Seismic Motion, Lithospheric Structures, Earthquake and Volcanic Sources: The Keiiti Aki Volume

Edited by
Yehuda Ben-Zion

2003

Springer Basel AG

Reprint from Pure and Applied Geophysics
(PAGEOPH), Volume 160 (2003), No. 3-4

Editors:

Yehuda Ben-Zion
University of Southern California
Department of Earth Sciences
Los Angeles CA, 90089-0740
USA

e-mail: benzion@terra.usc.edu

A CIP catalogue record for this book is available from the Library of Congress, Washington D.C., USA

Bibliographic information published by Die Deutsche Bibliothek
Die Deutsche Bibliothek lists this publication in the Deutsche Nationalbibliografie; detailed bibliographic data is available in the Internet at http://dnb.ddb.de.

ISBN 978-3-7643-7011-4   ISBN 978-3-0348-8010-7 (eBook)
DOI 10.1007/978-3-0348-8010-7

© 2003 Springer Basel AG
Originally published by Birkhäuser Verlag in 2003

Printed on acid-free paper produced from chlorine-free pulp

9 8 7 6 5 4 3 2 1   www.birkhauser-science.com

Contents

439   Introduction
        Y. Ben-Zion

445   A Review of the Discrete Wavenumber Method
        M. Bouchon

467   An Efficient Numerical Method for Computing Synthetic Seismograms for a
        Layered Half-space with Sources and Receivers at Close or Same Depths
        H.-m. Zhang, X.-f. Chen, S. Chang

487   Calculation of Synthetic Seismograms with Gaussian Beams
        R. L. Nowack

509   Wave Propagation, Scattering and Imaging Using Dual-domain One-way and
        One-return Propagators
        R.-S. Wu

541   Coda
        M. Fehler, H. Sato

555   Radiation from a Finite Reverse Fault in a Half Space
        R. Madariaga

579   Spontaneous Complex Earthquake Rupture Propagation
        S. Das

603   The Barrier Model and Strong Ground Motion
        A. S. Papageorgiou

635   Simulation of Ground Motion Using the Stochastic Method
        D. M. Boore

677   Characterization of Fault Zones
        Y. Ben-Zion, C. G. Sammis

717   Seismic Tomography of the Lithosphere with Body Waves
        C. H. Thurber

739   Volcano Seismology
        B. Chouet

789 Seismic Detection and Characterization of the Altiplano-Puna Magma Body,
Central Andes
G. Zandt, M. Leidig, J. Chmielowski, D. Baumont, X. Yuan

Pure appl. geophys. 160 (2003) 439–444
0033 – 4553/03/040439 – 6

Seismic Motion, Lithospheric Structures, Earthquake and Volcanic Sources: The Keiiti Aki Volume

Introduction

On March 16–18, 2000, the Department of Earth Science at the University of Southern California held a scientific meeting to honor the retirement of Professor Keiiti (Kei) Aki from academia. The meeting was attended by over 75 former students, postdoctoral fellows, close scientific associates and friends of Kei, representing universities, government agencies, and industry from the United States, Japan, China, France, New Zealand, England, Mexico, and Brazil. The scientific program consisted of 18 long review presentations and 12 additional short talks on theoretical and applied problems concerning earthquakes, faults, volcanoes, wave propagation, and engineering seismology. The 13 papers in this special volume are based on the long review talks given in the meeting and they provide examples of the broad range of topics that were pioneered, developed, and promoted by Aki over the last four decades.

Keiiti Aki was born on March 3, 1930, in Yokohama Japan to a family of engineers with a 100-year tradition of education and openness to the west. His grandfather was involved in construction of about 100 ports in Japan. After the 1923 great Tokyo earthquake he was in charge of rebuilding the port of Yokohama. He then built a house on top of a hill overlooking the port where Kei was later born. The house was burned in World War II but was rebuild. Kei's father was even more accomplished. He was a professor of hydrology at the University of Tokyo, a chairman of a national advisory committee to the Japanese government on utilization of natural resources, appeared often on television, worked together with Americans on various projects, and wrote scores of books. When it was time for Kei to attend university he wanted to study geophysics rather than continue the family tradition in engineering. His father agreed because he considered geophysics to be an important scientific field. The rest is history from which geophysics has greatly benefited. However, engineering did not lose! As is clear from Kei's career, he maintained an active interest in many engineering problems, and assisted in the development of an interface between seismology and earthquake engineering.

Kei obtained B.Sc. and Ph.D. degrees in geophysics from the University of Tokyo in 1952 and 1958. His Ph.D. dissertation under the supervision of Chuji Tsuboi was on space and time spectra in stationary stochastic waves with a special reference to microtremors. He was a postdoctoral fellow with Hewitt Dix and Frank Press at Caltech in the early 1960s and then an Associate Professor in the Earthquake Research Institute of the University of Tokyo from 1963 to 1966. When Frank Press left Caltech to head the MIT Department of Earth and Planetary Sciences, Aki was invited to join the MIT faculty where he began in 1966 his (U.S.) academic career as a Professor of Geophysics. In 1984, Kei moved to USC where he became the first W. M. Keck Foundation Professor of Earth Sciences, a position he held until his retirement in 2000. While at USC, Kei was largely responsible for the establishment of the Southern California Earthquake Center and he served as the first Science Director of the center.

Aki's scientific achievements cover source, path and site effects over the entire range of frequencies observed until about 1990. During the 1960 meeting of the IUGG in Helsinki, Kei was extremely impressed when theoretical spectral peaks of free oscillations predicted by Chaim Pekeris and observed peaks by three independent groups all matched. He was also very impressed by a dispersion analysis of Frank Press using seismic records from California generated by sources in the Pacific. These studies convinced Kei that it is possible to use deterministic methods for quantitative analysis of earthquakes, and inspired him to work on long-period seismology. That led to his fundamental 1966 paper on the seismic moment. Shortly after that paper was published, four talented young geophysicists developed important applications to the seismic moment: Hiroo Kanamori applied it to study of large earthquakes, Jim Brune related moment to plate tectonics, Boris Kostrov extended the concept to regional strain, and Adam Dziewonski developed systematic moment tensor inversion. With these and other seismologists working on long-period seismology, Kei decided to shift his focus to short-period seismology, a decision that eventually led him to return to California in 1984.

Kei's studies on high frequency waves include fundamental works on scaling laws of seismic spectra, scattering theory, analysis of coda waves, strong ground motion, seismic tomography, and frequency-dependent attenuation. Kei's group was instrumental in developing numerical methods for calculating seismic radiation from earthquake and volcanic sources, and imaging geophysical structures on different scales. Many of these works opened up new fields for seismologists and they played important roles in making quantitative seismology a mature science. One sign of a mature science is the existence of a standard comprehensive and rigorous textbook. This of course was provided in 1980 by the superb 2-volume book *Quantitative Seismology* of Aki and Richards. Another sign of a mature science is applications to various subjects. The papers in this special volume provide windows into such applications.

The first four papers deal with wave propagation techniques. Bouchon reviews the discrete wavenumber method of Bouchon and Aki (1977) for calculating Green functions for elastodynamic problems with irregular interfaces and finite sources. The method assumes spatial periodicity of the structure and source, extending earlier works of Rayleigh (1907) and Aki and Larner (1970) with a similar assumption on the structure. Using spectral representation and Fourier transform, the discrete wavenumber technique calculates seismic radiation as a superposition of plane waves. The method provides a simple yet powerful technique for computing realistic seismograms in a variety of applications. Computational schemes based on a spectral kernel function and wavenumber integration, like the discrete wavenumber and generalized reflection/transmission coefficient methods, converge very slowly when the spatial coordinates of the source and receiver in the direction that is transformed to the Fourier domain are similar. Zhang, Chen and Chang describe in the second paper an efficient remedy to this problem, using a peak and trough averaging of the oscillating kernel in the wavenumber domain integration. The technique is illustrated with examples associated with horizontally-layered structures.

In the third paper, Nowack covers the formulation, applications, and recent extensions of the Gaussian beam method for calculating synthetic seismograms in heterogeneous media. Following Popov (1981) and others, the method replaces rays (which are singular at caustics) with Gaussian beams. This allows a stable calculation of high-frequency elastic wavefield in various important cases. Additional types of beams and various implementation procedures are also discussed. Wu reviews in paper four the theoretical background, classical methods, and new techniques for wave propagation in heterogeneous media based on perturbation approaches. The review includes the Born, Rytov and De Wolf approximations, the phase-screen method for one-way propagation of elastic disturbances (neglecting reverberations and standing waves) implemented partially in the spatial and partially in the wavenumber domains, and several new generalized phase-screen methods. Applications of the new methods to seismic imaging are discussed and illustrated with 2-D and 3-D examples.

The well-known deterministic phases in seismograms, like P and S body waves, are followed by a decaying envelope of scattered waves. In a pioneering 1969 paper, Aki observed – without the benefit of modern broadband data and digital computers – some of the key properties of the seismic phases in that envelope which he called coda waves. Fehler and Sato review in the fifth paper analysis techniques, models, and applications of coda waves. The latter include imaging of crustal heterogeneities, scattering and attenuation of high-frequency waves, determination of site amplification factors, and more. Papers six to eight give results related to the seismic source. Madariaga clarifies conceptual and numerical subtleties associated with motion generated by a propagating rupture on a reverse dipping fault. The full solution has contributions from the rupture front and P, S, and surface wavefronts. The free surface modifies the frequency content of the solution as compared with full space results (in addition to adding surface waves). When the rupture does not reach the surface, the sum of all contributions gives finite motion although individual terms have singularities. When the rupture breaks the surface, a singularity at the surface-breaking tip remains in the full solution. Stable numerical calculations require regularization of the singularities. For shallow faulting, the rupture front produces strongly asymmetric disturbances across the fault with large motion in the hanging wall but not at the foot wall. In paper seven, Das reviews studies of dynamic rupture in the fracture mechanics and earthquake communities. The review is centered on her numerical works done first with Aki and then with Kostrov, and is focused on a discrete planar fault with possible strength/stress heterogeneities in a homogeneous isotropic medium. Simulation results include short rise time of earthquake slip, intersonic propagation, and spatially discontinuous ruptures. The discussion highlights important connections between material properties, rupture behavior, and radiated seismic fields, and the difficulty of inferring from available seismic data reliable information concerning properties and behavior at the earthquake source.

In paper eight, Papageorgiou discusses the specific barrier model of Papageorgiou and Aki (1983) and its use for calculations of strong ground motion. The model replaces the classical five-parameter (length and width of the fault, rupture velocity, final slip, and rise time) homogeneous kinematic framework of Haskell (1964, 1966, 1969) with a heterogeneous planar fault consisting of circular cracks separated by unbreakable barriers. Each circular crack fails at a random time following the passage of a uniformly propagating rupture front and sustains at failure a prescribed local stress drop. The model thus has a mixture of kinematic, dynamic, deterministic and stochastic ingredients. Calculations of seismic radiation involve specifying (or inverting for) six parameters, with the barrier interval, local stress drop, and maximum frequency replacing the final slip and rise time in the Haskell model. The interpretation of small-scale parameter values obtained by fitting model calculations to strong ground motion data remained controversial. Engineers typically simulate high frequency seismic waves with stochastic models. Boore gives in the ninth paper a comprehensive review of a stochastic method for ground motion calculations. The method combines a functional description of amplitude spectrum having contributions from source, path and site effects with random phase, using parameter choices related to the earthquake size and source-receiver distance. Detailed explanations and calculated examples illustrate the components needed to synthesize ground motion in a variety of seismological and engineering forms.

The final four papers of the special volume treat faults, volcanoes, and other lithospheric structures. Ben-Zion and Sammis discuss in paper ten conceptual frameworks and data on properties of earthquake fault zones. A wide variety of field, laboratory, and modeling results suggest that fault zones are formed as highly disordered structures characterized by strain hardening rheology and low mechanical efficiency. With increasing slip they appear to evolve toward strain weakening rheology, geometrical regularity and high mechanical efficiency. If this hypothesis is correct, it may be used to organize numerous geological and geophysical observations. The information available at present regarding the character of faults is, however, limited and the above hypothesis should be tested with future studies. In paper eleven, Thurber reviews teleseismic and local earthquake tomography, starting with the landmark 1976–1977 papers of Aki and coworkers on these topics and concluding with recent developments and applications. The discussed examples include tomographic imaging of sections of the San Andreas fault and the Valles caldera. Advances in computer power, 3-D ray-tracing techniques, inversion methods, and data availability (primarily through the establishment of IRIS and PASSCAL in the United States and similar initiatives in other countries) have led to significant improvements in the imaging capability of seismic tomography. This trend is likely to continue.

Chouet provides in paper twelve an extensive review of volcano seismology including theory, seismic observations, and lab data on source processes and structural properties. The physics of volcanic systems involves dynamic interactions between solid, liquid, and gas phases, and in that respect is more complex than the physics of tectonic earthquakes associated primarily with a solid state. Related to this, the diversity of volcanic sources and radiated seismic waves is larger than their counterparts in earthquake seismology and include, in addition to regular earthquakes and related waves, also tremor and long period events and waves. The latter types of waves are indicative of the eruption potential of volcanoes so quantitative predictive seismology is, perhaps paradoxically, more developed at present for volcanic activity than for tectonic earthquakes. Aki *et al.* (1977) developed the first quantitative description of volcanic tremor and long period events based on a fluid-driven crack model that is still in use today with some modifications. The pioneering tomographic 1976–1977 papers of Aki *et al.* provided the basis for imaging volcanic structures. After introducing these contributions, Chouet describes the state-of-the-art in volcano seismology using examples from his works with colleagues on tomographic imaging of the Kilauea volcano, source properties of tremor and long period events, and experiments with expanding gas-liquid flows. In the last paper of the volume, Zandt, Leidig, Chmielowski, Baumont and Yuan employ receiver functions to obtain a model of crustal stratigraphy and anisotropy in the Altiplano-Puna volcanic complex in the central Andes. The receiver functions provide information on impedance contrast interfaces by identifying P-to-S converted phases in seismograms. The method is especially useful for imaging narrow low velocity zones that have large impedance contrasts but produce small signals for tomographic studies. From teleseismic and local receiver functions, Zandt *et al.* infer on the existence of a thin sill-like regional magma body at a depth of about 20 km and overlying anisotropy in their study area.

Acknowledgments

Resources and facilities for the March 16–18, 2000, meeting were provided by the University of Southern California through the efforts of Tom Henyey, Charlie Sammis and Leon Teng. The administrative assistance of Sally Henyey, John McRaney, Michelle Smith and Shelly Werner made the meeting efficient and enjoyable. I thank the referees of the papers for critical reviews that improved the scientific quality of the volume. These include John Anderson, Rafael Benites, David Boore, Michel Bouchon, Michel Campillo, Bernard Chouet, Donna Eberhart-Phillips, Art Frankel, Hiroo Kanamori, Hitoshi Kawakatsu, Hiroshi Kawase, Brian Kennett, Geoff King, Peter Leary, Raul Madariaga, Chris Marone, Steve Martel, Bob Nowack, David Okaya, Tom Parsons, Jim Rice, Paul Richards, Francisco Sanchez-Sesma, Teruo Yamashita, Kiyoshi Yomogida, Yuehua Zeng, and Ru-Shan Wu.

Yehuda Ben-Zion
Department of Earth Sciences
University of Southern California
Los Angeles, CA, 90089-0740
U.S.A

Pure appl. geophys. 160 (2003) 445–465
0033–4553/03/040445–21

A Review of the Discrete Wavenumber Method

MICHEL BOUCHON[1]

Abstract — We present a review of the discrete wavenumber (DWN) method. The method, introduced by BOUCHON and AKI (1977), allows the simple and accurate calculation of the complete Green's functions for many problems in elastodynamics.

Key words: Wave propagation, synthetic seismograms, discrete wavenumber, earthquake ground motion.

Introduction

The evaluation of Green's functions for acoustic or elastic media is an important problem in fields such as seismology or acoustics. Since the pioneering work of LAMB (1904), many approaches have been proposed to evaluate the response of elastic solids to excitation by transient point sources. The methods devised for the calculation of the Green's functions are, however, often very complex or, in many cases, only provide approximate solutions. The discrete wavenumber method, introduced by BOUCHON and AKI (1977), provides a way to accurately calculate the complete Green's functions for many problems with a minimum amount of mathematics.

The principle of the method may be traced back to Rayleigh, who demonstrated that waves reflected by a sinusoidally corrugated surface propagate only at discrete angles that he referred to as the orders of the spectrum (RAYLEIGH, 1896, 1907). The existence of discrete orders in the horizontal wavenumber spectrum is an immediate consequence of the periodicity of the reflecting surface. AKI and LARNER, in 1970, extended Rayleigh's approach to study the scattering of plane waves in the vicinity of a periodic irregular surface with the use of complex frequency. In the same way, the discrete wavenumber (DWN) method introduces a spatial periodicity of sources to discretize the radiated wave field, and relies on the Fourier transform in the complex frequency domain to calculate the Green's functions.

[1] Laboratoire de Géophysique Interne et Tectonophysique, Université Joseph Fourier, BP 53, 38041 Grenoble, France. E-mail: Michel.Bouchon@ujf-grenoble.fr

Principle of the Method

We shall begin with a short consideration of the 2-D case, as the principle of the method is easiest to describe in this case. The steady-state radiation from a line source in an infinite homogeneous medium can be represented as a cylindrical wave or, equivalently, as a continuous superposition of homogeneous and inhomogeneous plane waves. Therefore, denoting by x and z the horizontal and vertical axes in the plane normal to the line source, any observable such as displacement or stress can be written in the form

$$F(x,z;\omega) = e^{i\omega t} \int\limits_{-\infty}^{\infty} f(k,z)e^{-ikx}\, dk \qquad (1)$$

where ω is the frequency and k is called the horizontal wavenumber. Equation (1) still holds for an extended two-dimensional source located in a medium which is homogeneous in any horizontal plane.

When the medium is finite or vertically heterogeneous, the integral kernel has poles and singularities, and the integration over the horizontal wavenumber becomes mathematically and numerically complicated. One simple way around these difficulties is to replace the single-source problem, whose solution is expressed by (1), by a multiple-source problem where sources are periodically distributed along the x axis. Then, equation (1) is replaced by:

$$G(x,z;\omega) = \int\limits_{-\infty}^{\infty} f(k,z)e^{-ikx} \sum_{m=-\infty}^{\infty} e^{ikmL}\, dk \qquad (2)$$

where L is the periodicity source interval and the $e^{i\omega t}$ time dependence is understood. Equation (2) reduces to:

$$G(x,z;\omega) = \frac{2\pi}{L} \sum_{n=-\infty}^{\infty} f(k_n,z)e^{-ik_n x} \qquad (3)$$

with

$$k_n = \frac{2\pi}{L}n$$

which in turn, if the series converges, can be approximated by the finite sum equation

$$G(x,z;\omega) = \frac{2\pi}{L} \sum_{n=-N}^{N} f(k_n,z)e^{-ik_n x} \ . \qquad (4)$$

In moving from equation (1) to equation (4), we have greatly reduced the calculation. In so doing, however, we have changed the problem from one of a single source, to one involving an infinite number of periodic sources, as illustrated in Figure 1. The DWN method calculates equation (4), that is $G(x, z; \omega)$, instead of evaluating equation (1).

The second stage of the method is to retrieve the single-source solution from the multiple-source problem that we have solved in the frequency domain. This would be straightforward if we could calculate the continuous Fourier transform of G, as we could then isolate the single source solution in the time domain, provided that we have chosen an appropriate value for L. In practice, however, we can only calculate G for a certain number of frequencies and use the discrete Fourier transform to obtain the time domain solution. Thus, on one hand we deal with a signal which has an infinite time response (because of the infinite set of sources), while on the other hand, we use the discrete Fourier transform, which yields a signal of finite duration $T = 2\pi/\Delta\omega$, where $\Delta\omega$ is the angular frequency sampling used in calculating G. This can indeed be accomplished by performing the Fourier transform in the complex frequency domain:

$$
g(x, z; t) = \int_{-\infty + i\omega_I}^{\infty + i\omega_I} G(x, z; \omega) e^{i\omega t} \, d\omega \tag{5}
$$

where ω_I denotes the constant imaginary part of the frequency and is chosen such that

$$
e^{\omega_I T} << 1 \; . \tag{6}
$$

This last equation insures the attenuation, over the time window T, of the previously infinite time response solution. Thus, provided that we have chosen L large enough so that no disturbance arrives at the receiver (x, z) from the next closest source in the time window of interest T, the time-domain single-source solution $f(x, z; t)$ is obtained from the frequency-domain multiple-source calculation $G(x, z; \omega)$ by

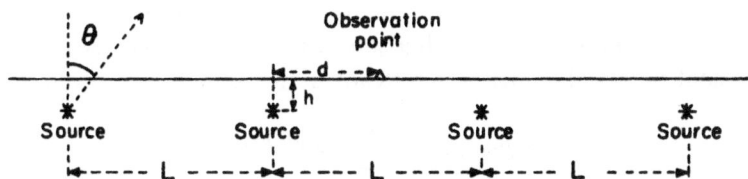

$$
L \sin\theta = n\lambda
$$

Figure 1

Physical interpretation of the DWN method. The single source is replaced by an infinite array of sources distributed horizontally at equal interval L. For a given radiation wavelength λ corresponding to a specific frequency of excitation, the elastic energy is radiated in discrete directions θ only.

$$f(x,z;t) = e^{-\omega_I t} \int_{-\infty}^{\infty} G(x,z;\omega)e^{i\omega_R t} \, d\omega_R \tag{7}$$

where the integral is computed by using the FFT.

Equation (6) shows that ω_I is only a function of the length of the time window T considered. The results should not be sensitive to the particular value of ω_I chosen, as long as it provides enough attenuation for the disturbances which arrive after the time window of interest T to be negligible. Values in the range:

$$\omega_I = \left[-\frac{\pi}{T}, -\frac{2\pi}{T} \right] \tag{8}$$

are recommended for most applications.

It is worth noting here that disturbances which arrive in the time range $[T, 2T]$ will be attenuated by $e^{\omega_I T}$, while disturbances in the time range $[2T, 3T]$ will be attenuated by $e^{2\omega_I T}$, and so on. The choice of ω_I may also be justified by the fact that the frequency spectrum $G(\omega)$ is not discrete, as would be the case with real frequencies, but is continuous with a bandwidth proportional to ω_I (LARNER, 1970). Choosing values in the range of relation (8) implies that the bandwidth of the spectral lines is of the order of the frequency interval. Thus, the calculated signals may also be considered as resulting from a nearly continuous sampling of the frequency domain.

In Figure 2, we present a comparison of the numerical results obtained through these equations with an analytical solution. The case considered involves an explosive line source in a half-space, as it is one of the rare cases where an analytical solution exists (GARVIN, 1956). The comparison shows the great accuracy of the DWN method.

Discretization in Various Coordinate Systems

The simplest type of elastic source in three-dimensions is an isotropic point source. The wave field radiated by such a source can be conveniently represented by the displacement potential, which, for a steady-state excitation, is given by:

$$\phi(R;\omega) = \frac{-V_S(\omega)}{4\pi R} e^{i\omega(t-R/\alpha)} \tag{9}$$

where V_S is the volume change at the source and α denotes the compressional wave velocity.

In the shallow earth, where boundaries are nearly horizontal and where the medium properties change primarily with depth, using this spherical wave representation would be most cumbersome, so we must express the wave field in more appropriate coordinate systems. One possibility is to use a Cartesian system with the

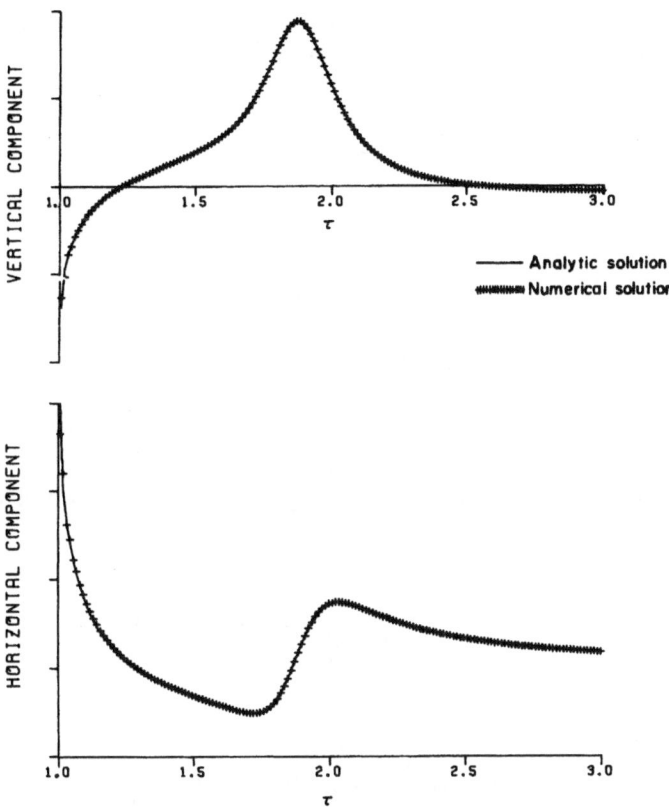

Figure 2
Comparison between numerical and analytical solutions for the surface displacement due to a buried explosive line source with step-function time dependence. Computations are made for a Poisson ratio of 0.25 and a ratio of distance R to source depth equal to 10. $\tau = t\alpha/R$, where t is time and α is the compressional wave velocity. The analytical displacements are infinite at the time of P-wave arrival ($\tau = 1$).

z axis running vertically. In such a system, the wave field is expressed as a double integral over the two components of the horizontal wavenumber, k_x and k_y, through the Weyl integral (LAMB, 1904; AKI and RICHARDS, 1980):

$$\phi(x, y, z; \omega) = \frac{iV_S(\omega)}{8\pi^2} \int\limits_{-\infty}^{\infty} \int\limits_{-\infty}^{\infty} \frac{1}{v} e^{-iv|z|} e^{-ik_x x} e^{-ik_y y} \, dk_x \, dk_y \tag{10}$$

with

$$v = \sqrt{\frac{\omega^2}{\alpha^2} - k_x^2 - k_y^2}, \qquad \text{Im}(v) < 0 \ ,$$

where the origin of the coordinate system is taken at the source, and the $e^{i\omega t}$ dependence is understood.

The generalization of the previous results from 2-D to 3-D is straightforward and leads to the following expressions (BOUCHON, 1979):

$$\phi(x,y,z;\omega) = \frac{iV_S(\omega)}{2L_xL_y} \sum_{n_x=-N_x}^{N_x} \sum_{n_y=-N_y}^{N_y} \frac{1}{v} e^{-iv|z|} e^{-ik_{nx}x} e^{-ik_{ny}y} \tag{11}$$

with

$$k_{nx} = \frac{2\pi}{L_x} n_x, \quad k_{ny} = \frac{2\pi}{L_y} n_y$$

for which the corresponding multiple-source problem is a periodic array of sources distributed at equal intervals L_x in the x direction, and L_y in the y direction.

In many wave propagation problems, the elastic wave field may also be conveniently expressed in a cylindrical coordinate system with z as the vertical axis. The wave field is then represented as an integral over the horizontal wavenumber through the Sommerfeld integral:

$$\phi(r,z;\omega) = \frac{iV_S(\omega)}{4\pi} \int_0^\infty \frac{k}{v} J_0(kr) e^{-iv|z|} \, dk \tag{12}$$

with

$$v = \sqrt{\frac{\omega^2}{\alpha^2} - k^2}, \quad \mathrm{Im}(v) < 0$$

and where J_0 denotes the zeroth order Bessel function.

The discretization of this equation can also be achieved by replacing the single-source by a periodic arrangement of sources which, in this case, consists of the original point source plus an infinite array of circular sources centered around the point source and distributed at equal radial interval L (BOUCHON, 1981). This physical arrangement leads to:

$$\phi(r,z;\omega) = \frac{iV_S(\omega)}{2} \sum_{n=0}^N \frac{k_n}{v_n} J_0(k_n r) e^{-iv_n|z|} \tag{13}$$

with

$$k_n = \frac{2\pi}{L} n \ .$$

The comparison between the two geometric source arrangements resulting in discretizations (11) and (13) is shown in Figure 3.

Once the source radiation has been decomposed, through equations (4), (11), or (13), into a superposition of waves propagating with discrete wavenumbers, the effect

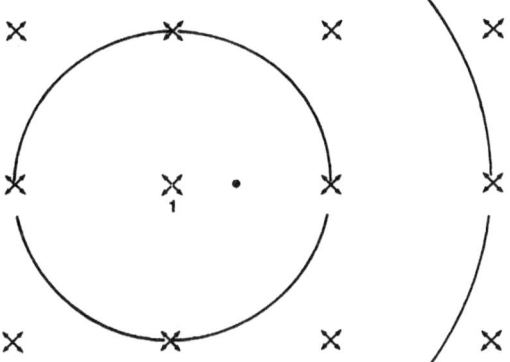

Figure 3
Geometries of source-receiver configurations leading to the discretization: a circular source array for the k discretization scheme and a rectangular network for the (k_x, k_y) discretization method. Source 1 is the original single-source problem. The black dot shows the receiver location.

of plane boundaries and flat layers is taken into account by using, for each horizontal wavenumber component, the corresponding plane-wave reflection and transmission coefficients at the medium surface and interfaces, and summing up all the wavenumber contributions. This is best done by calculating, for each wavenumber involved in the source radiation, the corresponding reflectivity and transmissivity matrices of the layered medium (KENNETT, 1974; KENNETT and KERRY, 1979; MÜLLER, 1985). The truncation of the wavenumber series is easily determined for each frequency by a simple convergence criterion which compares the new wavenumber contribution to the current sum of the series, and stops the calculation when the new contribution becomes negligible.

The accuracy of the two discretization schemes (11) and (13) can be measured by comparing synthetic seismograms obtained using these equations, as the two schemes are independent. This is done in Figure 4, where the similitude of the results demonstrates the accuracy of the DWN method. In most applications, the k discretization scheme will be preferred over the k_x, k_y scheme because it involves only one summation and the resulting calculation is faster. One such application is displayed in Figure 5.

For other types of problems, other schemes of discretization may be devised. For instance, in the case of a source in a borehole, common in exploration geophysics, it is convenient to use, for equation (9), the expression:

$$\phi(r, z; \omega) = \frac{-V_S(\omega)}{4\pi^2} \int_{-\infty}^{\infty} K_0(vr) e^{-ikz} \, dk \qquad (14)$$

with

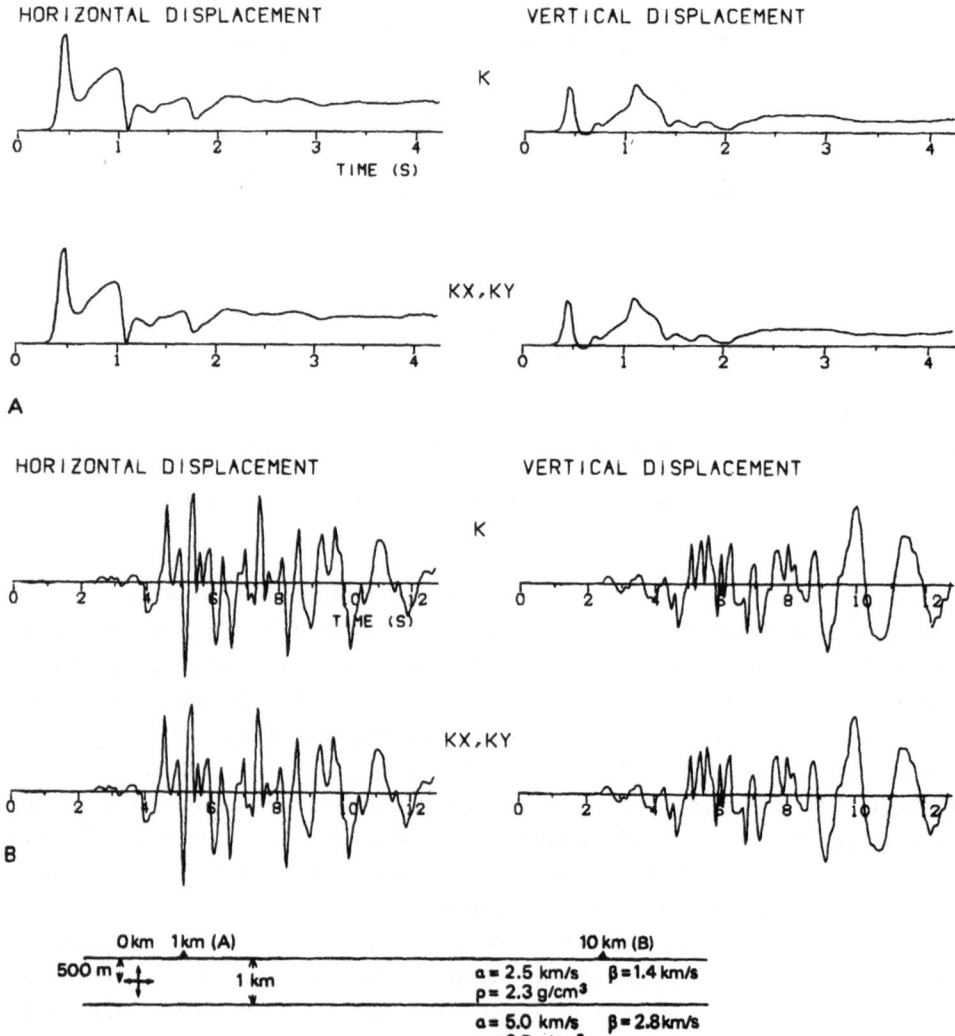

Figure 4
Comparison of surface displacements obtained using the k and (k_x, k_y) discretization schemes for an explosion in a layer over a half-space model. The source-time function is $\frac{1}{2}[1 + \tan h(t/t_0)]$ with $t_0 = 0.1$ s. First motions are up and away from the source.

$$v = \sqrt{\frac{\omega^2}{\alpha^2} - k^2}, \quad \mathrm{Im}(v) < 0 ,$$

where (r, z) are cylindrical coordinates centered at the source and z runs along the borehole axis, k is now the vertical wavenumber (in the case of a vertical borehole), and where K_0 denotes the zeroth-order modified Bessel function of the second kind.

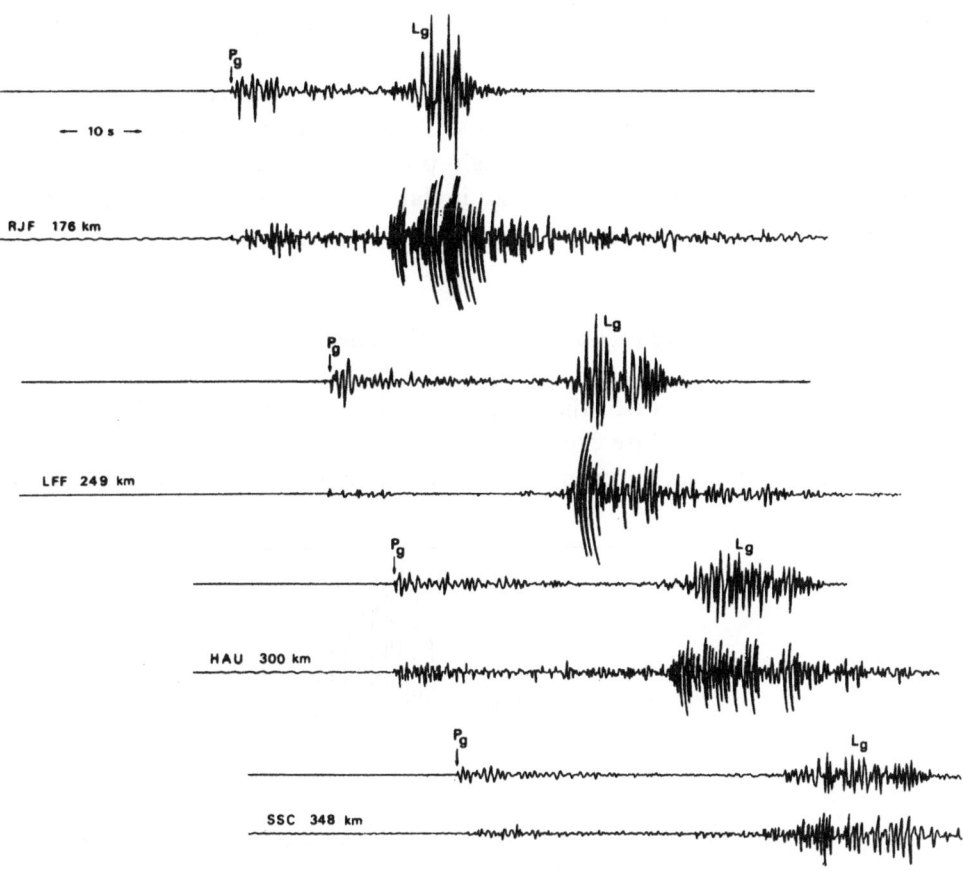

Figure 5
Comparison of the vertical short-period seismograms synthesized (upper trace) and observed (lower trace) at four stations for a small earthquake in central France. The epicentral distance of each station is indicated. The propagation model used in the calculation consists of four crustal layers overlaying a mantle half-space. The source is a double-couple point with the mechanism of the earthquake and located at a depth of 10 km. The slip time dependence is a ramp function with a rise time of 0.2 s (after BOUCHON, 1982a).

The discretization of this expression, which was introduced by CHENG and TOKSÖZ (1981), yields:

$$\phi(r, z; \omega) = \frac{-V_S(\omega)}{2\pi L} \sum_{n=-N}^{N} K_0(v_n r) e^{-ik_n z} \qquad (15)$$

with

$$k_n = \frac{2\pi}{L} n$$

and corresponds to a periodic arrangement of point sources distributed at interval L along the z axis.

Expression (15) is convenient to use in a borehole environment because, in this form, cylindrical boundaries of the borehole, tubing, mud casing, and/or borehole tool can be taken into account through propagator matrices or reflectivity/transmissivity matrices similar to the ones in flat layer media. An example of such a calculation is displayed in Figure 6.

Case of a Generalized and Extended Source

We now consider the case where the point source is a force with Cartesian components (F_x, F_y, F_z), and we express its radiation in a discretized form similar to (13). We assume again that the cylindrical coordinate system is centered at the source

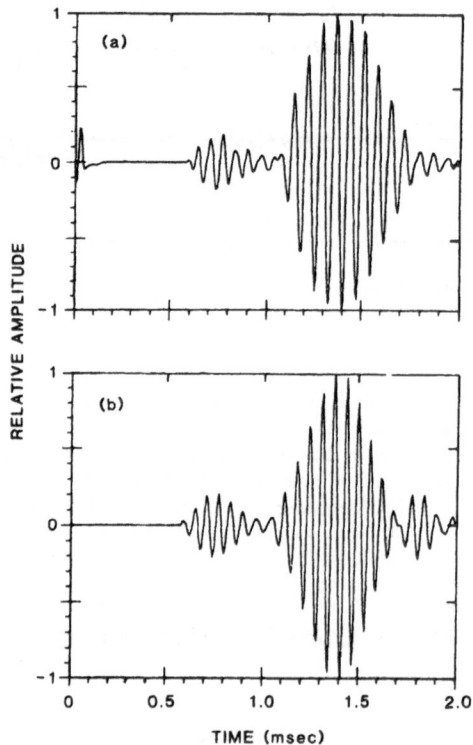

Figure 6

A comparison between (a) actual and (b) synthetic full waveform acoustic log microseismograms in a limestone formation. The source is a pressure point located in a fluid-filled cylindrical borehole. Parameters used are $\alpha = 5.95$ km/s, $\beta = 3.05$ km/s, $\rho = 2.3$ for the geological formation, and $\alpha = 1.83$ km/s, $\rho = 1.2$ for the fluid. The borehole radius is 6.7 cm. The synthetic microseismogram is calculated by discretizing the source radiation in the vertical wavenumber domain (after CHENG et al., 1982).

and that the z axis is vertical. We have for the compressional and rotational potentials:

$$\phi(r,\theta,z;\omega) = \frac{1}{2L\rho\omega^2}\left[\text{sgn}(z)F_z\sum_{n=0}^{N}k_nJ_0(k_nr)e^{-i\nu_n|z|}\right.$$

$$\left. -i(F_x\cos\theta + F_y\sin\theta)\sum_{n=0}^{N}\frac{k_n^2}{\nu_n}J_1(k_nr)e^{-i\nu_n|z|}\right]$$

$$\psi(r,\theta,z;\omega) = \frac{1}{2L\rho\omega^2}\left[-iF_z\sum_{n=0}^{N}\frac{k_n}{\gamma_n}J_0(k_nr)e^{-i\gamma_n|z|}\right.$$

$$\left. +\text{sgn}(z)(F_x\cos\theta + F_y\sin\theta)\sum_{n=0}^{N}J_1(k_nr)e^{-i\gamma_n|z|}\right] \qquad (16)$$

$$\chi(r,\theta,z;\omega) = i\frac{F_y\cos\theta - F_x\sin\theta}{2L\rho\beta^2}\sum_{n=0}^{N}\frac{1}{\gamma_n}J_1(k_nr)e^{-i\gamma_n|z|}$$

with

$$\gamma_n = \sqrt{\frac{\omega^2}{\beta^2} - k_n^2}, \quad \text{Im}(\gamma_n) < 0$$

and

$$\text{sgn}(z) = 1 \text{ for } z > 0, \quad \text{sgn}(z) = -1 \text{ for } z < 0.$$

where ρ is the density, β the shear-wave velocity, and J_1 is the Bessel function of the first order.

Any type of elastic source can be represented by a combination of point forces. In particular, a generalized point source is commonly represented in seismology by its moment tensor m_{ij} where m_{xx}, m_{yy}, and m_{zz} represent three force dipoles oriented along the Cartesian axes, while $m_{xy} = m_{yx}$, $m_{xz} = m_{zx}$, and $m_{yz} = m_{zy}$ are double couples with force oriented along the first axis index and arm along the second axis index. Expressions for the radiation from an arbitrary moment tensor source can then be obtained by linear operations on equations (16).

Of particular interest is the radiation from a double-couple source, as such a body source is equivalent to a point of shear dislocation. Denoting by (s_x, s_y, s_z) the components of the unit vector in the slip direction and by (n_x, n_y, n_z) those of the normal to the fault, the corresponding moment tensor components are:

$$m_{ij} = -\mu\,\text{slip}(\omega)\Delta S(s_in_j + s_jn_i) \qquad (17)$$

where μ is the rigidity and ΔS is the elementary fault surface on which slip occurs.

The simplest way to calculate the elastic radiation from an extended source is usually to represent the source by a superposition of elementary point sources.

Although analytical expressions of the radiation can sometimes be derived in the frequency-wavenumber domain for particular cases, the point-source superposition is generally more versatile. In the case of an earthquake, for instance, the fault can be discretized into a two-dimensional array of double-couple points distributed on the fault plane at a spacing smaller than the shortest wavelength considered in the problem. Each point radiates with a phase delay $e^{-i\omega t_r}$, where t_r denotes the time for rupture to propagate from the hypocenter to the particular fault location. Slip amplitude and duration may vary at each point. The summation of all the elementary contributions is done in the frequency-wavenumber domain, and does not affect the calculation of the reflection/transmission and reflectivity/transmissivity matrices. As an example, the simulation of the ground motion produced during the 1999 Izmit earthquake is presented in Figure 7. For this calculation, the 135 km long fault is represented by 10,800 double-couple points uniformly distributed at a spacing of 500 m in the horizontal and vertical directions. One important aspect of the DWN method, which is illustrated in this figure, is that the method calculates the complete elastic wave field, including both static and dynamic contributions.

Applications and Extensions of the Method

The DWN method has been successfully tested against analytical solutions and other techniques (e.g., YAO and HARKRIDER, 1983; BEN-ZION and AKI, 1990) and has been extensively used to check the accuracy of other methods like finite-differences, finite-elements, ray methods, mode summation, or pseudo-spectral techniques (e.g., STEPHEN et al., 1985; SAIKIA and HERRMANN, 1986; BEYDOUN and KEHO, 1987; MAUPIN, 1996; AOI and FUJIWARA, 1999; MOCZO et al., 1999).

It has been used to study a variety of problems in elastodynamics where the calculation of Green's functions is required. Many of the applications have been carried out using the numerical code of COUTANT (1990).

Applications include problems in seismic exploration (CHENG and TOKSÖZ, 1981; CHENG et al., 1982; DIETRICH and BOUCHON, 1985a,b; SCHMITT and BOUCHON, 1985; DIETRICH, 1988; SCHMITT, 1988b; CHENG, 1989; JEAN and BOUCHON, 1991; MEREDITH et al., 1993; GIBSON, 1994; FALK et al., 1996; HAARTSEN and PRIDE, 1997), earthquake seismology (CAMPILLO et al., 1984, 1985; SAIKIA and HERRMANN, 1987; GARIEL et al., 1990, 1991; OU and HERRMANN, 1990; CHIN and AKI, 1991; FUKUSHIMA et al., 1995; PLICKA and ZAHRADNIK, 1998; PLICKA et al., 1998), microseismicity studies (BERNARD and ZOLLO, 1989; GOT and FRÉCHET, 1993; JONGMANS and MALIN, 1995; ZOLLO et al., 1995; ZOLLO and IANNACCONE, 1996; THEODULIDIS et al., 1996), broadband modeling of local and regional seismograms (HERRMANN et al., 1980; EBERHART-PHILLIPS et al., 1981; BOUCHON, 1982a; CHRISTOFFERSSON et al., 1988; BERTIL et al., 1989; PAUL and NICOLLIN, 1989; ROBERTS and

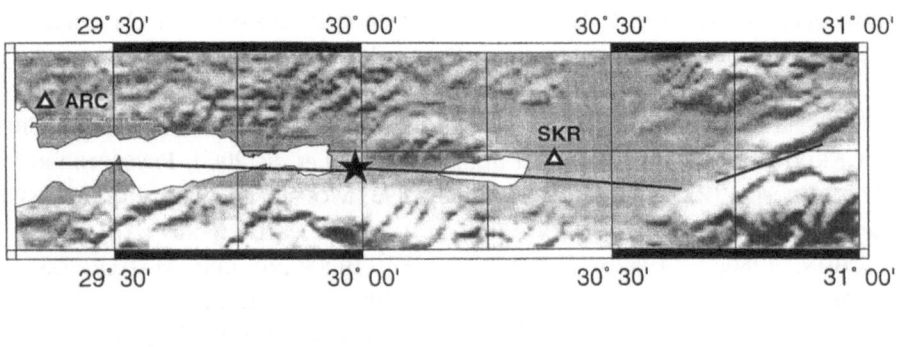

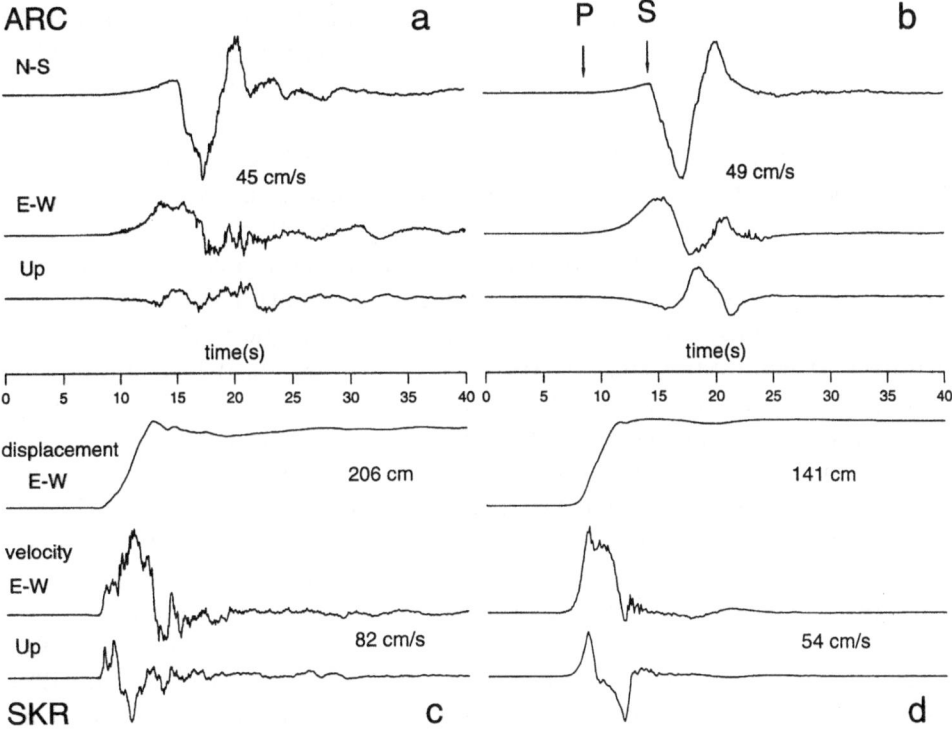

Figure 7

Comparison between recorded and calculated ground motion during the 1999 Izmit, Turkey, earthquake. (Top) Map of the surface rupture of the earthquake (solid line). The symbols indicate the location of the epicenter (star) and of the recording stations (triangles). (Middle) Ground velocity recorded (a) and calculated (b) at ARC. (Bottom) Displacement and velocity recorded (c) and calculated (d) at SKR. The numerical values indicated give the peak amplitudes of the observed/calculated velocity/displacement. All the traces start at the origin time of the rupture. The N-S component was inoperative at SKR. The fault is a vertically dipping strike-slip fault which follows the surface breaks shown on the map and extends from the surface down to 20 km. Rupture starts at the hypocenter, located at a depth of 17 km, and propagates toward the west at 3 km/s and toward the east at 4.7 km/s. Slip varies along the fault strike according to surface observations. Slip duration everywhere is 3 s. The lower than observed, peak values at SKR indicate a larger fault slip at depth near this station than the one observed at the surface.

CHRISTOFFERSSON, 1990; FUKUYAMA et al., 1991; PEDERSEN and CAMPILLO, 1991; TOKSÖZ et al., 1990; CAMPILLO and PAUL, 1992; TAKEO, 1992; CAMPILLO and ARCHULETA, 1993; STEIDL et al., 1996; SAIKIA and HELMBERGER, 1997; SINGH et al., 1997, 1999a,b; TSELENTIS and ZAHRADNIK, 2000), moment tensor inversion (CRUSEM and CARISTAN, 1992; NISHIMURA et al., 2000; SINGH et al., 2000; TEYSSONEYRE et al., 2001), scattering (ZENG et al., 1991; ZENG, 1993; MOINET and DIETRICH, 1998), fault zone effects (BEN-ZION, 1998), ground motion near earthquakes (AKI et al., 1978; BOUCHON, 1980a,b, 1982b; CAMPILLO, 1983; BERNARD and MADARIAGA, 1984; MENDEZ and LUCO, 1988; CAMPILLO et al., 1989; GARIEL and CAMPILLO, 1989; BARD et al., 1992; TAKEO and ITO, 1997; TAKEO and KANAMORI, 1997; PEYRAT et al., 2001), earthquake fault tomography (TAKEO, 1987, 1988; FUKUYAMA and MIKUMO, 1993; TAKEO et al., 1993; COTTON and CAMPILLO, 1994, 1995a,b; SEKIGUCHI et al., 1996, 2000; COTTON et al., 1996; IDE et al., 1996; MENDOZA and FUKUYAMA, 1996; IDE and TAKEO, 1996, 1997; COURBOULEX et al., 1997; NAKAYAMA and TAKEO, 1997; OGLESBY and ARCHULETA, 1997; HERNANDEZ et al., 1999; REBOLLAR et al., 1999; QUINTANAR et al., 1999), stress calculations (COTTON and COUTANT, 1997; BELARD-INELLI et al., 1999) and volcanology (CHOUET, 1981, 1982, 1985; AKI, 1984; CHOUET and JULIAN, 1985; TAKEO, 1990; NISHIMURA and HAMAGUCHI, 1993; GOLDSTEIN and CHOUET, 1994; UHIRA et al., 1994; NISHIMURA, 1995; NISHIMURA et al., 1995).

The DWN method has been extended to include anisotropic media (MANDAL and MITCHELL, 1986; MANDAL and TOKSÖZ, 1990, 1991; MANDAL, 1991) and two-phase media (BOUTIN et al., 1987; SCHMITT, 1988a, 1990; SCHMITT et al., 1988a, 1988b).

The discrete wavenumber formalism has also been extended to model wave propagation in 2-D or 3-D media through formulations based on boundary integral equations (BOUCHON, 1985; CAMPILLO and BOUCHON, 1985; CAMPILLO, 1987; PAUL and CAMPILLO, 1988; COUTANT, 1989; GAFFET and BOUCHON, 1989, 1991; BOUCHON et al., 1989, 1996; AXILROD and FERGUSON, 1990; CAMPILLO et al., 1993; CHAZALON et al., 1993; GAFFET et al., 1994; GIBSON and CAMPILLO, 1994; HAARTSEN et al., 1994; GAFFET, 1995; KARABULUT and FERGUSON, 1996; TAKENAKA et al., 1996; SHAPIRO et al., 1996), boundary elements (KAWASE, 1988; KAWASE and AKI, 1989, 1990; KIM and PAPAGEORGIOU, 1993; BOUCHON and COUTANT, 1994; DONG et al., 1995; PAPAGEORGIOU and PEI, 1998; ZHANG et al., 1998; FU and WU, 2001), or generalized reflection/transmission matrices (CHEN, 1990, 1995, 1996). Hybrid methods of calculation, combining the method with finite-difference or finite-element methods, have also been developed to study the propagation of seismic waves in complex geological structures (ZAHRADNIK, 1995; ZAHRADNIK and MOCZO, 1996; MOCZO et al., 1997; RIEPL et al., 2000).

REFERENCES

AKI, K. (1984), Evidence for Magma Intrusion during the Mammoth Lakes Earthquakes of May 1980 and Implications of the Absence of Volcanic (Harmonic) Tremor, J. Geophys. Res. 89, 7689–7696.

AKI, K. and LARNER K. (1970), *Surface Motion of a Layered Medium Having an Irregular Interface due to Incident Plane SH Waves*, J. Geophys. Res. *75*, 933–954.

AKI, K., BOUCHON, M., CHOUET, B., and DAS, S. (1978), *Quantitative Prediction of Strong Motion for a Potential Earthquake Fault*, Annali Geofisica *30*, 341–368.

AKI, K. and RICHARDS, P. G. *Quantitative Seismology: Theory and Methods* vol.1, (W. H. Freeman, San Francisco, Calif., 1980).

AOI, S. and FUJIWARA, H. (1999), *3D Finite-difference Method Using Discontinuous Grids*, Bull. Seismol. Soc. Am. *89*, 918–930.

AXILROD, H. D. and FERGUSON, J. F. (1990), *SH-wave Scattering from a Sinusoidal Grating: An Evaluation of Four Discrete Wavenumber Methods*, Bull. Seismol. Soc. Am. *80*, 643–655.

BARD, P. Y., BOUCHON, M., CAMPILLO, M., and GARIEL, J. C. *Numerical simulation of strong ground motion using the discrete wavenumber method: A review of main results*. In *Recent Advances in Earthquake Engineering and Structural Dynamics, 8th Ed.* (Nantes France, 1992) pp. 23–63.

BELARDINELLI, M. E., COCCO, M., COUTANT, O., and COTTON, F. (1999), *Redistribution of Dynamic Stress during Coseismic Ruptures: Evidence for Fault Interaction and Earthquake Triggering*, J. Geophys. Res. *104*, 14,925–14,945.

BEN-ZION, Y. (1998), *Properties of Seismic Fault Zone Waves and their Utility for Imaging Low-velocity Structures*, J. Geophys. Res. *103*, 12,567–12,585.

BEN-ZION, Y. and AKI, K. (1990), *Seismic Radiation from an SH Source in a Laterally Heterogeneous Planar Fault Zone*, Bull. Seismol. Soc. Am. *80*, 971–994.

BERNARD, P. and MADARIAGA, R. (1984), *A New Asymptotic Method for the Modeling of Near-field Accelerograms*, Bull. Seismol. Soc. Am. *74*, 539–557.

BERNARD, P. and ZOLLO, A. (1989), *Inversion of Near-source S Polarization for Parameters of Double-couple Point Sources*, Bull. Seismol. Soc. Am. *79*, 1779–1809.

BERTIL, D., BÉTHOUX, N., CAMPILLO, M., and MASSINON, B. (1989), *Modeling Crustal Phases in Southeast France for Focal Depth Determination*, Earth Planet. Sci. Lett. *95*, 341–358.

BEYDOUN, W. B. and KEHO, T. H. (1987), *The Paraxial Ray Method*, Geophysics *52*, 1639–1653.

BOUCHON, M. (1979), *Discrete Wavenumber Representation of Elastic Wave Fields in Three-space Dimensions*, J. Geophys. Res. *84*, 3609–3614.

BOUCHON, M. (1980a), *The Motion of the Ground during an Earthquake: 1. The Case of a Strike-slip Fault*, J. Geophys. Res. *85*, 356–366.

BOUCHON, M. (1980b), *The Motion of the Ground during an Earthquake: 2. The Case of a Dip-slip Fault*, J. Geophys. Res. *85*, 367–375.

BOUCHON, M. (1981), *A Simple Method to Calculate Green's Functions in Elastic Layered Media*, Bull. Seismol. Soc. Am. *71*, 959–971.

BOUCHON, M. (1982a), *The Complete Synthesis of Seismic Crustal Phases at Regional Distances*, J. Geophys. Res. *87*, 1735–1741.

BOUCHON, M. (1982b), *The Rupture Mechanism of the Coyote Lake Earthquake of 6 August 1979 Inferred from Near-field Data*, Bull. Seismol. Soc. Am. *72*, 745–757.

BOUCHON, M. (1985), *A Simple Complete Numerical Solution to the Problem of Diffraction of SH Waves by an Irregular Surface*, J. Acoust. Soc. Am. *77*, 1–5.

BOUCHON, M. and AKI, K. (1977), *Discrete Wave number Representation of Seismic Source Wave Fields*, Bull. Seismol. Soc. Am. *67*, 259–277.

BOUCHON, M., CAMPILLO, M., and GAFFET, S. (1989), *A Boundary Integral Equation – Discrete Wavenumber Representation Method to Study Wave Propagation in Multilayered Media Having Irregular Interfaces*, Geophysics *54*, 1134–1140.

BOUCHON, M. and COUTANT, O. (1994), *Calculation of Synthetic Seismograms in a Laterally-varying Medium by the Boundary Element—Discrete Wavenumber Method*, Bull. Seismol. Soc. Am. *84*, 1869–1881.

BOUCHON, M., SCHULTZ, C. A., and TOKSÖZ, M. N. (1996), *Effect of 3-D Topography on Seismic Motion*, J. Geophys. Res. *101*, 5835–5846.

BOUTIN, C., BONNET, G., and BARD, P. Y. (1987), *Green Functions and Associated Sources in Infinite and Stratified Poroelastic Media*, Geophys. J. Roy. Astr. Soc. *90*, 521–550.

CAMPILLO, M. (1983), *Numerical Evaluation of Near-field, High-frequency Radiation from Quasi-dynamic Circular Faults*, Bull. Seismol. Soc. Am. *73*, 723–734.

CAMPILLO, M. (1987), *Lg Wave Propagation in a Laterally Varying Crust and the Distribution of the Apparent Quality Factor in Central France*, J. Geophys. Res. *92*, 12,604–12,614.

CAMPILLO, M., BOUCHON, M., and MASSINON, B. (1984), *Theoretical Study of the Excitation, Spectral Characteristics, and Geometrical Attenuation of Regional Seismic Phases*, Bull. Seismol. Soc. Am. *74*, 79–90.

CAMPILLO, M. and BOUCHON, M. (1985), *Synthetic Seismograms in a Laterally Varying Medium by the Discrete Wavenumber Method*, Geophys. J. Roy. Astr. Soc. *83*, 307–317.

CAMPILLO, M., PLANTET, J. L., and BOUCHON, M. (1985), *Frequency-dependent Attenuation in the Crust beneath Central France from L_g Waves: Data Analysis and Numerical Modeling*, Bull. Seismol. Soc. Am. *75*, 1395–1411.

CAMPILLO, M., GARIEL, J. C., AKI, K., and SANCHEZ-SESMA, F. J. (1989), *Destructive Strong Ground Motion in Mexico City: Source, Path, and Site Effects during Great 1985 Michoacan Earthquake*, Bull. Seismol. Soc. Am. *79*, 1718–1735.

CAMPILLO, M. and PAUL, A. (1992), *Influence of the Lower Crustal Structure on the Early Coda of Regional Seismograms*, J. Geophys. Res. *97*, 3405–3416.

CAMPILLO, M. and ARCHULETA, R. J. (1993), *A Rupture Model for the 28 June 1992 Landers, California, Earthquake*, Geophys. Res. Lett. *20*, 647–650.

CAMPILLO, M., FEIGNIER, B., BOUCHON, B., and BÉTHOUX, N. (1993), *Attenuation of crustal waves across the Alpine range*, J. Geophys. Res. *98*, 1987–1996.

CHAZALON, A., CAMPILLO, M. GIBSON, R., and CARRENO, E. (1993), *Crustal Wave Propagation Anomaly across the Pyrenean Range. Comparison between Observations and Numerical Simulations*, Geophys. J. Int. *115*, 829–838.

CHEN, X. (1990), *Seismogram Synthesis for Multi-layered Media with Irregular Interfaces by Global Generalized Reflection/Transmission Matrices Method. 1. Theory of Two-dimensional SH case*, Bull. Seismol. Soc. Am. *80*, 1696–1724.

CHEN, X. (1995), *Seismogram Synthesis for Multi-layered Media with Irregular Interfaces by Global Generalized Reflection/Transmission Matrices Method. 2. Application for 2D SH Case*, Bull. Seismol. Soc. Am. *85*, 1094–1106.

CHEN, X. (1996), *Seismogram Synthesis for Multi-layered Media with Irregular Interfaces by Global Generalized Reflection/Transmission Matrices Method. 3. Theory of 2D P-SV Case*, Bull. Seismol. Soc. Am. *86*, 389–405.

CHENG, C. H. (1989), *Full Waveform Inversion of P Waves for V_s and Q_p*, J. Geophys. Res. *94*, 15,619–15,625.

CHENG, C. H. and TOKSÖZ, M. N. (1981), *Elastic Wave Propagation in a Fluid-filled Borehole and Synthetic Acoustic Logs*, Geophysics *46*, 1042–1053.

CHENG, C. H., TOKSÖZ, M. N., and WILLIS, M. E. (1982), *Determination of in situ Attenuation from Full Waveform Acoustic Logs*, J. Geophys. Res. *87*, 5477–5484.

CHIN, B. H. and AKI, K. (1991), *Simultaneous Study of the Source, Path, and Site Effects on Strong Ground Motion during the 1989 Loma Prieta Earthquake: A Preliminary Result on Pervasive Nonlinear Site Effects*, Bull. Seismol. Soc. Am. *81*, 1859–1884.

CHOUET, B. (1981), *Ground Motion in the Near-field of a Fluid-driven Crack and its Interpretation in the Study of Shallow Volcanic Tremor*, J. Geophys. Res. *86*, 5985–6016.

CHOUET, B. (1982), *Free Surface Displacements in the Near Field of a Tensile Crack Expanding in Three Dimensions*, J. Geophys. Res. *87*, 3868–3872.

CHOUET, B. (1985), *Excitation of a Buried Magmatic Pipe: A Seismic Source Model for Volcanic Tremor*, J. Geophys. Res. *90*, 1881–1893.

CHOUET, B. and JULIAN B. R. (1985), *Dynamics of an Expanding Fluid-filled Crack*, J. Geophys. Res. *90*, 11,187–11,198.

CHRISTOFFERSSON, A., HUSEBYE, E. S., and INGATE, S. F. (1988), *Wavefield Decomposition using ML-Probabilities in Modelling Single-site 3-component Records*, Geophys. J. *93*, 197–213.

COTTON, F. and CAMPILLO, M. (1994), *Application of Seismogram Synthesis to the Study of Earthquake Source from Strong Motion Records*, Ann. Geofis. *37*, 1539–1564.

COTTON, F. and CAMPILLO, M. (1995a), *Inversion of Strong Ground Motion in the Frequency Domain: Application to the 1992 Landers, California, Earthquake*, J. Geophys. Res. *100*, 3961–3975.

COTTON, F. and CAMPILLO, M. (1995b), *Stability of Rake during the 1992, Landers Earthquake. An Indication for a Small Stress Release?*, Geophys. Res. Lett. *22*, 1921–1924.

COTTON, F., CAMPILLO, M., DESCHAMPS, A., and RASTOGI, B. K. (1996), *Rupture History and Seismotectonics of the 1991 Uttarkashi, Himalaya Earthquake*, Tectonophysics, *258*, 35–51.

COTTON, F. and COUTANT, O. (1997), *Dynamic Stress Variations due to Shear Faults in a Plane Layered Medium*, Geophys. J. Int. *128*, 676–688.

COURBOULEX, F., SANTOYO, M. A., PACHECO, J. F., and SINGH, S. K. (1997), *The 14 September 1995 (M = 7.3) Copala, Mexico Earthquake: A Source Study Using Teleseismic, Regional, and Local Data*, Bull. Seism. Soc. Am. *87*, 999–1010.

COUTANT, O. (1989), *Numerical Study of the Diffraction of Elastic Waves by Fluid-filled Cracks*, J. Geophys. Res. *94*, 17,805–17,818.

COUTANT, O. (1990), *Programme de Simulation numérique AXITRA*, Rapport LGIT, Université Joseph Fourier, Grenoble, France.

CRUSEM, R. and CARISTAN, Y. (1992), *Moment Tensor Inversion, Yield Estimation, and Seismic Coupling Variability at the French Centre d'expérimentation du Pacifique*, Bull. Seismol. Soc. Am. *82*, 1253–1274.

DIETRICH, M. (1988), *Modeling of Marine Seismic Profiles in the $t − x$ and $\tau − p$ Domains*, Geophysics *53*, 453–465.

DIETRICH, M. and BOUCHON M. (1985a), *Synthetic Vertical Seismic Profiles in Elastic Media*, Geophysics *50*, 224–234.

DIETRICH, M. and BOUCHON, M. (1985b), *Measurements of Attenuation from Vertical Seismic Profiles by Iterative Modeling*, Geophysics *50*, 931–949.

DONG, W., BOUCHON, M., and TOKSÖZ, M. N. (1995), *Borehole Seismic-source Radiation in Layered Isotropic and Anisotropic Media: Boundary Element Modeling*, Geophysics *60*, 735–747.

EBERHART-PHILLIPS, D., RICHARDSON, R. M., SBAR, M. L., and HERRMANN, R. B. (1981), *Analysis of the 4 February 1976 Chino Valley, Arizona, Earthquake*, Bull. Seismol. Soc. Am. *71*, 787–801.

FALK, J., TESSMER, E., and GAJEWSKI, D. (1996), *Tube Wave Modeling by the Finite-difference Method with Varying Grid Spacing*, Pure. Appl. Geophys. *148*, 77–93.

FU, L. Y. and WU, R. S. (2001), *A Hybrid BE-GS Method for Modeling Regional Wave Propagation*, Pure Appl. Geophys. *158*, 1251–1277.

FUKUSHIMA, Y., GARIEL, J. C., and TANAKA, R. (1995), *Site-dependent Attenuation Relations of Seismic Motion Parameters at Depth Using Borehole Data*, Bull. Seismol. Soc. Am. *85*, 1790–1804.

FUKUYAMA, E., KINOSHITA, S., and YAMAMIZU, F. (1991), *Unusual High-stress Drop Subevent during the M 5.5 Earthquake, the Largest Event of the 1989 Ito-Oki Swarm Activity*, Geophys. Res. Lett. *18*, 641–644.

FUKUYAMA, E. and MIKUMO, T. (1993), *Dynamic Rupture Analysis: Inversion for the Source Process of the 1990 Izu-Oshima, Japan, Earthquake(M = 6.5)*, J. Geophys. Res. *98*, 6529–6542.

GAFFET, S. (1995), *Teleseismic Waveform Modeling Including Geometrical Effects of Superficial Geological Structures near to Seismic Sources*, Bull. Seismol. Soc. Am. *85*, 1068–1079.

GAFFET, S. and BOUCHON, M. (1989), *Effects of Two-dimensional Topographies Using the Discrete Wavenumber − Boundary Integral Equation Method in P-SV Cases*, J. Acoust. Soc. Am. *85*, 2277–2283.

GAFFET, S. and BOUCHON, M. (1991), *Source Location and Valley Shape Effects on the P-SV Displacement Field Using a Boundary Integral Equation − Discrete Wavenumber Representation Method*, Geophys. J. Int. *106*, 341–355.

GAFFET, S., MASSINON, B., PLANTET, J. L., and CANSI, Y. (1994), *Modelling Local Seismograms of French Nuclear Tests in Taourirt tan Afella Massif, Hoggar, Algeria*, Geophys. J. Int. *119*, 964–974.

GARIEL, J. C., ARCHULETA, R. J., and BOUCHON, M. (1990), *Rupture Process of an Earthquake with Kilometric Size Fault Inferred from the Modeling of Near-source Records*, Bull. Seismol. Soc. Am. *80*, 870–888.

GARIEL, J. C. and CAMPILLO, M. (1989), *The Influence of the Source on the High-frequency Behavior of the Near-field Acceleration Spectrum: A Numerical Study*, Geophys. Res. Lett. *16*, 279–282.

GARIEL, J. C., BARD, P. Y., and PITILAKIS, K. (1991), *A Theoretical Investigation of Source, Path and Site Effects during the 1986 Kalamata Earthquake (Greece)*, Geophys. J. Int. *104*, 165–177.

GARVIN, W. (1956), *Exact Transient Solution for the Buried Line Source Problem*, Proc. Roy. Soc. London, Ser. A *203*, 528–541.

GIBSON, R. L. (1994), *Radiation from Seismic Sources in Cased and Cemented Boreholes*, Geophysics *59*, 518–533.

GIBSON, R. L. and CAMPILLO, M. (1994), *Numerical Simulation of High- and Low-frequency Lg-Wave Propagation*, Geophys. J. Int. *118*, 47–56.

GOLDSTEIN, P. and CHOUET, B. (1994), *Array Measurements and Modeling of Sources of Shallow Volcanic Tremor at Kilauea Volcano, Hawaii*, J. Geophys. Res. *99*, 2637–2652.

GOT, J. L. and FRÉCHET, J. (1993), *Origins of Amplitude Variations in Seismic Doublets: Source or Attenuation Process*, Geophys. J. Int. *114*, 325–340.

HAARTSEN, M. W., BOUCHON, M., and TOKSÖZ, M. N. (1994), *A Study of Seismic Acoustic Wave Propagation Through a Laterally-varying Multilayered Medium using the Boundary-integral-equation – Discrete-wave-number Method*, J. Acoust. Soc. Am. *96*, 3010–3021.

HAARTSEN, M. W. and PRIDE, S. R. (1997), *Electroseismic Waves from Point Sources in Layered Media*, J. Geophys. Res. *102*, 24,745–24,769.

HERRMANN, R. B., DEWEY, J. W., and PARK, S. K. (1980), *The Dulce, New Mexico, Earthquake of 23 January 1966*, Bull. Seismol. Soc. Am. *70*, 2171–2183.

HERNANDEZ, B., COTTON, F., and CAMPILLO, M. (1999), *Contribution of Radar Interferometry to a Two-step Inversion of the Kinematic Process of the 1992 Landers Earthquake*, J. Geophys. Res. *104*, 13,083–13,099.

IDE, S. and TAKEO, M. (1996), *The Dynamic Rupture Process of the 1993 Kushiro-oki Earthquake*, J. Geophys. Res. *101*, 5661–5675.

IDE, S., TAKEO, M., and YOSHIDA, Y. (1996), *Source Process of the 1995 Kobe Earthquake: Determination of Spatio-temporal Slip Distribution by Bayesian Modeling*, Bull. Seismol. Soc. Am. *86*, 547–566.

IDE, S. and TAKEO, M. (1997), *Determination of Constitutive Relations of Fault Slip Based on Seismic Wave Analysis*, J. Geophys. Res. *102*, 27,379–27,391.

JEAN, P. and BOUCHON, M. (1991), *Cross-borehole Simulation of Wave Propagation*, Geophysics *56*, 1103–1113.

JONGMANS, D. and MALIN, P. E. (1995), *Microearthquake S-wave Observations from 0 to 1 km in the Varian Well at Parkfield, California*, Bull. Seismol. Soc. Am. *85*, 1805–1820.

KARABULUT, H. and FERGUSON, J. F. (1996), *SH-Wave Propagation by Discrete Wavenumber Boundary Integral Modeling in Transversely Isotropic Medium*, Bull. Seismol. Soc. Am. *86*, 524–529.

KAWASE, H. (1988), *Time-domain Response of a Semicircular Canyon for Incident SV, P, and Rayleigh Waves Calculated by the Discrete Wavenumber Boundary Element Method*, Bull. Seismol. Soc. Am. *78*, 1415–1437.

KAWASE, H. and AKI, K. (1989), *A Study on the Response of a Soft Basin for Incident S, P, and Rayleigh Waves with Special Reference to the Long Duration Observed in Mexico City*, Bull. Seismol. Soc. Am. *79*, 1361–1382.

KAWASE, H. and AKI, K. (1990), *Topography Effect at the Critical SV Wave Incidence: Possible Explanation of Damage Pattern by the Whittier Narrows, California, Earthquake of 1 October 1987*, Bull. Seismol. Soc. Am. *80*, 1–22.

KENNETT, B. L. N. (1974), *Reflections, Rays, and Reverberations*, Bull. Seismol. Soc. Am. *64*, 1685–1696.

KENNETT, B. L. N. and KERRY, N. J. (1979), *Seismic Waves in a Stratified Half-space*, Geophys. J. Roy. Astr. Soc. *57*, 557–583.

KIM, J. and PAPAGEORGIOU, A. S. (1993), *Discrete Wavenumber Boundary-element Method for 3-D Scattering Problems*, J. Eng. Mech. ASCE *119*, 603–624.

LAMB, H. (1904), *On the Propagation of Tremors at the Surface of an Elastic Solid*, Phil. Trans. Roy. Soc. London, Ser. A *203*, 1–42.

LARNER, K. L. (1970), *Near-receiver Scattering of Teleseismic Body Waves in Layered Crust-mantle Models Having Irregular Interfaces*, Ph.D. Thesis, Massachusetts Institute of Technology, Cambridge, Massachusetts.

MANDAL, B. (1991), *Reflection and Transmission Properties of Elastic Waves on a Plane Interface for General Anisotropic Media*, J. Acoust. Soc. Am. *90*, 1106–1118.

MANDAL, B. and MITCHELL, B. J. (1986), *Complete Seismogram Synthesis for Transversely Isotropic Media*, J. Geophys. *59*, 149–156.

MANDAL, B. and TOKSÖZ, M. N. (1990), *Computation of Complete Waveforms in General Anisotropic Media – Results from an Explosion Source in an Anisotropic Medium*, Geophys. J. Int. *103*, 33–45.

MANDAL, B. and TOKSÖZ, M. N., *Effects of an Explosive Source in an Anisotropic Medium*. In *Explosion Source Phenomenology, Geophys. Mono. 65, (Am. Geophys. Union 1991)* pp. 261–268.

MAUPIN, V. (1996), *The Radiation Modes of a Vertically Varying Half-space: A New Representation of the Complete Green's Function in Terms of Modes*, Geophys. J. Int. *126*, 762–780.

MENDEZ, A. J. and LUCO, J. E. (1988), *Near-source Ground Motion from a Steady-state Dislocation Model in a Layered Half-space*, J. Geophys. Res. *93*, 12,041–12,054.

MENDOZA, C. and FUKUYAMA, E. (1996), *The July 12, 1993, Hokkaido-Nansei-Oki, Japan, Earthquake: Coseismic Slip Pattern from Strong-motion and Teleseismic Recordings*, J. Geophys. Res. *101*, 791–801.

MEREDITH, J. A., TOKSÖZ, M. N., and CHENG, C. H. (1993), *Secondary Shear Waves from Source Boreholes*, Geophys. Prosp. *41*, 287–312.

MOCZO, P., BISTRICKY, E., KRISTEK, J., CARCIONE, J. M., and BOUCHON, M. (1997), *Hybrid Modeling of P-SV Seismic Motion at Inhomogeneous Viscoelastic Topographic Structures*, Bull. Seismol. Soc. Am. *87*, 1305–1323.

MOCZO, P., LUCKA, M., KRISTEK, J., and KRISTEKOVA, M. (1999), *3D Displacement Finite Differences and a Combined Memory Optimization*, Bull. Seismol. Soc. Am. *89*, 69–79.

MOINET, F. and DIETRICH, M. (1998), *Computation of Differential Seismograms for Point and Plane Scatterers in Layered Media*, Bull. Seismol. Soc. Am. *88*, 1311–1324.

MÜLLER, G. (1985), *The Reflectivity Method: A Tutorial*, J. Geophys. *58*, 153–174.

NAKAYAMA, W. and TAKEO, M. (1997), *Slip History of the 1994 Sanriku-Haruka-Oki, Japan, Earthquake Deduced from Strong-motion Data*, Bull. Seismol. Soc. Am. *87*, 918–931.

NISHIMURA, T. (1995), *Source Parameters of the Volcanic Eruption Earthquakes at Mount Tokashi, Hokkaido, Japan, and a Magma Ascending Model*, J. Geophys. Res. *100*, 12,465–12,473.

NISHIMURA, T. and HAMAGUCHI, H. (1993), *Scaling Law of Volcanic Explosion Earthquake*, Geophys. Res. Lett. *20*, 2479–2482.

NISHIMURA, T., HAMAGUCHI, H., and UEKI, S. (1995), *Source Mechanisms of Volcanic Tremor and Low-frequency Earthquakes Associated with the 1988–89 Eruptive Activity of Mt. Tokachi, Hokkaido, Japan*, Geophys. J. Int. *121*, 444–458.

NISHIMURA, T., NAKAMICHI, H., TANAKA, S., SATO, M., KOBAYASHI, T., UEKI, S., HAMAGUCHI, H., OHTAKE, M., and SATO, H. (2000), *Source Process of Very Long Period Seismic Events Associated with the 1998 Activity of Iwate Volcano, Northeastern Japan*, J. Geophys. Res. *105*, 19,135–19,147.

OGLESBY, D. D. and ARCHULETA, R. J. (1997), *A Faulting Model for the 1992 Petrolia Earthquake: Can Extreme Ground Acceleration be a Source Effect?*, J. Geophys. Res. *102* 11,877–11,897.

OU, G. B. and HERRMANN, R. B. (1990), *A Statistical Model for Ground Motion Produced by Earthquakes at Local and Regional Distances*, Bull. Seismol. Soc. Am. *80*, 1397–1417.

PAPAGEORGIOU, A. S. and PEI, D. (1998), *A Discrete Wavenumber Boundary Element Method for 2.5-D Elastodynamic Scattering Problems*, Earthq. Eng. Struct, Dyn. *27*, 619–638.

PAUL, A. and CAMPILLO, M. (1988), *Diffraction and Conversion of Elastic Waves at a Corrugated Interface*, Geophysics *53*, 1415–1424.

PAUL, A. and NICOLLIN, F. (1989), *Thin Crustal Layering in Northern France: Observations and Modelling of the PmP Spectral Content*, Geophys. J. Int. *99*, 229–246.

PEDERSEN, H. and CAMPILLO, M. (1991), *Depth Dependence of Q Beneath the Baltic Shield Inferred from Modeling Short-period Seismograms*, Geophys. Res. Lett. *18*, 1755–1758.

PEYRAT, S., OLSEN, K., and MADARIAGA, R. (2001), *Dynamic Modeling of the 1992 Landers Earthquake*, J. Geophys. Res. *106*, 26467–26482.

PLICKA, V. and ZAHRADNIK, J. (1998), *Inverting Seismograms of Weak Events for Empirical Green's Tensor Derivatives*, Geophys. J. Int. *132*, 471–478.

PLICKA, V., SOKOS, E., TSELENTIS, G. A., and ZAHRADNIK, J. (1998), *The Patras Earthquake (14 July 1993): Relative Roles of Source, Path and Site Effects*, J. Seismology *2*, 337–349.

QUINTANAR, L., YAMAMOTO, J., and JIMENEZ, Z. (1999), *Source Mechanism of Two Intermediate-depth-focus Earthquakes in Guerrero, Mexico*, Bull. Seismol. Soc. Am. *89*, 1004–1018.

RAYLEIGH, LORD (J. W. STRUTT), *The Theory of Sound*, vol. 2, 2nd ed., sect. 272 (Macmillan, London, 1896).

RAYLEIGH, LORD (J. W. STRUTT) (1907), *On the Dynamical Theory of Gratings*, Proc. Roy. Soc. London, Ser. A *79*, 399–416.

REBOLLAR, C. J., QUINTANAR, L., YAMAMOTO, J., and URIBE, A. (1999), *Source Process of the Chiapas, Mexico, Intermediate-depth Earthquake (M_w = 7.2) of 21 October 1995*, Bull. Seismol. Soc. Am. *89*, 348–358.

RIEPL, J., ZAHRADNIK, J., PLICKA, V., and BARD, P. Y. (2000), *About the Efficiency of Numerical 1-D and 2-D Modelling of Site Effects in Basin Structures*, Pure Appl. Geophys. *157*, 319–342.

ROBERTS, R. G. and CHRISTOFFERSSON, A. (1990), *Decomposition of Complex Single-station Three-component Seismograms*, Geophys. J. Int. *103*, 55–74.

SAIKIA, C. K. and HERRMANN, R. B. (1986), *Moment-tensor Solutions for three 1982 Arkansas Swarm Earthquakes by Waveform Modeling*, Bull. Seismol. Soc. Am. *76*, 709–723.

SAIKIA, C. K. and HERRMANN, R. B. (1987), *Determination of Focal Mechanism Solutions for four Earthquakes from Monticello, South Carolina, and Crustal Structure by Waveform Modelling*, Geophys. J. Roy. Astr. Soc. *90*, 669–691.

SAIKIA, C. K. and HEMBERGER, D. V. (1997), *Approximation of Rupture Directivity in Regional Phases Using Upgoing and Downgoing Wave Fields*, Bull. Seismol. Soc. Am. *87*, 987–998.

SCHMITT, D. P. (1988a), *Effects of Radial Layering when Logging in Saturated Porous Formations*, J. Acoust. Soc. Am. *84*, 2200–2214.

SCHMITT, D. P. (1988b), *Shear Wave Logging in Elastic Formations*, J. Acoust. Soc. Am. *84*, 2215–2229.

SCHMITT, D. P. (1990), *Acoustic Multiple Logging in Transversely Isotropic Poroelastic Formations*, J. Acoust. Soc. Am. *86*, 2397–2421.

SCHMITT, D. P. and BOUCHON, M. (1985), *Full Wave Acoustic Logging: Synthetic Microseismograms and Frequency-wavenumber Analysis*, Geophysics *50*, 1756–1778.

SCHMITT, D. P., BOUCHON, M., and BONNET, G. (1988a), *Full Wave Synthetic Acoustic Logs in Radially Semi-infinite Saturated Porous Media*, Geophysics *53*, 807–823.

SCHMITT, D. P., ZHU, Y., and CHENG, C. H. (1998b), *Shear-wave Logging in Semi-infinite Saturated Porous Formations*, J. Acoust. Soc. Am. *84*, 2230–2244.

SEKIGUCHI, H., IRIKURA, K., IWATA, T., KAKEHI, Y., and HOSHIBA, M. (1996), *Minute Locating of Fault Planes and Source Process of the 1995 Hyogo-ken Nanbu, Japan, Earthquake from the Waveform Inversion of Strong Ground Motion*, J. Phys. Earth *44*, 473–487.

SEKIGUCHI, H., IRIKURA, K., and IWATA, T. (2000), *Fault Geometry at the Rupture Termination of the 1995 Hyogo-ken Nanbu Earthquake*, Bull. Seismol. Soc. Am. *90*, 117–133.

SHAPIRO, N., BÉTHOUX, N., CAMPILLO, M., and PAUL, A. (1996), *Regional Seismic Phases across the Ligurian Sea: Lg Blockage and Oceanic Propagation*, Phys. Earth Planet. Inter. *93*, 257–268.

SINGH, S. K., PACHECO, J., COURBOULEX, F., and NOVELO, D. A. (1997), *Source Parameters of the Pinotepa Nacional, Mexico, Earthquake of 27 March, 1996 (M_w = 5.4) Estimated from Near-field Recordings of a Single Station*, J. Seismology *1*, 39–45.

SINGH, S. K., ORDAZ, M., DATTATRAYAM, R. S., and GUPTA, H. K. (1999a), *A Spectral Analysis of the 21 May 1997, Jabalpur, India, Earthquake (M_w = 5.8) and Estimation of Ground Motion from Future Earthquakes in the Indian Shield Region*, Bull. Seismol. Soc. Am. *89*, 1620–1630.

SINGH, S. K., DATTATRAYAM, R. S., SHAPIRO, N. M., MANDAL, P., PACHECO, J. F., and MIDHA, R.4K. (1999b), *Crustal and Upper Mantle Structure of Peninsular India and Source Parameters of the 21 May 1997, Jabalpur Earthquake (M_w = 5.8): Results from a New Regional Broadband Network*, Bull. Seismol. Soc. Am. *89*, 1631–1641.

SINGH, S. K., ORDAZ, M., PACHECO, J. F., and COURBOULEX, F. (2000), *A Simple Source Inversion Scheme for Displacement Seismograms Recorded at Short Distances*, J. Seismology *4*, 267–284.

STEIDL, J. H., TUMARKIN, A. G., and ARCHULETA, R. J. (1996), *What is a Reference Site?*, Bull. Seismol. Soc. Am. *86*, 1733–1748.

STEPHEN, R. A., CARDO-CASAS, F., and CHENG, C. H. (1985), *Finite-difference Synthetic Acoustic Logs*, Geophysics *50*, 1588–1609.

TAKENAKA, H., KENNETT, B. L. N., and FUJIWARA, H. (1996), *Effect of 2-D Topography on the 3-D Seismic Wavefield Using a 2.5-D Discrete Wavenumber-boundary Integral Equation Method*, Geophys. J. Int. *124*, 741–755.

TAKEO, M. (1987), *An Inversion Method to Analyse the Rupture Processes of Earthquakes using Near-field Seismograms*, Bull. Seismol. Soc. Am. *77*, 490–513.

TAKEO, M. (1988), *Rupture Process of the 1980 Izu-Hanto-Toho-Oki Earthquake Deduced from Strong Motion Seismograms*, Bull. Seismol. Soc. Am. *78*, 1074–1091.

TAKEO, M. (1990), *Analysis of Long-period Seismic Waves Excited by the November 1987 Eruption of Izu-Oshima Volcano*, J. Geophys. Res. *95*, 19,377–19,393.

TAKEO, M. (1992), *The Rupture Process of the 1989 Offshore Ito Earthquakes Preceding a Submarine Volcanic Eruption*, J. Geophys. Res. *97*, 6613–6627.

TAKEO, M., IDE, S., and YOSHIDA, Y. (1993), *The 1993 Kushiro-Oki, Japan, Earthquake: A High Stress-drop Event in a Subducting Slab*, Geophys. Res. Lett. *20*, 2607–2610.

TAKEO, M. and ITO, H. M. (1997), *What Can be Learned from Rotational Motions Excited by Earthquakes?*, Geophys. J. Int. *129*, 319–329.

TAKEO, M. and KANAMORI, H. (1997), *Simulation of Long-period Ground Motion Near a Large Earthquake*, Bull. Seismol. Soc. Am. *87*, 140–156.

TEYSSONEYRE, V., FEIGNIER, B., SILENY, J., and COUTANT, O. (2001), *Moment Tensor Inversion of Regional Phases: Application to a Mine Collapse*, Pure Appl. Geophys., *159*, 111–130.

THEODULIDIS, N., BARD, P. Y., ARCHULETA, R. J., and BOUCHON, M. (1996), *Horizontal-to-vertical Spectral Ratio and Geological Conditions: The Case of Garner Valley Downhole Array in Southern California*, Bull. Seismol. Soc. Am. *86*, 306–319.

TOKSÖZ, M. N., MANDAL, B., and DAINTY, A. (1990), *Frequency Dependent Attenuation in the Crust*, Geophys. Res. Lett. *17*, 973–976.

TSELENTIS, G. A. and ZAHRADNIK, J. (2000), *The Athens Earthquake of September 7, 1999*, Bull. Seismol. Soc. Am. *90*, 1143–1160.

UHIRA, K., YAMASATO, H., and TAKEO, M. (1994), *Source Mechanism of Seismic Waves Excited by Pyroclastic Flows Observed at Unzen Volcano, Japan*, J. Geophys. Res. *99*, 17,757–17,773.

YAO, Z. X. and HARKRIDER, D. G., (1983), *A Generalized Reflection-transmission Coefficient Matrix and Discrete Wavenumber Method for Synthetic Seismograms*, Bull. Seismol. Soc. Am. *73*, 1685–1699.

ZAHRADNIK, J. (1995), *A New Program Package for Modelling Seismic Ground Motions*, Proc. Fifth Int. Conf. on Seismic Zonation, Nice (France), *2*, 1221–1226.

ZAHRADNIK, J. and MOCZO, P. (1996), *Hybrid Seismic Modeling Based on Discrete Wavenumber and Finite-difference Methods*, Pure Appl. Geophys. *148*, 21–38.

ZENG, Y. (1993), *Theory of Scattered P- and S-wave Energy in a Random Isotropic Scattering Medium*, Bull. Seismol. Soc. Am. *83*, 1264–1276.

ZENG, Y., SU, F., and AKI, K. (1991), *Scattering Wave Energy Propagation in Random Isotropic Scattering Medium: 1. Theory*, J. Geophys. Res. *96*, 607–619.

ZHANG, B., PAPAGEORGIOU, A. S., and TASSOULAS, J. L. (1998), *A Hybrid Numerical Technique, Combining the Finite-element and Boundary-element Methods, for Modeling the 3D Response*, Bull. Seismol. Soc. Am. *88*, 1036–1050.

ZOLLO, A., DE MATTEIS, R., CAPUANO, P., FERULANO, F., and IANNACCONE, G. (1995), *Constraints on the Shallow Crustal Model of the Northern Apennines (Italy) from the Analysis of Microearthquake Seismic Records*, Geophys. J. Int. *120*, 646–662.

ZOLLO, A. and IANNACCONE, G. (1996), *Site and Propagation Effects on the Spectra of an S-to-S Reflected Phase Recorded from a Set of Microearthquakes in the Northern Apennines (Italy)*, Geophys. Res. Lett. *23*, 1163–1166.

(Received August 18, 2000, accepted April 5, 2001)

 To access this journal online:
http://www.birkhauser.ch

Pure appl. geophys. 160 (2003) 467–486
0033–4553/03/040467–20

Pure and Applied Geophysics

An Efficient Numerical Method for Computing Synthetic Seismograms for a Layered Half-space with Sources and Receivers at Close or Same Depths

HAI-MING ZHANG[1], XIAO-FEI CHEN[1] and SHYHONG CHANG[2]

Abstract — It is difficult to compute synthetic seismograms for a layered half-space with sources and receivers at close to or the same depths using the generalized R/T coefficient method (KENNETT, 1983; LUCO and APSEL, 1983; YAO and HARKRIDER, 1983; CHEN, 1993), because the wavenumber integration converges very slowly. A semi-analytic method for accelerating the convergence, in which part of the integration is implemented analytically, was adopted by some authors (APSEL and LUCO, 1983; HISADA, 1994, 1995). In this study, based on the principle of the Repeated Averaging Method (DAHLQUIST and BJÖRCK, 1974; CHANG, 1988), we propose an alternative, efficient, numerical method, the *peak-trough averaging method* (PTAM), to overcome the difficulty mentioned above. Compared with the semi-analytic method, PTAM is not only much simpler mathematically and easier to implement in practice, but also more efficient. Using numerical examples, we illustrate the validity, accuracy and efficiency of the new method.

Key words: Numerical integration method, layered half-space, synthetic seismograms, the peak-trough averaging method.

Introduction

Computing synthetic seismograms in a layered half-space is an important tool for investigating the interior structure of the earth as well as the dynamic process of seismic sources from well-recorded seismic data. Since the 1970s, many endeavors have been made regarding the calculation of Green's functions in seismological studies (FUCHS and MÜLLER, 1971; HELMBERGER, 1974; BOUCHON and AKI, 1977; BOUCHON, 1979; KENNETT and KERRY, 1979; WANG and HERRMANN, 1980; LUCO and APSEL; 1983; APSEL and LUCO, 1983; KENNETT, 1983; YAO and HARKRIDER, 1983; DRAVINSKI and MOSSESSIAN, 1988; CHEN, 1993, 1999; HISADA, 1994, 1995). According to the generalized R/T coefficient method (KENNETT, 1983; CHEN, 1999), the Green's function due to an arbitrary point seismic source buried in a layered

---

[1] Department of Geophysics, Peking University, Beijing 100871, China. E-mail: xfchen@pku.edu.cn
[2] Price Bargain Inc., Los Angeles, CA 90072, U.S.A.

half-space can be expressed as a summation of the products of radiation patterns and the following type of oscillatory integrals,

$$I_n(\omega) = \int_0^{+\infty} F(\omega, k) \cdot J_n(kr) \, dk \ , \tag{1}$$

where r is the epicentral distance, k is the horizontal wavenumber and ω is the circular frequency. $F(\omega, k)$ is the kernel function which includes a decaying factor $\exp(-\zeta^{(j)}|z^{(j)} - z^{(j-1)}|)$, $\exp(-\zeta^{(j)}|z - z^{(j)}|)$ or $\exp(-\zeta^{(j)}|z - z_s|)$, in which $\zeta^{(j)} = \sqrt{k^2 - (\omega/c^{(j)})^2}$ ($c^{(j)}$ is the wave velocity in the j-th layer) (see, CHEN, 1993). The depths of the receiver, the j-th interface and the source are represented by z, $z^{(j)}$, and z_s, respectively, and $J_n(kr)$ is the Bessel function of order n. The speed of convergence of the integrand in equation (1) is determined by both the kernel function and the Bessel function. When the sources and receivers are at close to or the same depths, both the kernel function and the Bessel function converge very slowly with k making the integral difficult to compute numerically. To explore the numerical convergence of integral (1), we introduce the following definite integral,

$$P_n(\omega, k) = \int_0^k F(\omega, \tilde{k}) \cdot J_n(\tilde{k}r) \, d\tilde{k} \ , \tag{1a}$$

where

$$I_n(\omega) = \lim_{k \to +\infty} \{P_n(\omega, k)\} \ .$$

We call $P_n(\omega, k)$ the partial integral of $I_n(\omega)$. Figures 1(a)–(c) show the kernel function $F(\omega, k)$, the integrand $F(\omega, k) \cdot J_n(kr)$ ($n = 2$), and the partial integral $P_n(\omega, k)$ respectively, versus wavenumber k for a particular layered media (Crust Model 2 in Table 3) at $f = 1.0$ Hz. Since the imaginary parts of the integrand and the partial integral converge more quickly than those of real parts (CHANG, 1988; HISADA, 1994), only the real parts are shown. We can see that both the integrand and partial integral converge very slowly. It is expected that the partial integral will converge at very large k, however, the corresponding computation is too expensive to implement. One remedy for this kind of problem is the semi-analytical method (abbreviated to SAM, see APSEL and LUCO, 1983; HISADA, 1994, 1995), in which the integral is broken into two parts; one of them is carried out analytically (we call them *numerical integral* and *analytic integral*, respectively, for the convenience in description). The prime merit of the semi-analytic method is that the integration interval of the numerical integral is greatly reduced. However, on the one hand, to what extent the integration interval is reduced depends heavily on the choice of the analytic integral. Usually such a choice is rather complicated mathematically (HISADA, 1994, 1995). On the other hand, since the result of the analytic integral is an

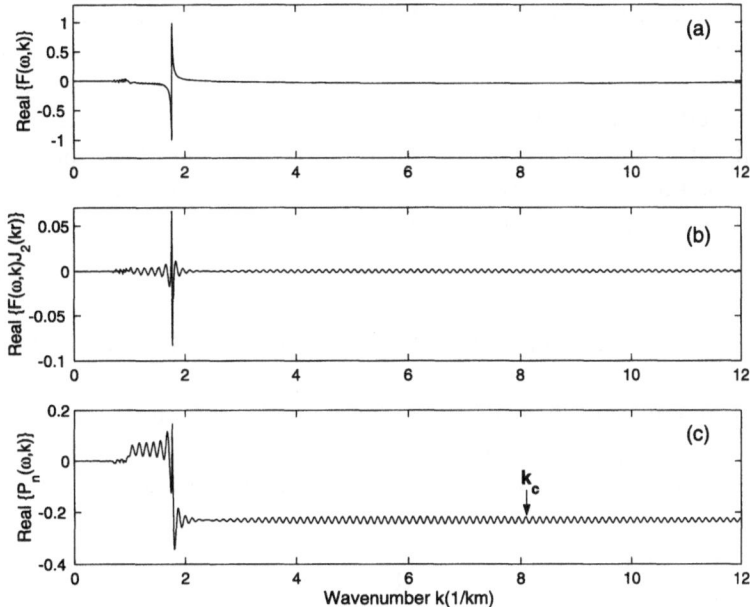

Figure 1

(a)–(c) Real parts of the kernel function $F(\omega, k)$, the integrand $F(\omega, k) \cdot J_n(kr)$ (for $n = 2$), and the partial integral $P_n(\omega, k)$ versus wavenumber k at a given frequency ($f = 1.0$ Hz). k_c in (c) is the critical wavenumber, beyond which the partial integral $P_n(\omega, k)$ becomes an oscillatory curve with a monotonically and smoothly decaying envelope. Crust Model 2 (see Table 3) is used. Here the epicentral distance is $r = 50$ km, the focal depth is $z_s = 0.3$ km, and the receiver depth is $z_0 = 0.0$ km. Notice that all traces here are normalized by the maximum amplitude of the top curve.

exact one, the integration step in the numerical integral, which is implemented numerically, must be very small to reduce the discrepancy in preciseness of the two integrals. Consequently, although the integration interval in the numerical integral is greatly reduced, considerable numerical computation is still involved.

In this study we propose an alternative method, the *Peak-Trough Averaging Method* (PTAM), based on the principle of the Repeated Averaging Method (DAHLQUIST and BJÖRCK, 1974; CHANG, 1988). The Repeated Averaging Method (RAM) is an efficient approach for evaluating the convergent values of slowly convergent alternating sequences (DAHLQUIST and BJÖRCK, 1974). As shown in Figure 1(c), beyond a critical wavenumber k_c, the partial integral $P_n(\omega, k)$ becomes an oscillatory curve with a monotonically and smoothly decaying envelope, indicating that the process is slowly approaching its convergent point, $I_n(\omega)$. The distribution of the peaks and troughs of this slowly convergent partial integral behaves like a monotonically decaying alternating sequence, thus it can be efficiently evaluated by applying RAM. Such an integral evaluation method is named the Peak-Trough Averaging Method. Compared with the semi-analytic approach, PTAM is much simpler mathematically and easier to implement in practice. By using a numerical test, we will demonstrate that the precision of PTAM is slightly higher than that of SAM with the same integration step and truncated upper limit, and PTAM is more efficient than SAM. In what follows we shall first briefly introduce the repeated averaging method, then describe its application to the evaluation of slowly convergent integrals of the kind of integral (1), i.e., introduce the peak-trough averaging method. Finally we will demonstrate the accuracy, efficiency and applicability of the new method by numerical examples.

Fundamentals of the Repeated Averaging Method

To explore our new efficient integration method, we shall first briefly introduce the *Repeated Averaging Method* (RAM) by considering the evaluation of the following slowly convergent alternating series (DAHLQUIST and BJÖRCK, 1974),

$$1 - \frac{1}{3} + \frac{1}{5} - \frac{1}{7} + \frac{1}{9} - \frac{1}{11} + \cdots + \frac{(-1)^n}{2n+1} + \cdots . \tag{2}$$

This series very slowly converges to its sum, $\pi/4$. We can define a corresponding sequence $\{M_0(n); n = 1, 2, \ldots\}$ as follows:

$$M_0(n) = \sum_{i=1}^{n} \frac{(-1)^{i-1}}{2i-1} . \tag{2a}$$

Obviously, the limit of this sequence equals the sum of the original series (2), namely,

$$\lim_{n \to +\infty} \{M_0(n)\} = \frac{\pi}{4} .$$

Hence, in what follows we shall discuss how to efficiently evaluate the limit of the sequence $\{M_0(n); n = 1, 2, \ldots\}$ rather than the sum of the original series (2). As shown in Figure 2(a), the partial sum sequence $M_0(n)$ converges to $\pi/4$ very slowly. $M_0(n)$ alternates around its limit $\pi/4$ with a monotonically and smoothly decaying envelope. To describe the convergence speed of the sequence $\{M_0(n)\}$, we define a relative error sequence $E_0(n)$ of $M_0(n)$ to its limit as follows,

$$E_0(n) = \frac{|M_0(n) - M_0(+\infty)|}{|M_0(+\infty)|} . \tag{3}$$

Obviously, $E_0(n) \to 0$ as $n \to +\infty$. Figure 2(b) displays the distribution of relative error sequence $E_0(n)$ that indicates the convergence speed of $M_0(n)$. It can be seen that the convergence speed of $M_0(n)$ is very slow. Accordingly, the usual direct evaluation of this kind of slowly convergent alternating sequence is very time-consuming and thus is less efficient. Fortunately, a method called the Repeated Averaging Method (RAM) described by DAHLQUIST and BJÖRCK (1974) can

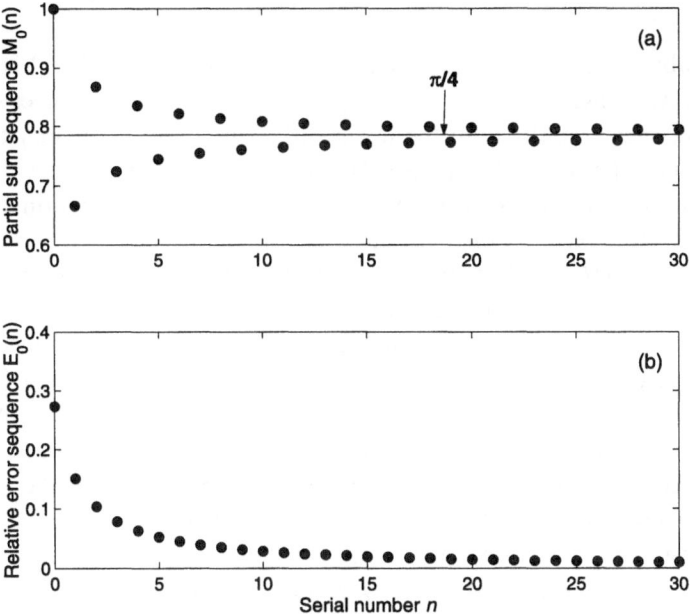

Figure 2

(a) The partial sum sequence $M_0(n)$ (see equation (2a) in the text) alternates around its limit point $\pi/4$ with a monotonously and smoothly decaying envelope. (b) The distribution of the relative error sequence $E_0(n)$, which indicates the speed of convergence of $M_0(n)$.

dramatically improve the evaluation efficiency of this kind of slowly convergent alternating sequence.

To explain the basic idea of RAM clearly, we first define the i-th order *reduced sequence* $M_i(n)$ and the corresponding *relative error sequence* $E_i(n)$ as follows,

$$M_i(n) = \frac{1}{2}[M_{i-1}(n+1) + M_{i-1}(n)] \quad (i = 1, 2, 3 \ldots) \tag{4a}$$

$$E_i(n) = \frac{|M_i(n) - M_i(+\infty)|}{|M_i(+\infty)|} \quad (i = 1, 2, 3 \ldots) \ , \tag{4b}$$

where $M_0(n)$ is defined in equation (2a). Since the limit of the sequence $M_0(n)$ does exist, we can show that all the reduced sequences $M_i(n)$ $(i = 1, 2, 3 \ldots)$ converge to the same limit as that of $M_0(n)$ as follows.

For $i = 1$,

$$\lim_{n \to +\infty} M_1(n) = \frac{1}{2}\left[\lim_{n \to +\infty} M_0(n+1) + \lim_{n \to +\infty} M_0(n)\right]$$
$$= \frac{1}{2}[M_0(+\infty) + M_0(+\infty)] = M_0(+\infty) \ . \tag{5a}$$

Likewise, using the iterative relation (4a) one can easily show the results for $i > 1$ below,

$$M_i(+\infty) = M_{i-1}(+\infty) = \cdots = M_1(+\infty) = M_0(+\infty) \ . \tag{5b}$$

Another notable property of the reduced sequence $M_i(n)(i = 1, 2, 3 \ldots)$ is its speed of convergence. As shown in Figure 3, the higher order reduced sequences converge to the limit value more rapidly than the lower order ones, and the higher the order is, the more rapidly the reduced sequence $M_i(n)$ converges to the limit value than the original sequence $M_0(n)$. Accordingly, the relative error function $E_i(n)$ more rapidly goes to zero. For instance, $E_0(n) < 10^{-6}$ for $n > 3 \times 10^5$, i.e., $\{|M_0(n) - M_0(\infty)|/|M_0(\infty)|\} < 10^{-6}$ for $n > 3 \times 10^5$. This means that with an accuracy of 10^{-6}, the usual direct evaluation of infinite series (2) needs at least 3×10^5 terms. However, as seen from Figure 3, the sixth-order reduced sequence $\{M_6(n)\}$ converges so fast that n is only required to be greater than 5 for the same accuracy. Thus $\{|M_6(n) - M_6(\infty)|/|M_6(\infty)|\} < 10^{-6}$ for $n > 5$, which is five orders of magnitude faster than the usual direct evaluation of the sum of infinite series (2) or the limit of sequence $\{M_0(n); n = 1, 2, \ldots\}$. On the other hand, all the higher order reduced sequences converge to the *same* limit as demonstrated in equation (5). Therefore we can evaluate the limit of higher order reduced sequence $\{M_i(n); n = 1, 2, \ldots\}$, rather than directly evaluate the limit of the original sequence $\{M_0(n); n = 1, 2, \ldots\}$, so that the computation efficiency will be dramatically

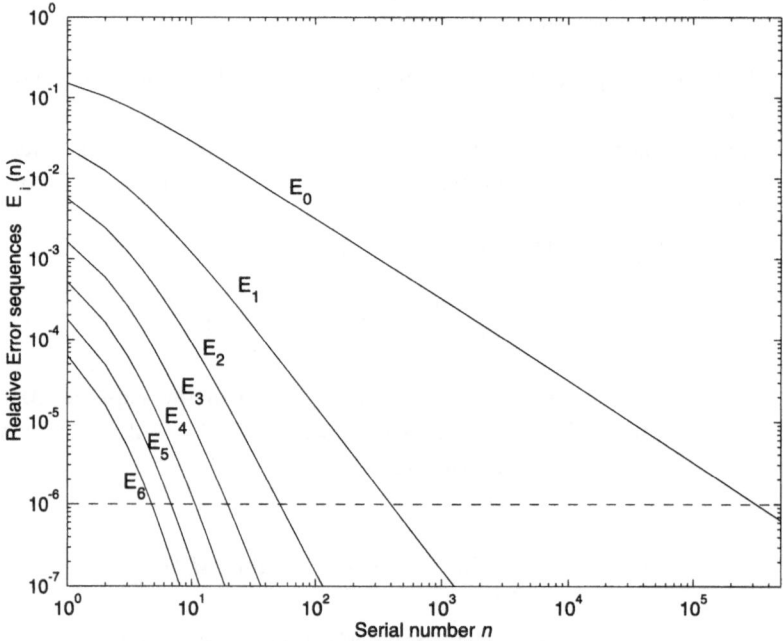

Figure 3
Distribution of relative error sequences $E_i(n)$ of different orders. The higher order reduced sequences converge to the limit value more rapidly than the lower order ones.

improved as shown above. Such an evaluation is not only mathematically efficient, but also easy to implement for computer calculation. Table 1 illustrates the arithmetic evaluation process, in which each term of the higher order reduced sequence is simply generated by the arithmetic mean of the corresponding pair of neighbor terms of the previous lower order reduced sequence as indicated by the short lines. It should be recognized that to generate the n-th term of the i-th reduced sequence $M_i(n)$, the first $(n + i)$ terms of the original infinite series are needed, according to the definition of $M_i(n)$ given in equation (4a). In summary, RAM is not only very efficient and accurate but also a very simple algorithm to evaluate an alternating series or the limit of an alternating sequence (DAHLQUIST and BJÖRCK, 1974).

The Peak-trough Averaging Method

We now consider how to efficiently evaluate the integral (1) by applying RAM. As shown in Figure 1(c), beyond a critical k_c, the partial integral $P_n(\omega, k)$ becomes an oscillatory function with a monotonically and smoothly decaying envelope, indicating the process of slowly approaching its limit, $I_n(\omega)$. The behavior of the peaks and troughs of this slowly convergent function acts like an alternating sequence, which suggests adopting RAM to efficiently evaluate the slowly convergent integral (1) by the following procedures.

First, determine the critical k_c beyond which the partial integral $P_n(\omega, k)$ becomes an oscillatory function with a monotonically and smoothly decaying envelope. The critical wavenumber k_c can be determined by an empirical formula

$$k_c = 1.5 \times |\omega|/v_{\min} , \tag{6}$$

where $v_{\min}$ is the minimum velocity of the structure model, ω is the circular frequency. When the discrete wavenumber integration method (see e.g., BOUCHON and AKI, 1977) is used, an imaginary frequency is introduced to depress the influence

Table 1

Arithmetic evaluation process of the Repeated Averaging Method. The bits which are different in higher order reduced sequences M_i $(i = 1, 2, \ldots, 6)$ are underlined

n	M_0	M_1	M_2	M_3	M_4	M_5	M_6
5	0.744012	0.782474	0.785038	0.785340	0.785387	0.785396	0.785398
6	0.820935	0.787602	0.785641	0.785434	0.785405	0.785400	
7	0.754268	0.783680	0.785228	0.785376	0.785395		
8	0.813092	0.786776	0.785523	0.785414			
9	0.760460	0.784270	0.785305				
10	0.808079	0.786340					
11	0.764601						

of fictitious sources and structures. In such a case, $\omega = \omega_R + i\omega_I$, where ω_R and ω_I is the real and imaginary circular frequency, respectively. As mentioned earlier, the kernel function $F(\omega, k)$ in the j-th layer includes a decaying factor $\exp(-\zeta^{(j)}|z - z_s|)$, with $\zeta^{(j)} = \sqrt{k^2 - (\omega/c^{(j)})^2}$ ($c^{(j)}$ is the wave velocity in the j-th layer). Obviously, $\exp(-\zeta^{(j)}|z - z_s|)$ behaves as an oscillatory function when $k < |\omega|/c^{(j)}$; however, it becomes a decaying factor when $k > |\omega|/c^{(j)}$. Consequently, $|\omega|/v_{\min}$ is a critical point, beyond which the partial integral $P_n(\omega, k)$ decays with the increase of upper limit k (see, Figure 1(c)). Usually in practice, an empirical coefficient, such as 1.5 in equation (6), is added to ensure the decaying property of $P_n(\omega, k)$.

Second, determine the raw peaks and troughs of the curve of $P_n(\omega, k)$ versus k in the range of k greater than k_c by the following simple screening process as the integral evaluation continues. As k ($k > k_c$) increases with a fixed step-size Δk (in discrete wavenumber integration method, $\Delta k = 2\pi/L$, see, BOUCHON and AKI, 1977), we record every three successive integral sampling points k_i and the corresponding values of the partial integral S_i ($= P_n(\omega, k_i)$), i.e., $(k_i, S_i)(i = 1, 2, 3)$. If S_2 is larger or smaller than both S_1 and S_3, then (k_2, S_2) is a rough location of a peak or trough. If not, there is no peak or trough found in this interval. Regardless of the case, move forward one point and form a new three successive integral sampling points and the partial integral values $(k_i', S_i')(i = 1, 2, 3)$, then repeat the above screening procedure until enough peaks and troughs are found.

Finally, regard the sequential peaks and troughs as an alternating sequence, and apply RAM to evaluate the limit value that equals $I_n(\omega)$. As shown in Figure 1(c), the sequential distribution of those peaks and troughs behaves like an alternating sequence (hereafter we call it a *peak-trough sequence*) whose limit is believed to be the same as the convergent value of the partial integral $P_n(\omega, k)$, i.e., $I_n(\omega)$. This allows us to adopt the efficient algorithm, RAM, to evaluate the integral $I_n(\omega)$ through evaluating the limit value of the *peak-trough sequence*. Therefore, we designate this efficient integration method as the *Peak-Trough Averaging Method* (PTAM).

It is noted that the accuracy of this approach heavily depends on the accuracy of the values of those peaks and troughs. However, the raw values of peaks and troughs obtained by the above screening process are not accurate enough, and must be further refined. This can be achieved by a simple quadratic interpolation technique. We construct the following quadratic interpolation polynomial

$$S(k) = a_1 \left(\frac{k - k_1}{k_3 - k_1}\right)^2 + a_2 \left(\frac{k - k_1}{k_3 - k_1}\right) + a_3 , \qquad (7)$$

where $a_1 = 2S_3 - 4S_2 + 2S_1$, $a_2 = 4S_2 - S_3 - 3S_1$, $a_3 = S_1$, and $S_i = S(k_i)$ ($i = 1, 2, 3$), and the coordinates of the refined peak (when $a_1 < 0$) or trough (when $a_1 > 0$) can be easily found as

$$\left(k_1 - \frac{a_2}{2a_1}(k_3 - k_1), \ S_1 - \frac{a_2^2}{4a_1} \right) .$$

Applying the above process to all raw peaks and troughs, we can obtain a set of refined peaks and troughs. Figure 4 shows an example of refining the location and peak value from a raw location and peak value (k_2, S_2). The dashed line is a simple sine function; the analytic form of which is assumed to be unknown in a practical computation, and the cross and asterisk are the accurate peak and refined peak obtained by using the above technique, respectively. It can be seen that the refined peak (k^*, S^*) is decidedly more accurate than the raw one; actually, the absolute error of S^* to the accurate value is less than 10^{-5}, whereas the error of the raw one is about 2.5×10^{-3}. Having these accurate peaks and troughs, we can directly apply the RAM to evaluate the limit value of the related peak-trough sequence, and thus obtain the integration value of equation (1).

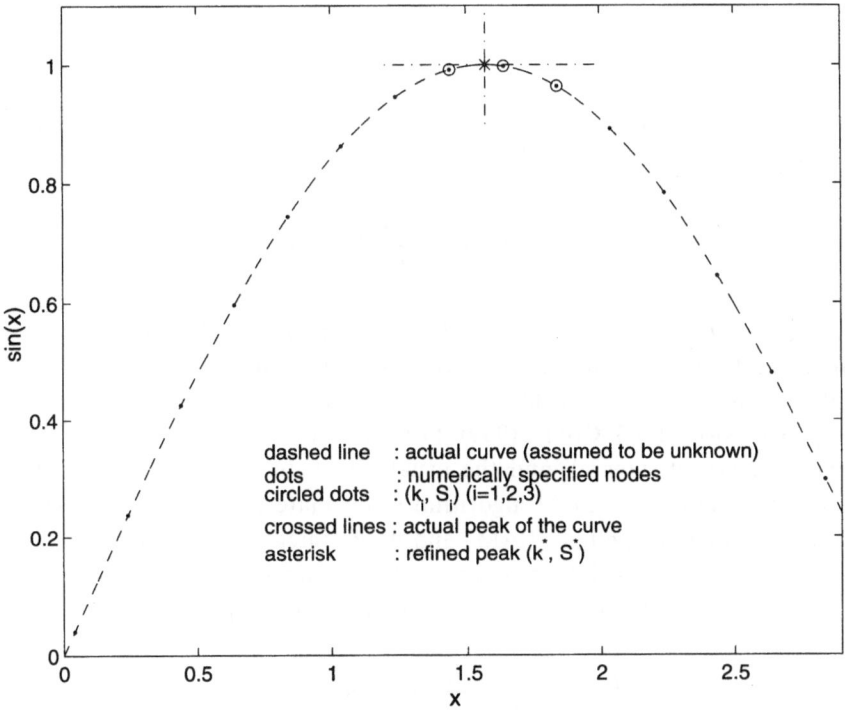

Figure 4

An example of refining the location and peak value from a raw peak (k_2, S_2). Here the dashed line is $\sin(x)$, the analytic form of which is assumed to be unknown in practical numerical computation. The dots represent the numerically specified nodes, the circled dots represent the three successive points $(k_i, S_i)(i = 1, 2, 3)$ to be interpolated, and the asterisk and cross represent the refined peak (k^*, S^*) and the accurate value, respectively. The absolute error of S^* to the accurate value is less than 10^{-5}, while the error of the raw one is about 2.5×10^{-3}.

Validation and Application of PTAM

Validity of PTAM

We shall consider two well-known examples to test the validity of PTAM. The first example considered is the calculation of synthetic scalar wave field in an infinite homogeneous and isotropic media. This problem has the following analytic solution (see, e.g., AKI and RICHARDS, 1980):

$$\phi(x, y, z, t) = \frac{f(t - R/c)}{2\pi R} \ , \tag{8}$$

where $R = \sqrt{z^2 + r^2}$, and c is the velocity of the wave. In the frequency domain, the above solution has the form of

$$\bar{\phi}(x, y, z, \omega) = G(\mathbf{x}, \omega) F(\omega) \ ,$$

where $F(\omega)$ is the spectrum of the source-time function $f(t)$ and

$$G(\mathbf{x}, \omega) = \frac{\exp(-i\omega R/c)}{R} \ . \tag{9}$$

Notice that $G(\mathbf{x}, \omega)$ can be represented by the superposition of a set of cylindrical waves via the following Sommerfeld integral (see, e.g., AKI and RICHARDS, 1980):

$$G(\mathbf{x}, \omega) = \int\limits_0^{+\infty} \frac{\exp[-\gamma(k)|z|]}{\gamma(k)} J_0(kr) k \, dk \ , \tag{10}$$

where $\gamma(k) = \sqrt{k^2 - (\omega/c)^2}$ with $\text{Re}\{\gamma(k)\} \geq 0$. The reason for taking this problem as an example is not only its analytic solution, but also its similarity to the formulation of the general problem of seismic wave propagation in layered media (see, e.g., KENNETT, 1983; CHEN, 1999). For the case of $|z| \sim 0$, integral (10) becomes a slowly convergent integral, and the usual direct integration method will be less efficient. We shall apply PTAM algorithm to evaluate this integral, and compare it with the semi-analytic method (SAM) and direct integration method.

Due to the pole on the integration path, the Sommerfeld integral in equation (10) is an improper integral. However, it can be converted to a proper integral as follows:

$$\int\limits_0^{+\infty} \frac{\exp[-\gamma(k)|z|]}{\gamma(k)} J_0(kr) k \, dk = I_1 + I_2 \ , \tag{11}$$

where,

$$I_1 = -2i \int\limits_0^{\sqrt{\frac{\omega}{c}}} \frac{\exp[-i \cdot q\sqrt{2\omega/c - q^2}|z|]}{\sqrt{2\omega/c - q^2}} \left(\frac{\omega}{c} - q^2\right) J_0\left[\left(\frac{\omega}{c} - q^2\right)r\right] dq \ , \tag{11a}$$

and

$$I_2 = 2 \int_0^{+\infty} \frac{\exp[-q\sqrt{2\omega/c + q^2}|z|]}{\sqrt{2\omega/c + q^2}} \left(\frac{\omega}{c} + q^2\right) J_0\left[\left(\frac{\omega}{c} + q^2\right)r\right] dq \ . \tag{11b}$$

Notice that there are no poles on the integration paths of I_1 and I_2, and the integrand in I_2 is a purely real function. Since the integration interval of I_1 is limited, it can be calculated by using a standard numerical integration technique. Thus we will focus on the calculation of I_2.

In SAM, the integral in equation (10) can be rewritten as

$$\int_0^{+\infty} \frac{\exp[-\gamma(k)|z|]}{\gamma(k)} J_0(kr)k\,dk = \int_0^{+\infty} \left\{\frac{\exp[-\gamma(k)|z|]}{\gamma(k)}k - \exp[-k|z|]\right\} J_0(kr)dk$$

$$+ \int_0^{+\infty} \exp[-k|z|]J_0(kr)\,dk \ , \tag{12}$$

where

$$\int_0^{+\infty} \exp[-k|z|]J_0(kr)\,dk = 1/R = 1/\sqrt{z^2 + r^2} \ . \tag{12a}$$

(GRADSHTEYN and RYZHIK, 1980). Equation (12) can also be converted to a proper integral similar to equation (11).

The integrand in I_2 versus variable q and I_2 versus the truncated upper limit q_{max} for $z = 0.01$ km and $f = 3.0$ Hz are shown (solid lines) in Figures 5(a) and (b), respectively. For comparison, the corresponding integrand in the numerical integral of SAM is shown (dot-dash line) in Figure 5(a). We can see that the integrand of SAM converges more rapidly than that of PTAM. In Figure 5(b), the calculated integral value of I_2 using PTAM is indicated as a straight line, and the peaks and troughs are represented as asterisks. The accurate value of $I_1 + I_2$ and those of PTAM and SAM and the corresponding relative errors are listed in Table 2, in which the integration step and truncated upper limit of SAM are the same as those of PTAM. As we can see, both methods have a high precision.

To demonstrate the result in time domain, we choose the Ricker wavelet (RICKER, 1977) as the source time function:

$$f(t) = \frac{\sqrt{\pi}}{2}\left(\frac{u^2}{4} - \frac{1}{2}\right)\exp\left(-\frac{u^2}{4}\right) \tag{13}$$

with $u = 2\sqrt{6}(t - t_0)/t_b$, where t_0 and t_b are the shift time and the width between the two peaks, respectively. The comparison of PTAM, SAM and the accurate value and

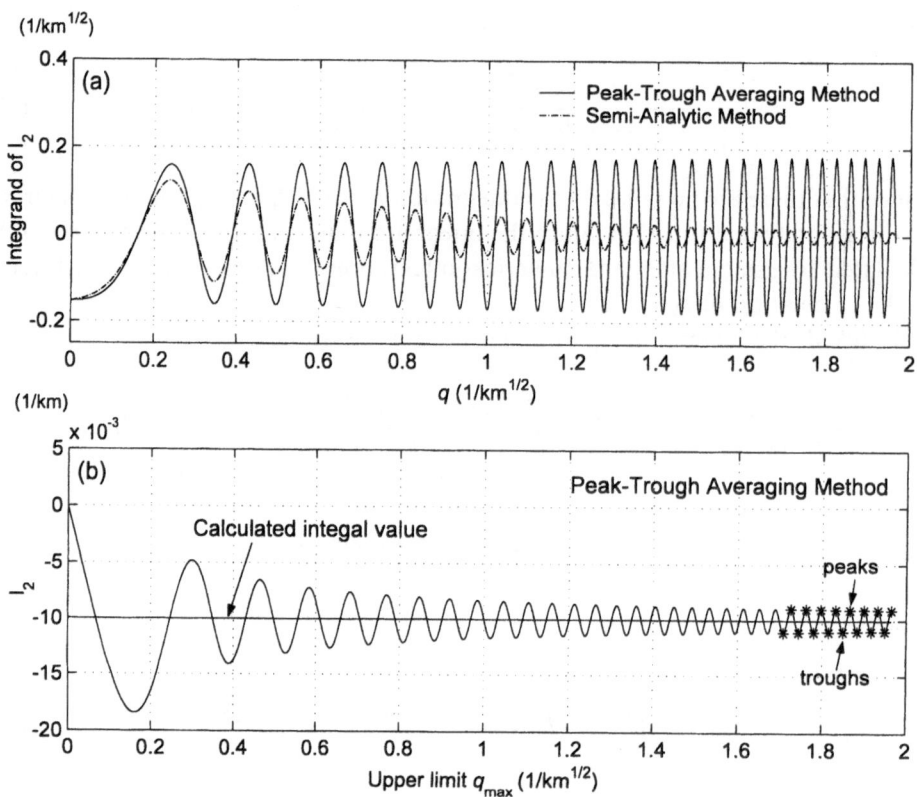

Figure 5

(a) The integrand in I_2 versus variable q $(1/\text{km}^{1/2})$ for $z = 0.01$ km and $f = 3.0$ Hz is shown by a solid line. The corresponding integrand in the numerical integral of the semi-analytic method is also shown by a dot-dash line. (b) I_2 versus the truncated upper limit $q_{\max}$. The calculated integral value of I_2 using PTAM is indicated as a straight line, and the peaks and troughs are represented as asterisks.

Table 2

Comparison of results and the relative errors by using PTAM and SAM

Accurate value (1/km)		Real part	Imaginary part
		-9.999088×10^{-3}	1.732102×10^{-2}
PTAM	Value (1/km)	-1.001710×10^{-2}	1.732054×10^{-2}
	Relative Error	0.180096%	0.028169%
SAM	Value (1/km)	-9.954172×10^{-3}	1.732054×10^{-2}
	Relative Error	0.449199%	0.028169%

the relative errors of PTAM and SAM are shown in Figures 6(a) and (b), respectively. Here $t_0 = 5.0$ s, $t_b = 3.0$ s. The relative error is defined as the ratio of the absolute error and the maximum amplitude of $\phi(x, y, z, t)$. The maximum relative errors of PTAM and SAM are both less than 0.1%, implying that both methods are

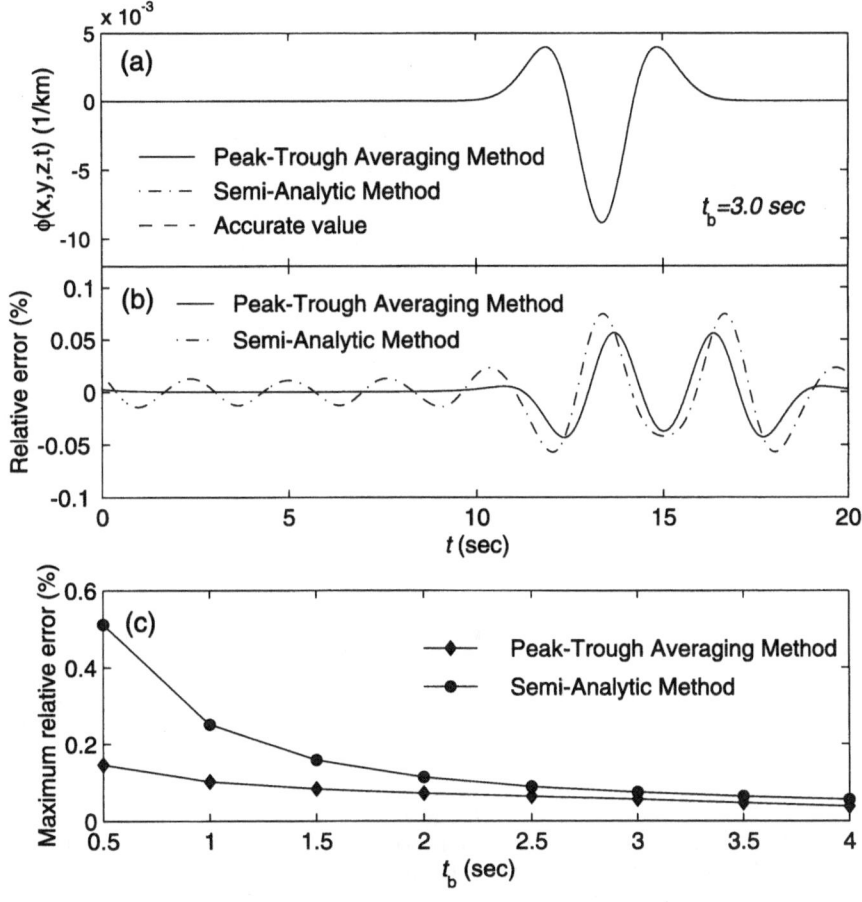

Figure 6
(a) Comparison of $\phi(x,y,z,t)$ by using PTAM, SAM and the accurate solution ($t_0 = 5.0$ s, $t_b = 3.0$ s).
(b) Relative errors of PTAM and SAM. (c) Comparison of the maximum relative errors of PTAM (line
with diamonds) and SAM (line with circles) versus t_b.

accurate. Figure 6(c) illustrates the comparison of the maximum relative errors of
PTAM and SAM versus t_b. Note that the integration steps and the truncated upper
limits for both methods are the same. As shown, the larger t_b is, the smaller the
maximum relative error is, and for the same t_b, the precision of PTAM is slightly
higher than that of SAM.

Having proved the accuracy, we proceed to demonstrate the efficiency of PTAM.
Figure 7 shows the comparison of the computation time of PTAM, SAM and direct
integration method for various parameter z, which is a crucial parameter for the
convergence of integral (10). For each z, we use PTAM with six peaks and six
troughs, the computation time is shown as a line with diamonds. For the same
precision as PTAM at each z, SAM and the direct integration is performed and the

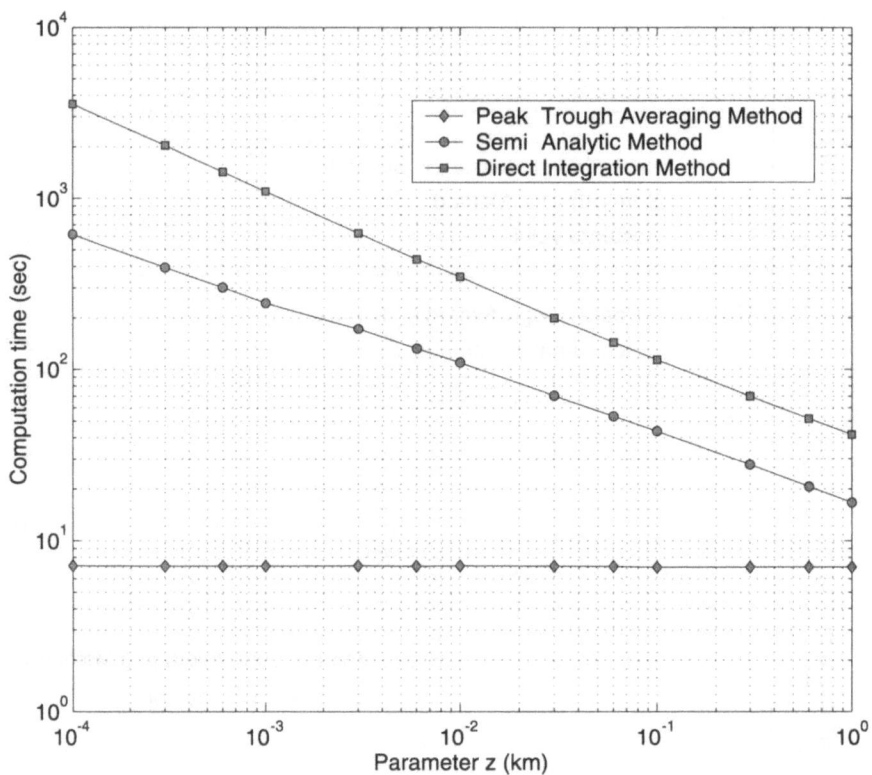

Figure 7

Comparison of the computation time of PTAM (line with diamonds), SAM (line with circles) and the direct integration method (line with squares) for different parameter z.

corresponding computation time is plotted as lines with circles and squares, respectively. It is obvious that with the decrease of z, the computation time of SAM and the direct integration method increase dramatically, whereas the computation time of PTAM remains unchanged. Although the integration interval is greatly reduced and considerable computation time is saved compared with the direct integration method, SAM proves to be much less efficient than PTAM when z tends to zero.

The second example for the validity of PTAM is the classical Lamb's problem (LAMB, 1904). The original Lamb's problem involved the computation of the response at surface of a semi-infinite isotropic elastic solid due to a vertical point force applied at the surface. A closed form solution of the integral solution was given by LAMB (1904). We compute the response using the generalized R/T coefficient method (KENNETT, 1983; CHEN, 1999) with PTAM and Lamb's formulae, respectively. A Poisson solid with a P-wave velocity of 5 km/s is used. The epicentral distance is 67 km, and the source time function is $p/(t^2 + p^2)$ with $p = 0.15$ s.

Figure 8 presents the comparison of the two results. As shown, the waveforms with the two completely different methods are in very good agreement, which again indicates that the PTAM is accurate.

Application to Calculation of Synthetic Seismograms

Having illustrated the validity and efficiency of PTAM using numerical examples, we now apply it to the calculation of synthetic seismograms.

We first consider a two-layer crust model (Crust Model 1, see Table 3). The source is a strike-slip double-couple point source at the surface, and the receivers are also at the surface and aligned from 30 km to 200 km set apart with a step of 5 km and an azimuth of 30°. The source time function here is a smoothed ramp function (BOUCHON, 1982) with a rise time 0.2 s. The synthetic three-component displacement seismograms are shown in Figure 9, where reduced time $t - R/\alpha_1$ is used (R is the epicentral distance and α_1 is the P-wave velocity in the top layer). The main theoretical arrival time curves are also plotted. The waveforms of vertical and radial

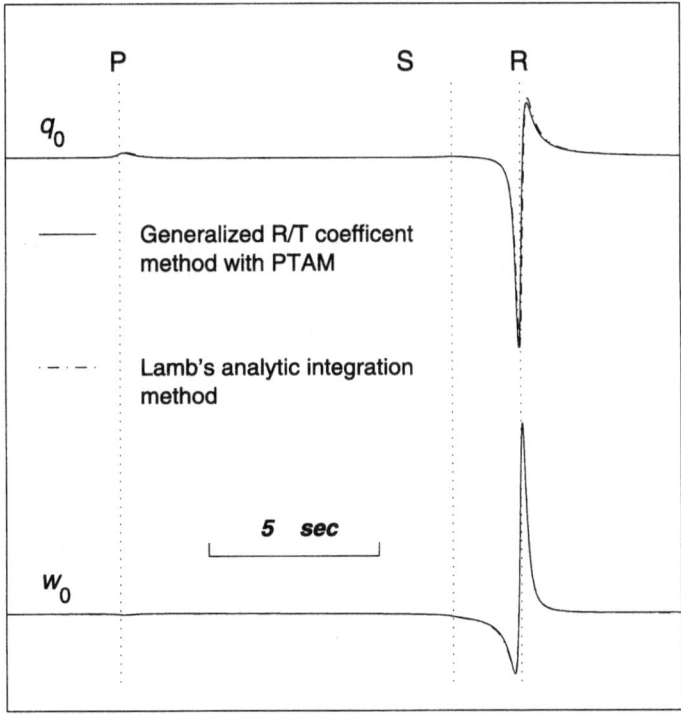

Figure 8
Comparison of the results of Lamb's problem by using the generalized R/T coefficient method with PTAM and Lamb's analytic integration method.

Table 3

Layered crust models

Layer Thickness (km)	P-wave Velocity (km/s)	S-wave Velocity (km/s)	Density (g/cm³)	Q_P	Q_S
Crust Model 1					
30.0	6.30	3.65	2.90	2000.0	2000.0
∞	8.20	4.70	3.30	2000.0	2000.0
Crust Model 2					
18.0	6.00	3.50	2.80	2000.0	2000.0
6.0	6.30	3.65	2.90	2000.0	2000.0
6.0	6.70	3.90	3.10	2000.0	2000.0
∞	8.20	4.70	3.30	2000.0	2000.0

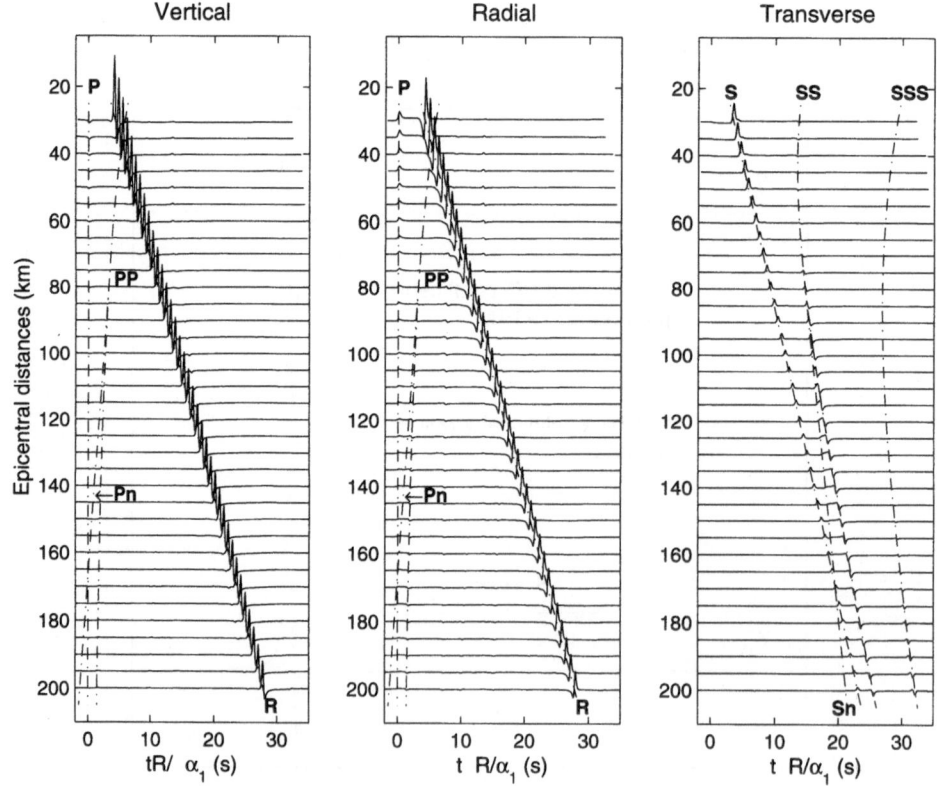

Figure 9

The synthetic three-component displacement seismograms for different epicentral distances, where reduced time $t - R/\alpha_1$ is used (R is the epicentral distance, and α_1 is the P velocity in the top layer). Here, Crust Model 1 (see Table 3) is used, and the source and receivers are all at the free surface.

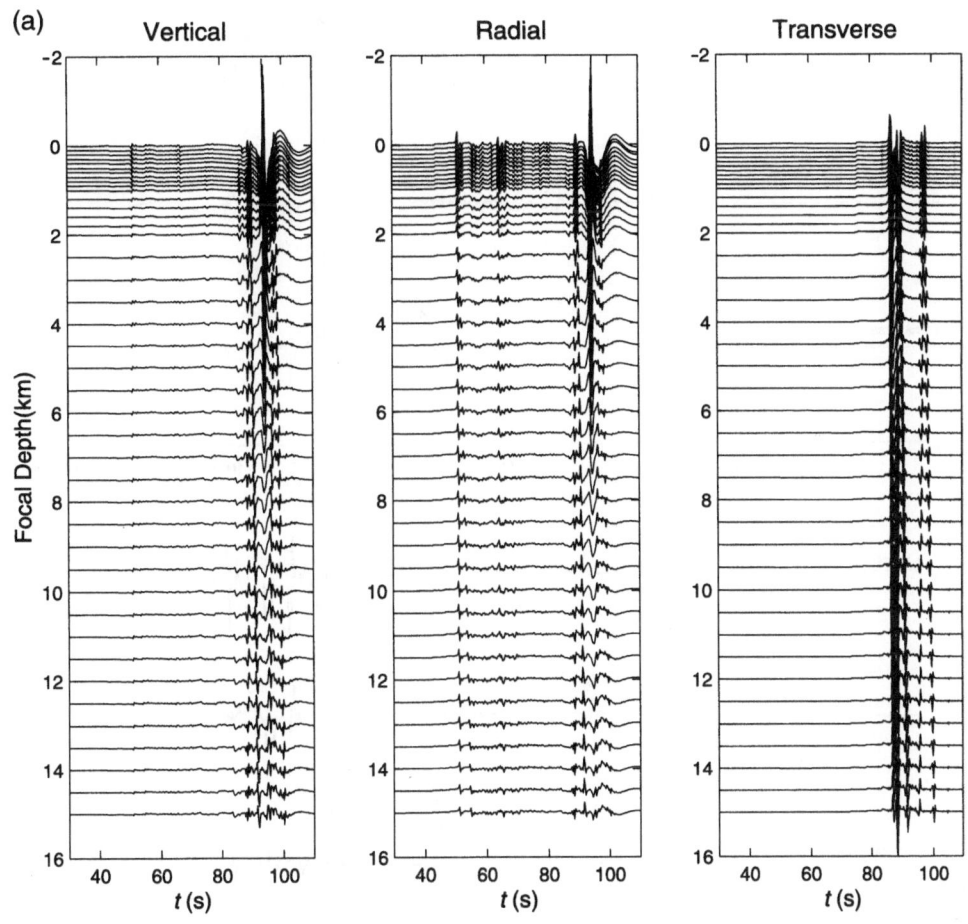

Figure 10
(a) The synthetic three-component displacement seismograms with different focal depths (0–15 km). Here, different focal depth steps are used for different depth ranges: 0.1 km for 0 to 1 km, 0.2 km for 1 to 2 km, and 0.5 km for 2 to 15 km. (b) and (c) The synthetic seismograms with a focal depth of 15 km obtained in this study and those in BOUCHON (1982), respectively. Crust Model 2 (see Table 3) is used, and the source a strike-slip double-couple point force. The receivers are at the surface with an epicentral distance of 300 km and an azimuth of 18°.

components are somewhat like the Lamb pulses, and we can see strong Rayleigh waves in both components, whereas for the transverse component, S, SS and SSS can be identified clearly.

We next consider a more complicated four-layer crust model (Crust Model 2, see Table 3). Here the source is also a strike-slip point double couple, and placed away from the free surface at a depth of 15 km. Different focal depth steps are used for different depth ranges: 0.1 km for 0 to 1 km, 0.2 km for 1 to 2 km, and 0.5 km for 2 to 15 km. The receiver is at the free surface with an epicentral distance of 300 km and

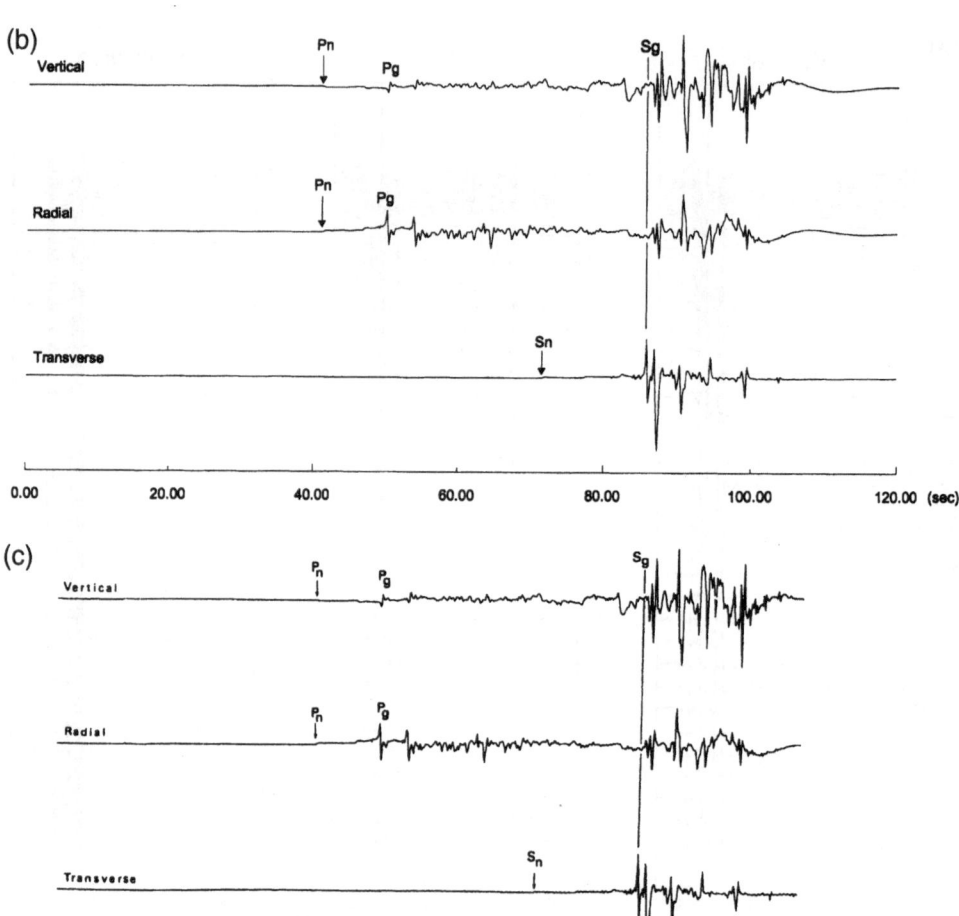

Figure 10b,c

an azimuth of 18°. The source time function matches the previous one. The result at a focal depth of 15.0 km obtained in this study and that in BOUCHON (1982) are shown in Figures 10(b) and (c), respectively. The excellent agreement between the two indicates that the result obtained by using our code is reliable. As shown in Figure 10(a), the synthetic displacement seismograms change continuously as focal depth decreases, which indicates that the results at shallow focal depths are reasonable.

Conclusions

To efficiently calculate synthetic seismograms for a layered half-space with close to or the same depths, we proposed a simple while efficient method—the *Peak-*

Trough Averaging Method (PTAM), which is built upon the Repeated Averaging Method (DAHLQUIST and BJÖRCK, 1974; CHANG, 1988). Compared with the semi-analytic method (APSEL and LUCO, 1983; HISADA, 1994, 1995), this method is not only for simpler mathematically and easier to implement, but also more efficient. Using numerical examples, we illustrate that this method is accurate and efficient. Therefore, it is expected to be very useful in the computation of synthetic seismograms.

Acknowledgements

The basic idea of this work was originally proposed by Dr. Shyhong Chang. Drs. Xiaofei Chen and Shyhong Chang would like to express their deepest gratitude to Prof. K. Aki for his guidance throughout the years of graduate study at USC. We thank Dr. B. Nowack and an anonymous reviewer for their critical comments and for the enhancement of English in the manuscript. This work is supported by the National Natural Sciences Foundation of China under grant of DYF49625406 and 40074008, and partially by the China National Fundamental Research Project (G1998040702).

REFERENCES

AKI, K. and RICHARDS, P. Q., *Quantitative Seismology* (Freeman, San Francisco, 1980).

APSEL, R. J. and LUCO, J. E. (1983), *On the Green's Functions for a Layered Half-space. Part II*, Bull. Seismol. Soc. Am. *73*, 931–951.

BOUCHON, M. (1979), *Discrete Wavenumber Representation of Elastic Wave Fields in Three Space Dimensions*, J. Geophys. Res. *84*, 3604–3614.

BOUCHON, M. (1982), *The Complete Synthesis of Seismic Crustal Phases at Regional Distance*, J. Geophys. Res. *87*(B3), 1735–1741.

BOUCHON, M. and AKI, K. (1977), *Discrete Wavenumber Representation of Seismic Source Wave Fields*, Bull. Seismol. Soc. Am. *67*, 259–277.

CHANG, S. H. (1988), *Complete Wave-field Modeling and Seismic Inversion for Lossy-elastic Layered Half-space due to Surface Force*, Ph. D. Thesis, University of Southern California, Los Angeles.

CHEN, X. F. (1993), *A Systematic and Efficient Method of Computing Normal Modes for Multi-layered Half-space*, Geophys. J. Int. *115*, 391–409.

CHEN, X. F. (1999), *Seismograms Synthesis in Multi-layered Half-space Media. Part I. Theoretical Formulations*, Earthquake Res. in China *13*, 149–174.

DAHLQUIST, G. and BJÖRCK, Å., *Numerical Methods* (Prentice-Hall Inc., Englewood Cliffs, N. J., 1974).

DRAVINSKI, M. and MOSSESSIAN, T. K. (1988), *On Evaluation of the Green Function for Harmonic Line Loads in an Elastic Half-space*, J. Num. Meth. Engng. *26*, 823–841.

FUCHS, K. and MÜLLER, G. (1971), *Computation of Synthetic Seismograms with Reflectivity Method and Comparison with Observations*, Geophys. J. R. Astr. Soc. *23*, 417–433.

GRADSHTEYN, I. S. and RYZHIK, I. M., *Table of Integrals, Series, and Products* (Academic Press, New York, 1980).

HELMBERGER, D. V. (1974), *Generalized Ray Theory for Shear Dislocations*, Bull. Seismol. Soc. Am. *64*, 45–64.

HISADA, Y. (1994), *An Efficient Method for Computing Green's Functions for a Layered Half-space with Sources and Receivers at Close Depths*, Bull. Seismol. Soc. Am. *84*, 1457–1472.

HISADA, Y. (1995), *An Efficient Method for Computing Green's Functions for a Layered Half-space with Sources and Receivers at Close Depths (Part 2)*, Bull. Seismol. Soc. Am. *85*, 1080–1093.

KENNETT, B. L. N. *Seismic Wave Propagation in Stratified Media* (Cambridge University Press, New York, 1983).

KENNETT, B. L. N. and KERRY, N. J. (1979), *Seismic Waves in a Stratified Half space*, Geophys. J. R. Astr. Soc. *57*, 557–583.

LAMB, H. (1904), *On the Propagation of Tremors over the Surface of an Elastic Solid*, Phil. Trans. Roy. Soc. (London) A *203*, 1–42.

LUCO, J. E. and APSEL, R. J. (1983), *On the Green's Functions for a Layered Half-space. Part I*, Bull. Seismol. Soc. Am. *73*, 909–929.

RICKER, N. H., *Transient Waves in Visco-elastic Media* (Elsevier Scientific Publishing Co., Amsterdam, Holland, 1977).

WANG, C. Y. and HERRMANN, R. B. (1980), *A Numerical Study of P-, SV-, and SH-wave Generation in Plane-layered Medium*, Bull. Seismol. Soc. Am. *70*, 1015–1036.

YAO, Z. X. and HARKRIDER, D. G. (1983), *A Generalized Reflection-transmission Coefficient Matrix and Discrete Wavenumber Method for Synthetic Seismograms*, Bull. Seismol. Soc. Am. *73*, 1685–1699.

(Received December 11, 2000, accepted April 10, 2001)

 To access this journal online:
http://www.birkhauser.ch

Pure appl. geophys. 160 (2003) 487–507
0033–4553/03/040487–21

Ⅰ Pure and Applied Geophysics

Calculation of Synthetic Seismograms with Gaussian Beams

Robert L. Nowack[1]

Abstract — In this paper, an overview of the calculation of synthetic seismograms using the Gaussian beam method is presented accompanied by some representative applications and new extensions of the method. Since caustics are a frequent occurrence in seismic wave propagation, modifications to ray theory are often necessary. In the Gaussian beam method, a summation of paraxial Gaussian beams is used to describe the propagation of high-frequency wave fields in smoothly varying inhomogeneous media. Since the beam components are always nonsingular, the method provides stable results over a range of beam parameters. The method has been shown, however, to perform better for some problems when different combinations of beam parameters are used. Nonetheless, with a better understanding of the method as well as new extensions, the summation of Gaussian beams will continue to be a useful tool for the modeling of high-frequency seismic waves in heterogeneous media.

Key words: Synthetic seismograms, Gaussian beams.

Introduction

The Gaussian beam method is an asymptotic method for the computation of wave fields in smoothly varying inhomogeneous media, and was proposed by Popov (1981, 1982) based on an earlier work of Babich and Pankratova (1973). The method was first applied by Popov et al. (1980), Katchalov and Popov (1981) and Červený et al. (1982) to describe high-frequency seismic wave fields by the summation of paraxial Gaussian beams. One of the advantages of the method is that the individual Gaussian beams have no singularities either at caustics in the spatial domain or at pseudo-caustics in the wavenumber domain. Although caustics and pseudo-caustics are generally at different locations, methods such as the Maslov method require them to be well separated. The lack of singularities of the individual beams assures that the summation of Gaussian beams is regular everywhere. Another advantage of the Gaussian beam method is that it naturally introduces smoothing effects and is therefore not as sensitive to model parameterizations as the ray method. Finally, the Gaussian beam method does not require two-point ray tracing, unlike the ray method.

---

[1] Department of Earth and Atmospheric Sciences, Purdue University, West Lafayette, IN 47907.
E-mail: nowack@purdue.edu

An overview of the Gaussian beam method is first given, including a description of paraxial Gaussian beams, the superposition of Gaussian beams for the construction of more general wave fields, and the selection of beam parameters. Next, a representative list of applications of the method is presented. Finally, several extensions of the original method that have been proposed are described.

Rays and Linearized Rays

A high-frequency wave solution connected to a given ray can be written in ray centered coordinates (q_1, q_2, s) as

$$\vec{u}(q_I, s) = \vec{U}(s)e^{-i\omega(t-\tau(q_I,s))} \ , \tag{1}$$

where $I = 1, 2$ and the phase time up to second order is

$$\tau(q_I, s) = \tau(s) + \frac{1}{2}\vec{q}^T M(s)\vec{q} \ , \tag{2}$$

where $M(s)$ is a 2×2 matrix related to the local curvature of the wavefront by $K(s) = vM(s)$, and v is the medium velocity along the ray. $M(s)$ is obtained by solving a matrix Riccati equation along the ray, and can be decomposed into two submatrices as $M(s) = P(s)Q^{-1}(s)$ (POPOV and PŠENČÍK, 1978; ČERVENÝ and HRON, 1980).

The matrices $P(s)$ and $Q(s)$ are 2×2 matrices which are solutions of the first-order dynamic or linearized ray equations which can be written as

$$\frac{dX}{ds} = AX, \quad \text{where} \quad A = \begin{pmatrix} 0 & v\delta_{IJ} \\ -v^{-2}v_{,IJ} & 0 \end{pmatrix}, \quad X(s) = \begin{pmatrix} Q(s) \\ P(s) \end{pmatrix} \ , \tag{3}$$

where I, J range from 1, 2, s is the coordinate along the ray, δ_{IJ} is the Kronecker delta symbol, and $v_{,IJ} = \partial^2 v/\partial q_I \partial q_J$ is the second-derivative matrix of the velocity field transverse to the ray. As an alternative, the dynamic ray equations can be written in Cartesian or other coordinate systems (ČERVENÝ, 1985b; 2001), but the ray-centered coordinates are convenient for the discussion here. $Q(s)$ and $P(s)$ are solutions for the entire ray bundle in the vicinity of the central ray, and for specific initial conditions can be related to small offsets and angle changes of a linearized ray about the central ray. One of the initial applications of dynamic ray equations was to the computation of geometric spreading about a given central ray, with the ray amplitude related to $(\det Q(s))^{-1/2}$. For $Q(s_0)$ specified for a given source, then $Q(s)$ at the receiver can be computed by the dynamic ray equations.

The solution $X(s)$ of the dynamic ray equations can be written in terms of a 4×4 fundamental matrix $\pi(s, s_0)$ as

$$X(s) = \pi(s, s_0)X(s_0) = \begin{pmatrix} Q_1 & Q_2 \\ P_1 & P_2 \end{pmatrix} X(s_0) \ , \tag{4}$$

where Q_1, Q_2, P_1 and P_2 are 2×2 sub-matrices of $\pi(s, s_0)$. The fundamental matrix has the properties

$$\pi(s_0, s_0) = \begin{pmatrix} \delta_{IJ} & 0 \\ 0 & \delta_{IJ} \end{pmatrix} \quad \text{and} \quad \det[\pi(s, s_0)] = 1 , \tag{5}$$

where I, J range from 1, 2 and δ_{IJ} is the Kronecker delta symbol. Two primary initial conditions to the dynamic ray equations are

$$X(s_0) = \begin{pmatrix} Q(s_0) = 0 \\ P(s_0) \neq 0 \end{pmatrix}, \quad X(s_0) = \begin{pmatrix} Q(s_0) \neq 0 \\ P(s_0) = 0 \end{pmatrix} , \tag{6}$$

and are related to a point source with initial angle variations of linearized rays about the central ray, and to a planar source with initial displacement variations of linearized rays about the central reference ray, respectively.

Paraxial Gaussian Beams

A paraxial Gaussian beam is an asymptotic solution of a one-way parabolic wave equation in ray-centered coordinates about the central ray. The solution is similar in form to the high-frequency ray solution given in Eqn. (1), except that now $M(s)$ is complex with Im $(M(s))$ positive definite. It can be constructed from the linearized ray solution by using complex initial conditions $X(s_0)$ in Eqn. (4). Since the matrix $M(s)$ does not change if both $Q(s)$ and $P(s)$ are multiplied by a nonsingular matrix, the initial conditions to the dynamic ray equations can be written as

$$X(s_0) = \begin{pmatrix} I \\ M(s_0) \end{pmatrix} . \tag{7}$$

Since Im $M(s_0)$ is positive-definite, the Gaussian beam solution will be a combination of the initial point source and plane wave solutions in Eqn. (6) which will exponentially decay away from the central ray. For a Gaussian beam solution, if $Q(s)$ is regular at one point along the ray, it is regular along the entire ray (ČERVENÝ, 1985b). Based on the regularity of $Q(s)$ for the Gaussian beam solution, then the amplitude factor $(\det Q(s))^{-1/2}$ will be nonsingular for all points along the ray.

In general, there will be three complex or six real parameters that are needed to specify $M(s)$ for a given point along the ray. The dynamic ray equations can then be used to find $M(s)$ at other points on the ray. For the special case of a circular beam at a given point on the ray, then $M(s)$ can be written as

$$M(s) = \text{Re } M(s) + i \text{ Im } M(s) = \left[v^{-1} K(s) + \frac{i}{\pi L^2(s)} \right] \delta_{IJ} , \tag{8}$$

where v is the velocity along the ray, $K(s)$ is the beam-front curvature and $L(s)$ is the beam half-width at a frequency of 1 Hz. Therefore, the real part of $M(s)$ describes the curvature properties of the phase-front of the beam and the imaginary part of $M(s)$ describes the beam-width, where a smaller Im $M(s)$ corresponds to a larger beamwidth. Extensions of high-frequency Gaussian beams to propagation in elastic media were described by ČERVENÝ and PŠENČÍK (1983a,b, 1984), and to anisotropic media by HANYGA (1986).

Expansion of High-frequency Wave Fields into Gaussian Beams

A high-frequency expansion of a wave field into Gaussian beams was proposed by POPOV (1981, 1982), and initially applied by POPOV et al. (1980), KATCHALOV and POPOV (1981) and ČERVENÝ et al. (1982). An early overview of the method was also given by ČERVENÝ (1981). The expansion of a time-harmonic wave field into Gaussian beams and evaluated at a position S in the medium can be written as

$$\vec{u}(S,\omega) = \iint_D \Phi(\gamma_I)\vec{u}_{\gamma_I}^{GB}(S,\omega,M(s_b))d^2\gamma \ , \tag{9}$$

where $\Phi(\gamma_I)$ is the weighting function, $\vec{u}_{\gamma_I}^{GB}$ are the beam solutions, and $M(s_b)$ are the beam parameters for a specified position s_b along the ray. The ray parameters γ_I $(I = 1, 2)$ specify a given central ray along the initial wavefront, and the domain D depends on the type of source to be decomposed into beams. For an initial point source, the ray parameters γ_I can be specified by initial angles at the source, and for a planar wavefront, the γ_I can be specified by positions along the initial wave front. Since the central rays of the individual Gaussian beams need only be in the vicinity of the receiver, no two-point ray tracing is required for the method. Also by summing over Gaussian beams, Eqn. (9) naturally results in some smoothing of the wave field, and is therefore less sensitive to model parameterizations than the ray method.

The individual Gaussian beams in Eqn. (9) can be written in ray-centered coordinates (q_1, q_2, s) as

$$\vec{u}_{\gamma_I}^{GB}(S,\omega,M(s_b)) = \frac{\vec{U}^N(s)}{\det(Q(s))^{1/2}} e^{-i\omega(t-\tau(s)-\frac{1}{2}\vec{q}^T M(s)\vec{q})} \ , \tag{10}$$

where $M(s) = \operatorname{Re} M(s) + i \operatorname{Im} M(s)$, $\operatorname{Im} M(s)$ is positive-definite, $\det[Q(s)]^{-1/2} \neq 0$, and the beam parameters are given by $M(s_b)$ at a specified position along the ray. The values of $M(s)$ and $Q(s)$ are then computed using the dynamic ray equations above. $\vec{U}^N(s)$ is spreading-free ray amplitude given by $\vec{U}^R(s)\det(Q^R(s))^{1/2}$ where $\vec{U}^R(s)$ is the complete ray amplitude factor.

The weighting function for the asymptotic expansion of a wave field into Gaussian beams in Eqn. (9) can be written as

$$\Phi(\gamma_I) = \frac{\omega}{2\pi} \left[-\det\big(Q^T(s)(M(s) - M^R(s))Q^R(s)\big)\right]^{1/2}$$

$$= \frac{\omega}{2\pi} \left[-\det\big(P^T Q^R - Q^T P^R\big)\right]^{1/2} \tag{11}$$

where $Q^T(s)$ and $P^T(s)$ are the transposes of the complex 2×2 matrices $Q(s)$ and $P(s)$. The values $Q^R(s)$, $P^R(s)$ and $M^R(s)$ are the corresponding ray values for the given type of source. For example, if we specify $Q^R(s) = I$, $P^R(s) = 0$, and $M^R(s) = 0$ at the source, then the resulting weighting function is the same as that for the expansion of a plane wave into Gaussian beams as given by ČERVENÝ (1982). Similarly, the weighting function for the expansion of an initial point source into Gaussian beams is given by ČERVENÝ (1985b). The weighting function for the asymptotic expansion of an arbitrary initial wavefront into Gaussian beams can also be obtained (KLIMEŠ,1984a; ČERVENÝ, 1985a,b). Based on the properties of the solutions of the dynamic ray equations, the weighting factor $\Phi(\gamma_I)$ is an invariant and can be evaluated at any point along the ray (ČERVENÝ, 1985b).

If we specify the weighting function at the endpoint of the ray, the complete amplitude term for the integrand in Eqn. (9) can be written as

$$\Phi(\gamma_I)\frac{\vec{U}^N(s)}{\det(Q(s))^{1/2}} = \frac{\omega}{2\pi} \left|\det(Q^R(s))\right|^{1/2} \left[-\det\big(M(s) - M^R(s)\big)\right]^{1/2}\vec{U}^N(s) \tag{12}$$

where $\mathrm{Re}\,[-\det(M(s) - M^R(s))]^{1/2} > 0$, and all parameters other than $M(s)$ are the corresponding ray values. The specification of $M(s)$ at the endpoint of the ray gives added phase stability to the Gaussian beam summations.

Different methods for performing the Fourier transform of the Gaussian beam summation to the time-domain have then been given by ČERVENÝ (1983, 1985a,b). A direct frequency domain approach can be written as

$$\vec{u}(S,t) = \frac{1}{\pi}\mathrm{Re} \int\limits_0^\infty F(\omega)\vec{u}(S,\omega)e^{-i\omega t}d\omega \ , \tag{13}$$

where $F(\omega)$ is the Fourier transform of source-time function $f(t)$, and $\vec{u}(S, \omega)$ is the frequency response given in Eqn. (9). This is the most convenient form for the incorporation of additional frequency-dependent effects such as causal attenuation. The incorporation of a general moment tensor source function has been described by ČERVENÝ et al. (1987).

A convolutional approach can also be used where the impulse response of the Gaussian beam summation is convolved with the source-time function. For a complex phase factor $\tau(q_i, s)$ of the form given in Eqn. (2), this can be written as

$$\vec{u}(S,t) = f(t)^* \frac{1}{\pi} \iint\limits_D d\gamma^2 \mathrm{Im}\left(\left\{\frac{\Phi(\gamma_I)\vec{U}^N(s)\det(Q(s))^{-1/2}}{t - \tau(q_I, s)}\right\}\right) . \tag{14}$$

Finally, for particular choices of $f(t)$, elementary signals in the time domain can be asymptotically obtained. The most common choice is the Gabor wavelet which is a Gaussian weighted cosine function. This choice for $f(t)$ results in the fastest computational approach and can be written as

$$\vec{u}(S, t) = \iint\limits_{D} d\gamma^2 \vec{u}_{\gamma_I}^{GW}(S, t) \ , \tag{15}$$

where $\vec{u}_{\gamma_I}^{GW}(S, t)$ are the individual Gabor wave packets which are now Gaussian in both space and time.

For numerical calculations, the Gaussian beam integrals in Eqns. (9), (14) and (15) must be discretized to form discrete summations of Gaussian beams. The error in performing this discretization has been described by ČERVENÝ (1985a) and KLIMEŠ (1986), and the minimization and bounding of the discretization error have been used as one selection criterion for the beam parameters.

Choices of Beam Parameters

The complex beam parameters $M(s)$ are commonly specified at either the source or receiver, although other choices, such as at interfaces, are possible as well. For the special case of a circular beam at a given position along the ray, then $M(s)$ can be written in terms of the wavefront curvature $K(s)$ and the beam half-width $L(s)$ as in Eqn. (8). However, after propagation, $M(s)$ will generally no longer represent a circular beam.

We first describe several common choices of beam parameters at the source point s_0. For Re $M(s_0) \to 0$ and Im $M(s_0) \to 0$ this generates large planar beams, and results in a plane-wave expansion at the source. A second choice of the beam parameters at the source for a chosen Re $M(s_0)$ is to specify Im $M(s_0)$ to produce the smallest beamwidth at the receiver. For this case, Im $M(s_0)$ can be written as

$$\text{Im } M(s_0) = C\left\{[Q_2^{-1}(s)Q_1(s) - \text{Re}\,M(s_0)]^2 + A^2\right\}^{1/2} \ , \tag{16}$$

where C is a constant, A is a constant 2×2 matrix and Q_1 and Q_2 are subcomponent matrices of the fundamental ray matrix at the receiver given in Eqn. (4) (ČERVENÝ, 1985b). Choosing Re $M(s_0) \to 0$ with $C = 1$ and $A = 0$ gives the so-called optimal beam choice for planar beams at the source that results in the smallest beams at the receiver (ČERVENÝ et al., 1982).

The beam parameters can also be chosen at the receiver. For Re $M(s) \to 0$ and Im $M(s) \to 0$, this gives large planar beams at the receiver and results in Chapman-Maslov seismograms (KLIMEŠ, 1984b; ČERVENÝ, 1985b). Alternatively, for velocity gradients or topography at the receiver, then Re $M(s)$ can be specified to give effective planar beams at the receiver, providing stable summation results.

Approximate ray synthetic seismograms can be obtained by choosing Re $M(s)$ at the receiver to be equal to $M^R(s)$, and Im $M(s)$ very large to give small beamwidths at the receiver (ČERVENÝ, 1985a). For the practical summation of Gaussian beams for this case, then Im $M(s)$ needs to be chosen in relation to the ray-sampling interval (ČERVENÝ, 1985a).

Another choice at the receiver for a specified Re $M(s)$ is

$$\text{Im } M(s) = C\left\{ [M^R(s) - \text{Re } M(s)]^2 + A^2 \right\}^{1/2} , \tag{17}$$

where C is a constant, A is a constant 2×2 matrix, and $M^R(s)$ is related to the ray-curvature matrix. Similar to the specification in Eqn. (16) at the source, this gives narrow beams at the receiver for short ray paths and wider beams for longer ray paths. Specifying Re $M(s) \to 0$ results in planar beams at the receiver. ČERVENÝ (1985a,b) gave modifications of Re $M(s)$ to incorporate lateral heterogeneities or a curved interface at the receiver. With the specification of beam parameters at the receiver and these heterogeneity corrections, very stable results for the summation of Gaussian beams can be obtained. The choice in Eqn. (17) can also be derived by minimizing the discretization error from replacing a continuous Gaussian beam integral with a discrete summation (ČERVENÝ, 1985a).

An alternate approach to the selection of beam parameters is to choose broad Gaussian beams which also limit the discretization error of the Gaussian beam summation. For a specified Re $M(s)$ in the 2-D case, ČERVENÝ (1985a) specified Im $M(s)$ which results in broad beams at the receiver which still limits the discretization error given by KLIMEŠ (1986). ČERVENÝ (1985a) found that this choice gave very stable results for vertically inhomogeneous media.

In many cases, it was found that stable results can be obtained for a large range of beam parameters used in the Gaussian beam summation (ČERVENÝ et al., 1982; NOWACK and AKI, 1984). For example, Fig. 1 shows the individual beams and the resulting Gaussian beam summation for a layer over a linear gradient. The model has a velocity of 5.6 km/s down to 15 km and then increases linearly to 8.0 km/s at 40 km. Two geometric arrivals occur at a distance range of 140 km as shown by the stationary phase points of the individual beam contributions in Figure 1. Figure 1A presents the results for the optimal beam parameter choice of ČERVENÝ et al. (1982) as given by Eqn. (16) with Re $M(s_0) = 0$, $C = 1$ and $A = 0$. Figure 1B displays the corresponding results for beams that are 16 times wider at the source than in Figure 1A. The resulting beam summations can be seen to be very similar for both beam parameter choices. Nevertheless, certain choices of beam parameters will give better results for a given type of source. For example, NOWACK and AKI (1984) found that wide planar beams at the source give better results for the expansion of a point source, while narrow planar beams at the source give better results for planar sources.

For structures with strong lateral velocity variations, larger values of Im $M(s)$ are required along the path in order to ensure validity of the individual beam elements.

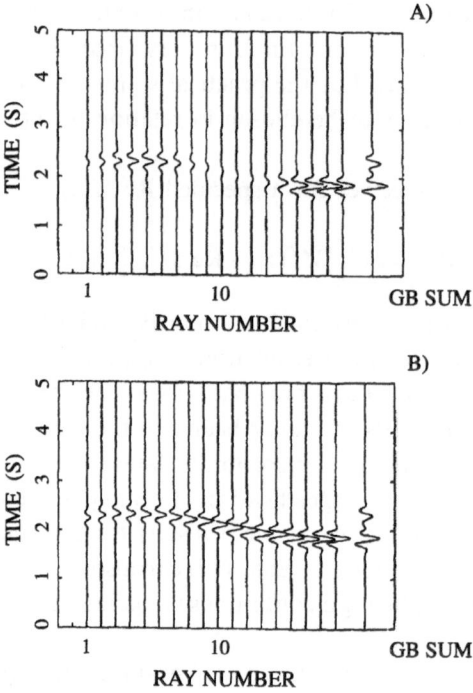

Figure 1
Gaussian beam summation for a layer over a gradient model with a 5.6 km/s layer down to 15 km and a
velocity gradient from 5.6 km/s at 15 km to 8 km/s at 40 km. The individual Gaussian beam contributions
are shown in the time domain using a Gabor wavelet, and the Gaussian beam summation is shown for a
receiver at 140 km. A) The Gaussian beam contributions and summation are shown using the optimal
beam choice of ČERVENÝ et al. (1982). B) The Gaussian beam contributions and summation are shown for
beams that are 16 times wider at the source than in A).

For example, Fig. 2 shows a random velocity layer with a thickness of 120 km. The
layer is spline interpolated with velocities of 8 km/s ± 3% and a heterogeneity scale
of 15 km. Gaussian beam seismograms are shown in Figure 2B for two different
pulse frequencies of 1 Hz and 2 Hz, and with beam parameters specified at the source
with Re $M(s_0) = 0$, $C = 1$ and $A = 0$ in Eqn. (16). These results compared well with
results obtained using a parabolic finite-difference approach (NOWACK and AKI,
1984). Nonetheless, smaller scale features or multi-scale structures could be more
problematic for the Gaussian beam method.

   Additional choices of beam parameters have been given by MADARIAGA (1984),
MÜLLER (1984), WEBER (1988a), and KLIMEŠ (1989b). KLIMEŠ (1989a) describes the
beam superposition in terms of Gaussian packets with a summation along the ray as
well. For specific problems, additional choices of beam parameters may give more
optimized results, whereas some beam parameters could lead to poor results
(NOWACK and AKI, 1984; ČERVENÝ, 1985a; WHITE et al., 1987). This dependence on
the beam parameters was also noted by FELSEN (1984) and NORRIS (1986) and

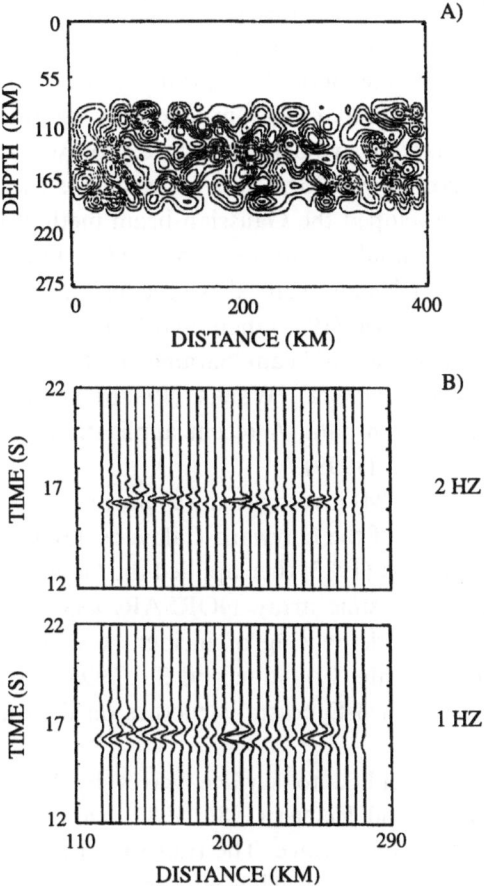

Figure 2

The Gaussian beam method applied to the propagation of a plane wave through a random velocity layer.
A) Contour plot of a 120 km thick layer with randomly fluctuating velocities with velocities of
8 km/s ± 3% and a scale length of 15 km. B) Gaussian beam wave fields for a vertically incident plane
wave propagated through the random velocity layer in A) for pulse center frequencies of 1 and 2 Hz (from
NOWACK and AKI, 1984).

resulted in various extensions to the original method as described below. Nonetheless, the original Gaussian beam method provides very stable, nonsingular results over a wide range of beam parameters for smoothly varying media.

Applications of the Gaussian Beam Method

In this section, an overview is given of representative applications of the Gaussian beam method. In addition to applications of the method by the original Czech and Russian groups, groups at MIT led by K. Aki and later at USC, as well as groups in

France, Germany and elsewhere performed early studies of the Gaussian beam method. As an example of early work from the MIT group, NOWACK and AKI (1984) performed a number of validity tests of the method with different beam parameters. They then applied the Gaussian beam method to random, smoothly varying media (see Figure 2), as well as to focusing effects from a volcanic structure. NOWACK and AKI (1986) applied the method to the inversion of waveform data for velocity structure.

MADARIAGA (1984) developed the Gaussian beam method for vertically varying media and used modified initial conditions expressed in terms of WKB and point source solutions. Gaussian beams were also specified in geographic coordinates. MADARIAGA and PAPADIMITRIOU (1985) then used Gaussian beams for the modeling of upper mantle phases. Different beam parameter choices were investigated by MÜLLER (1984) and WEBER (1988a). WEBER (1988b) applied the method to the modeling of regional refraction data. A recent application to upper mantle phases was given by LEBORGNE et al. (1999).

CORMIER and SPUDICH (1984) investigated waveform complexity from focusing in the heterogeneous fault zone of the Hayward-Calaveras fault system using Gaussian beams. NOWACK and CORMIER (1985) then applied the Gaussian beam method to the 3-D structure beneath the seismic array NORSAR. CORMIER (1987) applied the method to the focusing and defocusing of incident teleseismic waves by the 3-D structure at the Nevada test site. CORMIER and SU (1994) used the Gaussian beam method to study the effects of 3-D crustal structure on the estimation of fault slip history and ground motion.

Gaussian beams were applied to surface waves by YOMOGIDA (1985, 1987), YOMOGIDA and AKI (1985) and JOBERT (1986, 1987) using vertical adiabatic modes and horizontal beams along the surface. The transformation of JOBERT and JOBERT (1983) was used to perform 2-D ray tracing on a sphere. FRIEDERICH (1989) directly propagated Gaussian beams for long-period surface waves on a sphere. YOMOGIDA and AKI (1987) used the Gaussian beam method to invert surface wave amplitude and phase data for velocity anomalies in the Pacific Ocean basin. A research group lead by K. Aki and T.L. Teng at USC performed further surface wave studies using Gaussian beams. For example, ZHENG et al. (1989) used surface waves to map the crust and upper mantle in the Arctic region, KATO et al. (1993) studied surface wave propagation in sedimentary basins in Japan, QU et al. (1994) applied Gaussian beams for short-period surface wave propagation in Southern California, and CHEN et al. (1998) studied short-period surface waves in Taiwan.

CORMIER (1989) applied the Gaussian beam method to the diffraction of seismic pulses from downgoing subducted slabs. WEBER (1990) and SEKIGUCHI (1992) then used Gaussian beams to investigate the influence on P-wave travel times and amplitudes of heterogeneous subduction zones. CORMIER (1995) performed time-domain modeling of PKIKP precursors for lower mantle heterogeneities. Studies of the lower mantle using Gaussian beams were also conducted by WEBER and DAVIS (1990) and WEBER (1993).

Further applications by Russian and Czech groups included KATCHALOV *et al.* (1983), GRIKUROV and POPOV (1983), KATCHALOV and POPOV (1985, 1988). Modeling in 3-D was performed by ČERVENÝ and KLIMEŠ (1984). The relation between the Gaussian beam method and the Maslov method was investigated by KLIMEŠ (1984b). ČERVENÝ *et al.* (1987) applied the Gaussian beam method to the modeling of extended earthquake sources in laterally varying structures and found good agreement with results from finite-element modeling and the isochron method. An overview of the Gaussian beam method was given by BABICH and POPOV (1989) along with additional references up to that time.

Studies of the range of validity of rays and beams were conducted by BEN-MENAHEM and BEYDOUN (1985) and BEYDOUN and BEN MENAHEM (1985). Applications of the Gaussian beam method in other fields have included that of PORTER and BUCKER (1987) who applied the method to ocean acoustics (see also, JENSEN *et al.*, 1994). The method was applied to atmospheric acoustics by GABILLET *et al.* (1992).

Perturbation methods were used to compute approximate rays and beams in complicated media from results in more simple media by FARRA and MADARIAGA (1987). NOWACK and LUTTER (1988) applied ray perturbation methods for the investigation of linearized rays and their influence on the inversion of travel times and amplitudes. NOWACK (1990) then used perturbation methods for the calculation of Gaussian beam seismograms in a laterally varying perturbed medium using results computed in a laterally homogeneous medium.

As a recent example of the Gaussian beam method, NOWACK and STACY (2002) applied the method to the calculation of interference head waves for an interface with a velocity gradient beneath. Interference waves result from multiple bounces on the underside of an interface, as well as wide-angle reflected and head waves. In an earlier study, CORMIER and RICHARDS (1977) used a full waveform technique to sum gallery phases for a gradient beneath an interface and applied this to the inner core boundary (AKI and RICHARDS, 1980). In Fig. 3, synthetic seismograms are shown using the Gaussian beam and reflectivity methods for a 6 km/s layer from 0 to 25 km over a velocity gradient going from 8 km/s at 25 km to 8.57 km/s at 40 km depth. The source depth was 6 km and the source-time function was a Gabor wavelet with a dominant frequency of 2.77 Hz.

Reflectivity synthetics for this model are displayed in Figure 3A where the first arriving phases are the direct and interference P wave phases from the interface at 25 km. Later phases such as the surface reflected P waves and S waves are also seen. The Gaussian beam synthetics are shown in Figure 3B and include only the direct P waves, interference waves and wide-angle reflections from the interface. For the interference waves produced by the gradient, the semi-automatic choice of ČERVENÝ (1985a) was used which specifies broad beams at the receiver that also limits the discretization error of the Gaussian beam summation. This choice was specified in order to produce stable results and also avoid caustics for the interference waves. To

REFLECTIVITY

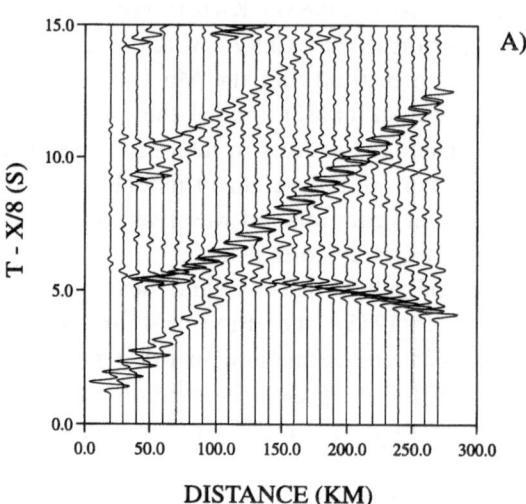

DISTANCE (KM)

GAUSSIAN BEAM

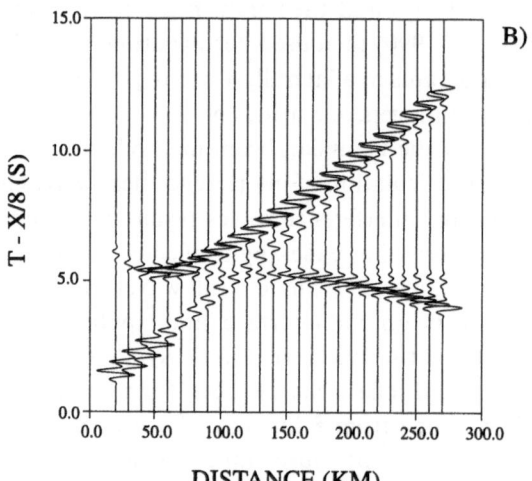

DISTANCE (KM)

Figure 3
Synthetic seismograms are shown for a model with a layer of 6 km/s over a velocity gradient from 8 km/s at 25 km to 8.57 km/s at 40 km. Reflectivity synthetics are shown in A) and Gaussian beam synthetics are shown in B). Only the direct P wave and the interference and wide-angle phases from the interface at 25 km are shown for the Gaussian beam synthetics (from Nowack and Stacy, 2002).

model the wide-angle reflections somewhat broader beams were used to ensure that the direct head-wave contribution was obtained. For the direct wave, the Gaussian beam method was run in ray mode since the direct wave is standard for this case. The final Gaussian beam synthetics in Figure 3B are the summation of results for these different phases. Figure 3 shows that a good agreement between the results from the Gaussian beam and reflectivity methods was obtained for the first-arrival and wide-angle *P*-wave phases. In addition, the Gaussian synthetics were obtained for a small fraction of the computing time as the reflectivity results.

The relation between the Gaussian beam method and the use of complex source points was presented by FELSEN (1984) and WU (1985) (see also, DESCHAMPS, 1971; KELLER and STREIFER, 1971), and extensions of the Gaussian beam method using complex source points are described below. FELSEN (1984) also recommended propagation corrections for lateral beam shifts at interfaces as a way to better use narrow paraxial beams in beam summations. However, this correction was later incorporated by GAO *et al.* (1990) and was shown to be insufficient to provide the head-wave contribution when using only paraxially narrow beams in the summation.

Extensions of the Gaussian Beam Method

Although the original Gaussian beam method has been successful in modeling smoothly varying media, structures with discontinuities or corner points have presented difficulties for the method. This was identified by NOWACK and AKI (1984), KONOPASKOVA and ČERVENÝ (1984a,b), and WHITE *et al.* (1987) who demonstrated that broad beams are required to obtain the head waves at interfaces. Also, GEORGE *et al.* (1987) showed that corner points present difficulties when paraxially approximated Gaussian beams are used. For structures with strong heterogeneities, WHITE *et al.* (1987) showed that the Gaussian beam method depends on the chosen beam parameters. Numerous studies have attempted to extend the Gaussian beam method, and these have come under several categories including Gaussian beams as building blocks in other methods such as the boundary integral method, more exact beam elements and more complete beam superposition algorithms.

An expansion of a time-harmonic point source was performed by NORRIS (1986) using a distribution of exact complex source points. LU *et al.* (1987) also used complex source points to model reflections from an interface using a complex Huygen's principle, and HEYMAN (1989) extended this to time-domain point sources. Further applications of summation of complex source points are described by DEZHONG (1995) and NORRIS and HANSEN (1997). For complex sources in general heterogeneous media, complex rays must be used for ray tracing. An overview of complex rays is given by KRAVSTOV *et al.* (1999) (see also, THOMSON, 1997). As an alternative to complex ray tracing, ZHU and CHUN (1994a,b) used ray perturbation methods to obtain complex rays.

Another extension of the Gaussian beam method has been to construct superpositions of either complex source points or paraxial beam elements for extended sources. EINZIGER *et al.* (1986) used a Gabor expansion described below to study an extended aperture by the superposition of beam waves, where each of the elementary beams was specified on a phase-space lattice of shifted and tilted beams. The propagation of waves away from extended apertures by phase-space beam summations was investigated by several authors including MARCIEL and FELSEN (1989) and STEINBERG *et al.* (1991a). Self-consistent Gaussian beam superpositions using Gabor expansions were investigated by FELSEN *et al.* (1991) for 2-D applications and KLOSNER *et al.* (1992) for 3-D applications. The use of pulsed-time signals instead of time-harmonic signals for the beam summations was also considered (see for example, STEINBERG *et al.*, 1991b; MELAMED, 1997).

The Gabor expansion was initially described by GABOR (1946) and is related to windowed Fourier transforms. It can be written as

$$u(x) = \sum_n \sum_m \Phi(n,m)w(x - mL)e^{in\Omega x} \ , \tag{18}$$

where L is the sampling in position, Ω is the sampling in wavenumber, $\Phi(n,m)$ is the weighting function, and $w(x)$ is the window function. An example of a window function is the Gaussian window, $w(x) = (2^{1/2}/L)^{1/2}e^{-\pi x^2/L^2}$. The sampling parameter L also determines the width of the Gaussian window functions. The Gabor expansion was investigated by BASTIAANS (1980) who derived a set of biorthogonal expansion coefficients $\Phi(n,m)$ for a given function $u(x)$ in the case of critical sampling, where critical sampling results when the sampling in position L and the sampling in wavenumber Ω are related by $\Omega L = 2\pi$.

If the phase-shifted Gaussian window functions are considered as initial Gaussian beams in the aperture plane, then Eqn. (18) can be interpreted as a decomposition of an initial wave field into shifted and tilted Gaussian beams. Figure 4A presents an example of a 2-D phase-space lattice representing the wave field in the aperture (from FELSEN *et al.*, 1991). Figure 4B displays an example of a shifted and tilted beam element, where $\sin \theta$ is related to the wavenumber. Assuming that the initial Gaussian beams can either be exactly or asymptotically propagated away from the aperture, then the resulting wave field can be written

$$u(x,z) = \sum_n \sum_m \Phi(n,m)u_{n,m}(x,z) \ , \tag{19}$$

where $u_{n,m}(x,z)$ are the individual beams for $z > 0$ and $\Phi(n,m)$ are the weight functions. Several applications of Gabor expansions in optics have been described by BASTIAANS (1998).

Although, this approach provides a self-consistent selection criterion for the summation of Gaussian beams, it was observed by DAUBECHIES (1990) that the Gabor expansion coefficients of BASTIAANS (1980) at critical sampling are only marginally stable. Nonetheless, a useful localization can still be obtained for a Gabor expansion when oversampled frames are used, where $\Omega L < 2\pi$. Although the oversampled frames are nonunique, a minimum norm solution for the coefficients can be obtained (WEXLER and RAZ, 1990; QIAN and CHEN, 1993; ZIBULSKI and

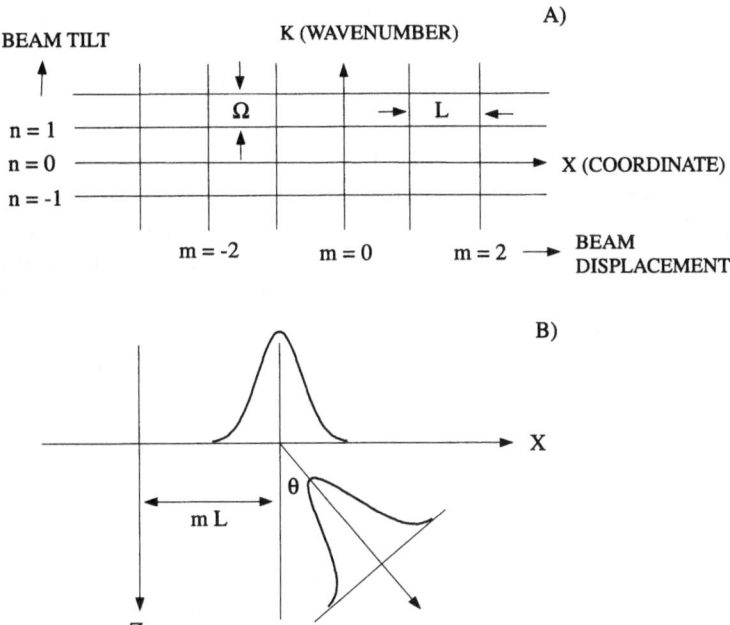

Figure 4

Summation of beams on a phase-space lattice. A) A phase-space lattice is shown for lateral beam displacement and beam tilt. The beam sampling is L and the wavenumber sampling is Ω which is related to the beam tilt. B) An example of a laterally displaced and tilted beam (from FELSEN et al., 1991).

ZEEVI, 1993; BASTIAANS and GEILEN, 1996). An example of an oversampled Gabor expansion of an initial wave field propagated away from an aperture was given by LUGARA and LETROU (1998).

In geophysics, there have been several applications which have attempted to increase the spectral content of the Gaussian beam expansion. These have included the studies by WANG and WALTHAM (1995a,b), who used the stability and versatility of Gaussian beam tracking in smoothly heterogeneous media along with edge and tip diffracted waves to generalize the Gaussian beam decomposition providing more spectral content in the decomposition. Another approach has been to expand the wave field into so-called coherent states, and this method has been described by FOSTER and HUANG (1991) and THOMSON (2001) for geophysical applications based on the work of KLAUDER (1987a,b). This approach is also related to a Gaussian-windowed Fourier transform of the wave field.

BENITES and AKI (1989) used Gaussian beams as building blocks within the framework of a boundary integral method, and thus used beams in a different type of expansion of the wave field. BENITES and AKI (1994) then used this boundary integral–Gaussian beam approach for the calculation of ground motions in sedimentary basins with velocity gradients and in models with surface topography.

An important development in exploration geophysics has been the use of Gaussian beam decompositions for the imaging of seismic reflection data (COSTA, et al., 1989; LAZARATOS and HARRIS, 1990; HILL, 1990). In these studies, reflection data are decomposed into local slant or beam stacks (RAZ, 1987) and matched to the seismic data at the surface. These local slant-stacks are then used within Gaussian beam migration algorithms. These approaches were found to be very successful in imaging steep dips using Gaussian beam propagation and superposition. An example of steep dip imaging of a salt dome by Gaussian beam migration was presented by HILL et al. (1991), and a numerical overview of the method was provided by HALE (1992a,b). Gaussian beam migration has been applied to anisotropic media by ALKHALIFAH (1995), and the use of Gaussian beam summations for prestack reflection data has been described by HILL (2001).

Conclusions

The Gaussian beam method has been shown to be a very stable asymptotic method for the computation of high-frequency wave fields in smoothly varying inhomogeneous media. One of the advantages of the method is that individual Gaussian beam components have no singularities along their paths. This assures the summation of Gaussian beams to be regular everywhere. The Gaussian beam method also introduces smoothing. Therefore, the method is not as sensitive to model parameterizations as the ray method. Another advantage is that the Gaussian beam method does not require two-point ray tracing. A number of successful applications of the method have been presented. However, the selection of beam parameters is still a topic of ongoing research and different extensions of the method have been reviewed. Nonetheless, the advantages of Gaussian beam methods will continue to make them useful for the modeling of high-frequency seismic waves in heterogeneous media.

Acknowledgments

I thank Keiiti Aki for his initial and ongoing interest in Gaussian beam methods. His interest has motivated a number of application studies in this area. Also, his ability to investigate new problems clearly and rationally has been an inspiration for his many students and colleagues.

REFERENCES

AKI, K. and RICHARDS, P. G., *Quantitative Seismology* (Freeman, San Francisco, 1980).
ALKHALIFAH, T. (1995), *Gaussian Beam Depth Migration for Anisotropic Media*, Geophysics 60, 1474–1484.

BABICH, V. M. and PANKRATOVA, T. F, *On discontinuities of Green's function of the wave equation with variable coefficient*, In *Problems of Mathematical Physics Vol. 6* (ed. Babich, V.M.) (Leningrad Univ. Press, Leningrad, 1973), pp. 9–27.

BABICH, V. M. and POPOV, M. M. (1989), *Gaussian Summation Method (Review)*, Izvestiya Vysshikh Uchebnykh Zavedenii, Radiofizika *32*, 1447–1466 (translated in Radiophysics and Quantum Electronics *32*, 1063–1081, 1990).

BASTIAANS, M. J. (1980), *Gabor's Expansion of a Signal into Gaussian Elementary Signals*, Proc. IEEE *68*, 538–539.

BASTIAANS, M. J. and GEILEN, M. C. (1996), *On the Discrete Gabor Transform and the Discrete Zak Transform*, Signal Proc. *49*, 151–166.

BASTIAANS, M. J., *Gabor's Signal Expansion in Optics*. In *Gabor Analysis and Algorithms* (eds. Feichtinger, H.G. and Strohmer, T.) (Birkhäuser, Boston, 1998) pp. 427–451.

BEN-MENAHEM, A. and BEYDOUN, W. B. (1985), *Range of Validity of Seismic Ray and Beam Methods in General Inhomogeneous Media-I. General Theory*, Geophys. J. R. Astr. Soc. *82*, 207–234.

BENITES, R. and AKI, K. (1989), *Boundary Integral-Gaussian Beam Method for Seismic Wave Scattering: SH Waves in Two-Dimensional Media*, J. Acoust. Soc. Am. *86*, 375–386.

BENITES, R. and AKI, K. (1994), *Ground Motion at Mountains and Sedimentary Basins with Vertical Seismic Velocity Gradient*, Geophys. J. Int. *116*, 95–118.

BEYDOUN, W. B. and BEN-MENAHEM, A. (1985), *Range of Validity of Seismic Ray and Beam Methods in General Inhomogeneous Media-II. A Canonical Problem*, Geophys. J. R. Astr. Soc. *82*, 235–262.

ČERVENÝ, V., *Seismic Wave Fields in Structurally Complicated Media (Ray and Gaussian Beam Approaches)* (Lecture Notes, Universiteit Utrecht, Vening-Meinesz laboratory, Utrecht, 1981).

ČERVENÝ, V. (1982), *Expansion of a Plane Wave into Gaussian Beams*, Studia Geophys. Geod. *26*, 120–131.

ČERVENÝ, V. (1983), *Synthetic Body Wave Seismograms for Laterally Varying Layered Structures by the Gaussian Beam Method*, Geophys. J. R. Astr. Soc. *73*, 389–426.

ČERVENÝ, V. (1985a), *Gaussian Beam Synthetic Seismograms*, J. Geophys. *58*, 44–72.

ČERVENÝ, V., *The application of ray tracing to the numerical modeling of seismic wave fields in complex structures*. In *Seismic Shear Waves* (ed. Dohr, G.) (Geophysical Press, London, 1985b) pp. 1–124.

ČERVENÝ, V. (2001), *Seismic Ray Theory*, Cambridge Univ. Press.

ČERVENÝ, V. and HRON, F. (1980), *The Ray Series Method and Dynamic Ray Tracing Systems in 3-D Inhomogeneous Media*, Bull. Seismol. Soc. Am. *70*, 47–77.

ČERVENÝ, V. and KLIMEŠ, L. (1984), *Synthetic Body Wave Seismograms for Three-dimensional Laterally Varying Media*, Geophys. J. R. Astr. Soc. *79*, 119–133.

ČERVENÝ, V., PLEINEROVÁ, J., KLIMEŠ, L., and PŠENČÍK, I. (1987), *High-frequency Radiation from Earthquake Sources in Laterally Varying Layered Structures*, Geophys. J. R. Astr. Soc. *88*, 43–79.

ČERVENÝ, V., POPOV, M. M. and PŠENČÍK, I. (1982), *Computation of Wave Fields in Inhomogeneous Media – Gaussian Beam Approach*, Geophys. J. R. Astr. Soc. *70*, 109–128.

ČERVENÝ, V. and PŠENČÍK, I. (1983a), *Gaussian Beams in Two-dimensional Elastic Inhomogeneous Media*, Geophys. J. R. Astr. Soc. *72*, 417–433.

ČERVENÝ, V. and PŠENČÍK, I. (1983b), *Gaussian Beams and Paraxial Ray Approximation in Three-dimensional Elastic Inhomogeneous Media*, J. Geophys. *53*, 1–15.

ČERVENÝ, V. and PŠENČÍK, I. (1984), *Gaussian Beams in Elastic 2-D Laterally Varying Layered Structures*, Geophys. J. R. Astr. Soc. *78*, 65–91.

CHEN, C. H., TENG, T. L., and GUNG, Y. C. (1998) *Ten-second Love-wave Propagation and Strong Ground Motions in Taiwan*, J. Geophys. Res. *103*, 21,253–21,273.

CORMIER, V. F. (1987), *Focusing and Defocusing of Teleseismic P Waves by Known Three-dimensional Structure Beneath Pahute Mesa, Nevada Test Site*, Bull. Seismol. Soc. Am. *77*, 1688–1703.

CORMIER, V. F. (1989), *Slab Diffraction of S Waves*, J. Geophys. Res. *94*, 3006–3024.

CORMIER, V. F. (1995), *Time-domain modeling of PKIKP Precursors for Constraints on the Heterogeneity in the Lower Most Mantle*, Geophys. J. Int. *121*, 725–736.

CORMIER, V. F. and RICHARDS, P. G. (1977), *Full-wave Theory Applied to a Discontinuous Velocity Increase: The Inner Core Boundary*, J. of Geophys. *43*, 3–31.

CORMIER, V. F. and SPUDICH, P. (1984), *Amplification of Ground Motion and Waveform Complexity in Fault Zones: Examples from the San Andreas and Calaveras Faults*, Geophys. J. R. Astr. Soc. *79*, 135–152.

CORMIER, V. F. and SU. W. J. (1994), *Effects of Three-dimensional Crustal Structure on the Estimated Slip History and Ground Motion of the Loma Prieta Earthquake*, Bull. Seismol. Soc. Am. *84*, 284–294.

COSTA, C., RAZ, S., and KOSLOFF, D. (1989), *Gaussian Beam Migration*, 59th Ann. Internat. Mtg. Soc. Expl. Geophys., Expanded Abstracts, 1169–1171.

DAUBECHIES, I. (1990), *The Wavelet Transform, Time Frequency Localization and Signal Analysis*, IEEE Trans. Info. Theory *36*, 961–1005.

DESCHAMPS, G. A. (1971), *Gaussian Beam as a Bundle of Complex Rays*, Electron. Lett. *7*, 684.

DEZHONG, Y. (1995), *Study of Complex Huygens Principle*, Int. J. Infrared and Millimeter Waves *16*, 831–838.

EINZIGER, P. D., RAZ, S., and SHAPIRA, M. (1986), *Gabor Representation and Aperture Theory*, J. Opt. Soc. Am. A *3*, 508–522.

FARRA, V. and MADARIAGA, R. (1987), *Seismic Waveform Modeling in Heterogeneous Media by Ray Perturbation Theory*, J. Geophys. Res. *92*, 2697–2712.

FELSEN, L. B. (1984), *Geometrical Theory of Diffraction, Evanescent Waves, Complex Rays and Gaussian Beams*, Geophys. J. R. Astr. Soc. *79*, 77–88.

FELSEN, L. B., KLOSNER, J. M., LU, I. T., and GROSSFELD, Z. (1991), *Source Field Modeling by Self-consistent Gaussian Beam Superposition (Two-dimensional)*, J. Acoust. Soc. Am. *89*, 63–72.

FOSTER, D. J. and HUANG, J. I. (1991), *Global Asymptotic Solutions of the Wave Equation*, Geophys. J. Int. *105*, 163–171.

FRIEDERICH, W. (1989), *A New Approach to Gaussian Beams on a Sphere: Theory and Application to Long-Period Surface Wave Propagation*, Geophys. J. Int. *99*, 259–271.

GABILLET, Y., SCHROEDER, H., DAIGLE, G., and L'ESPERANCE, A. (1992), *Application of the Gaussian Beam Approach to Sound Propagation in the Atmosphere: Theory and Experiments*, J. Acoust. Soc. Am. *93*, 3105–3116.

GABOR, D. (1946), *Theory of Communication*, J. Inst. Elec. Eng. *93-III*, 429–457.

GAO, X. J., FELSEN, L. B., and LU, I. T., *Spectral options to improve the paraxial narrow Gaussian beam algorithm for critical reflection and head waves*. In *Computational Acoustics* (eds. Lee, D., Cakmak, R., and Vichnevetsky, R.) (Elsevier, New York, 1990) pp. 149–166.

GEORGE, Th., VIRIEUX, J., and MADARIAGA, R. (1987), *Seismic Wave Synthesis by Gaussian Beam Summation: A Comparison with Finite Differences*, Geophysics *52*, 1065–1073.

GRIKUROV, V. E. and POPOV, M. M. (1983), *Summation of Gaussian Beams in a Surface Waveguide*, Wave Motion *5*, 225–233.

HANYGA, A. (1986), *Gaussian Beams in Anisotropic Elastic Media*, Geophys. J. R. Astr. Soc. *85*, 473–503.

HALE, D. (1992a), *Migration by the Kirchhoff, Slant Stack, and Gaussian Beam Methods*, CWP-126, Center for Wave Phenomena, Colorado School of Mines.

HALE, D. (1992b), *Computational Aspects of Gaussian Beam Migration*, CWP-127, Center for Wave Phenomena, Colorado School of Mines.

HEYMAN, E. (1989), *Complex Source Pulsed Beam Representation of Transient Radiation*, Wave Motion *11*, 337–349.

HILL, N. R. (1990), *Gaussian Beam Migration*, Geophysics *55*, 1416–1428.

HILL, N. R. (2001), *Prestack Gaussian Beam Depth Migration*, Geophysics, *66*, 1240–1250.

HILL, N. R., WATSON, T. H., HASSLER, M. H., and SISEMORE, L. K. (1991), *Salt-flank Imaging Using Gaussian Beam Migration*, 61st Ann. Internat. Mtg. Soc. Expl. Geophys., Expanded Abstracts, 1178–1180.

JENSEN, F. B., KUPERMAN, W. A., PORTER, M. B., and SCHMIDT, H., *Computational ocean acoustics*. In *AIP Series in Modern Acoustics and Signal Processing* (ed. Beyer, R.T.), (AIP Press, New York, 1994) pp. 149–202.

JOBERT, N. (1986), *Mantle Wave Propagation Anomalies on Laterally Heterogeneous Global Models of the Earth by Gaussian Beam Synthesis*, Ann. Geophys. *4*, 261–270

JOBERT, N. (1987), *Mantle Wave Deviations from "Pure-Path" Propagation on Aspherical Models of the Earth by Gaussian Beam Waveform Synthesis*, Phys. Earth Planet Int. *47*, 253–266.

JOBERT, N. and JOBERT, G. (1983), *An Approximation of Ray Theory to the Propagation of Waves along a Laterally Heterogeneous Spherical Surface*, Geophys. Res. Lett. *10*, 1148–1151.

KATCHALOV, A. P. and POPOV, M. M. (1981), *Application of the Method of Summation of Gaussian Beams for Calculation of High-frequency Wave Fields*, Sov. Phys. Dokl. *26*, 604–606.

KATCHALOV, A. P. and POPOV, M. M. (1985), *Application of the Gaussian Beam Method to Elasticity Theory*, Geophys. J. R. Astr. Soc. *81*, 205–214.

KATCHALOV, A. P. and POPOV, M. M. (1988), *Gaussian Beam Methods and Theoretical Seismograms*, Geophys. J. *93*, 465–475.

KATCHALOV, A. P., POPOV, M. M., and PŠENČÍK, I. (1983), *Applicability of the Gaussian Beams Summation Method to Problems with Angular Points on the Boundaries*, Zapiski Nauchnykh Seminarov Leningradskogo Otdeleniya Matematicheskogo Instituta im. V. A. Steklova AN SSSR *128*, 65-71 (translated in J. Sov. Math., 2406–2410, 1985).

KATO, K., AKI, K. and TENG, T. L. (1993), *3-D Simulations of Surface Wave Propagation in the Kanto Sedimentary Basin, Japan-Part 1: Application of the Surface Wave, Gaussian Beam Method*, Bull. Seismol. Soc. Am. *83*, 1676–1699.

KELLER, J. B. and STREIFER, W. (1971), *Complex Rays with an Application to Gaussian Beams*, J. Opt. Soc. Am. *61*, 40–43.

KLAUDER, J. R., (1987a), *Global, Uniform, Asymptotic Wave-equation Solutions for Large Wavenumbers*, Ann. Phys. *180*, 108–151.

KLAUDER, J. R. *Some recent results on wave equations, path integrals, and semiclassical approximations.* In *Random Media* (ed. Papanicolau, G.) (Springer-Verlag, New York, 1987b) pp. 163–182.

KLIMEŠ, L. (1984a), *Expansion of a High-frequency Time-harmonic Wave Field Given on an Initial Surface into Gaussian Beams*, Geophys. J. R. Astr. Soc. *79*, 105–118.

KLIMEŠ, L. (1984b), *The Relation Between Gaussian Beams and Maslov Asymptotic Theory*, Studia Geoph. et Geod. *28*, 237–247.

KLIMEŠ, L. (1986), *Discretization Error for the Superposition of Gaussian Beams*, Geophys. J. R. Astr. Soc. *86*, 531–551.

KLIMEŠ, L. (1989a), *Gaussian Packets in the Computation of Seismic Wave Fields*, Geophys. J. Int. *99*, 421–433.

KLIMEŠ, L. (1989b), *Optimization of the Shape of Gaussian Beams of a Fixed Length*, Studia Geoph. et Geod. *33*, 146–163.

KLOSNER, J. M., FELSEN, L. B., LU, I. T., and GROSSFELD, H. (1992), *Three-dimensional Source Field Modeling by Self-consistent Gaussian Beam Superposition*, J. Acoust. Soc. Am. *91*, 1809–1822.

KONOPASKOVA, J. and ČERVENÝ, V. (1984a), *Numerical Modelling of Time-harmonic Seismic Wave Fields in Simple Structures by the Gaussian Beam Method. Part I.*, Studia Geoph. et Geod. *28*, 19–35.

KONOPASKOVA, J. and ČERVENÝ, V. (1984b), *Numerical Modelling of Time-harmonic Seismic Wave Fields in Simple Structures by the Gaussian Beam Method. Part II.*, Studia Geoph. et Geod. *28*, 113–128.

KRAVTSOV, Y. A., FORBES, G. W., and ASATRYAN, A. A., *Theory and applications of complex rays.* In *Progress in Optics* (ed. Wolf, E.) (Elsevier, New York, 1999), pp. 2–62.

LAZARATOS, S. K. and HARRIS, J. M. (1990), *Radon Transform/Gaussian Beam Migration*, 60th Ann. Internat. Mtg. Soc. Expl. Geophys., Expanded Abstracts, 1178–1180.

LEBORGNE, S., MADARIAGA, R., and FARRA, V. (1999), *Body Waveform Modeling of East Mediterranean Earthquakes at Intermediate Distance (17°–30°) with a Gaussian Beam Summation Method*, J. Geophys. Res. *104*, 28,813–28,828.

LU, I. T., FELSEN, L. B. and RUAN, Y. Z. (1987), *Spectral Aspects of the Gaussian Beam Method: Reflection from a Homogeneous Half-space*, Geophys. J. R. Astr. Soc. *89*, 915–932.

LUGARA, D. and LETROU, C. (1998), *Alternative to Gabor's Representation of Plane Aperture Radiation*, Electron. Lett. *34*, 2286–2287.

MARCIEL, J. J. and FELSEN, L. B., (1989), *Systematic Study of Fields Due to Extended Apertures by Gaussian Beam Discretization*, IEEE Trans. on Antennas and Propagation *37*, 884–892.

MADARIAGA, R. (1984), *Gaussian Beam Synthetic Seismograms in a Vertically Varying Medium*, Geophys. J. R. Astr. Soc. *79*, 589–612.

MADARIAGA, R. and PAPADIMITRIOU, P. (1985), *Gaussian Beam Modelling of Upper Mantle Phases*, Ann. Geophys. *3*, 799–812.

MELAMED, T. (1997), *Phase-space Beam Summation: A Local Spectrum Analysis of Time Dependent Radiation*, J. Electromagnetic Waves and Appl. *11*, 739–773.

MÜLLER, G. (1984), *Efficient Calculation of Gaussian-beam Seismograms for Two-dimensional Inhomogeneous Media*, Geophys. J. R. Astr. soc. *79*, 153–166.

NORRIS, A. N. (1986), *Complex Point-source Representation of Real Point Sources and the Gaussian Beam Summation Method*, J. Opt. Soc. Am. A *3*, 2005–2010.

NORRIS, A. N. and HANSEN, T. B., (1997), *Exact Complex Source Representations of Time-harmonic Radiation*, Wave Motion *25*, 127–141.

NOWACK, R. L. (1990), *Perturbation methods for rays and beams*. In *Computational Acoustics*, (eds. Lee, D., Cakmak, R. and Vichnevetsky, R.) (Elsevier, New York, 1990) pp. 167–180.

NOWACK, R. L. and AKI, K. (1984), *The Two-dimensional Gaussian Beam Synthetic Method: Testing and Application*, J. Geophys. Res. *89*, 7797–7819.

NOWACK, R. L. and AKI, K. (1986), *Iterative Inversion for Velocity Using Waveform Data*, Geophys. J. R. Astr. Soc. *87*, 701–730.

NOWACK, R. L. and CORMIER, V. F. (1985), *Computed Amplitudes Using Ray and Beam Methods for a Known 3-D Structure*, EOS Trans. Am. Geophys. Un. *66*, 980.

NOWACK, R. L. and LUTTER, W. J. (1988), *Linearized Rays, Amplitude and Inversion*, Pure Appl. Geophys. *128*, 401–421.

NOWACK, R. L. and STACY, S. (2002), *Synthetic Seismograms and Wide-angle Seismic Attributes from the Gaussian Beam and Reflectivity Methods for Models with Interfaces and Gradients*, Pure Appl. Geophys., *159*, 1447–1464.

POPOV, M. M. (1981), *A New Method of Computing Wave Fields in the High-frequency Approximation*, Zapiski Nauchnykh Seminarov Leningradskogo Otdeleniya Matematicheskogo Instituta im. V. A. Steklova AN SSSR *104*, 195–216 (translated in J. of Sov. Math. *20*, 1869–1882, 1982).

POPOV, M. M. (1982), *A New Method of Computation of Wave Fields Using Gaussian Beams*, Wave Motion *4*, 85–97.

POPOV, M. M. and PŠENČÍK, I. (1978), *Computation of Ray Amplitudes in Inhomogeneous Media with Curved Interfaces*, Studia Geophys. Geod. *22*, 248–258.

POPOV, M. M., PŠENČÍK, I., and ČERVENÝ, V. (1980), *Uniform Ray Asymptotics for Seismic Wave Fields in Laterally Inhomogeneous Media (Abstract)*, Prog. Abstr. XVII General Assembly of the European Seismological Commission, Hungarian Geophysical Society, Budapest, p. 143.

PORTER, M. B. and BUCKER, H. P. (1987), *Gaussian Beam Tracing for Computing Ocean Acoustic Fields*, J. Acoust. Soc. Am. *82*, 1349–1359.

QIAN, S. and CHEN, D. (1993), *Discrete Gabor Transform*, IEEE Trans. Signal Proc. *41*, 2429–2438.

QU, J., TENG, T. L., and WANG, J. (1994), *Modeling of Short-period Surface-wave Propagation in Southern California*, Bull. Seismol. Soc. Am. *84*, 596–612.

RAZ, S. (1987), *Beam Stacking: A Generalized Preprocessing Technique*, Geophysics *52*, 1199–1210.

SEKIGUCHI, S. (1992), *Amplitude Distribution of Seismic Waves for Laterally Heterogeneous Structures Including a Subducting Slab*, Geophys. J. Int. *111*, 448–464.

STEINBERG, B. Z., HEYMAN, E., and FELSEN, L.B. (1991a), *Phase-space Beam Summation for Time-harmonic Radiation from Large Apertures*, J. Opt. Soc. Am. A *8*, 41–59.

STEINBERG, B. Z., HEYMAN, E., and FELSEN, L.B. (1991b), *Phase-space Beam Summation for Time Dependent Radiation from Large Apertures: Continuous Parameterization*, J. Opt. Soc. Am. A *8*, 943–958.

THOMSON, C. J. (1997), *Complex Rays and Wavepackets for Decaying Signals in Inhomogeneous, Anisotropic and Anelastic Media*, Studia Geophys. Geod. *41*, 345–381.

THOMSON, C. J. (2001), *Seismic Coherent States and Ray Geometrical Spreading*, Geophys. J. Int. *144*, 320–342.

WANG, X. and WALTHAM, D. (1995a), *The Stable-beam Seismic Modeling Method*, Geophys. Prosp. *43*, 939–961.

WANG, X. and WALTHAM, D. (1995b), *Seismic Modeling Over 3-D Homogeneous Layered Structure—Summation of Gaussian Beams*, Geophys. J. Int. *122*, 161–174.

WEBER, M. (1988a), *Computation of Body-wave Seismograms in Absorbing 2-D Media Using the Gaussian Beam Method: Comparison with Exact Methods*, Geophys. J. *92*, 9–24.

WEBER, M. (1988b), *Application of the Gaussian Beam Method in Refraction Seismology – Urach Revisited*, Geophys. J. *92*, 25–31.

WEBER, M. (1990), *Subduction Zones—Their Influence on Traveltimes and Amplitudes of P Waves*, Geophys. J. Int. *101*, 529–544.

WEBER, M. (1993), *P-wave and S-wave reflections from Anomalies in the Lower Most Mantle*, Geophys. J. Int. *115*, 183–210.

WEBER, M. and DAVIS, J. P. (1990), *Evidence of a Laterally Variable Lower Mantle Structure from P Waves and S Waves*, Geophys. J. Int. *102*, 231–255.

WEXLER, J. and RAZ, S. (1990), *Discrete Gabor Expansions*, Signal Proc. *21*, 207–221.

WHITE, B. S., NORRIS, A., BAYLISS, A., and BURRIDGE, R. (1987), *Some Remarks on the Gaussian Beam Summation Method*, Geophys. J. R. Astr. Soc. *89*, 579–636.

WU, R. S. (1985), *Gaussian Beams, Complex Rays, and the Analytic Extension of the Green's Function in Smoothly Inhomogeneous Media*, Geophys. J. R. Astr. Soc. *83*, 93–110.

YOMOGIDA, K. (1985), *Gaussian Beams for Surface Waves in Laterally Slowly-varying Media*, Geophys. J. R. Astr. Soc. *82*, 511–533.

YOMOGIDA, K. (1987), *Gaussian Beams for Surface Waves in Transversely Isotropic Media*, Geophys. J. R. Astr. Soc. *88*, 297–304.

YOMOGIDA, K. and AKI, K. (1985), *Waveform Synthesis of Surface Waves in a Laterally Heterogeneous Earth by the Gaussian-beam Method*, J. Geophys. Res. *90*, 7665–7688.

YOMOGIDA, K. and AKI, K. (1987), *Amplitude and Phase Data Inversion for Phase Velocity Anomalies in the Pacific Ocean Basin*, Geophys. J. R. Astr. Soc. *88*, 161–204.

ZHENG, Y., TENG, T. L., and AKI, K. (1989), *Surface-wave Mapping of the Crust and Upper Mantle in the Arctic Region*, Bull. Seismol. Soc. Am. *79*, 1520–1541.

ZHU, T. and CHUN, K. Y. (1994a), *Understanding Finite-frequency Wave Phenomena: Phase-ray Formulation and Inhomogeneity Scattering*, Geophys. J. Int. *119*, 78–90.

ZHU, T. and CHUN, K. Y. (1994b), *Complex Rays in Elastic and Anelastic Media*, Geophys. J. Int. *119*, 269–276.

ZIBULSKI, M. and ZEEVI, Y. Y. (1993), *Oversampling in the Gabor Scheme*, IEEE Trans. on Signal Proc. *41*, 2679–2687.

(Received August 16, 2000, accepted April 25, 2001)

To access this journal online:
http://www.birkhauser.ch

Pure appl. geophys. 160 (2003) 509–539
0033–4553/03/040509–31

© Birkhäuser Verlag, Basel, 2003

▌Pure and Applied Geophysics

Wave Propagation, Scattering and Imaging Using Dual-domain One-way and One-return Propagators

RU-SHAN WU[1]

Abstract — Dual-domain one-way propagators implement wave propagation in heterogeneous media in mixed domains (space-wavenumber domains). One-way propagators neglect wave reverberations between heterogeneities but correctly handle the forward multiple-scattering including focusing/defocusing, diffraction, refraction and interference of waves. The algorithm shuttles between space-domain and wavenumber-domain using FFT, and the operations in the two domains are self-adaptive to the complexity of the media. The method makes the best use of the operations in each domain, resulting in efficient and accurate propagators. Due to recent progress, new versions of dual-domain methods overcame some limitations of the classical dual-domain methods (phase-screen or split-step Fourier methods) and can propagate large-angle waves quite accurately in media with strong velocity contrasts. These methods can deliver superior image quality (high resolution/high fidelity) for complex subsurface structures. One-way and one-return (De Wolf approximation) propagators can be also applied to wave-field modeling and simulations for some geophysical problems. In the article, a historical review and theoretical analysis of the Born, Rytov, and De Wolf approximations are given. A review on classical phase-screen or split-step Fourier methods is also given, followed by a summary and analysis of the new dual-domain propagators. The applications of the new propagators to seismic imaging and modeling are reviewed with several examples. For seismic imaging, the advantages and limitations of the traditional Kirchhoff migration and time-space domain finite-difference migration, when applied to 3-D complicated structures, are first analyzed. Then the special features, and applications of the new dual-domain methods are presented. Three versions of GSP (generalized screen propagators), the hybrid pseudo-screen, the wide-angle Padé-screen, and the higher-order generalized screen propagators are discussed. Recent progress also makes it possible to use the dual-domain propagators for modeling elastic reflections for complex structures and long-range propagations of crustal guided waves. Examples of 2-D and 3-D imaging and modeling using GSP methods are given.

Key words: Wave propagation, scattering, seismic imaging, modeling, one-way propagation, depth migration.

1. Introduction

Perturbation approach is one of the well-known approaches for wave propagation, scattering and imaging (see Ch. 9 of MORSE and FESHBACH, 1953; Ch. 13 of AKI and RICHARDS, 1980; WU, 1989). Traditionally, perturbation methods are used only

[1] Modeling and Imaging Laboratory, Institute of Geophysics and Planetary Physics, University of California, Santa Cruz, California, U.S.A. E-mail: wrs@es.ucsc.edu

for weakly inhomogeneous media and short propagation distance. However, recent progress in this direction has led to the development of iterative perturbation solutions in the form of one-way marching algorithm for scattering and imaging problems in strongly heterogeneous media. In this article, a historical review and theoretical analysis regarding the perturbation approach, including the Born, Rytov, and De Wolf approximations, are given in section 2. The relative strong and weak points of the Born and Rytov approximations are analyzed. Since the Born approximation is a weak scattering approximation, it is not suitable for large volume or long-range numerical simulations. The Rytov approximation is a smooth scattering approximation, which works well for long-range small-angle propagation problems, but is not applicable to large-angle scattering and backscattering. Then the De Wolf approximation (multiple-forescattering-single-backscattering, or "one-return approximation") is introduced to overcome the limitations of the Born and Rytov approximations in long-range forward propagation and backscattering calculations, which can serve as the theoretical basis of the new dual-domain propagators. A review of classical dual-domain propagators (phase-screen or split-step Fourier method) is also given, followed by a summary and analysis of the new dual-domain propagators, in section 3. The iterative perturbation approach has been developed in parallel to the operator splitting approach, and has led to the generalization of phase-screen approximation for scalar waves to the elastic wave screen propagators (see section 3.3). Three versions of GSP (generalized screen-propagators): the hybrid pseudo-screen, the wide-angle Padé-screen, and the higher-order generalized screen propagators are presented in section 3.4 as examples of the newly developed wide-angle dual-domain propagators. The applications of the new propagators to seismic imaging are reviewed in section 4. The advantages and limitations of the traditional Kirchhoff migration and time-space domain finite-difference migration, when applied to imaging of 3-D complicated structures, are first analyzed. Then the special features and applications of the new dual-domain methods are presented. Examples of 2-D and 3-D imaging (post- and pre-stack depth migrations) using synthetic data sets from the Marmousi model and SEG-EAEG salt model are given. Further progress also makes it possible to use the dual-domain propagators for modeling elastic reflections for complex structures and long-range propagation of crustal guided waves. These applications are briefly discussed in section 5. Conclusion is given in section 6.

2. From Born, Rytov to De Wolf

2.1 Born Approximation and Rytov Approximation: Their Strong and Weak Points

In recollection of my study and collaboration with Professor K. Aki, I would like to briefly digress to communicate my work on elastic Born scattering when I was a graduate student at MIT.

When we started to work on elastic Born scattering, we were not aware of GUBERNATIS et al.'s (1977a,b) work. We started from the basic principle and referred to MORSE and FESHBACH (1953) for our derivation. After a few months, I showed my derivation and part of the results to professor Aki. He was very delighted by the elegance of the theory and the practical implications of the results. He told me that I was very lucky to have achieved such a nice result on such a fundamental problem. But he added, "Such a neat result on this kind of fundamental problems should have been solved a long time ago. Somehow people neglected this spot and left some nice thing there. You are very lucky to pick up this stuff!" It turned out later, however that I was not as lucky as it appeared. When circulating our results to other colleagues, it was pointed out that similar results have been published in the Journal of Applied Physics (GUBERNATIS et al., 1977a,b). Of course, there were differences. Gubernatis et al.'s results regard a uniform elastic inclusion; while ours pertain to an arbitrarily heterogeneous body, and we had nice expressions for the velocity-type and impedance-type heterogeneities. Nevertheless, the general form of elastic Born scattering was published in that paper. In the beginning, I felt embarrassed and was very disappointed, like a defeated hero. Later I recovered from that mode of debacle. I comforted myself by looking at the event from a different perspective. I looked it as a test of my ability and good fortune. I said to myself that I can attack such problems and perhaps have luck in my future work. Thus I extended my work on elastic Born scattering to more general cases and to random media, with professor Aki's assistance, and published two papers in Geophysics and Journal of Geophysical Research (WU and AKI, 1985a,b), respectively. It turned out to be a prelude to my endeavor concerning the research of seismic wave propagation and scattering.

For the sake of simplicity, we consider the scalar wave case as an example. The scalar wave equation in inhomogeneous media can be written as

$$\left(\nabla^2 + \frac{\omega^2}{c^2(\vec{r})}\right)u(\vec{r}) = 0 \ , \tag{1}$$

where ω is the circular frequency, $\vec{r}$ is the position vector, and $c(\vec{r})$ is wave velocity at $\vec{r}$. Define c_0 as the background velocity of the medium, resulting in

$$(\nabla^2 + k^2)u(\vec{r}) = -k^2\varepsilon(\vec{r})u(\vec{r}) \ , \tag{2}$$

where $k = \omega/c_0$ is the background wavenumber and

$$\varepsilon(\vec{r}) = \frac{c_0^2}{c^2(\vec{r})} - 1 \tag{3}$$

is the perturbation function (dimensionless force). Set

$$u(\vec{r}) = u^0(\vec{r}) + U(\vec{r}) \ , \tag{4}$$

where $u^0(\vec{r})$ is the unperturbed wave field or "incident wave field" (field in the homogeneous background medium), and $U(\vec{r})$ is the scattered wave field. Substitute (4) into (2) and note that $u^0(\vec{r})$ satisfies the homogeneous wave equation, resulting in

$$u(\vec{r}) = u^0(\vec{r}) + k^2 \int\limits_V d^3\vec{r}' g(\vec{r};\vec{r}')\varepsilon(\vec{r}')u(\vec{r}') \ , \tag{5}$$

where $g(\vec{r};\vec{r}')$ is the Green's function in the background medium and the integral is over the entire volume of the medium. This is the Lippmann-Schwinger integral equation. Since the field $u(\vec{r})$ under the integral is the total field which is unknown, equation (5) is not an explicit solution but an integral equation.

Born Approximation

Approximating the total field under the integral with the incident field $u^0(\vec{r}')$, we obtain the Born Approximation

$$u(\vec{r}) = u^0(\vec{r}) + k^2 \int\limits_V d^3\vec{r}' g(\vec{r};\vec{r}')\varepsilon(\vec{r}')u^0(\vec{r}') \tag{6}$$

In general, the Born approximation is only valid when the scattered field is much smaller than the incident field, which implies that the heterogeneities are weak and the propagation distance is short. However, the valid regions of Born approximation are very different for forward scattering and for backscattering. Forward-scattering divergence or catastrophe is the weakest point of Born approximation. For simplicity, we will use "forescattering" to stand for "forward scattering." As can be seen from (6), the total scattering field is the sum of scattered fields from all parts of the scattering volume. Each contribution is independent from other contributions since the incident field is not updated by the scattering process. In the forward direction, the scattered fields from each part propagate with the same speed as the incident field, so they will be coherently superposed, leading to the linear increase of the total field. The Born approximation has no energy conservation. The energy increase will be fastest in the forward direction, resulting in a catastrophic divergence for long distance propagation. This can be demonstrated schematically as shown in Figure 1a. The medium is divided into blocks each represented by a concentrated scatterer at its center. It also can be considered as a discrete scattering medium. At the observation point, the total field will be the sum of the incident field (without any correction) and the scattered fields from all parts of the scattering volume. On the contrary, backscattering behaves quite differently from forescattering. As shown in Figure 1b, since there is no incident wave in the backward direction, the total observed field is the sum of all the backscattered fields from all the scatterers. However, the size of coherent stacking for backscattered waves is about $\lambda/4$ because of the two-way travel-time difference. Beyond this coherent region, all other contributions will be cancelled out. For this reason, backscattering does not have the

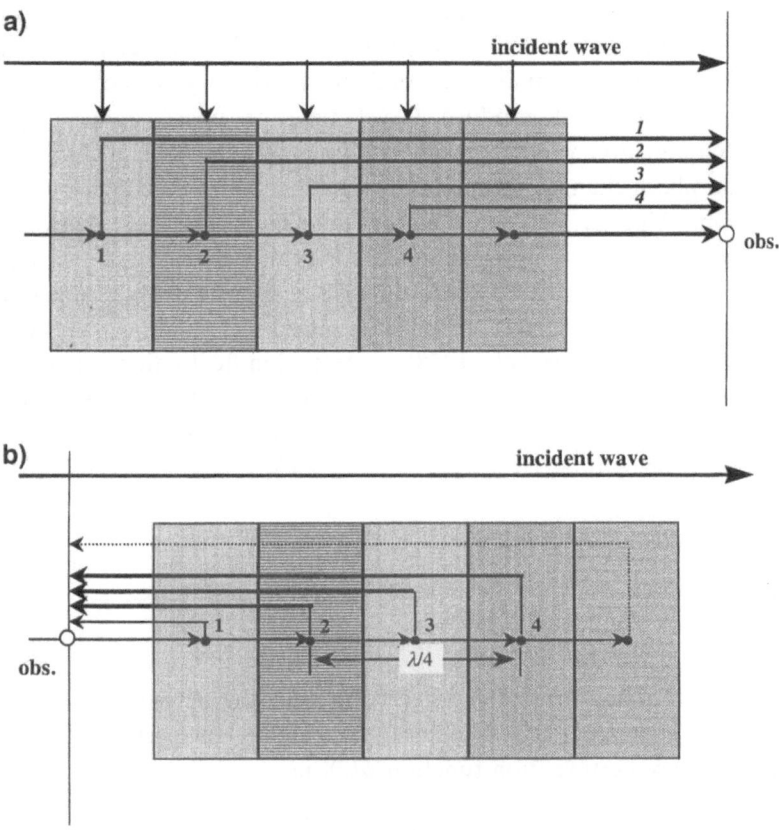

Figure 1
(a) Schematic demonstration of the forward scattering catastrophe of the Born approximation.
(b) Schematic demonstration of the size of the coherent response for backscattering.

catastrophic divergence even when Born approximation is used. This can be further explained with the spectral responses of heterogeneities to scatterings with different scattering angles.

From the analysis of scattering characteristics, we know that the forescattering is controlled by the d.c. component of the medium spectrum $W(0)$, but the backscattering is determined by the spectral component at spatial frequency $2k$, i.e., $W(2k)$, where k is wavenumber of the wave field in the background medium (see, Wu and Aki, 1985b; Wu, 1989a). The d.c. component of the medium spectrum linearly increases with the propagation distance in general, while $W(2k)$ is usually considerably smaller and increases much slower than $W(0)$. The validity condition for the Born approximation is the smallness of the scattered field compared with the incident field. Therefore the region of validity of the Born approximation for backscattering is much larger than that for forescattering. The other difference between backscattering and forescattering is their responses to different types of heterogeneities. The backscattering is sensitive to the impedance type of heterogeneities, while forescattering mainly responds to the velocity type of heterogeneities. Velocity perturbation will produce travel time or phase change, which can accumulate to quite large values, causing the breakdown of the Born approximation. This kind of phase-change accumulation can be easily handled by the Rytov transformation. This is why the Rytov approximation has decidedly better performance than the Born approximation for forescattering and has been widely used for long distance propagation with only forescattering or small-angle scattering involved, such as the line-of-sight propagation of optical or radio waves (CHERNOV, 1960; TATARSKII, 1971; ISHIMARU, 1978), transmission fluctuations of seismic waves at arrays (AKI, 1973; FLATTÉ and WU, 1988; WU and FLATTÉ, 1990), diffraction tomography (DEVANEY, 1982, 1984; WU and TOKSÖZ, 1987), and seismic imaging using one-way propagators (HUANG et al., 1999a,b).

Rytov Approximation

Let $u^0(\vec{r})$ be the solution in the absence of perturbations, i.e.,

$$(\nabla^2 + k^2)u^0 = 0 \tag{7}$$

and the perturbed wave field after interaction with the heterogeneity as $u(\vec{r})$. We normalize $u(\vec{r})$ by the unperturbed field $u^0(\vec{r})$ and express the perturbation of the field by a complex phase perturbation function $\psi(\vec{r})$, i.e.,

$$u(\vec{r})/u^0(\vec{r}) = e^{\psi(\vec{r})} \ . \tag{8}$$

This is the Rytov Transformation (see TATARSKII, 1971; or ISHIMARU, 1978, Ch. 17, p. 349). $\psi(\vec{r})$ denotes the phase- and log-amplitude deviations from the incident field:

$$\psi = \log u - \log u^0 = \log\left[\frac{A}{A^0}\right] + i(\phi - \phi^0) \ , \tag{9}$$

where A is the amplitude and ϕ is phase angle. Combining (2), (7) and (8) yields

$$2\nabla u^0 \cdot \nabla\psi + u^0\nabla^2\psi = -u^0(\nabla\psi \cdot \nabla\psi + k^2\varepsilon) \tag{10}$$

The simple identity

$$\nabla^2(u^0\psi) = \psi\nabla^2 u^0 + 2\nabla u^0 \cdot \nabla\psi + u^0\nabla^2\psi$$

together with (7) results in

$$2\nabla u^0 \cdot \nabla\psi + u^0\nabla^2\psi = (\nabla^2 + k^2)(u^0\psi) \ . \tag{11}$$

From (10) and (11) we obtain

$$(\nabla^2 + k^2)(u^0\psi) = -u^0(\nabla\psi \cdot \nabla\psi + k^2\varepsilon) \ . \tag{12}$$

The solution of (12) can be expressed as an integral equation:

$$u^0(\vec{r})\psi(\vec{r}) = \int_V d^3\vec{r}' g(\vec{r};\vec{r}')u^0(\vec{r}')\left[\nabla\psi(\vec{r}') \cdot \nabla\psi(\vec{r}') + k^2\varepsilon(\vec{r}')\right], \qquad (13)$$

where $g(\vec{r};\vec{r}')$ is the Green's function for the background medium, $u^0(\vec{r}')$ and $u^0(\vec{r})$ are the incident field at $\vec{r}'$ and $\vec{r}$, respectively.

Equation (13) is a nonlinear (Ricatti) equation. Assuming $|\nabla\psi \cdot \nabla\psi|$ is small with respect to $k^2|\varepsilon|$, we can neglect the term $\nabla\psi \cdot \nabla\psi$ and obtain a solution known as the Rytov approximation:

$$\psi(\vec{r}) = \frac{k^2}{u^0(\vec{r})} \int_V d^3\vec{r}' g(\vec{r};\vec{r}')\varepsilon(\vec{r}')u^0(\vec{r}'). \qquad (14)$$

Now we discuss the relation between the Rytov and Born approximations, and their strong and weak points, respectively. By expanding e^ψ into power series, the scattered field can be written as

$$u - u^0 = u^0(e^\psi - 1) = u^0\psi + \tfrac{1}{2}u^0\psi^2 + \cdots, \qquad (15)$$

When $\psi \ll 1$, i.e., the accumulated phase change is less than one radian (corresponding to about one sixth of the wave period), the terms of ψ^2 and higher terms can be neglected, and

$$u - u^0 = u^0\psi = k^2 \int_V d^3\vec{r}' g(\vec{r};\vec{r}')\varepsilon(\vec{r}')u^0(\vec{r}') \qquad (16)$$

which is the Born approximation. This indicates that when $\psi \ll 1$, Rytov approximation reduces to Born approximation. In case of large phase-change accumulation, for which Born approximation is no longer valid, Rytov approximation still holds as long as the condition $|\nabla\psi \cdot \nabla\psi| \ll k^2|\varepsilon|$ is satisfied.

Let us look at the implication of the condition $|\nabla\psi \cdot \nabla\psi| \ll k^2|\varepsilon|$ for the Rytov approximation. Assume that the observed total field after the wave interacted with the heterogeneities is nearly a plane wave:

$$u = Ae^{i\vec{k}\cdot\vec{r}}$$

which could be the refracted wave in the forward direction, or the backscattered field. Since the incident wave is

$$u^0 = A_0 e^{i\vec{k}_0\cdot\vec{r}}$$

the complex phase field ψ can be written as

$$\psi = \log(A/A_0) + i(\vec{k} - \vec{k}_0) \cdot \vec{r} \qquad (17)$$

and

$$\nabla \psi = \nabla \log(A/A_0) + i(\vec{k} - \vec{k}_0) \tag{18}$$

$$\nabla \psi \cdot \nabla \psi = |\nabla \log(A/A_0)|^2 - |\vec{k} - \vec{k}_0|^2 + 2i(\vec{k} - \vec{k}_0) \cdot \nabla \log(A/A_0) . \tag{19}$$

Normally wave amplitudes vary much slower than the phases, so the major contribution to $\nabla \psi \cdot \nabla \psi$ in (19) is from the phase term $|\vec{k} - \vec{k}_0|^2$. Therefore, the condition for the Rytov approximation can be approximately stated as

$$|\vec{k} - \vec{k}_0|^2 = 4k_0^2 \sin^2 \frac{\theta}{2} \ll k_0^2 |\varepsilon| \tag{20}$$

where θ is the scattering angle. Therefore the Rytov approximation is only valid when the scattering angle (deflection angle) is small enough to satisfy

$$\sin \frac{\theta}{2} \ll \sqrt{\frac{1}{4} \varepsilon} = \frac{1}{2} \sqrt{\frac{c_0^2 - c^2(\vec{r})}{c^2(\vec{r})}} . \tag{21}$$

This is a point-to-point analysis for the contributions from different terms (for example, the terms in the differential equation (12)). For the integral equation (13), one needs to estimate the integral effects of $\nabla \psi \cdot \nabla \psi$ and $k^2 \varepsilon$. The heterogeneities need to be smooth enough to guarantee the smallness of the integral of $\nabla \psi \cdot \nabla \psi$ which is related to scattering angles, in comparison with the total scattering contribution $k^2 \varepsilon$. Regardless, the Rytov approximation is totally inappropriate for backscattering. In the exactly backward direction, $\theta = 180°$ and $\sin \theta/2 = 1$, inequality (21) is hardly to be satisfied. Therefore, although not explicitly specified, the Rytov approximation is a somewhat small-angle approximation. Together with the parabolic approximation, they formed a set of analytical tools widely used for the forward propagation and scattering problems, such as the line-of-sight propagation problem (e.g., FLATTÉ, 1979; ISHIMARU, 1978; TATARSKII, 1971). The Rytov approximation is also used in modeling transmission fluctuation for seismic array data (WU and FLATTÉ, 1990), diffraction tomography (DEVANEY, 1982, 1984; WU and TOKSÖZ, 1987). TATARSKII (1971, Ch. 3B) discussed the relation of the Rytov approximation and parabolic approximation.

2.2 De Wolf Approximation

We see the limitations of both the Born and Rytov approximations. Even in weakly inhomogeneous media we need better tools for wave modeling and imaging for long distance propagation. Higher order terms of the Born series (defined later in this section) may help in some cases. However, for strong scattering media, Born series will either converge very slow, or become divergent. That is because the Born series is a global interaction series, each term of which is global in nature. The first term (the Born approximation) is a global response and the higher terms are just global corrections. If it makes an undue error in the first step, it will be hard to correct later. One solution to the divergence of the scattering series is the renormalization procedure. Renormalization methods try to split the operations so that the scattering series can be reordered into many sub-series. We hope some sub-series can be summed up theoretically so that the divergent elements of the series can be removed. De Wolf approximation splits the scattering potential into forescattering and backscattering parts and renormalizes the incident field and Green's function into the forward propagated field and forward propagated Green's function (forward propagator), respectively (DE WOLF, 1971, 1985). The forward propagated field u_f is the sum of an infinite sub-series which includes all the multiply forescattered fields. The forward propagator G_f is the sum of a similar sub-series which includes multiple forescattering corrections to the Green's function. The De Wolf approximation is also called "one-return approximation" (WU, 1996; WU and HUANG, 1995; WU et al., 2000a,b), since it is a multiple-forescattering-single-backscattering (MFSB) approximation. It is also somewhat of a local Born approximation with both the incident field and Green's function (propagator) calculated by one-way forward propagators. From previous sections we know that Born approximation works well for backscattering locally. With the renormalized incident field and Green's function the local Born (MFSB) proved to work surprisingly well for many practical applications. The key is to have good forward propagators. RINO (1988) has obtained better approximation than MFSB in the wavenumber domain and pointed out the error of De Wolf approximation in the calculation of backscattering enhancement. The error (overestimation) is again due to the violation of energy conservation law by the Born approximation. Even with forescattering correction, the backscattered energy is still not removed from the forward propagated waves for the local Born approximation. However, for short propagation distances in exploration seismology, the errors in reflection amplitudes may not become a serious problem.

In the appendix, we give a brief derivation of De Wolf approximation using formal operator algebra (DE WOLF, 1985). In this section, we will adopt an intuitive approach of derivation to discerns the physical meaning of the approximation. De Wolf approximation bears similarity to the Twersky approximation for discrete scatterers (TWERSKY, 1964; ISHIMARU, 1978). The Twersky approximation includes all the multiple scattering except the reverberations between pairs of scatterers, which excludes the paths which connect the two neighboring scatterers more than once. The Twersky approximation has less restrictions and therefore a wider range of applications than the De Wolf approximation. The latter needs to define the split of forward and back scatterings. We define the scattering to the forward hemisphere as forescattering and its complement as backscattering.

The Lippmann-Schwinger equation (5) can be written symbolically as

$$u = u^0 + G_0 \varepsilon u \ , \tag{22}$$

where ε is a diagonal operator in space domain, and G_0 is a nondiagonal integral operator. If the reference medium is homogeneous, G_0 will be the volume integral with the Green's function $g_0(\vec{r}; \vec{r}')$ as the kernel. Formally (22) can be expanded into an infinite scattering series (Born series)

$$u = u^0 + G_0 \varepsilon u^0 + G_0 \varepsilon G_0 \varepsilon u^0 + \cdots \ . \tag{23}$$

If we split the scattering potential into the forescattering and backscattering parts

$$\varepsilon = \varepsilon_f + \varepsilon_b \tag{24}$$

and substitute it into (23), we can have all combinations of multiple forescattering and backscattering. We neglect the multiple backscattering (reverberations), i.e., drop all the terms containing two or more backscattering potentials, resulting in a multiple scattering series which contains terms with only one ε_b.

The general term will look like

$$G_0 \varepsilon_f G_0 \varepsilon_f \cdots G_0 \varepsilon_b G_0 \varepsilon_f \cdots G_0 \varepsilon_f u^0 \ . \tag{25}$$

The multiple forescattering on the left side of ε_b can be written as

$$G_f^m = [G_0 \varepsilon_f]^m G_0 \tag{26}$$

and on its right side,

$$u_f^n = [G_0 \varepsilon_f]^n u^0 \ . \tag{27}$$

Collecting all the terms of G_f^m and u_f^n respectively, we have

$$G_f^M = \sum_{m=0}^{M} [G_0 \varepsilon_f]^m G_0$$

$$u_f^N = \sum_{n=0}^{N} [G_0 \varepsilon_f]^n u^0 \ . \tag{28}$$

Let M and N go to infinite, then the renormalized G_f (forward propagator) and u_f (forescattering corrected incident field) are:

$$G_f = \sum_{m=0}^{\infty} [G_0 \varepsilon_f]^m G_0$$

$$u_f = \sum_{n=0}^{\infty} [G_0 \varepsilon_f]^n u^0 \tag{29}$$

and De Wolf approximation becomes

$$u = u_f + G_f \varepsilon_b u_f \ . \tag{30}$$

The observed total field u in (30) is different for different observation geometries. For transmission problems, the backscattering potential has no effect under the De Wolf approximation,

$$u_{\text{transmission}} = u_f \ . \tag{31}$$

On the other hand, for reflection measurement, that is, when the observations are at the same level as or behind the source with respect to the propagation direction, there is no u_f in the total field (30),

$$u_{\text{reflection}} = G_f \varepsilon_b u_f \ . \tag{32}$$

Write it into integral form, (30) becomes

$$u(\vec{r}) = u_f(\vec{r}) + \int_V d^3\vec{r}' g_f(\vec{r},\vec{r}')\varepsilon_b(\vec{r}')u_f(\vec{r}') \ . \tag{33}$$

Note that both the incident field and the Green's function have been renormalized by the multiple forescattering process through the multiple interactions with the forward-scattering potential ε_f.

3. Dual-domain One-way Propagators for Scalar, Acoustic and Elastic Waves

3.1 Classical Scalar-wave Dual-domain Propagators

The first use of a dual-domain propagator can be traced back to CHANDRA-SEKHAR's work (1952) on the calculations of amplitude fluctuations of light (scintillations) passing through the atmosphere using the thin-phase-screen. The field fluctuation outside the extended medium containing irregularities in refractive index, can be calculated as if produced by a thin phase-changing screen (CHANDRASEKHAR, 1952; BRAMLEY, 1954, 1977). The early use of phase-screen is a single screen for the whole inhomogeneous layer (atmosphere for light or ionosphere for radio waves). The interaction with heterogeneities is concentrated at the screen: a phase-changing operation in the space domain; the propagation is in the wavenumber domain. The formulation is simple with an exponential transformation and is similar to the Rytov transformation. However, in the phase-screen approximation, the phase-function is real, not complex as in the Rytov transformation. Therefore the phase-screen only imposes phase modulation to waves passing through it. Although simple, it has advantages over the formulation using Born approximation (BOOKER and GORDON, 1950), since for strong fluctuation the accumulated phase error by Born approximation may become significant. The approach also has been used for wave propagation through a single turning point (MERCIER, 1962; SALPETER, 1967; FLATTÉ, 1979, ch. 11) and for strong fluctuation theory (see e.g., ISHIMARU, 1978, ch. 20). Later the method was extended to multi-screen to accommodate long-range propagation (HERMANN and BRADLEY, 1971; BROWN, 1973; FLECK et al., 1976; FEIT and FLECK, 1978; RINO, 1978, 1982; KNEPP, 1983; MARTIN and FLATTÉ, 1988). The interaction between the heterogeneities and wavefield is through phase screens at each step along the propagation path. It is widely used for laser propagation through the atmosphere and later through optical fibers. Random media are modeled through a series of random phase-screens (ibid).

The method was introduced to ocean acoustics by HARDIN and TAPPERT (1973), TAPPERT (1974), FLATTÉ and TAPPERT (1975), and McDANIEL (1975), and was called split-step Fourier method as a purely numerical method for solving parabolic wave equations.

The original phase-screen propagator is derived from the parabolic wave equation, and the free propagator suffered the parabolic approximation. A "wide-angle" split-step propagator has been obtained, based on the symmetric splitting of the square-root operator (FLECK et al., 1976; FEIT and FLECK, 1978; THOMSON and CHAPMAN, 1983), in which the free propagator is an accurate one. The accuracy of this improved one-way propagator has been analyzed in those papers and more recently by HUANG and FEHLER (1998). The other approach to improve the phase-screen propagator was to match its travel time with the ray equation (TOLSTOY et al., 1985; BERMAN et al., 1989). BERMAN et al. (1989) changed the phase correction term of the screen into $\log n$, where n is the refraction index of the medium. However, all the improvement is kept in the realm of classical phase-screen correction.

Dual-domain one-way propagation methods were introduced to exploration seismology early in the 90s, with methods such as the split-step Fourier method (STOFFA et al., 1990; LEE et al., 1991), or the phase-screen method (WU and HUANG, 1992; LIU and WU, 1994) as alternatives to the time-space finite-difference solutions. These methods operate in the frequency domain and use the dual-domain implementation with operations shuttling between space and wavenumber domains by Fast Fourier Transform. Free propagation is accomplished in the wavenumber domain through a homogeneous medium which has some reference velocity. This reference velocity can vary with depth. Wave-medium interaction is done in the space domain that accounts for the effects of the heterogeneity to the wavefront. These methods have no grid dispersion and are unconditionally stable.

3.2 Wide-angle Dual-domain Propagators

The abovementioned phase-screen or split-step methods can be viewed as classical dual-domain propagator methods. These methods are accurate only for small-angle waves and cannot correctly handle large-angle waves. This severely limits its practical applications. Recently, significant progress has been made in improving the large-angle accuracy of the dual-domain methods and in extending them to acoustic and elastic waves (WU, 1994, 1996; WU and XIE, 1994; RISTOW and RUHL, 1994; WU and HUANG, 1995; HUANG and WU, 1996; HUANG and FEHLER, 1998,

2000; WU and JIN, 1997; XIE and WU, 1998, 1999, 2000; JIN and WU, 1999a,b; JIN *et al.*, 1998, 1999, 2000; HUANG *et al.*, 1999a,b; DE HOOP *et al.*, 2000; LE ROUSSEAU and DE HOOP, 2000). Various modifications and extensions have been introduced to improve the wide-angle response of the dual-domain propagators with different names for the propagators. Approximate propagators based on the use of local Born and local Rytov approximations (local Born and local Rytov propagators) have been developed by HUANG *et al.*, (1999a,b) and HUANG and FEHLER (2000a) (see also the early work of WU and HUANG, 1995). Fourier finite-difference methods developed by RISTOW and RUHL (1994), HUANG and FEHLER (2000b) and other authors are hybrid methods, which adopt the finite-difference calculations for wide-angle corrections to the phase-screen propagators. Generalized screen methods including pseudo-screen, complex screen, windowed screen, Padé-screen, and higher-order generalized screens have been developed and applied to synthetic and field data (WU, 1994; WU and HUANG, 1995; WU and JIN, 1997; JIN and WU, 1999a,b; JIN *et al.*, 1999, 2000; XIE and WU, 1998, 1999, 2000; DE HOOP *et al.*, 2000; LE ROUSSEAU and DE HOOP, 2000).

Based on the De Wolf approximation, WU (1994) derived an elastic one-way propagator, the complex-screen propagator. In the limiting case (null shear rigidity) he derived a new one-way propagator for scalar waves (WU, 1994, §5). For small angles it reduces to the classical phase-screen ("wide-angle" of FEIT and FLECK, 1978). But for large angles it keeps the form of the local Born approximation, which has better accuracy than the phase-screen solution. The new propagator is named "generalized phase-screen propagator". Along this direction, DE HOOP *et al.*, (2000) formulated a new type of acoustic one-way propagator based on the Hamilton path-integral and pseudo-differential operator theory. Wu's new propagator coincides with the first-order expansion of the new class of propagators identified as generalized screen propagators (GSP). The first-order approximation "generalized phase-screen propagator" was renamed as "pseudo-screen propagator" (see next section). The original form of pseudo-screen propagator has a singularity in the wavenumber domain and numerical instability. The problems were solved by the introduction of the Taylor expansion around the singularity in the extended local Born Fourier method and by the use of the Rytov approximation in the extended local Rytov Fourier method (HUANG *et al.*, 1999a,b). JIN *et al.*, (1998, 1999) solved the problems by using the Padé expansion and implementing the wide-angle corrections with an implicit finite-difference algorithm. For further development of the new propagators see the examples in the next section.

3.3 Acoustic and Elastic Screen Propagators

Dual-domain methods have also been developed for modeling elastic wave propagation in heterogeneous media (WU, 1994, 1996; WILD and HUDSON, 1998; WILD *et al.*, 2000) and for modeling primary reflections (XIE and WU, 1995, 1999, 2000; WU, 1996; WILD and HUDSON, 1998; WU and WU, 1998, 1999). Methods for wave propagation using dual-domain propagators for regional seismic waves in complicated Crustal structures (half-space screen propagators) have been developed and tested by comparing the results to finite difference solutions (WU *et al.*, 2000a,b). Dual-domain propagators for modeling acoustic wave reflections were developed in similar time (WU and HUANG, 1995; WU *et al.*, 1995; DE HOOP *et al.*, 2000).

3.4 Examples of Wide-angle Dual-domain Propagators

As we have discussed, there are many different versions of dual-domain propagators (DDP). Here we examine some generalized screen propagators (GSP) as examples. Different versions of GSP with various approximations can be derived through different approaches. The early derivation used the local Born approximation and the De Wolf approximation (WU, 1994, 1996). Later the one-way wave propagation with GSP was more rigorously cast into a Hamilton (phase space) path-integral formulation (DE HOOP *et al.*, 2000), which forms a mathematical basis for accuracy analysis and further development of screen propagators. However, for the path integral in exact form, the vertical slowness symbol is hard to solve and the implementation would be very involved even if we could find the exact form. Therefore different approximations must be invoked for practical use of the method. In the following, we discuss three versions of GSP from the viewpoint of path-integral formulation and the approximation of vertical slowness symbol: pseudo-screen approach, generalized screen series expansion, and Padé expansion approach.

   Pseudo-screen propagator starts with the weak scattering assumption, so that the vertical slowness symbol can be decomposed into background and perturbation parts. The perturbation part can be derived with a local Born approximation. However, the Born approximation is basically a low-frequency approximation, and has severe phase errors for strong contrast and high-frequencies, especially for large-angle waves. In order to have better phase accuracy, which is important for imaging (migration), some high-frequency asymptotic phase-matching has been applied to the local Born solution. Even zero-order matching leads to a solution better than the classic phase-screen method (spit-step Fourier method). The term "pseudo-screen" first appeared in WU and DE HOOP (1996) and HUANG and WU (1996) to distinguish the new form of screen propagator from the classic phase-screen propagator. Phase-screen has operations only in the space domain so that the phase correction is accurate only for small-angle waves; while pseudo-screen has operations in both the space and wavenumber domains to improve the accuracy for large-angle waves. The operations of pseudo-screen for heterogeneity correction have deviated from the function of a physical "screen"; and the phase-delay is angle-dependent. That is why the correction is termed "pseudo-screen." The asymptotic phase-matching method used by JIN *et al.*, (1998, 1999) in the hybrid pseudo-screen propagator with a wavenumber filter in the form of continued fraction expansion can improve the large-angle wave response significantly. In the method, the wide-angle correction is implemented with an implicit finite-difference scheme and the expansion coefficients are optimized by phase-matching.

Generalized screen series expansion expands the perturbation part of the vertical slowness symbol into a series (DE HOOP et al., 2000) in terms of both the smallness parameter and the smoothness parameter of the perturbations. The first-order term in smallness is in parallel with the local Born solution. Higher-order terms can improve the wide-angle performance but involve more calculations. LE ROUSSEAU et al., (2000) extended the scalar GSP to transversely isotropic media with a vertical symmetry axis.

Wide-angle Padé-screen propagator starts with a smooth approximation for the vertical slowness symbol in the very beginning. The approach does not require the weak perturbation assumption and therefore can handle strong contrast media more naturally. However, the approximation applied corresponds to the local homogeneity approximation in the traditional way of expanding the square-root operator. Large errors may exist around sharp boundaries. The expansion of the symbol in terms of h–f series can be found in FISHMAN and MCCOY (1984). Retaining only the leading term leads to a simple form of $\gamma_{(0)}^{HF} = \{\alpha^2(X_T) - \alpha_T^2\}^{1/2}$ which is the principal part of the symbol (DE HOOP et al., 2000), where γ is the vertical slowness, $\alpha(X_T)$ is the local slowness (inverse velocity) at a transverse position X_T and α_T is the horizontal slowness. After the smooth approximation (h–f asymptotics), the vertical slowness symbol is expanded into a Padé series and a finite-difference (FD) scheme is used to implement the wide-angle corrections (XIE and WU, 1998; XIE et al., 2000).

As we pointed out, the FD implementation of wide-angle corrections in this approach makes the method resembling the Fourier finite-difference method (RISTOW and HUHL, 1994; HUANG and FEHLER, 2000b). They are both based on the local homogeneity approximation. In HUANG and FEHLER (2000b) the coefficients of the first Padé expansion are globally optimized.

4. Imaging Using Dual-domain Propagators

Two milestones in the development of seismic imaging were the introduction of the one-way (parabolic) wave equation finite-difference algorithm (CLEARBOUT, 1970, 1976) and the introduction of the Kirchhoff integral method (SCHNEIDER, 1978). Both approaches have been widely used in the industry. The application of the dual-domain technique emerged in exploration seismology only in the beginning of the 90s and is relatively new to the industry. I will give first a brief overview of the abovementioned two approaches and their limitations, followed by a summary of the features of the newly developed dual-domain propagators.

4.1 Limitations of the Kirchhoff Migration Method

The widely used Ray-Kirchhoff imaging (depth migration) method (or simply "Kirchhoff migration") is a ray-theory based method. The method uses the Kirchhoff integral with a ray-theory approximated Green function. The process consists of ray-tracing from both the source and receiver down to the imaging point and then the pickup and stack of the wavelets from all the seismic traces according to the corresponding travel times. This approach has been successfully used in areas with relatively simple structures. However, it has two fundamental limitations which cause problems in applications to complicated regions, especially in 3-D cases.

One limitation of the Kirchhoff method is its high-frequency approximation (ray approximation) of the Green's function. The Fresnel radius of a ray can be viewed as more or less the distance from the center of the ray within which the wave field has less than 180 degrees of phase difference. The Fresnel radius increases with propagation distance and hence increases with depth leading to low resolution (lateral resolution) and poor image quality for deep targets in complex region. In contrast, wave equation migration methods are based on the wave theory which includes all the frequency-dependent properties of the wave field. The decrease in resolution and image quality of wave-theory based methods with depth is significantly less severe than that of ray-theory based methods.

The other limitation of the Kirchhoff method is the low fidelity of the amplitude information carried in the imaging process and for the final image. It is difficult to obtain amplitude information for rays propagating through complex structures because of the presence of ray caustics, multiple arrivals, and interference. Wave equation migration methods maintain the true amplitude information and thus provide high-fidelity images.

Other problems with the ray-theory based Kirchhoff method include the difficulties in dealing with multiple arrival interference, caustics, chaotic rays, sensitivity to velocity structures especially those with irregular sharp interfaces, and the calculation and storage of large travel-time tables for 3-D imaging.

The fundamental limitations of the Kirchhoff method severely limit its applications for the high resolution/high fidelity imaging in complicated regions. Nevertheless it will remain to be quite useful and convenient for some industrial applications because of its flexibility in target-oriented imaging and straightforward implementation.

4.2 Space-domain One-way Wave Finite-difference Migration

The time-space (t-x) domain finite-difference algorithm is a one-way wave-equation based method that has many advantages over the ray-Kirchhoff method. Since the introduction of the method by Clearbout by the end of 60s and the beginning of 70s (CLAERBOUT, 1970, 1976, 1985) many improvements have been made in various aspects. The method keeps the basic features of wave-theory based imaging, and overcomes the fundamental limitations of ray-theory based methods. In addition, it has abundantly faster speed of computation compared to full-wave equation methods.

Despite its success, the t-x domain finite difference approach also has certain intrinsic difficulties, especially for 3-D imaging. One is the grid dispersion problem which originates from the rectangular discretization of space that results in different propagation speeds for waves of different angles. This dispersion will cause errors and artifacts for imaging. Suppression of these artifacts usually leads to severe attenuation of large-angle waves which are important for imaging steep structures. Other problems with the t-x finite-difference approach include the numerical anisotropy for 3-D geometry and difficulty in formulating a midpoint-offset domain approach for more efficient migration. Because of these problems, the Kirchhoff method has dominated the exploration industry for a long period despite its fundamental limitations.

4.3 Features of Dual-domain Propagators (DDP)

The dual-domain propagators, which are wide-angle one-way propagators, neglect wave reverberations between heterogeneities but correctly handle the forward multiple-scattering including focusing/defocusing, diffraction, refraction and interference of waves. Due to recent progress, new versions of dual-domain methods can propagate large-angle waves quite accurately in strong contrast media, resulting in superior image quality for complex geological regions.

Dual-domain methods are self-adaptive to the complexity of the medium. In homogeneous regions, the algorithm will automatically perform wavenumber domain operations that are accurate up to 90°; while in heterogeneous regions properly weighted space-domain operations will be added according to the strength of the heterogeneities. The adaptive phase-space (dual-domain) manipulation makes the best use of the operations of each domain, resulting in efficient and accurate propagators.

The wide-angle capability of these new propagators can be seen from the propagating wave fronts in strongly perturbed media. Figure 2 exhibits comparisons of wavefronts calculated using different propagators. The reference velocity used in calculating wave propagation in each case was chosen to be a factor of two different from the real velocity, so we can investigate how well the propagators correct for the difference between the chosen reference velocity and the medium velocity. Figure 2a shows the wavefront calculated using the Split-Step Fourier (SSF) (phase-screen) method. Figure 2b presents the result using the hybrid pseudo-screen propagator, and Figure 1c is that by a traditional 65-degree finite-difference method implemented in the frequency-space domain. We observe that the SSF wavefront is only accurate for small-angle waves. While the F-X finite difference responded better for large-angle waves; the dispersion and artifacts are quite conspicuous. In contrast, the dual-domain Hybrid Pseudo-Screen Propagator performs quite well for large-angle waves.

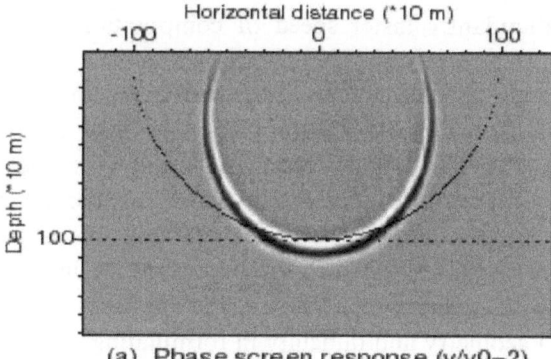

(a) Phase screen response (v/v0=2)

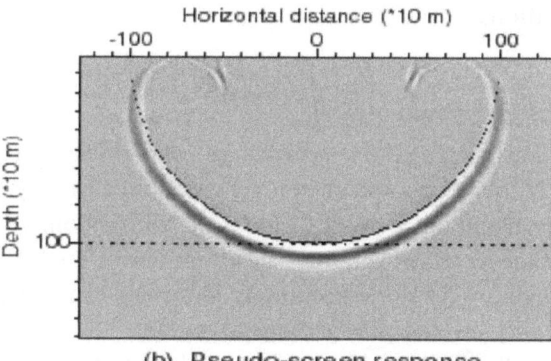

(b) Pseudo-screen response

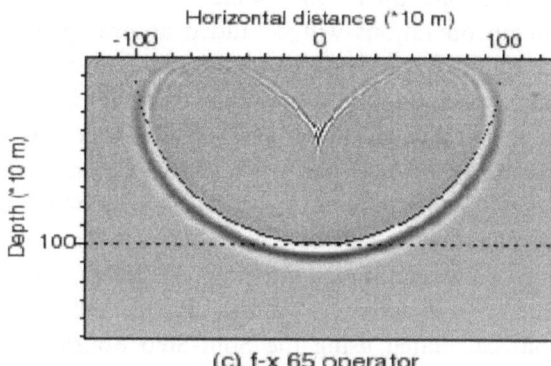

(c) f-x 65 operator

Figure 2

Impulse responses of three different migration operators: (a) phase-screen propagator; (b) hybrid pseudo-screen propagator; (c) 65 degree finite-difference propagator.

The amplitude-preserving feature of the new dual-domain propagators is demonstrated in Figure 3. Events with a dip angle up to 60° can be migrated with little distortion of amplitude. Even for the structures dipping up to 70°, only the deepest portion of the migrated image has significant distortion. In the lower panel of Figure 3, it is also shown that the amplitude is preserved even in the case of strong lateral velocity variations (100% perturbations).

The efficiency of the dual-domain wave-theory based methods is no longer a weak point in the 3-D case compared with the ray-theory based methods, such as Kirchhoff migration. For 2-D imaging, the dual-domain methods are generally a few times slower than the Kirchhoff method. However, for 3-D imaging, the situation is different: While the time for ray-tracing required in Kirchhoff migration increase as N^4 (N is the number of points in one dimension), the time for dual-domain methods increases as only $2N^2 \log_2 N$. When N is large in the 3-D case, the dual-domain methods are not necessarily slower than the ray methods. Further, the dual-domain methods can be formulated in the midpoint-offset coordinate system without difficulty (JIN and WU, 1999b), while a finite difference solution for this system has not been found. For marine data, the number of offsets is considerably smaller than the number of sources; thus there is a gain in efficiency when using an offset-domain migration formulation with dual-domain propagators.

4.4 Examples of GSP Migrations Applied to Different Data Sets

We present migration examples for 2-D and 3-D models to demonstrate the features and excellent performance of this approach. The first example is the 2-D prestack depth migration for the A-A′ profile of the SEG-EAEG salt model. The profile crosses many of the difficult structural elements in the model including steep, irregular shallow salt flanks, abrupt dip changes where the faults are located, strong velocity contrasts between the salt body and the surrounding medium (3–4 time differences). These pose an immense challenge to the conventional imaging methods. Figure 4 shows the imaging results of prestack depth migration using our Padé-screen propagator method (XIE and WU, 2000). We see that not only the salt body but also the subsalt structures were imaged clearly. The lateral and vertical resolutions are excellent.

The next example is the 2-D prestack migration image for the Marmousi model using the offset-domain pseudo-screen propagator (for theory see JIN and WU, 1999). The model contains very complicated geological features, especially shallow steep faults and an underlying high velocity lateral salt body intrusion. In addition, the model contains salt structure related traps and a reservoir structure beneath this complex geology. Figure 5c shows the result of ProMAX prestack Kirchhoff depth migration using finite-difference eikonal traveltimes. As expected, the multiple arrivals generated by this model cause the mislocation of reflections in complicated regions. Figure 5a is the image by 70° explicit finite-difference migration. Figure 5b

a) ## GSP : Migration Amplitudes

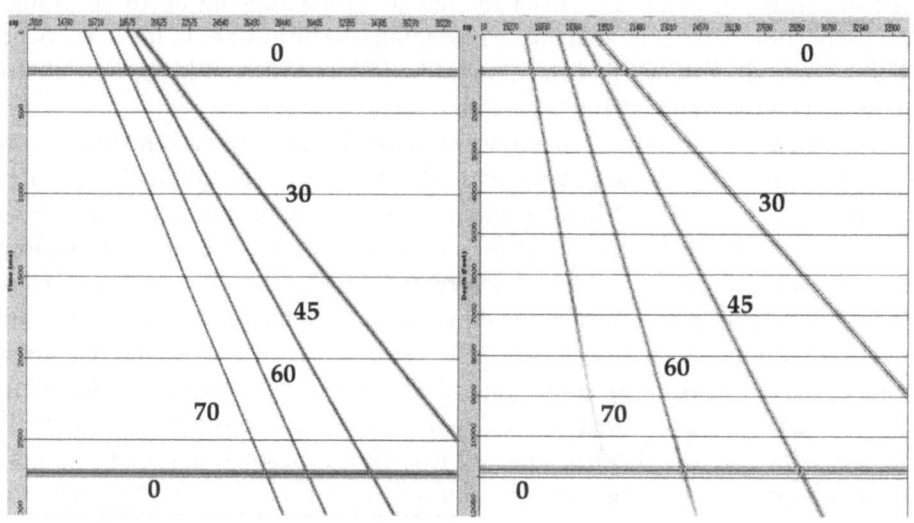

Migration input: unit amplitude   GSP migration output

b) ## GSP : Migration Amplitudes

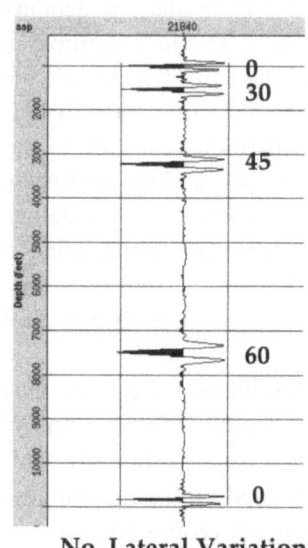
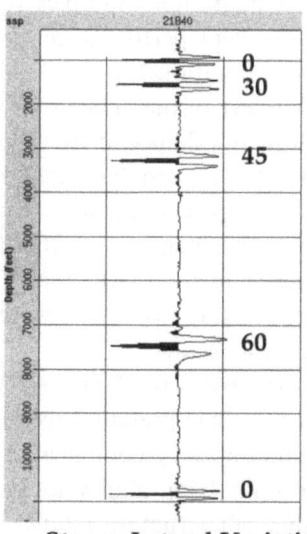

No Lateral Variation   Strong Lateral Variation

Figure 3
Amplitude preserving property of the wide-angle dual-domain migration methods. Top: Migration input and output for events with different dip-angles; Bottom: Traces showing the detailed amplitude and waveform information for the cases of no lateral variation and with strong lateral variation.

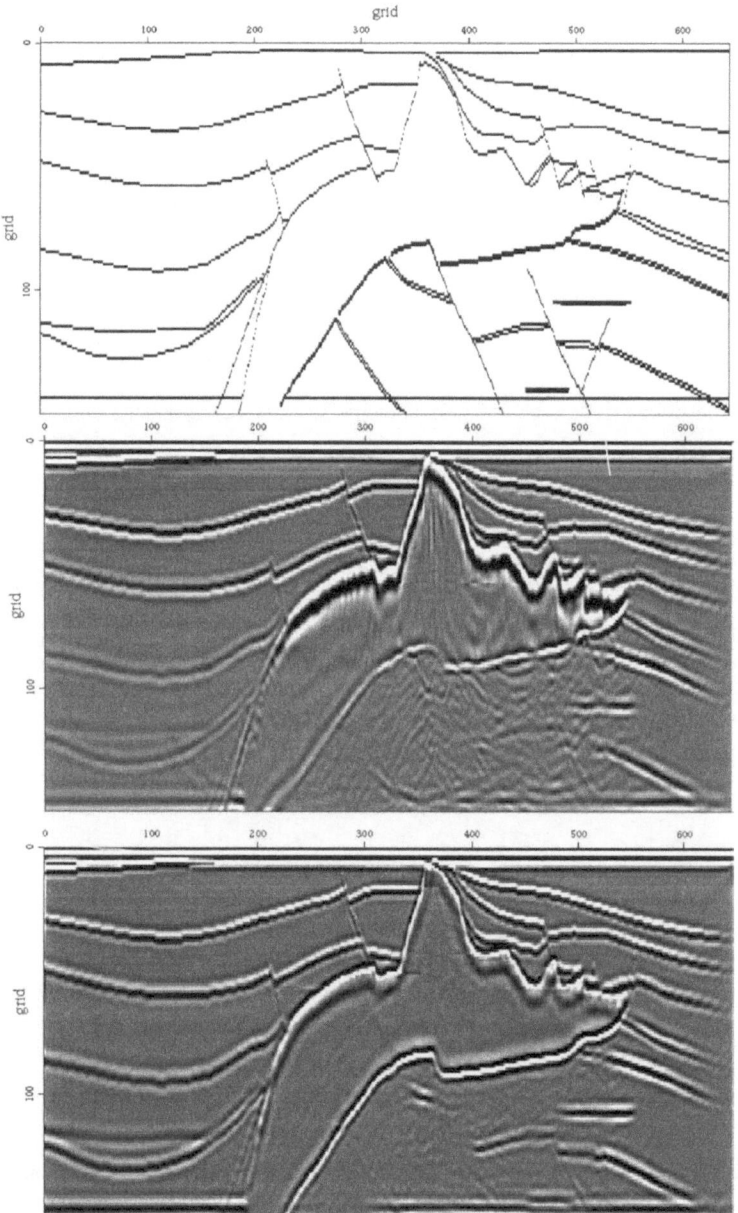

Figure 4
Prestack depth migration for A-A′ profile of the SEG-EAGE salt model using the wide-angle Padé-screen:
Top: Reflection model; Middle: Image by phase-screen; Bottom: Image by Padé-screen.

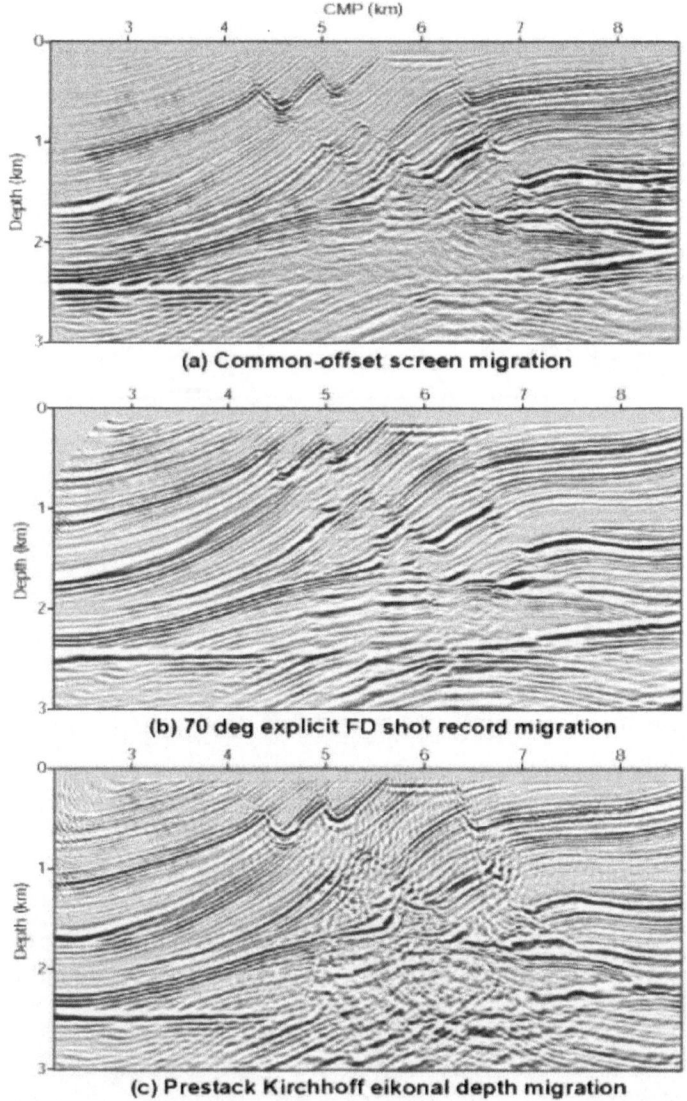

Figure 5
Prestack depth migration of Marmousi data set: (a) Offset-domain pseudo-screen migration. Forty-eight offset sections are used. (b) Image by 70 degree explicit finite-difference shot record migration. (c) Image by Kirchhoff migration.

shows the result of offset-domain pseudo-screen depth migration. Forty-eight offset sections with full offset range between 200 m and 2550 m were used in the test. The superiority of the image quality is apparent and it can delineate faults and reservoirs well enough to identify the major features.

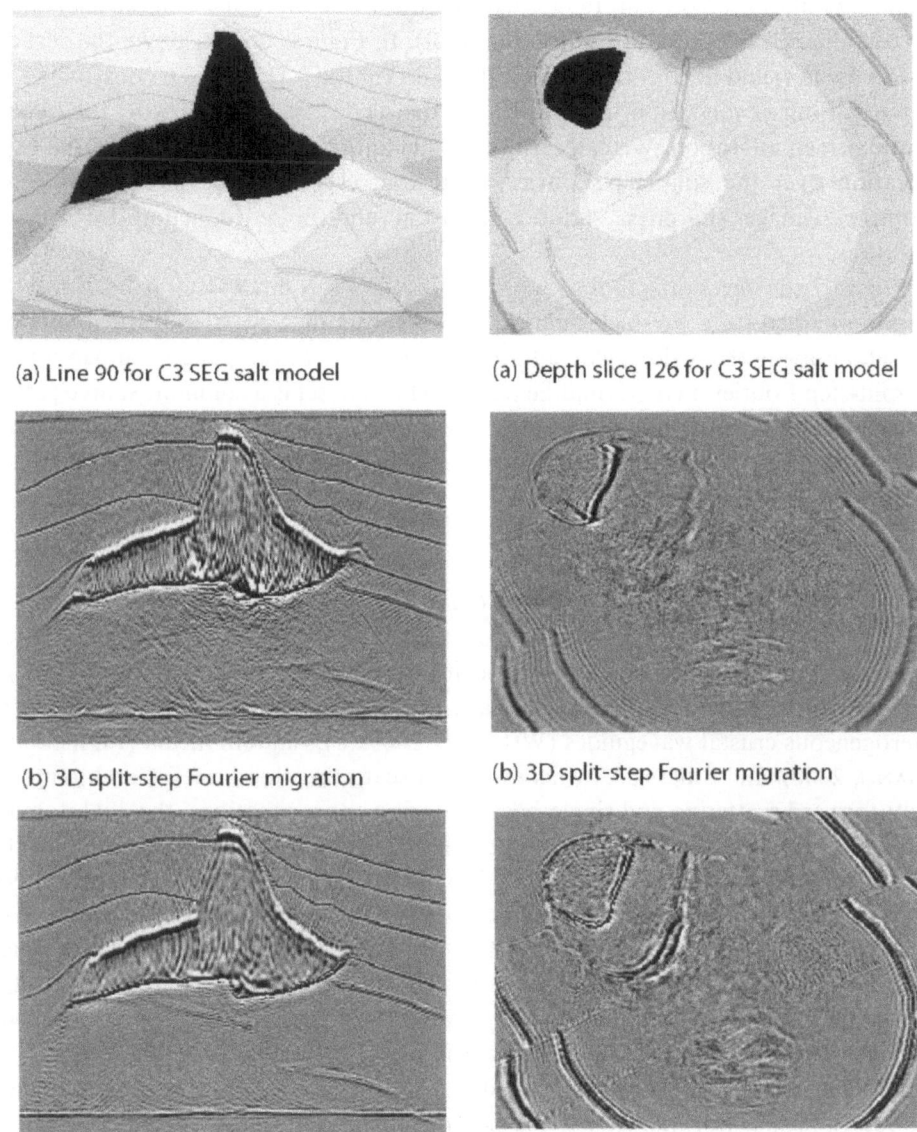

(a) Line 90 for C3 SEG salt model   (a) Depth slice 126 for C3 SEG salt model

(b) 3D split-step Fourier migration   (b) 3D split-step Fourier migration

(c) 3D GSP migration   (c) 3D GSP migration

Figure 6

Comparison between images from different methods for the C3 subset of 3-D SEG Salt model. On the left
is a vertical slice at line 90 and on the right is a horizontal slice at depth 126. Panels A, B and C are velocity
models, images from the split-step Fourier method and images from the GSP migration, respectively.

Figure 6 provides an example of poststack 3-D migration for the C3 subset of the
SEG-EAEG salt model. The synthetic data were generated by ARCO for a
decimated model of 250×250 with a 40 m spacing using a finite-difference

exploding reflector algorithm. Figure 6a presents a vertical profile (on the left) and a horizontal slice at depth grid 126 (on the right). In Figures 6b are shown the vertical cut (on the left) and horizontal slice (on the right) of the 3-D images reconstructed by split-step Fourier migration. Figures 6c are the reconstructed images by the hybrid pseudo-screen migration. We can see clearly the improved image quality of the GSP migration over the split-step Fourier migration, especially the improvements of resolution, image sharpness, fault delineation and noise reduction for subsalt structures.

Figure 7 shows a horizontal slice of the 3-D SEG salt model (top panels) and the images obtained from prestack migration of a subset of a portion of the numerical dataset, applying the wide-angle Padé-screen method (bottom panel) compared with the split-step Fourier method (middle panel). The data set is a common-source gather with a total of 45 shots. It can be seen clearly that the new wide-angle dual-domain method performs substantially better in imaging the faults and defining the sharp structural boundaries.

5. Modeling and Simulation Using Dual-domain Propagators

Because of the super wide-angle capacity of the new dual-domain propagators, they can be applied to modeling the wave propagation in complex media such as heterogeneous crustal waveguides (WU et al., 2000a,b), random media (FEHLER and HUANG, 2000), and other cases where forward scattering dominates. Combining the multi-forward scattering and single-backscattering approximations, the dual-domain propagators can be used to model the primary reflections in complex elastic media (WU, 1994, 1996; XIE and WU, 1995, 1996, 1998, 1999; WILD and HUDSON, 1998; WILD et al., 2000; WU and WU, 1998, 1999). Figure 8 illustrates an example of synthetic reflection seismograms using the thin-slab operator for the elastic French Model (top panel). The dark model has -20% perturbations relative to the surrounding medium for both P- and S-wave velocities. In the figure, the solid reference lines were calculated using a full-wave finite-difference algorithm. The dotted lines are the results of the thin-slab approximation, which is a dual-domain one-return propagator. Direct arrivals are not shown in the figure. It can be seen that the thin-slab operator can accurately calculate the backscattered waves extending to quite large angles.

6. Conclusions

Wide-angle dual-domain propagators, including the generalized screen propagators (GSP) have adaptive phase-space manipulations. In homogeneous regions the wavenumber-domain operation dominates; while in heterogeneous regions, it turns

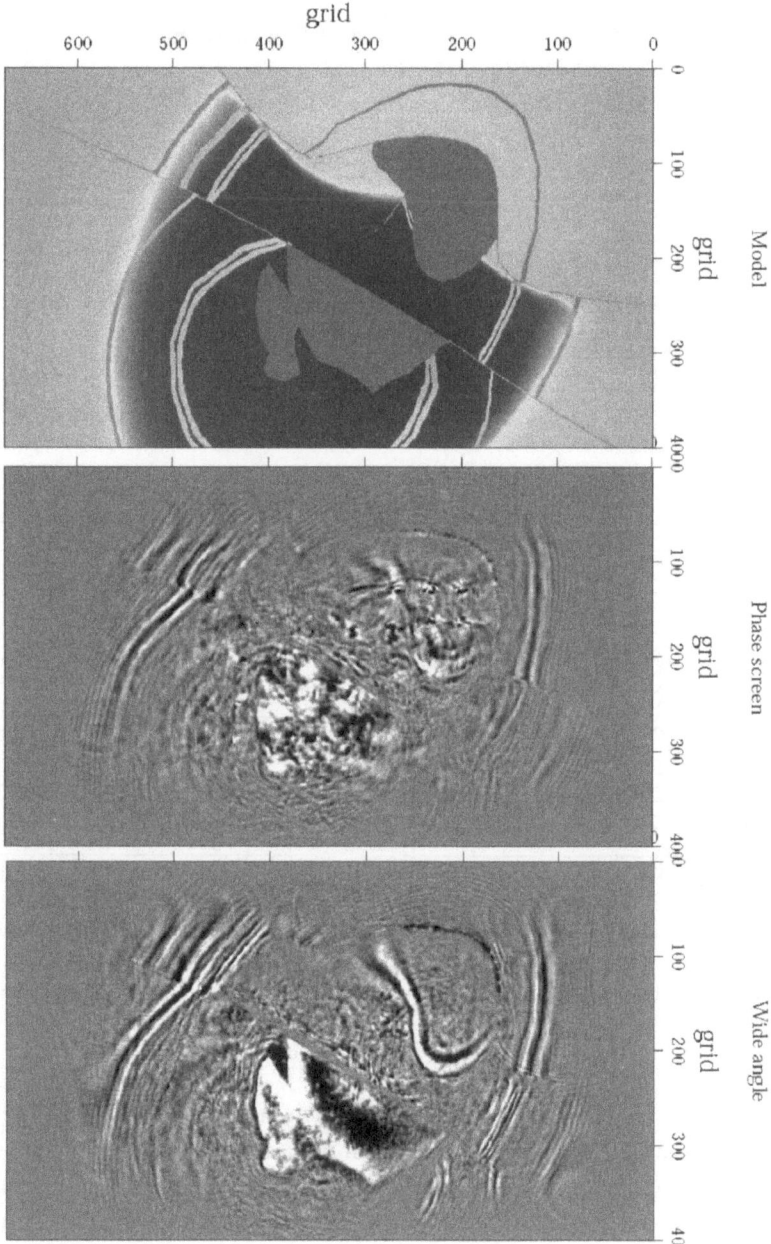

Figure 7

Comparison of 3-D prestack migration images using different methods. The horizontal slice is located at $Z = 2100$ m. From top to bottom are velocity model, migration image using the phase-screen method, and image using the wide-angle Padé-screen method, respectively.

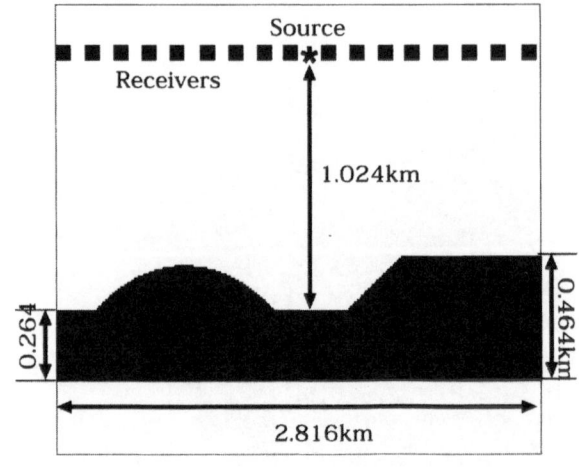

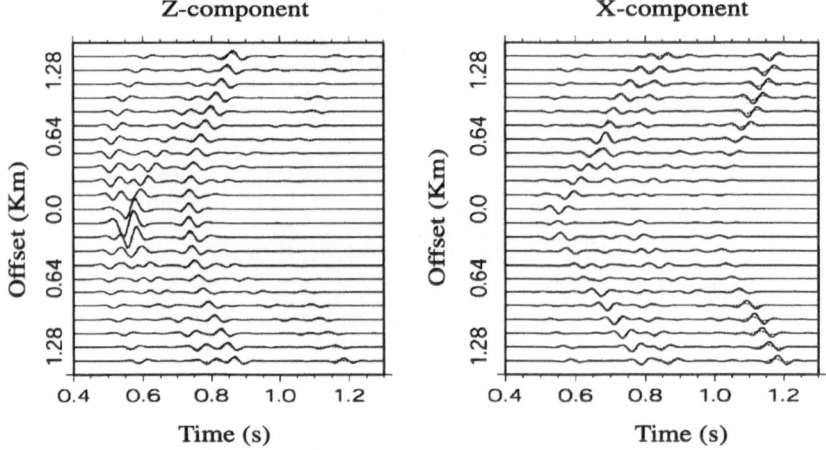

Figure 8

Synthetic reflection seismograms calculated by the thin-slab approximation (dotted lines) compared with those by finite-difference calculation (solid lines) for an elastic French model with an explosion source.

into weighted mixed domain operations. The weight is proportional to the strength of heterogeneity and space-domain operation may dominate. In summary, the wide-angle dual-domain propagators have the following features: **High-resolution** and **High-fidelity**: imaging based on wave theory; **High-speed**: one-way approximation + FFT implementation; **Super-wide-angle performance**: dual-domain adaptability to heterogeneities; **Reduced grid-dispersion**: hybrid FT-FD; **Midpoint-offset domain imaging capability**: high efficiency; **Dual-domain information available** at each step: convenient for imaging **velocity analysis** and **AVA** (amplitude versus angle) analysis.

Acknowledgements

I am grateful to Professor Keiiti Aki for introducing me to the study of scattering theory, and his encouragement and support of my research in this field. The development of new dual-domain propagators, especially the generalized screen propagators (GSP), grows out of collaborative efforts between many institutions, scientists and industrial partners. I thoroughly enjoyed working with M. Fehler, M. de Hoop, L.J. Huang, X.B. Xie, S. Jin, C. Peng, C. Mosher, C. Burch, R. Cook, H. Guan, X. Wu, F. Mignet, among others. I appreciate their contributions and collaborations. The support from the DOE/BES grants and the WTOPI Research Consortium at the University of California at Santa Cruz are highly appreciated. Contribution No. 430 of the Center for Study of Imaging and Dynamics of the Earth, IGPP, University of California, Santa Cruz.

Appendix

Renormalization of Scattering Series and the De Wolf Approximation

The Lippmann-Schwinger equation

$$u = u^0 + k^2 \int_V d^3\vec{r}' g_0(\vec{r};\vec{r}')\varepsilon(\vec{r}')u(\vec{r}') \ , \tag{A.1}$$

where $\varepsilon(\vec{r}')$ is the equivalent body force for scattering or scattering potential as identified in the scattering theory, can be written symbolically as

$$u = u^0 + G_0\varepsilon u \tag{A.2}$$

where ε is a diagonal operator in space domain, and G_0 is a nondiagonal integral operator. If the reference medium is homogeneous, G_0 will be the volume integral with the Green's function $g_0(\vec{r};\vec{r}')$ as the kernel. A formal "solution" of (A.2) is

$$u = [1 - G_0\varepsilon]^{-1}u^0 \ . \tag{A.3}$$

If we expand (A.3) into a series by iteration, it will become an infinite scattering series which is the familiar Born series. The Born series may converge very slowly or become divergent for strong scattering. Now let us split the operations (the interaction between the medium and the wave) so that we can resort the scattering series and sum up certain sub-series theoretically to remove the divergent elements in the Born series. This is the intent of renormalization. Here we split the scattering potential operator into a forescattering and a backscattering operator: $\varepsilon = \varepsilon_f + \varepsilon_b$, such that $G_0\varepsilon_f$ and $G_0\varepsilon_b$ correspond to the forescattered and backscattered Born solutions, respectively. It can be seen from straightforward operator algebra that

$$u = [1 - G_f \varepsilon_b]^{-1} u_f \tag{A.4}$$

with

$$u_f = [1 - G_0 \varepsilon_f]^{-1} u_0$$
$$G_f = [1 - G_0 \varepsilon_f]^{-1} G_0$$

or in the form of

$$u = u_f + G_f \varepsilon_b u \tag{A.5}$$

with

$$u_f = u^0 + G_0 \varepsilon_f u_f$$
$$G_f = G_0 + G_0 \varepsilon_f G_f \ .$$

In explicit form, (A.5) can be written as

$$u(\vec{r}) = u_f(\vec{r}) + \int_V d^3 \vec{r}' g_f(\vec{r}; \vec{r}') \varepsilon_b(\vec{r}') u(\vec{r}') \ . \tag{A.6}$$

Equation (A.6) thus expresses $u(\vec{r})$ in terms of a renormalized forward propagated field u_f, and a scattered component due to a medium with scattering potential ε_b, and an effective forward propagator $g_f(\vec{r}; \vec{r}')$ instead of the background Green's function g_0. Equation (A.6) can be solved with an iterative procedure. The first iteration will be the De Wolf approximation (33).

References

AKI, K. (1973), *Scattering of P Waves under the Montana Lasa*, J. Geophys. Res. *78*, 1334–1346.
AKI, K. and RICHARDS, P., *Quantitative Seismology*, vol. II (W.H. Freeman, San Francisco 1980).
BERMAN, D. H., WRIGHT, E. B., and BAER, R. N. (1989), *An Optimal PE-type Wave Equation*, J. Acoust. Soc. Am. *86*, 228–233.
BOOKER, H. G. and GORDON, W. E. (1950), *A Theory of Radio Scattering in the Troposphere*, Proc. Inst. Radio Eng. *38*, 401.
BRAMLEY, E. N. (1954), *The Diffraction of Waves by Irregular Refracting Medium*, Proc. R. Soc. A*225*, 515.
BRAMLEY, E. N. (1977), *The Accuracy of Computing Ionospheric Radio-wave Scintillation by the Thin-phase-screen Approximation*, J. Atmos. and Terres. Phys. *39*, 367–373.
BROWN, W. P. (1973), *High Energy Laser Propagation*, Hughes Research Laboratory, Rept. on contract N00014-B-C-0460.
CHANDRASEKHAR, S. (1952), Mon. Not. R. Ast. Soc. *112*, 475.
CHERNOV, L. A., *Wave Propagation in a Random Medium* (McGraw-Hill, New York 1960).
CLAERBOUT, J. F. (1970), *Coarse Grid Calculations of Waves in Inhomogeneous Media with Applications to Delineation of Complicated Seismic Structure*, Geophys. *35*, 407–418.
CLAERBOUT, J. F., *Fundamentals of Geophysical Data Processing* (McGraw-Hill, 1976).
CLAERBOUT, J. F., *Imaging the Earth's Interior* (Blackwell Scientific Publications, 1985).
DE HOOP, M., LE ROUSSEAU, J., and WU, R.-S. (2000), *Generalization of the Phase-screen Approximation for the Scattering of Acoustic Waves*, Wave Motion *31*, 43–70.

DE WOLF, D. A. (1971), *Electromagnetic Reflection from an Extended Turbulent Medium: Cumulative Forward-scatter Single-backscatter Approximation*, IEEE Trans. Ant. and Prop. *AP-19*, 254–262.

DE WOLF, D. A. (1985), *Renormalization of EM fields in Application to Large-angle Scattering from Randomly Continuous Media and Sparse Particle Distribution*, IEEE Trans. Ant. and Prop. *AP-33*, 608–615.

DEVANEY, A. J. (1982), *A Filtered Back Propagation Algorithm for Diffraction Tomography*, Ultrasonic Img. *4*, 336–350.

DEVANEY, A. J. (1984), *Geophysical Diffraction Tomography*, Trans. IEEE GE-*22*, 3–13.

FEHLER, M., SATO, H., and HUANG, L.-J. (2000), *Envelope Broadening of Outgoing Waves in 2-D Random Media: A Comparison between the Markov Approximation and Numerical Simulations*, Bull. Seismol. Soc. Am., in press.

FEIT, M. D. and FLECK, J. A., Jr. (1978), *Light Propagation in Graded-index Optical Fibers*, Appl. Opt. *17*, 3990–3998.

FISHMAN, L. and McCOY, J. J. (1984), *Derivation and Application of Extended Parabolic Wave Theories II. Path Intergral Representations*, J. Math. Phys. *25*, 297–308.

FLATTÉ, S. M., *Sound Transmission through a Fluctuating Ocean* (Cambridge University Press, 1979).

FLATTÉ, S. M. and TAPPERT, F. D. (1975), *Calculation of the Effect of Internal Waves on Oceanic Sound Transmission*, J. Acoust. Soc. Am. *58*, 1151–1159.

FLATTÉ, and WU (1988), *Small-scale structure in the lithosphere and asthenosphere deduced from arrival-time and amplitude fluctuations at NORSAR*, J. Geophys. Res. *93*, 6601–6614.

FLECK, J. A., Jr., MORRIS, J. R., and FEIT, M. D. (1976), *Time-dependent Propagation of High Energy Laser Beams through the Atmosphere*, Appl. Phys. *10*, 129–160.

GUBERNATIS, J. E., DOMANY, E., and KRUMHABSL, J. A. (1977a), *Formal Aspects of the Theory of the Scattering of Ultrasound by Flaws in Elastic Materials*, J. Appl. Phys. *48*, 2804–2811.

GUBERNATIS, J. E., DOMANY, E., KRUMHABSL, J. A., and HUBERMAN, M. (1977b), *The Born Approximation in the Theory of the Scattering of Elastic Waves by Flaws*, J. Appl. Phys. *48*, 2812–2819.

HARDIN, R. H. and TAPPERT, F. D. (1973), *Applications of the Split-step Fourier Method to the Numerical Solution of Nonlinear and Variable Coefficient Wave Equation*, SIAM Rev. *15*, 423.

HERMANN, J. and BRADLEY, L. C. (1971), *Numerical Calculation of Light Propagation*, MIT Lincoln Laboratory, Cambridge, Mass., Rept. LTP-10.

HUANG, L.-J. and FEHLER, M. (1998), *Accuracy Analysis of the Split-step Fourier Propagator: Implications for Seismic Modeling and Migration*, Bull. Seismol. Soc. Am. *88*, 18–29.

HUANG, L.-J. and FEHLER, M. (2000a), *Quasi-born Fourier Migration*, Geophys. J. Int. *140*, 521–534.

HUANG, L.-J. and FEHLER, M. (2000b), *Globally Optimized Fourier Finite-difference Method*, Geophys., submitted.

HUANG, L.-J., FEHLER, M., ROBERTS, P., and BURCH, C. C. (1999a), *Extended Local Rytov Fourier Migration Method*, Geophysics *64*, 1535–1545.

HUANG, L.-J., FEHLER, M., and WU, R.-S. (1999b), *Extended Local Born Fourier Migration Method*, Geophys. *64*, 1524–1534.

HUANG, L-J. and WU R.-S. (1996), *Prestack Depth Migration with Acoustic Pseudo-screen Propagators*, Mathematical Methods in Geophysical Imaging IV, SPIE *2822*, 40–51.

ISHIMARU, A., *Wave Propagation and Scattering in Random Media*, vol. II (Academic Press, New York 1978).

JIN, S. and WU, R.-S. (1999a), *Depth Migration with a Windowed Screen Propagator*, J. Seismic Exploration *8*, 27–38.

JIN, S. and WU, R.-S. (1999b), *Common-offset Pseudo-screen Depth Migration*, Expanded Abstracts, SEG 69th Annual Meeting, 1516–1519.

JIN, S., MOSHER, C. C., and WU, R.-S. (2000), *3-D Prestack Wave Equation Common-offset Pseudo- screen Depth Migration*, Expanded abstracts, SEG 70th Annual Meeting, 842–845.

JIN, S., WU, R.-S., and PENG, C. (1998), *Prestack Depth Migration Using a Hybrid Pseudo-screen Propagator*, Expanded Abstracts, SEG 68th Annual Meeting, 1819–1822.

JIN, S., WU, R.-S., and PENG, C. (1999), *Seismic Depth Migration with Screen Propagators*, Comput. Geosci. *3*, 321–335.

LIU, Y. B. and WU, R.-S. (1994), *A Comparison between Phase-screen, Finite Difference and Eigenfunction Expansion Calculations for Scalar Waves in Inhomogeneous Media*, Bull. Seismol. Soc. Am. *84*, 1154–1168.

KNEPP, D. L. (1983), *Multiple Phase-screen Calculation of the Temporal Behavior of Stochastic Waves*, Proc. IEEE *71*, 722–727.

LE ROUSSEAU, J. H. and DE HOOP, M. V. (2000), *Scalar Generalized-screen Algorithms in Transversely Isotropic Media with a Vertical Symmetry Axis*, Geophysics, submitted.

LEE, D., MASON, I. M., and JACKSON, G. M. (1991), *Split-step Fourier Shot-record Migration with Deconvolution Imaging*, Geophysics 56, 1786–1793.

MARTIN, J. M. and FLATTÉ, S. M. (1988), *Intensity Images and Statistics from Numerical Simulation of Wave Propagation in 3-D Random Media*, Appl. Opt. *17*, 2111–2126.

MCDANIEL, S. T. (1975), *Parabolic Approximations for Underwater Sound Propagation*, J. Acoust. Soc. Am. *58*, 1178–1185.

MERCIER, R. P. (1962), *Diffraction by a Screen Causing Large Random Phase Fluctuations*, Proc. Cambridge Phil. Soc. *58*, 382–400.

MILES, J. W. (1960), *Scattering of Elastic Waves by Small Inhomogeneities*, Geophys. *25*, 643–648.

MORSE, P. M. and FESHBACH, H., *Methods of Theoretical Physics* (McGraw-Hill Book Comp., New York 1953).

RINO, C. L. (1978), *Iterative Methods for Treating Multiple Scattering of Radio Waves*, J. Atmos. Terr. Phys. *40*, 1101–1118.

RINO, C. L. (1982), *On the Application of Phase-screen Models to the Interpretation of Ionospheric Scintillation Data*, Radio Sci. *17*, 855–867.

RINO, C. L. (1988), *A Spectral-domain Method for Multiple Scattering in Continuous Randomly Irregular Media*, IEEE Tr. Anten. and Prop. *36*, 1114–1128.

RISTOW, D. and RUHL, T. (1994), *Fourier Finite-difference Migration*, Geophys. *59*, 1882–1893.

SALPETER, E. E. (1967), *Interplanetary Scintillations*, Astrophys. J. *147*, 433–448.

SCHNEIDER, W. A. (1978), *Integral Formulation in Two and Three Dimensions*, Geophys. *43*, 49–76.

STOFFA, P., FOKKEMA, J., DE LUNA FREIRE, R., and KESSINGER, W. (1990), *Split-step Fourier Migration*, Geophys. *55*, 410–421.

TAPPERT, F. D. (1974), *Parabolic Equation Method in Underwater Acoustics*, J. Acoust. Soc. Am. *55*, Supplement 34(A).

TATARSKII, V. L., *The Effects of the Turbulent Atmosphere on Wave Propagation* (translated from Russian) (National Technical Information Service, 1971).

THOMSON, D. J. and CHAPMAN, N. R. (1983), *A Wide Angle Split-step Algorithm for the Parabolic Equation*, J. Acoust. Soc. Am. *74*, 1848–1854.

TOLSTOY, A., BERMAN, D. H., and FRANCHI, E. R. (1985), *Ray Theory versus the Parabolic Equation in a Long-range Ducted Environment*, J. Acoust. Soc. Am. *78*, 176–189.

TWERSKY, V. (1964), *On Propagation in Random Media of Discrete Scatterers*, Proc. Am. Math. Soc. Soc. Symp. Stochas. Proc. Math. Phys. Eng. *16*, 84–116.

WILD, A. J. and HUDSON, J. A. (1998), *A Geometrical Approach to the Elastic Complex-screen*, J. Geophys. Res. *103*, 707–726.

WILD, A. J., HOBBS, R. W., and FRENJE, L. (2000), *Modeling Complex Media: An Introduction to the Phase-screen Method*, Phys. Earth and Planet. Interiors *120*, 219–226.

WU, R.-S. (1989a), *The Perturbation Method in Elastic Wave Scattering*, Pure Appl. Geophys. *131*, 605–637.

WU, R.-S. (1989b), *Seismic wave scattering*. In *Encyclopedia of Geophysics* (ed., D. James, Van Norstrand Reinhold and Comp.), 1166–1187.

WU, R.-S. (1994), *Wide-angle Elastic Wave One-way Propagation in Heterogeneous Media and an Elastic Wave Complex-screen Method*, J. Geophys. Res. *99*, 751–766.

WU, R.-S. (1996), *Synthetic Seismograms in Heterogeneous Media by One-return Approximation*, Pure Appl. Geophys. *148*, 155–173.

WU, R.-S. and AKI, K. (1985a), *Scattering Characteristics of Elastic Waves by an Elastic Heterogeneity*, Geophys. *50*, 582–595.

Wu, R.-S. and Aki, K. (1985b), *Elastic Wave Scattering by a Random Medium and the Small-scale Inhomogeneities in the Lithosphere*, J. Geophys. Res. *90*, 10,261–10,273.

Wu, R.-S. and De Hoop, M. V. (1996), *Accuracy analysis and numerical tests of screen propagators for wave extrapolation*, Mathematical Methods in Geophysical Imaging IV, SPIE *2822*, 196–209.

Wu, R.-S. and Flatté S. M. (1990), *Transmission Fluctuations across an Array and Heterogeneities in the Crust and Upper Mantle*, Pure Appl. Geophys. *132*, 175–196.

Wu, R.-S. and Huang, L.-J. (1992), *Scattered Field Calculation in Heterogeneous Media Using Phase-screen Propagator*, Expanded Abstracts of the Technical Program, SEG 62nd Annual Meeting, 1289–1292.

Wu, R.-S. and Huang, L.-J. (1995), *Reflected wave modeling in heterogeneous acoustic media using the De Wolf approximation*, Mathematical Methods in Geophysical Imaging III, SPIE *2571*, 176–186.

Wu, R.-S., Huang, L.-J., and Xie, X.-B. (1995), *Backscattered Wave Calculation Using the De Wolf Approximation and a Phase-screen Propagator*, Expanded Abstracts, SEG 65th Annual Meeting, 1293–1296.

Wu, R.-S. and Jin, S. (1997), *Windowed GSP (Generalized Screen Propagators) Migration Applied to SEG-EAEG Salt Model Data*, Expanded abstracts, SEG 67th Annual Meeting, 1746–1749.

Wu, R.-S., Jin, S., and Xie, X.-B. (2000a), *Seismic Wave Propagation and Scattering in Heterogeneous Crustal Waveguides Using Screen Propagators: I SH Waves*, Bull. Seismol. Soc. Am. *90*, 401–413.

Wu, R.-S., Jin, S., and Xie, X.-B. (2000b), *Energy Partition and Attenuation of Lg Waves by Numerical Simulations Using Screen Propagators*, Phys. Earth and Planet. Inter. *120*, 227–244.

Wu, R.-S. and Toksöz, M. N. (1987), *Diffraction Tomography and Multisource Holography Applied to Seismic Imaging*, Geophys. *52*, 11–25.

Wu, R.-S. and Xie, X.-B. (1993), *A Complex-screen Method for Elastic Wave One-way Propagation in Heterogeneous Media*, Expanded Abstracts of the 3rd International Congress of the Brazilian Geophysical Society.

Wu, R.-S. and Xie, X.-B. (1994), *Multi-screen backpropagator for fast 3D elastic prestack migration*, Mathematical Methods in Geophysical Imaging II, SPIE, *2301*, 181–193.

Wu, R.-S., Xie, X.-B., and Wu, X.-Y. (1999), *Lg wave simulations in heterogeneous crusts with irresular topography using half-space screen propagators*, Proceedings of the 21th Annual *Seismic Research Symposium on Monitoring a Comprehensive Test Ban Treaty*, pp. 683–693.

Wu, X.-Y. and Wu, R.-S. (1998), *An Improvement to Complex Screen Method for Modeling Elastic Wave Reflections*, Expanded abstracts, SEG 68th Annual Meeting, 1941–1944.

Wu, X.-Y. and Wu, R.-S. (1999), *Wide-angle Thin-slab Propagator with Phase Matching for Elastic Wave Modeling*, Expanded abstracts, SEG 69th Annual Meeting, 1867–1870.

Xie, X.-B. and Wu, R.-S. (1995), *A Complex-screen Method for Modeling Elastic Wave Reflections*, Expanded Abstracts, SEG 65th Annual Meeting, 1269–1272.

Xie, X.-B. and Wu, R.-S. (1996), *3-D Elastic Wave Modeling Using the Complex Screen Method*, Expanded Abstracts of the Technical Program, SEG 66th Annual Meeting, 1247–1250.

Xie, X.-B. and Wu, R.-S. (1998), *Improve the Wide Angle Accuracy of Screen Method under Large Contrast*, Expanded abstracts, SEG 68th Annual Meeting, 1811–1814.

Xie, X.-B. and Wu, R.-S. (1999), *Improving the Wide Angle Accuracy of the Screen Propagator for Elastic Wave Propagation*, Expanded abstracts, SEG 69th Annual Meeting, 1863–1866.

Xie, X.-B. and Wu, R.-S. (2000), *Modeling Elastic Wave Forward Propagation and Reflection Using the Complex-screen Method*, J. Acoust. Soc. Am., accepted.

Xie, X.-B., Mosher, C. C. and Wu, R.-S. (2000), *The Application of Wide Angle Screen Propagator to 2-D and 3-D Depth Migrations*, Expanded abstracts, SEG 70th Annual Meeting, 878–881.

(Received December 6, 2000, accepted June 7, 2001)

To access this journal online:
http://www.birkhauser.ch

Pure appl. geophys. 160 (2003) 541–554
0033–4553/03/040541–14

© Birkhäuser Verlag, Basel, 2003

| **Pure and Applied Geophysics**

Coda

MICHAEL FEHLER[1] and HARUO SATO[2]

Abstract—Observations and analysis of seismic scattering in the heterogeneous earth have grown from the initial observations of Aki in the 1960s into a well-developed subfield of seismology. The area presents many challenging and interesting problems for seismologists today and there are many areas of fruitful research. We focus on a small subset of areas of research within the general area of, "coda study," that can be most directly tied to Kei's early work in this field: scattering coefficient, coda Q, coda normalization method, and the radiative transfer approach. These are the most useful tools in the interpretation of high-frequency seismograms. In each of these areas, Kei provided initial inspiration through insightful observation and well-thought out models for his observations. The results of ongoing work in these areas have provided insight into the complexity of wave propagation in the earth and have yielded new insights into the character of the earth's lithosphere. They have also provided reliable means to obtain practical information like relative site amplification factors and relative source radiation as a function of frequency.

Key words: Scattering, coda, seismology, heterogeneity, attenuation, Q.

Introduction

In 1969, Keiiti Aki first called attention to the continuous wavetrains in the tail portion of seismograms (AKI, 1969). Kei called these wavetrains "coda waves" and this term has been used since to describe the tail portion of regional seismograms. He observed that early portions of seismograms seem to be composed of waves that were propagating away from the source and decrease with amplitude with increasing propagation distance. Conversely, he observed that the coda had similar amplitude at all stations independent of epicentral distance and had similar spectral content among stations. He also noted that these waves were not simply correlated from station to station but rather appeared as an incoherent series of waves. Aki proposed that the appearance of coda waves was caused by the incoherent waves scattered from random heterogeneity in the earth's lithosphere. His original proposal was that these waves were surface waves. Characteristic of Kei, he argued that these waves,

[1] Los Alamos Seismic Research Center, MS D443; Los Alamos National Laboratory; Los Alamos, NM 87545, U.S.A. E-mail: fehler@lanl.gov
[2] Department of Geophysics, Graduate School of Science, Tohoku University, Aoba-ku, Sendai-shi, 980-8578, Japan. E-mail: sato@zisin.geophys.tohoku.ac.jp

while they may seem complicated, were a tremendous resource because they provided a means to estimate source excitation characteristics (AKI, 1969). In that paper, Kei proposed to use coda spectra for the estimation of source spectra from seismograms in which the direct arrivals are clipped. Figure 1 shows traces of an earthquake recorded at two local stations in Japan. The P and S arrivals and P and S codas are labeled.

Following the idea of random heterogeneity influencing seismic observations, AKI (1973) investigated travel time and amplitude fluctuations at the seismic array in Montana (LASA), using predictions in CHERNOV's (1960) book on wave propagation in a random medium. In this paper, Kei argued that observed travel time and amplitude fluctuations could be explained by considering the effects on both of scattering from random heterogeneity.

Following on the observations of some of Kei's very early work (for example, AKI, 1956; AKI and TSUJIURA, 1959), and observations made by others (NIKOLAYEV and TREGUB, 1970; DAINTY et al., 1974; CAPON, 1974; NAKAMURA et al., 1970, Kei and Bernard Chouet wrote a paper that outlined their combined ideas regarding coda waves (AKI and CHOUET, 1975). They concluded that at low frequencies, less than about 10 Hz, coda waves are dominated by scattered surface waves but that higher frequency coda waves are comprised mostly of scattered body waves. They also outlined the following basic characteristics of coda waves at epicentral distances of less than about 100 km and at times greater than about twice the S-wave travel time: (1) the spectrum of the coda waves for one earthquake is the same at all recording

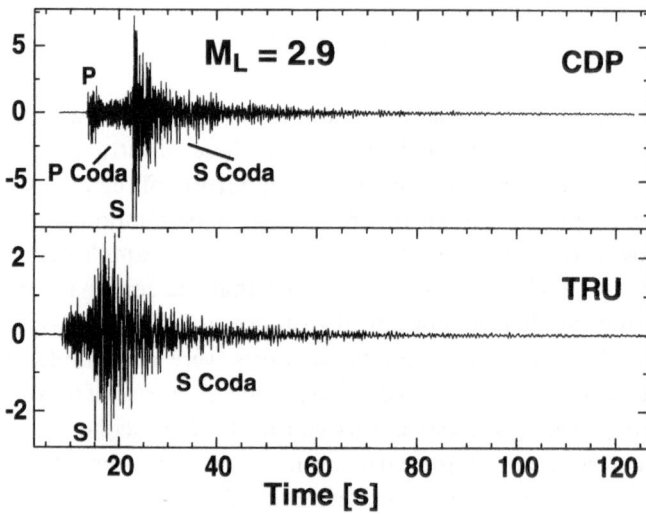

Figure 1
Traces recorded at two stations of an earthquake in Japan. Direct P and S arrivals are labeled. P and S codas are also labeled.

stations, (2) the coda length can be used as a reliable measure of earthquake magnitude (TSUMURA, 1967), (3) the power spectra of coda waves decays as a function of time in the same manner at all stations and for all events within a given region (RAUTIAN and KHALTURIN, 1978), (4) the temporal decay shape described in (3) is independent of earthquake magnitude for events with magnitude less than about 6, and (5) coda amplitude varies with local geology at a recording site. AKI and CHOUET (1975) introduced the parameter coda Q as a measure of the decay rate of the coda within a given frequency band, and they showed that this decay rate was independent of recording site and event location, provided that observations were made within approximately 100 km of the epicenter of an event. They also proposed to estimate the scattering power of random heterogeneity of the earth medium from coda excitation of local earthquakes.

Observations by TSUJIURA (1978) showed that the site amplification determined for coda waves agreed well with that determined from direct S waves but differed from that for P waves. This observation, combined with an apparent agreement between the coda Q and measurements of shear wave Q, led Kei to propose that coda waves are dominated by S waves and that attenuation in the earth is dominated by scattering (AKI, 1980).

Later work by Kei, his students, and collaborators was conducted to both refine observations of coda waves and improve our models for coda waves and/or high-frequency seismogram envelopes. Many of these studies are thoroughly discussed in a recent book (SATO and FEHLER, 1998), whose completion is a monument to Kei's influence on many scientists.

Since there are so many areas of study of coda waves, both observational and theoretical, we have chosen to limit our focus. We choose to discuss observational methods that can be most directly related to Kei's early observations and ideas, namely coda Q and the coda normalization method. We introduce some justification for the assertion by Kei in his 1969 paper that at a given time, the power spectrum of coda waves of a local earthquake is nearly independent of epicentral distance. Finally, we mention a more recent area of investigation, the use of the radiative transfer theory, which was introduced into seismology by one of Kei's students (WU, 1985). Later this approach was developed to correctly adopt the nonstationary process for the case of earthquake source radiation by Kei's students and collaborators. This approach has offered a way to synthesize the whole seismogram envelope and has provided a reliable method for estimating the amount of attenuation caused by scattering and intrinsic mechanisms.

Scattering Coefficient and Coda Q

The conceptual model of coda excitation was introduced by AKI and CHOUET (1975). Their single backscattering model is proposed to explain phenomenologically the coda envelopes of local earthquakes. They wrote the coda power as a convolution of the source, scattering and site amplification as

$$\left\langle \dot{u}_{ij}^{S\,Coda}(t;f)^2 \right\rangle_T = \frac{W_i^S(f)g_0(f)N_j^S(f)^2}{2\pi\beta_0^2 t^2} e^{-Q_C^{-1}2\pi f t} , \tag{1}$$

where $\dot{u}_{ij}^{S\,Coda}(t;f)$ is the S-coda particle velocity at site j at center frequency f, $\langle\ldots\rangle_T$ means average over some time window T, $W_i^S(f)$ is the S-wave source power for the ith earthquake, $g_0(f)$ is the total scattering coefficient representing the scattering power per unit volume, $N_j^S(f)$ is the site amplification factor, and β_0 is the background S-wave velocity. They introduced coda Q, Q_C as a parameter to account for anelastic loss of energy from the wavefield. Later, JANNAUD et al. (1991) numerically confirmed the validity of relation (1).

Measurements of g_0 and Q_c are made by applying equation (1) to bandpass-filtered seismic data. Some investigators used an empirical relationship between radiated energy and local earthquake magnitude to estimate $W_i^S(f)$. Others used joint analysis of direct S-wave and S-coda envelopes. By first multiplying the filtered seismic trace data by t and then taking the logarithm and plotting it vs. lapse time t, Q_c can be determined from the resulting slope. Measurements are made using data over a given lapse time range that generally begins long after the S-wave onset at each station.

The total scattering coefficient $g_0(f)$ has been measured in many regions worldwide. Roughly, this parameter is the ratio of S-coda power to radiated S-wave source power. SATO and FEHLER (1998, Figure 3.10) compiled reported values of g_0, where they included reported total scattering coefficients estimated by assuming isotropic scattering. We find that total scattering coefficient g_0 for S-to-S wave scattering is of the order of 10^{-2} km^{-1} for frequencies 1–30 Hz and the scatter is a factor of two for individual measurements of g_0; however, regional differences related to tectonic activity have not yet been quantitatively clarified.

Observations of Q_c have been made by numerous authors using data from networks at many locations worldwide. Results confirm the conclusions of AKI and CHOUET (1975) that Q_c is relatively constant within a given region. Q_c varies with frequency, lapse time interval used in the observations, and tectonic region. It has generally been found to be lower in tectonically active regions and higher in tectonically stable regions. Changes in Q_c have also been suggested as a precursor to earthquakes although some have criticized precursory studies. In general, Q_C^{-1} is about 10^{-2} at 1 Hz and decreases to about 10^{-3} at 20 Hz. The frequency dependence within a region can be written as $Q_C^{-1} \propto f^{-n}$ for $f > 1$ Hz, where the power n ranges between 0.5 and 1. The variation from region to region is more than a factor of 10. Summaries of observations of Q_c can be found in SATO and FEHLER (1998) and SATO et al. (2002).

The relation between Q_c and intrinsic and scattering Q is not obvious. It is clear from reading papers of Aki that he struggled with this issue. He alternatively speculated that Q_c is the intrinsic loss (AKI and CHOUET, 1975) and equivalent to total attenuation which he said is dominated by scattering loss (AKI, 1980). FRANKEL and WENNERBERG (1987) developed the energy-flux model for coda waves that was based on the idea that energy is uniformly distributed within some region surrounding an earthquake at some lapse time after an earthquake and that Q_c is intrinsic loss. GUSEV (1995) demonstrated that coda decay is quantitatively well explained if the total scattering coefficient decreases with depth, when the leakage of scattered energy to the bottom cannot be discriminated from intrinsic loss. Later, MAYEDA *et al.* (1992) examined the relation among scattering Q and intrinsic Q determined using the radiative transfer theory and Q_c. They concluded that there is no simple relation between Q_c and scattering and intrinsic Q. It is necessary for us to develop an interpretation of coda Q that includes the depth dependence of the total scattering coefficient and intrinsic absorption in addition to velocity structure.

Coda Normalization Method

The coda waves provide a reliable way to isolate and quantify the seismic source radiation and receiver site amplification. It also allows the investigation of propagation effects. The fundamental empirical base of the coda normalization method is that there is a uniform distribution of coda energy in some volume surrounding the source at some lapse time. The key observation in support of this method is that coda envelopes have a common decay curve that is independent of the source-receiver distance (RAUTIAN and KHALTURIN, 1978; AKI and CHOUET, 1975). Coda amplitudes vary with source size and recording site amplification.

Site Factors

When the lapse time t_c is large enough, the relative coda amplitude at two different sites j and k should be the same except for the influence of the near-recording site amplification since the source factor is common:

$$N_j^S(f)/N_k^S(f) = \sqrt{\left\langle \dot{u}_{ij}^{S\,\text{Coda}}(t_c;f)^2 \right\rangle_T \Big/ \left\langle \dot{u}_{ik}^{S\,\text{Coda}}(t_c;f)^2 \right\rangle_T}. \qquad (2)$$

TSUJIURA (1978) reported that the frequency dependence of site amplification factors estimated by the coda normalization method is more stable than those estimated from direct wave analysis. He also found that the amplification determined by averaging observations of coda amplification found for many earthquakes was similar to that obtained by averaging measurements of direct S waves. This conclusion was important to Aki's later assertion that coda waves are composed predominately of scattered S waves (AKI, 1980). Analyzing local earthquakes in northern California, PHILLIPS and AKI (1986) determined site amplifications for 150 stations. SU et al. (1992) extended the study of PHILLIPS and AKI (1986) and found that site amplification in central California is related to the geologic age of rocks at the receiver site. SU and AKI (1995) used coda waves to determine site amplifications of 158 stations in Central and Southern California and confirmed the relation between geologic age and site amplification found by SU et al. (1992). FEHLER et al. (1992) developed a map of site amplification factors for the Kanto–Tokai region of Japan. They showed that the site amplification factors at 6 Hz agreed well with local magnitude residuals, which were obtained from maximum amplitude measurements for the same stations.

Detailed comparisons of the relative site amplification factors determined using S waves and coda waves have led to the conclusion that there can be differences between the two, particularly in structurally complex areas (e.g., BONILLA et al., 1997); however, other studies indicate that the differences are small (e.g., KATO et al., 1995). However, the differences can be reduced with an appropriate choice of reference site or when the average response of multiple sites is used as a reference site.

Source Size

From equation (1), we ascertain relative source radiation as a function of frequency by dividing the coda amplitude of the seismogram recorded at one site j for a given earthquake i by the amplitude at the same site for a different earthquake k taken at the same absolute lapse time t_c at the same site j:

$$W_i^S(f)/W_k^S(f) = \left\langle \dot{u}_{ij}^{S\,\text{Coda}}(t_c;f)^2 \right\rangle_T \Big/ \left\langle \dot{u}_{kj}^{S\,\text{Coda}}(t_c;f)^2 \right\rangle_T . \tag{3}$$

We know that coda wave excitation is mostly insensitive to radiation-pattern differences. Using seismic moments from two well-analyzed earthquakes in Alaska, BISWAS and AKI (1984) developed a coda-amplitude vs. seismic-moment scale. That relationship can be used to make a fast and reliable determination of the size of an earthquake using data from only one station. DEWBERRY and CROSSON (1995) performed a detailed analysis of the source spectrum of earthquakes in the U.S. Pacific Northwest using data from coda waves. HARTSE et al. (1995) determined source radiation as a function of frequency for earthquakes and nuclear explosions in the U.S.A.

Single Station Attenuation Measurements

The single station method for measuring attenuation was proposed by AKI (1980). By normalizing direct S-wave amplitude by S-coda amplitude, this method corrects for source size and site amplification, thus allowing data to be combined from many earthquakes to find a more stable estimate of attenuation. The direct S-wave particle velocity amplitude at station j at frequency f for local earthquake i is

$$\left| \dot{u}_{ij}^{S\ \text{Direct}}(f) \right| \propto \frac{1}{r_{ij}} \sqrt{W_i^S(f)} N_j^S(f) e^{-Q_S^{-1}\pi f\ r_{ij}/\beta_0} , \qquad (4)$$

where r_{ij} is the source-receiver distance, β_0 is S-wave velocity, and Q_S is the Q of direct S waves. Taking the logarithm of the ratio of the product of hypocentral distance and the direct S-wave amplitude to the averaged coda amplitude given by equation (1), the common site amplification and source terms cancel, we get

$$\ln \frac{r_{ij}\left| \dot{u}_{ij}^{S\ \text{Direct}}(f) \right|}{\sqrt{\left\langle \dot{u}_{ij}^{S\ \text{Coda}}(t_c;f)^2 \right\rangle_{\text{T}}}} = -(Q_S^{-1}(f)\pi f/\beta_0)r_{ij} + \text{Const} . \qquad (5)$$

We smooth out the radiation pattern differences when the measurements are made over a large enough number of earthquakes. Plotting the left-hand side against hypocentral distance, the gradient gives the attenuation per travel distance. AKI (1980) first applied this method to seismograms of local earthquakes recorded in Kanto, Japan. His work stimulated measurements of attenuation in many areas of the world during the 1980s. Later, YOSHIMOTO et al. (1993) extended the conventional coda-normalization method to measure Q_P^{-1}.

Qualitative Discussion of the Coda Normalization Method

The foundation of the coda normalization method was stated near the beginning of AKI (1969) where he stated that "the power spectrum of coda waves at a given time measured from the earthquake origin time appears to be nearly independent of the epicentral distance." This observation is critical to the success of the coda normalization method. Stated another way; after some time, if energy is uniformly distributed within some region surrounding the source, all stations within that region will have trace amplitude that is proportional to the size of the earthquake. Variations in amplitude observed at differing stations are caused by local site amplification effects. These statements are implicit in equation (1).

The uniform distribution of coda energy within some region surrounding the source for lapse times greater than some lapse time is predicted by the single isotropic scattering model (see for example Figure 3.6 of SATO and FEHLER, 1998). Results calculated using the radiative transfer equation for models that include multiple scattering, the effects of nonisotropic scattering, and nonisotropic source radiation also indicate that energy is uniformly distributed within a medium after sufficient time has passed from the initial source excitation time (see for example Figure 7.24 of SATO and FEHLER, 1998).

Uniform distribution of coda energy was also observed in results of numerical simulations by FRANKEL and CLAYTON (1986). To examine this in more detail, we show results of 2-D numerical simulations of scalar wave propagation in a medium that is characterized by a von Kármán autocorrelation function with $\kappa = 0.1$. A medium with a von Kármán autocorrelation function is extremely rich in short wavelength heterogeneity. The power spectrum density (PSDF) of the medium heterogeneity is shown in Figure 2. Numerical modeling for a medium with a Gaussian autocorrelation function was discussed in FEHLER *et al.* (2000). They discussed a case in which forward scattering dominates because the dominant wavelength of the source excitation (2 km) was much less than the dominant wavelength of the medium heterogeneity (5 km). In the simulations of FEHLER *et al.* (2000) no coda are evident in the vicinity of the source. In addition, after the initial wave packet had passed a given observation distance, there is no coda at that distance.

Our numerical modeling was conducted using a finite-difference simulation of the 2-D scalar wave equation in the same manner as described in FEHLER *et al.* (2000). We used a medium with a correlation length of 5 km and a RMS fractional fluctuation in the velocity of 5%. We note that the wavenumber corresponding to 2 Hz is about 3.1/km, which is in the power-law range of the von Kármán type PSDF with $\kappa = 0.1$ for the background velocity of 4 km/s. Due to the presence of small-

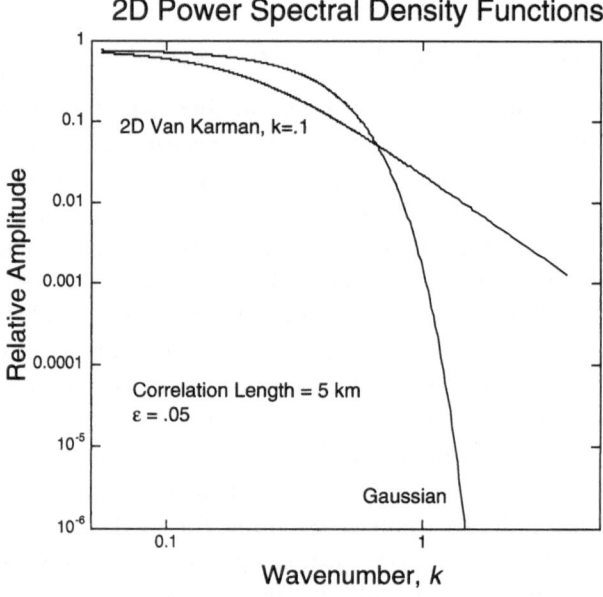

Figure 2

PSDF for random media with 2-D Gaussian and von Kármán autocorrelation functions. Correlation length of both media is 5 km and RMS velocity fluctuation is 5%. Note that the medium with von Kármán autocorrelation function has more short wavelength inhomogeneity; e.g., more power in large wavenumbers, than the medium characterized by a Gaussian autocorrelation function.

scale heterogeneity, coda waves are formed by backscattering near the source. Figure 3 displays ensemble-average RMS envelopes at distances ranging from 25 to 200 km from the source, constructed from 50 numerical simulations. The procedure for calculating the ensemble average envelopes is described in FEHLER *et al.* (2000). Contrary to the results obtained from modeling propagation in a Gaussian random medium where correlation distance is larger than the dominant wavelength, the envelopes here show well-developed coda with a long tail at short distances. Note also that at large lapse times, e.g., times much later than the direct arrival time at a given distance, the coda amplitude reaches a level that is independent of observation distance. The constant level of coda amplitude for the case of random media possessing spectra rich in short-wavelength components is in agreement with the phenomenological observation that the coda normalization method is reliable.

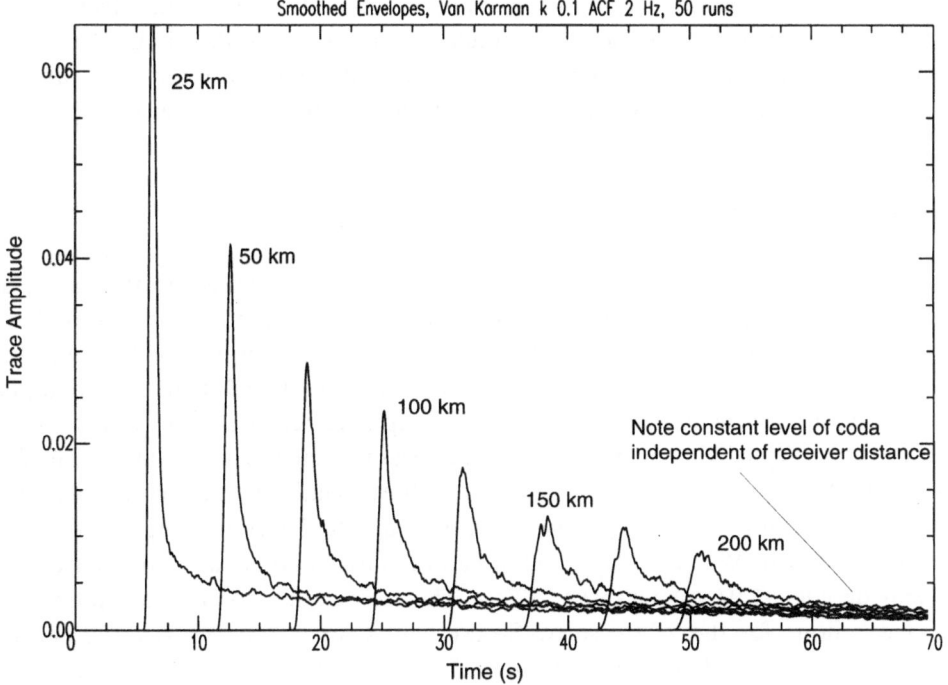

Figure 3

Ensemble-average RMS envelopes calculated for media with von Kármán autocorrelation function whose PSDF is shown in Figure 2. Since source was a 2 Hz. Ricker wavelet, envelopes should be considered to be envelopes of seismic waveforms that have been band-passed in a frequency range from about 1–4 Hz. Note coda of envelopes at all distances whose amplitude at large lapse-time is nearly the same, independent of distance from the source.

Power-law Spectra

Analyzing the frequency dependence of *S*-wave attenuation and scattering coefficient revealed from coda excitation of local earthquakes, WU and AKI (1985) found that inhomogeneities in the lithosphere have power-law characteristics. This means that the spectral content of random inhomogeneities of the earth is rich in short wavelength components compared with the Gaussian spectra. Direct evidence was found from the study of well log data. SATO (1979) reported that the autocorrelation of a borehole acoustic velocity log in crystalline rock in Kanto, Japan is not Gaussian but exponential-like, peaking sharply at lag distance zero. Analyzing sonic log data of the KTB deep-borehole in Germany, WU *et al.* (1994) reported that the spectrum of heterogeneity obeys a power-law for a wide wavelength range. LEARY and ABERCROMBIE (1994) also reported that the PSDF of sonic velocity log at the Cajon Pass borehole in California obeys a power-law.

Analyzing coda of a microearthquake recorded by a borehole seismometer located at 2.5 km depth in the Cajon Pass hole, LEARY and ABERCROMBIE (1994) found that the spectral ratio of coda spectra to *S*-wave source spectra is proportional to the spectra of acoustic reflectivity log data for a broad frequency range from 10 to 200 Hz. As discussed in the previous simulation, the uniform distribution of coda energy might be related with such spectral content. It is necessary for us to further study seismic wave scattering through random media with such power-law spectra.

Radiative Transfer Theory

Aki's initial models explaining seismic coda waves were based on the use of a single scattering approximation. The single scattering assumption worked well to explain seismic observations even though there were concerns raised about the importance of multiple scattering (see e.g., AKI and CHOUET, 1995). Several attempts were made to investigate the importance of multiple scattering as observations were made that indicated that it might be an important factor. The radiative transfer theory has been an attractive theory because it allows relatively tractable calculations of effects of multiple scattering. The radiative transfer theory was first introduced into seismology by WU (1985) although it had been used in other areas of physics for several decades (CHANDRASEKHAR, 1960). The formulation for the nonstationary state was done for the 2-D case by SHANG and GAO (1988). ZENG *et al.* (1991) beautifully formulated the non-stationary multiple scattering process in 3-D. The diffusion model, which was used for the study of lunar seismograms (DAINTY *et al.*, 1974), is considered as an extreme limit of the multiple scattering process.

Here, we introduce the equation that governs radiative transfer for media with multiple isotropic scattering. We imagine a 3-D scattering medium with the background propagation velocity V_0, in which point-like isotropic scatterers of cross section σ_0 are randomly and homogeneously distributed with number density n, where $g_0 \equiv n\sigma_0$ is the total scattering coefficient characterizing the scattering power per unit volume. We assume an impulsive radiation of energy W at time zero from a source located at the origin. The energy density due to the propagation of coherent waves from the source is $We^{-(V_0 g_0 + \eta)t} \times \delta(t - r/V_0)/4\pi V_0 r^2$. We include the exponential decay term $V_0 g_0$ to account for scattering loss and η for intrinsic attenuation per time. Generation of scattered energy per unit time from a unit volume at the last scattering point $(\mathbf{x}', t')$ is a product of g_0, V_0, and energy density $E(\mathbf{x}', t')$. Including geometrical spreading and the time lag due to propagation, we derive the energy-flux density at the receiver at $(\mathbf{x}, t)$ due to scattered waves from a unit volume. Integrating the scattered energy-flux density over the entire 3-D space and dividing by V_0, we derive the total contribution of scattered energy density. Adding the energy density of direct propagation of coherent waves from the source to that of the scattered waves, we obtain

$$E(\mathbf{x}, t) = W \, G_E(\mathbf{x}, t) + g_0 V_0 \int\limits_{-\infty}^{\infty} \int\limits_{-\infty}^{\infty} \int\limits_{-\infty}^{\infty} \int\limits_{-\infty}^{\infty} G_E(\mathbf{x} - \mathbf{x}', t - t') E(\mathbf{x}', t') \, d\mathbf{x}' dt' \quad (6)$$

where the direct propagation of coherent wave energy is given by

$$G_E(\mathbf{x}, t) = \frac{1}{4\pi V_0 r^2} H(t) \delta\left(t - \frac{r}{V_0}\right) e^{-(g_0 V_0 + \eta)t} \ . \quad (7)$$

Solutions of equations (6) with (7) yield information about the distribution of energy in space and time. The solution in the time domain gives the mean-square envelope of the entire wavefield. The calculations are conducted in terms of energy since the wavefield is assumed to be composed of incoherently scattered waves. Solutions of the radiative transfer equation which have forms similar to equation (6) for various assumptions pertaining to source and media are a fruitful frontier area of research. For example, the effects of multiple scattering on the polarization of S waves have recently been studied (BAL and MOSCOSO, 2000) and Monte Carlo simulations have been used to investigate radiative transfer in an elastic medium (MARGERIN et al., 2000).

The radiative transfer approach forms the basis of a viable analysis method for making independent estimates of the amount of scattering that is caused by intrinsic and scattering mechanisms. An initial approach for making these measurements was proposed by WU and AKI (1988). WU and AKI (1988) used their stationary-state solution to analyze data. However, application of this solution required the measurement of the total energy in a seismogram, which may be underestimated if significant portions of energy are buried in noise in later portions of the seismogram. The method was later modified with the introduction of the Multiple Lapse Time Window analysis Method (FEHLER et al., 1992; MAYEDA et al., 1992), which does not require the total energy in the seismograms to be measured. Results obtained using the radiative transfer theory show that at frequencies near 1 Hz, scattering is the dominant mechanism of attenuation. However, as frequency increases to 10 Hz, scattering looses importance as an attenuation mechanism and intrinsic (nonelastic) attenuation mechanisms dominate (Fig. 7.12, SATO and FEHLER, 1998). These results provide useful information for developing models of seismic attenuation and scattering.

Conclusions

Kei Aki once remarked that of all the work he had done in seismology, he was most proud of his work on coda waves because if he had not done the work, it is likely that no one else would have done so. With his remarkable insight, beginning with his observations in the 1950s, Kei has provided seismologists with a wealth of tools and ideas regarding wave propagation in the heterogeneous earth. He has inspired many of his students and other investigators to work on related topics. To his credit, many of the ideas and observations he made early in his career using very little data that were poorly recorded by today's standards are still viable today. Some of these observations still provide challenges to us today as we attempt to formalize the physics that explains his observations.

Coda waves have proved to be a valuable tool. With no formal theory to explain their formation, the coda normalization method is a valuable tool for making quantitative measurements of source radiation and site amplification effects. Enhancements to allow the measurement of seismic attenuation using a single station method as developed by AKI (1980) or using the radiative transfer theory have provided us with results that would not be achievable by other methods.

The number of researchers who are building careers out of using various aspects of the ideas initiated by Keiiti Aki for dealing with coda waves and wave propagation in random media is enormous. Each has added to our body of knowledge relating to the earth and each has gained enormously from the inspiration of Kei Aki.

Acknowledgments

We gratefully thank Kei and all of his students who have worked in the area of coda and scattering for their discussions and support. We thank editor Yehuda Ben-Zion and reviewers Yuehua Zeng and Peter Leary for comments that facilitated our enhancement of the manuscript. Work at Los Alamos National Laboratory was supported by the United States Department of Energy through contract W-7405-ENG-36 from the Office of Basic Energy Sciences headed by Nick Woodward.

REFERENCES

AKI, K. (1956), *Correlogram Analyses of Seismograms by Means of a Simple Automatic Computer*, J. Phys. Earth 4, 71–79.

AKI, K. (1969), *Analysis of Seismic Coda of Local Earthquakes as Scattered Waves*, J. Geophys. Res. 74, 615–631.

AKI, K. (1973), *Scattering of P Waves under the Montana LASA*, J. Geophys. Res. 78, 1334–1246.

AKI, K. (1980), *Attenuation of Shear Waves in the Lithosphere for Frequencies from 0.05 to 25 Hz*, Phys. Earth Planet. Inter. 21, 50–60.

AKI, T. and TSUJIURA, M. (1959). *Correlation Study of Near-earthquake Waves*, Bull. Earthquake Res. Inst. Tokyo Univ. 37, 207–232.

AKI, K. and CHOUET, B. (1975), *Origin of Coda Waves: Source, Attenuation and Scattering Effects*, J. Geophys. Res. 80, 3322–3342.

BAL, G. and MOSCOSO, M. (2000), *Polarization Effects of Seismic Waves on the Basis of Radiative Transfer Theory*, Geophys. J. Int. 142, 571–585.

BISWAS, N. N. and AKI, K. (1984), *Characteristics of Coda Waves: Central and South Central Alaska*, Bull. Seismol. Soc. Am. 74, 493–507.

BONILLA, L., STEIDL, J., LINDLEY, G., TUMARKIN, A., and ARCHULETA, R. (1997), *Site Amplification in the San Fernando Valley, California: Variability of Site-effect Estimation Using the S-wave, coda, and H/V Methods*, Bull. Seismol. Soc. Am. 87, 710–730.

CAPON, J. (1974), *Characterization of Crust and Upper Mantle Structure under LASA as a Random Medium*, Bull. Seismol. Soc. Am. 64, 235–266.

CHANDRASEKHAR, S., *Radiative Transfer* (Oxford University Press, Cambridge 1960).

CHERNOV, L. A., *Wave Propagation in a Random Medium* (Engl. trans. by R. A. Silverman) (McGraw-Hill, New York (1960)).

DAINTY, A., TOKSÖZ, M., ANDERSON, K., PINES, P., NAKAMURA, Y., and LATHAM, G. (1974), *Seismic Scattering and Shallow Structure of the Moon in Oceanus Procellarum*, Moon 9, 11–29.

DEWBERRY, S. R. and CROSSON, R. S. (1995), *Source Scaling and Moment Estimation for the Pacific Northwest Seismograph Network Using S-coda Amplitudes*, Bull. Seismol. Soc. Am. 85, 1309–1326.

FEHLER, M., HOSHIBA, M., SATO, H., and OBARA, K. (1992), *Separation of Scattering and Intrinsic Attenuation for the Kanto-Tokai Region, Japan, Using Measurements of S-wave Energy versus Hypocentral Distance*, Geophys. J. Int. 108, 787–800.

FEHLER, M., SATO, H., and HUANG, L.-J. (2000), *Envelope Broadening of Outgoing Waves in 2-D Random Media: A Comparison between the Markov Approximation and Numerical Simulations*, Bull. Seismol. Soc. Am. 90, 914–928.

FRANKEL, A. and CLAYTON, R. W. (1986), *Finite-Difference Simulations of Seismic Scattering: Implications for the Propagation of Short-period Seismic Waves in the Crust and Models of Crustal Heterogeneity*, J. Geophys. Res. 91, 6465–6489.

FRANKEL, A. and WENNERBERG, L. (1987), *Energy-flux Model of Seismic Coda: Separation of Scattering and Intrinsic Attenuation*, Bull. Seismol. Soc. Am. 77, 1223–1251.

GUSEV, A. A. (1995), *Vertical Profile of Turbidity and Coda Q*, Geophys. J. Int. 123, 665–672.

HARTSE, H., PHILLIPS, W. S., FEHLER, M., and HOUSE, L. (1995), *Single-station Spectral Discrimination Using Coda Waves*, Bull. Seismol. Soc. Am. 85, 1464–1474.

JANNAUD, L. R., ADLER, P. M., and JACQUIN C. G. (1991), *Spectral Analysis and Inversion of Codas*, J. Geophys. Res. 96, 18,215—18,231.

KATO, K., AKI, K., and TAKEMURA, M. (1995) *Site Amplification from Coda Waves: Validation and Application to S-wave Site Response*, Bull. Seismol. Soc. Am. 85, 467–477.

LEARY, P. and ABERCROMBIE R. (1994), *Frequency-dependent Crustal Scattering and Absorption at 5–160 Hz from Coda Decay Observed at 2.5 km Depth*, Geophys. Res. Lett. 21, 971–974.

MARGERIN, L., CAMPILLIO, M., and TIGGELEN, B. (2000), *Monte Carlo Simulation of Multiple Scattering of Elastic Waves*, J. Geophys. Res. 105, 7873–7893.

MAYEDA, K., KOYANAGI, S., HOSHIBA, M., AKI, K., and ZENG, Y. (1992), *A Comparative Study of Scattering, Intrinsic, and Coda Q^{-1} for Hawaii, Long Valley and Central California between 1.5 and 15 Hz*, J. Geophys. Res. 97, 6643–6659.

NAKAMURA, Y., LATHAM, G., EWING, M., and DORMAN, L. (1970), *Lunar Seismic Energy Transmissions (abstract)* EOS. Trans. Am Geophys. Un. *51*, 776.

NIKOLAYEV, A. and TREGUB, F. (1970), *A Statistical Model for the Earth's Crust: Method and Results*, Tectonophys. *10*, 573–578.

PHILLIPS, W. S. and AKI, K. (1986), *Site Amplification of Coda Waves from Local Earthquakes in Central California*, Bull. Seismol. Soc. Am. *76*, 627–648.

RAUTIAN, T. G. and KHALTURIN, V. I. (1978), *The Use of the Coda for Determination of the Earthquake Source Spectrum*, Bull. Seismol. Soc. Am. *68*, 923–948.

SATO, H. (1979), *Wave Propagation in One-dimensional Inhomogeneous Elastic Media*, J. Phys. Earth *27*, 455–466.

SATO, H. and FEHLER, M. (1998), *Seismic Wave Propagation and Scattering in the Heterogeneous Earth* (AIP Press/Springer Verlag, New York 2000).

SATO, H., FEHLER, M., and WU, R.-S. (2002), *Scattering and attenuation of seismic waves in the lithosphere*, Chapter 13 in *International Handbook of Earthquake and Engineering Seismology* (eds. P. Jennings, H. Kanamori, and W. Lee), 195–208.

SHANG, T. and GAO, H. (1988), *Transportation Theory of Multiple Scattering and its Application to Seismic Coda Waves of Impulsive Source*, Scientia Sinica (series B, China) *31*, 1503–1514.

SU, F., AKI, K., TENG, L., ZENG, Y., KOYANAGI, S., and MAYEDA, K. (1992), *The Relation between Site Amplification Factor and Surface Geology in Central California*, Bull. Seismol. Soc. Am. *82*, 580–602.

SU, F. and AKI, K. (1995), *Site Amplification Factors in Central and Southern California Determined from Coda Waves*, Bull. Seismol. Soc. Am. *85*, 452–466.

TSUJIURA, M. (1978), *Spectral Analysis of the Coda Waves from Local Earthquakes*, Bull. Earthq. Inst. Univ. Tokyo *53*, 1–48.

TSUMURA, K. (1967), *Determination of Earthquake Magnitude from Duration of Oscillation*, Zisin (in Japanese) *20*, 30–40.

WU, R.-S. (1985), *Multiple Scattering and Energy Transfer of Seismic Waves—Separation of Scattering Effect from Intrinsic Attenuation—I. Theoretical Modeling*, Geophys. J. R. Astron. Soc. *82*, 57–80.

WU, R.-S. and AKI, K. (1985), *The Fractal Nature of the Inhomogeneities in the Lithosphere Evidenced from Seismic Wave Scattering*, Pure Appl. Geophys. *123*, 805–818.

WU, R.-S. and AKI, K. (1988), *Multiple Scattering and Energy Transfer of Seismic Waves—Separation of Scattering Effect from Intrinsic Attenuation. II. Application of the Theory to Hindu–Kush Region*, Pure Appl. Geophys. *128*, 49–80.

WU, R.-S., XU, Z., and LI, X. P. (1994), *Heterogeneity Spectrum and Scale-anisotropy in the Upper Crust Revealed by the German Continental Deep-Drilling (KTB) Holes*, Geophys. Res. Lett. *21*, 911–914.

YOSHIMOTO, K., SATO, H., and OHTAKE, M. (1993), *Frequency-dependent Attenuation of P and S Waves in the Kanto Area, Japan, Based on the Coda-normalization Method*, Geophys. J. Int. *114*, 165–174.

ZENG, Y., SU, F., and AKI, K. (1991), *Scattering Wave Energy Propagation in a Random Isotropic Scattering Medium 1. Theory*, J. Geophys. Res. *96*, 607–619.

(Received August 31, 2000, accepted March 19, 2001)

To access this journal online:
http://www.birkhauser.ch

Pure appl. geophys. 160 (2003) 555–577
0033–4553/03/040555–23

| Pure and Applied Geophysics

Radiation from a Finite Reverse Fault in a Half Space

RAÚL MADARIAGA[1]

Abstract — Using a set of well-known results for the seismic field radiated by a simple dip-slip dislocation in a half space, we study interesting details of the motion at the surface of the half space. The static solution for a dislocation in a half space was found by FREUND and BARNETT in 1976. The corresponding elastodynamic solution was solved exactly in the Fourier and Laplace domain by several authors about 20 years ago, however its properties remained unexplored because of analytical difficulties. We remove these difficulties and show that the solution contains three important phenomena: Seismic wave fronts of P, S and SP type; the near-field pulse associated with the propagation of the dislocation front; and the long-time elastic response that converges toward the static solution of Freund and Barnett. Based on these results we show that solutions to all these problems are self-similar and homogeneous in x/h and $\alpha t/h$ so that when the fault depth h approaches 0, the solutions become concentrated near the origin and around the P, S and surface wave travel times. This explains several paradoxes in the radiation from dip-slip faults; among these the most notable are the presence of a point force singularity at the tip of a surface breaking fault and the reduction in high frequency radiation near the surface.

Key words: Earthquakes, elastic wave propagation, dislocations.

1. Introduction

In 1968 Aki published his classical study of the Parkfield earthquake where he computed the near field of strike-slip fault and used this result to accurately fit observed near field displacement on the perpendicular component of station 2 (AKI, 1968). Since then the techniques to compute the near field from strike-slip earthquakes have become increasingly complex and accurate. It is currently common to invert accelerograms for slip distribution and rupture history for large shallow strike-slip earthquakes (see, e.g., ARCHULETA, 1984; WALD and HEATON, 1994; COTTON and CAMPILLO, 1995 or COHEE and BEROZA, 1994). More recently, dynamic models of specific earthquakes have been reported in the literature (OLSEN et al., 1997; PEYRAT et al., 2001).

For shallow dip-slip earthquake, the situation is different because few events of this type have been well recorded in the near field. A substantial effort to model dip-slip events and the strong motion they generate has been made by OGLESBY et al.

[1] Ecole Normale Supérieure, 24 rue Lhomond, 75231 Paris Cedex 05, France.
E-mail: madariag@geologie.ens.fr

(1998, 2000a, 2000b), SHI *et al.* (1998). These works raised a number of interesting questions regarding the effect of the free surface above the shallow dipping fault that need careful study. The recent Chichi earthquake of 1999 in Taiwan (see, e.g., KAO and CHEN, 2000) provided a wealth of information that can be used to address a number of questions concerning the effects of the free surface on the strong motion near fault break outs.

In this paper I will review numerous results for kinematic dislocation models that are scattered in the literature in order to clarify a certain number of problems identified by J. Brune (personal communication, 2000). This work will be entirely based on papers published in the late seventies and early eighties by several authors. One of the earliest works on kinematical models of shallow faulting is that of BOORE and ZOBACK (1974) reviewed in chapter 14 of AKI and RICHARDS (1980). The two-dimensional kinematic problem of a dip-slip fault buried in a half space was precisely solved by NIAZI (1975) using the Cagniard-de Hoop method. Unfortunately his solution could not be simply computed. BOUCHON and AKI (1977) and BOUCHON (1978) proposed a numerical spectral method for this problem, nonetheless their discrete frequency wavenumber does not work properly when the fault breaks the free surface due to a singularity of the wavefield at infinite wavenumber. In 1980 I published a different solution to the two-dimensional dip-slip fault in a half space that could be computed exactly using the Cagniard de Hoop method. In the following I will present a shorter derivation of the solution and invert it to space and time domains. I will give exact solutions for ground velocities and the stress component that is different from zero on the free surface. Then I will exploit these solutions to obtain several properties of the effect of the free surface.

2. Basic Formulation

We study the propagation of a dislocation along an inclined straight fault using the Cagniard de Hoop method (AKI and RICHARDS, 1980). We solve the elastic wave equation:

$$\rho \frac{\partial^2 \mathbf{u}}{\partial t^2} = (\lambda + \mu)\nabla(\nabla \cdot \mathbf{u}) + \mu \nabla^2 \mathbf{u} \tag{1}$$

in a half space containing a dislocation source. λ and μ are the elastic constants, ρ is density and elastic wave speeds will be designated α and β for P and S waves, respectively.

As shown in Figure 1, let a dislocation run at speed v along a fault inclined at an angle θ with respect to the horizontal. Rupture starts at position $(0, h)$ and propagates with constant rupture speed v for a finite time L/v in the direction θ stopping at a distance L. Let ξ be a coordinate along the dislocation line. On the fault located between $0 < \xi < L$ slip is constant and equal to D.

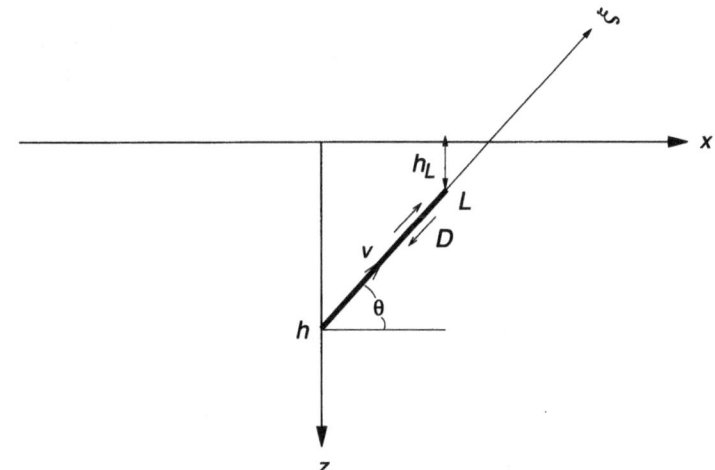

Figure 1

Geometry of a finite fault of dip θ buried in a half space. Rupture starts at the depth h and propagates along the fault for a finite distance L.

Since we are going to use the Cagniard de Hoop method for our calculations we introduce the double Laplace transformed solution defined by

$$u_i(x,z,t) = \frac{1}{2\pi i} \int_{C_s} \frac{1}{2\pi i} \frac{1}{\alpha} \int_{C_p} \bar{u}_i(p,z,s) e^{s(t-\frac{p}{\alpha}x)} \, dp \, s ds \tag{2}$$

where C_p and C_s are appropriate Bromwich contours. We note that the convention in (2) is different from that of MADARIAGA (1980). The problem is formally solved by the determination of the doubly transformed displacement $\bar{u}_i(p,z,s)$.

We look for solutions of (1) by the method of potentials. We define as usual $\mathbf{u} = \nabla\phi + \nabla \times (\psi \mathbf{e}_y)$. These potentials satisfy the P- and the S-wave equation, respectively. In order to satisfy appropriate radiation conditions at infinity we choose different forms of the potential in the upper and lower half spaces separated by the line $z = h$, the depth at which the source is located.

$$\bar{\phi} = \Phi^+(s,p)e^{-\frac{s}{\alpha}q_p(z-h)} + \Phi^-(s,p)e^{+\frac{s}{\alpha}q_p(z-h)} \tag{3}$$

$$\bar{\psi} = \Psi^+(s,p)e^{-\frac{s}{\alpha}q_s(z-h)} + \Psi^-(s,p)e^{+\frac{s}{\alpha}q_s(z-h)} \tag{4}$$

where $q_p = \sqrt{1 - p^2}$, $q_s = \sqrt{\kappa^2 - p^2}$, and $\kappa = \alpha/\beta$, the ratio between P- and S-wave speeds. Signs are chosen so that these fields vanish at infinity. Thus, Φ^+ and Ψ^+ are the appropriate solutions in the half space $z < h$; Φ^- and Ψ^- must be used in $z > h$.

Particle velocities and stresses can be easily computed from (3) and (4) in the Laplace transformed domain. They are written here for reference in a Haskell matrix notation:

$$
\begin{bmatrix} \dot{u}_x/\alpha \\ \dot{u}_z/\alpha \\ \sigma_{xz}/\mu \\ \sigma_{zz}/\mu \\ \sigma_{xx}/\mu \end{bmatrix} = \frac{s^2}{\alpha^2} \begin{bmatrix} -p & -q_s & -p & q_s \\ -q_p & p & q_p & p \\ 2pq_p & (\kappa^2 - 2p^2) & -2pq_p & (\kappa^2 - 2p^2) \\ (\kappa^2 - 2p^2) & -2pq_s & (\kappa^2 - 2p^2) & 2pq_s \\ (\kappa^2 - 2q_p^2) & 2pq_s & (\kappa^2 - 2q_p^2) & -2pq_s \end{bmatrix} \begin{bmatrix} \Phi^+ e^{-s/\alpha q_p(z-h)} \\ \Psi^+ e^{-s/\alpha q_s(z-h)} \\ \Phi^- e^{+s/\alpha q_p(z-h)} \\ \Psi^- e^{+s/\alpha q_s(z-h)} \end{bmatrix}.
$$

$$(5)$$

3. Radiation from a Point Dislocation

Let us first provide the solution for the simple problem of a point double-couple source buried in an infinite homogeneous elastic medium. In the following section we will use this solution as a Green function to compute radiation from a finite fault. Let us assume that the source is a point double-couple of Moment M_0 and source time function $H(t)$, where H is Heaviside's function. The point source is located at $(x = 0, z = h)$, the dislocation has a dip angle θ and slip is oriented as in Figure 1 so that the fault is a thrust.

We can write the transformed displacement Green function for a point double-couple source in the general form

$$
G_i^\ell(p, z, s | 0, h, 0) = \frac{M_0}{2\mu\kappa^2 s} \mathscr{D}_i^\ell(p) \mathscr{R}^\ell(p, \theta) \frac{e^{-\frac{s}{\alpha} q_\ell(h-z)}}{q_\ell} , \tag{6}
$$

where the notation $G_i^\ell(p, z, s | 0, h, 0)$ indicates that the source is at position $(0, h, 0)$ in space-time, and the indices are i, the component of displacement which takes values x or z, and ℓ, which stands for either P or S waves.

$$
\mathscr{D}_i^\ell(p) = \begin{bmatrix} -p & q_s \\ q_p & p \end{bmatrix} \tag{7}
$$

is the matrix that converts from potential to displacements; and the "plane-wave" radiation patterns $\mathscr{R}^\ell$ are defined by

$$
\begin{aligned}
\mathscr{R}^P(p, \theta) &= (1 - 2p^2) \sin 2\theta + 2pq_p \cos 2\theta \\
\mathscr{R}^S(p, \theta) &= 2pq_s \sin 2\theta - (\kappa^2 - 2p^2) \cos 2\theta .
\end{aligned} \tag{8}
$$

By the same procedure described above for the displacement we can now compute stresses associated with the Green function. Using (5) we get

$$
\sigma_{ij}^\ell(p, z, s) = \frac{M_0}{2\kappa^2\alpha} \mathscr{S}_{ij}^\ell(p) \mathscr{R}^\ell(p, \theta) \frac{e^{-\frac{s}{\alpha} q_\ell(h-z)}}{q_\ell} \tag{9}
$$

where the coefficients $\mathscr{S}^\ell_{ij}$ can be derived from (5):

$$
\begin{bmatrix} \mathscr{S}^\ell_{xz} \\ \mathscr{S}^\ell_{zz} \\ \mathscr{S}^\ell_{xx} \end{bmatrix} = \begin{bmatrix} -2pq_p & (\kappa^2 - 2p^2) \\ (\kappa^2 - 2p^2) & 2pq_s \\ (\kappa^2 - 2q_p^2) & -2pq_s \end{bmatrix}.
$$

4. Radiation from a Finite Fault

We consider as in Figure 1 a rupture front starting from the hypocenter at $(x = 0, z = h)$ and propagating along a fault with dip angle θ. At time $t = 0$, rupture starts moving along the fault with constant speed v, and after a certain time $t = L/v$ it stops, leaving a final slip zone of length L. Slip of the fault is uniform and equal to D. As usual, we can compute the radiation from this finite fault by the superposition of point double-couples of type (6).

Let us introduce the coordinate ξ along this line. Then the source located at a point of coordinate ξ along the line will have coordinates $(x = \xi \cos \theta, z = h - \xi \sin \theta)$ and will be activated at time $t = \xi/v$. It will produce a radiation that we can compute from (6):

$$
G^\ell_i(p, z, s | \xi \cos \theta, h - \xi \sin \theta, \xi/v) = \frac{D}{2\kappa^2 s} \mathscr{D}^\ell_i(p) \mathscr{R}^\ell(p, \theta) \frac{e^{-\frac{s}{\alpha} q_\ell (h - z)}}{q_\ell}
$$
$$
\times e^{-\frac{s}{\alpha}(\Gamma - p \cos \theta - q_\ell \sin \theta)\xi} d\xi \tag{10}
$$

where $\Gamma = \alpha/v$ is the ratio between P and rupture speed. In general, for rupture speeds less than the shear-wave speed, $\Gamma > \kappa$. All the other coefficients are independent of ξ.

The radiation from the finite fault is readily computed in the Laplace domain by integration of (10) for a series of identical sources distributed continuously from $(0 \leq \xi \leq L)$. Carrying out the integration we find that the ground velocity produced by the finite source can be written on the form:

$$
\bar{u}^\ell_i(p, z, s, L) = \bar{u}^\ell_i(p, z, s) - \bar{u}^\ell_i(p, z, s) e^{\frac{s}{\alpha} \chi^\ell(p, \theta)L} \tag{11}
$$

where $\bar{u}^\ell_i$ will be defined below, and the factor

$$
\chi^\ell(p, \theta) = p \cos \theta + q_\ell \sin \theta - \Gamma \tag{12}
$$

is the directivity due to rupture propagation along the fault.

Each term in (11) represents the radiation from a semi-infinite dislocation that moves along the fault plane with speed v. The velocity field produced by this semi-infinite dislocation is given by:

$$
\bar{u}^\ell_i(p, z, s) = \frac{\alpha D}{2\kappa^2 s} \frac{\mathscr{D}^\ell_i(p) \mathscr{R}^\ell(p, \theta)}{\chi^\ell(p, \theta)} \frac{e^{-\frac{s}{\alpha} q_\ell (h - z)}}{q_\ell} \tag{13}
$$

and for the stress field we find a similar expression:

$$\bar{\sigma}_{ij}^{\ell}(p,z,s) = \frac{\mu D}{2\kappa^2 s} \frac{\mathscr{S}_{ij}^{\ell}(p)\mathscr{R}^{\ell}(p,\theta)}{\chi^{\ell}(p,\theta)} \frac{e^{-\frac{s}{v}q_{\ell}(h-z)}}{q_{\ell}} \tag{14}$$

Equations (13) and (14) represent the waves radiated by a rupture front that starts at the hypocenter and propagates along a fault of dip θ. The rupture front propagates at constant speed v without ever stopping. Behind the rupture front slip is constant and equal to D.

Solution (11) is the sum of a positive rupture front of slip D that starts running at speed v from the hypocenter $(0, h)$ at time $t = 0$; and a second dislocation of the opposite sign that is triggered at the instant when the main rupture reaches the end point L of the fault. The appropriate time delay for the second rupture is given by the phase in the exponential term. Computing the effect of finite faults by addition and subtraction of appropriately delayed dislocations is a well-known technique that has been applied to numerous problems, including the computation of finite sources in static and dynamic problems (YOFFE, 1960; COMMINOU and DUNDURS, 1975). This method is exact for uniform slip faults, but it can be used as an approximation for more realistic slip distributions.

4.1 Effect of the Free Surface

The solution we found in the previous section is for a fault line embedded in an infinite medium. The free surface has a very strong effect on this solution; and several authors tried to find approximate expressions for the effect of the surface (BOORE and ZOBACK, 1974; NIELSEN, 1998; OGLESBY et al., 2000a,b). It is very common to assume that the free surface doubles the displacements that would have recorded in a full space. BOUCHON (1998) has given strong support for this approximation. NIELSEN (1998) tried to obtain a similar expression for the only stress component that is not zero on the free surface (σ_{xx}).

Let us consider stress conditions near the free surface. Since both σ_{xz} and σ_{zz} are zero, σ_{xx} can be directly computed from ϵ_{xx} (see also SAVAGE, 1983). On the free surface

$$\epsilon_{zz} = -\frac{\lambda}{\lambda + 2\mu} \epsilon_{xx} \tag{15}$$

so that

$$\sigma_{xx} = \frac{4\mu(\lambda + \mu)}{\lambda + 2\mu} \epsilon_{xx} = \frac{2\mu}{1 - v} \epsilon_{xx} \, , \tag{16}$$

where v is Poisson's modulus. Thus, in general there is little one can say about the horizontal stress on the surface except that the elastic response on the free surface is that of plane stress instead of plane strain.

We can compute the effect of the free surface exactly using the standard seismological method for plane waves. This is a standard problem that need not be repeated here. The effect of the free surface is to multiply the incident fields by a so-called surface response factor. Our final solution on the free surface ($z = 0$) is

$$\bar{u}_i^\ell(p,0,s) = \frac{\alpha D}{2\kappa^2 s} \frac{\mathcal{T}_i^\ell(p)\mathcal{R}^\ell(p,\theta)}{R(p)\chi^\ell(p,\theta)} \frac{e^{-\frac{s}{\alpha}q_\ell h}}{q_\ell} \tag{17}$$

where

$$R(p) = (\kappa^2 - 2p^2)^2 + 4p^2 q_P q_S$$

is Rayleigh's function, and the surface response functions are

$$\mathcal{T}_i^\ell(p) = \begin{bmatrix} -4\kappa^2 q_p q_s p & 2\kappa^2(\kappa^2 - 2p^2)q_s \\ 2\kappa^2(\kappa^2 - 2p^2)q_p & 4\kappa^2 p q_p q_s \end{bmatrix} .$$

Expression (17) is the exact solution to our problem in the Laplace transform domain. It can be verified against similar expressions found by BOUCHON (1978) in the Fourier domain using the transformations $s \rightarrow i\omega$ and $sp/\alpha \rightarrow ik$.

On the free surface the only stress component that is different from zero is σ_{xx} which is given by the relatively complex expression:

$$\bar{\sigma}_{xx}^\ell(p,0,s) = \frac{\mu D}{2\kappa^2 s} \frac{S_{xx}^\ell(p)\mathcal{R}^\ell(p,\theta)}{\chi^\ell(p,\theta)} \frac{e^{-\frac{s}{\alpha}q_\ell h}}{q_\ell} \tag{18}$$

where the surface responses are

$$S_{xx}^P(p) = \frac{16(\kappa^2 - 1)p^2 q_p q_s}{R(p)}$$

$$S_{xx}^S(p) = -\frac{8(\kappa^2 - 1)(\kappa^2 - 2p^2)pq_s}{R(p)} .$$

These factors are complex in the range $1 < p < \kappa$, become infinite at the Rayleigh pole and become constant at infinity. I consider it very unlikely that a simple relation between stresses with and without free surface may be found.

5. Inversion to the Time Domain

The Laplace transformed solutions (17) and (18) may be inverted to the time and space domain using the Cagniard-de Hoop method. Details of this method are relegated to the Appendix. For running dislocations there are a couple of nontrivial problems with the Cagniard-de Hoop method because the denominators of (17) and (18) have poles in the p complex plane. The residues at these poles must be taken into account in the inversion of either displacements or stresses. There are three poles:

1. A pole at the zero of $R(p)$. This pole is located on the real axis at the Rayleigh wave slowness. It only affects numerical computations when the depth of the hypocenter $h \to 0$.

2. A pole at infinity. This is a very frequent problem with the Cagniard-de Hoop method: the transforms for particle velocities for P and S waves diverge at infinity like p, but their difference converges like p^{-1}. For the stress field the divergence of individual waves is $\mathcal{O}(p^2)$ but the difference converges to a constant value (the residual stress). The main problem created by the pole at infinity is that computations of velocities and stresses for long times become unstable. Convergence can be improved using high precision arithmetic.

3. A pole due to zeroes of the directivity factor, χ^{ℓ}, in the denominator. This is the most important pole because it is associated with the jump in displacement across the fault. It produces the so-called rupture front wave (see AKI and RICHARDS, 1980). Fortunately, as shown by MADARIAGA (1980), the contribution of the pole at the free surface cancels out if the fault is of finite length L such that it does not break the free surface. If the fault breaks the free surface, both velocities and stresses on the free surface become singular at the exit point of the fault, as will be shown later. For practical numerical computations with a finite sampling interval, the free surface contribution can be computed by letting the fault stop just short of the free surface.

Using the results of the Appendix, we get for the P wave the exact result

$$\dot{u}_i^p(x, 0, t) = \frac{\alpha D}{2\pi\kappa^2 r} \Re\left[\frac{\mathcal{T}_i^p(p(t, \phi)) \mathcal{R}^p(p(t, \phi), \theta)}{R(p(t, \phi)) \chi(p(t, \phi), \theta)} \right] \frac{H(t - r/\alpha)}{\sqrt{\tau^2 - 1}} \tag{19}$$

and

$$\dot{u}_i^s(x, 0, t) = \frac{\alpha D}{2\pi\kappa^2 r} \Re\left[\frac{\mathcal{T}_i^s(p(t, \phi)) \mathcal{R}^p(p(t, \phi), \theta)}{R(p(t, \phi)) \chi(p(t, \phi), \theta)} \frac{1}{\sqrt{\tau^2 - \kappa^2}} \right] H(t - t_{sp}) \tag{20}$$

for S waves. $\Re$ denotes the real part. Here $t_{sp} = \min[r/\beta, |x|/\alpha + \sqrt{1 - \kappa^{-2}} h/\beta]$ is the arrival time of the SP wave, the S wave diffracted by the free surface (see BOUCHON, 1978 for a full discussion of the role of this wave in shallow dip-slip faults). Similar expression can be found for the σ_{xx} component of stress. Equations (19) and (20) are the complete solution for a semi-infinite dislocation moving at constant speed along a fault of dip θ. They can be easily computed by any computer algebra package and compared with the same solution when there is no free surface.

A very important property of these solutions is that particle velocity is a homogeneous function of order -1 of the depth h. This means that we can rewrite (19) in the compact form:

$$\dot{u}_i^{\ell}(x, 0, t) = \frac{\alpha D}{h} V_i^{\ell}\left(\frac{x}{h}, \theta, \frac{\alpha t}{h}\right) \tag{21}$$

where the function $V_i^{\ell}(x/h, \theta, \alpha t/h)$ is homogeneous of order 0; therefore, if we multiply depth h by a certain constant, and the time t and the position x change proportionally, the velocity increases by a factor of $1/h$. This is a general property of self-similar solutions to the elastic wave equation. It is perhaps important to remark that only self-similar problems can be integrated exactly by the method of Cagniard.

Now we can understand what happens with the radiation from the two ends of the fault. We write the complete time-space solution (11) for the displacement of the free surface due to a fault of length $L < h/\cos\theta$ in the compact form

$$\ddot{u}_i^\ell(x,0,t,L) = \alpha D \left[\frac{1}{h} V_i^\ell\left(\frac{x}{h},\theta,\frac{\alpha t}{h}\right) - \frac{1}{h_L} V_i^\ell\left(\frac{x-L\cos\theta}{h_L},\theta,\frac{\alpha(t-L/v)}{h_L}\right) \right] \qquad (22)$$

where $h_L = h - L\sin\theta$ is the depth of the second source (see Fig. 2). Thus, the stopping phase due radiation from rupture arrest at L is simply an appropriately scaled and delayed version of the radiation from the starting phase.

In order to show the properties of the velocity field we study a simple model. Since all quantities in (19), (20) and (22) are scaled by depth h, slip D and the P-wave speed α, we choose $h = 1, \alpha = 1$, and $D = 1$. We model a fault of dip $\theta = 30°$ and rupture speed $\Gamma = v/\alpha = 0.4$. The fault length is $L = 1.9$ so that rupture stops just below the free surface at depth $h_L = 0.05$. We choose a finite depth for the end of the fault in order to avoid problems with the singularities due to the Rayleigh wave and the rupture front. The exit point of the fault is $x = 1.732$.

In Figure 3 we show a seismogram section for the horizontal particle velocity produced by the starting phase (the first term in 22). We plot seismograms from $x = 0$ to $x = 3$ every $\Delta x = 0.05$. Figure 4 shows the second term in (22). This represents the contribution from the stopping phase at $L = 1.9$.

In Figure 3 we can easily identify the P and S waves emitted by the initiation of rupture: these are the hyperbolic-shaped arrivals that arrive at the free surface at

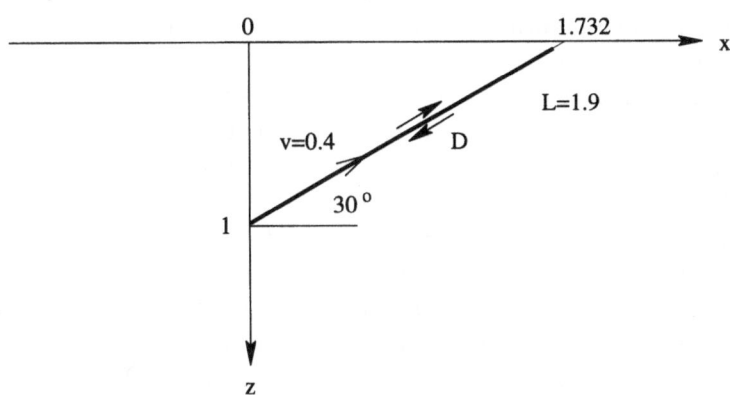

Figure 2
Model used for the simulations. A shallow dipping fault of angle $\theta = 30°$ and length L $= 1.9$. Rupture start from the hypocenter at time $t = 0$ and propagates at speed $\Gamma = v/\alpha = 0.4$.

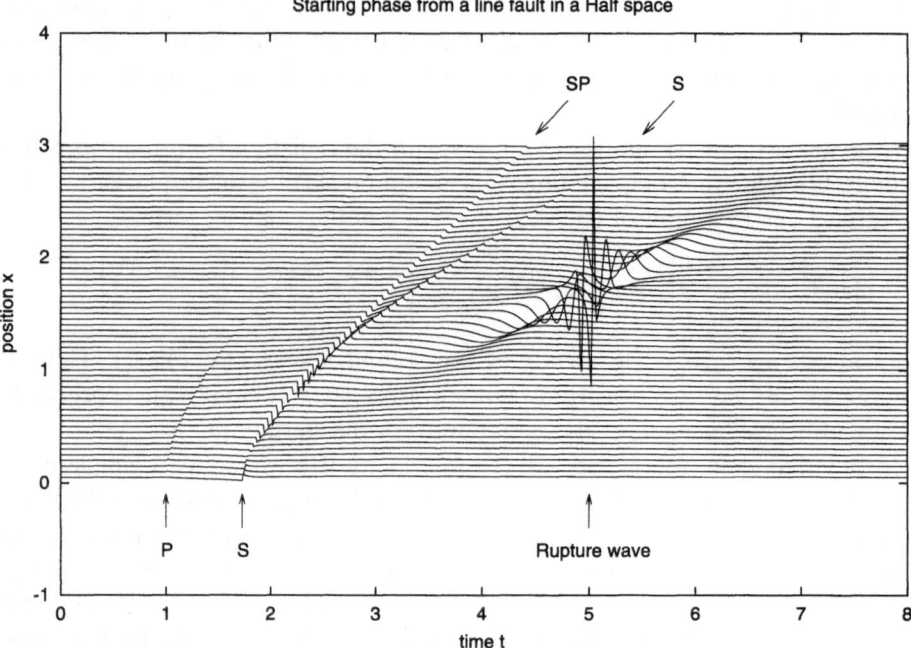

Figure 3

Velocity field produced by the shallow dip-slip dislocation model of Figure 2. This figure shows the starting phases due to the initiation of rupture and its propagation at constant speed along the fault in Figure 2.

point $x = 0$ at times $t = 1$ and $t = 1.73$, respectively. SP waves are also clearly identified in this figure, as the straight lined arrival that splits off the S wave at point $x = 0.707$ and time $t_{sp} = 2.122$. The critical distance for the SP wave can be easily computed from the Vp/Vs ratio κ. The most important arrival in this Figure is the strong rupture phase that appears as the large waves near $x = 1.732$, $t = 2/0.4 = 5$. This is the very well-known rupture front wave associated to the propagating dislocation. The properties of these waves in infinite elastic media were extensively discussed by BOORE and ZOBACK (1974) and AKI and RICHARDS (1980, Chapter 14). We shall study these waves in more detail in a later section.

Figure 4 shows the stopping phases emitted when the dislocation stops at $h_L = 0.9$. These waves were computed using only the second term in (22) except for the minus sign. Although this figure looks very different from the previous one, it is essentially the same as Figure 3 but scaled as indicated in (22). Signals are larger and it is considerably easier to identify the arrivals of P and S waves which decrease slowly as they move away from the tip of the fault. In the forward direction we see a large Rayleigh wave that propagates without geometrical spreading. Behind the Rayleigh wave in the forward direction, we can see the stopping phase of the rupture front wave. This wave has exactly the same shape, phase and arrival time as that shown in Figure 3. As will be discussed in a later section, when we subtract these two

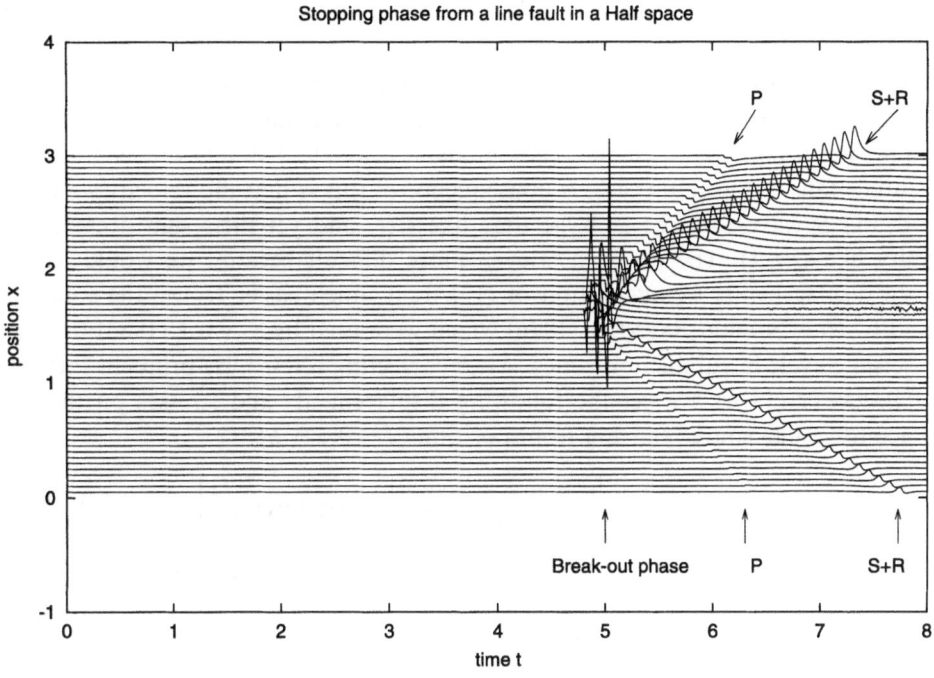

Figure 4

Velocity field produced by the shallow dip-slip dislocation model of Figure 2. This figure shows the stopping (or break-out) phases emitted when the rupture reaches the free surface in Figure 2.

figures to produce the radiation from a finite buried fault, the rupture wave disappears in the forward direction. Some noise is observed in the later part of the seismograms at $x = 1.75$. This is due to the poor numerical canceling of the contribution from P and S waves. These two waves diverge as a function of time, although their sum converges with time as t^{-1}.

As we mentioned earlier, if the rupture reaches the free surface, i.e., if $h_L \to 0$, (19) and (20) can still be used but a number of precautions have to be taken in order to avoid singularities. The Cagniard contour become flat and wraps around the singularities of the Rayleigh wave, the rupture front and the pole at infinity. In this case it is necessary to carefully smooth computations near each of these singularities. In particular, the Cagniard-de Hoop solutions become singular and blow up at $x = 1.73$, this is reasonable since particle velocity is infinite at the point where the fault breaks the surface. The worst instability that occurs when $h_L \to 0$ is due to the poor estimation of the static solution. The small oscillations observed in Figure 4 at $x = 1.75$ become large and the long-time computation blows up. The static solution can be obtained with great difficulty from the Cagniard-de Hoop expressions, however it will be inaccurate and cannot be integrated to displacement without numerical smoothing.

6. Static Displacement and Seismic Moment

As we just showed, retrieving the static solution from the Cagniard contour solutions is a very difficult exercise. The reason is that the solution (17) has been written in terms of separate P and S potentials. It is well known that the separation in potentials fails in the static case. A consequence of this is that the velocity solutions for P and S waves diverge at large time like $O(t)$ and yet their sum decreases at infinity like $O(t^{-1})$. The problem is even more difficult because we would like to obtain displacements by integration of (17). It is much easier to use the solutions found by FREUND and BARNETT (1976) which were used by SAVAGE (1983) to study the seismic cycle in subduction zones. After correcting a few miss-prints in FREUND and BARNETT (1976), we get the following displacements and stresses on the free surface for a line source starting from position $(x = 0, z = h)$

$$u_x(x,0) = \frac{D}{\pi}\left[\frac{h(h\sin\theta + x\cos\theta)}{r^2} - \cos\theta\arctan\frac{x}{h}\right]$$

$$u_z(x,0) = \frac{D}{\pi}\left[\frac{h(h\cos\theta - x\sin\theta)}{r^2} - \sin\theta\arctan\frac{x}{h}\right] \qquad (23)$$

$$\sigma_{xx}(x,0) = -\frac{4\mu D}{\pi(1-v)}\frac{hx(h\sin\theta + x\cos\theta)}{r^4}$$

where $r = \sqrt{x^2 + h^2}$ is the distance from the hypocenter to the observation point at $(x, 0)$. The other two components of stress are zero on the free surface.

Once again the displacements in (23) are homogeneous of order 0, i.e., we can write them in the general form

$$u_i(x,0) = DU_i\left(\frac{x}{h},\theta\right) \ .$$

This solution is self-similar, so that the distribution of displacement along the surface is stretched or compressed depending on the value of h, the depth of the hypocenter. Strictly, solution (23) was computed by Freund and Barnett, for a fault that extends from the hypocenter downward to infinity. They have been used by many authors for a fault that breaks the surface using a particular restriction regarding how the angle $\arctan x/h$ is computed. While this is correct for the computation of displacements, it is inaccurate for stresses as, we show next.

In order to properly compute the displacement and stresses produced by a fault of dip θ and finite length L, we subtract a second dislocation located at the end of the fault $(x = L\cos\theta, z = h_L)$, where $h_L = h - L\cos\theta$:

$$u_x(x,0,L) = D\left[U_x\left(\frac{x}{h},\theta\right) - U_x\left(\frac{x - L\cos\theta}{h_L},\theta\right)\right] \ . \qquad (24)$$

Because of self-similarity, the effect of the second term is simply a scaled version of the first. We can now see what happens when the fault elongates until it breaks the free surface, i.e., when $h_L \to 0$. In this case the second term becomes increasingly concentrated in a domain of width h_L around the tip of the fault. When h_L is strictly zero the displacement $U_x((x - L\cos\theta)/h_L) \to \cos\theta\, H(x - L/\cos\theta)$. This represents a jump of amplitude $\cos\theta$ at the tip of the fault. This term is just the displacement discontinuity at the tip of the surface breaking fault. In previous uses of the FREUND and BARNETT (1976) solution, the second term in (24) has not been explicitly computed because it can be integrated into the definition of the branch cut for the calculation of arctan. For instance, by defining the range $-\pi + \theta < \arctan x/h < \theta$. These two procedures are equivalent but I think that the formulation (24) is clearer.

The stress field near the free surface is computed using the same procedure. We consider first a finite fault of length L and dip θ, then stress field on the free surface $z = 0$ is:

$$\sigma_{xx}(x, 0, L) = \frac{4\mu D}{\pi(1 - v)} \left[\frac{1}{h} S\left(\frac{x}{h}, \theta\right) - \frac{1}{h_L} S\left(\frac{x - L\cos\theta}{h_L}, \theta\right) \right] \tag{25}$$

where S is the non-dimensional function

$$S(u, \theta) = -\frac{u(\sin\theta + u\cos\theta)}{(u^2 + 1)^2} \; . \tag{26}$$

The last term in (25) was not included in the expression for ϵ_{xx} by SAVAGE (1983) because he used the stress field for an infinite fault. If the fault is finite of length L, the second term cannot be neglected. If finite fault breaks the free surface, the second term in (25) becomes zero except near the crack tip, where it is singular. Its value can be computed if we consider stress near the fault tip as a distribution which is the limit of a series of regular functions that depend on the single parameter h_L. We find that

$$\lim_{h_l \to 0} \frac{1}{h_L} S\left(\frac{x - L\cos\theta}{h_L}, \theta\right) = -\frac{\pi}{2}\cos\theta\,\delta(x - L\cos\theta) \; . \tag{27}$$

The amplitude of the delta function can be found by integration of (25) from $-\infty$ to ∞ and then letting $h_L \to 0$. Inserting this term into (25), we observe that there is a point force located exactly at the tip of the fault. The strength of the point force is proportional to the jump in displacement across the fault D projected along the surface, multiplied by the plane stress elastic constant $2\mu/(1 - v)$ (see 16).

Let us now examine the moment tensor representation of a surface-breaking fault. From a straightforward analysis of the way FREUND and BARNETT (1976) found the static solution for a semi-infinite fault, we find, as shown on Figure 5, that displacements and stresses (25) and (24) are due to the combined effect of a finite fault in the lower half space, plus an image fault in the upper virtual half space, and a correction term on the surface. When the fault breaks the surface, as in our example, the set of equivalent forces is that shown in Figure 5. These forces are obviously in equilibrium of force and moment as they should be, otherwise our solution would be mechanically incorrect. Thus the stress singularity of σ_{xx} is crucial to understand the

Equivalent forces for a fault that breaks the free surface

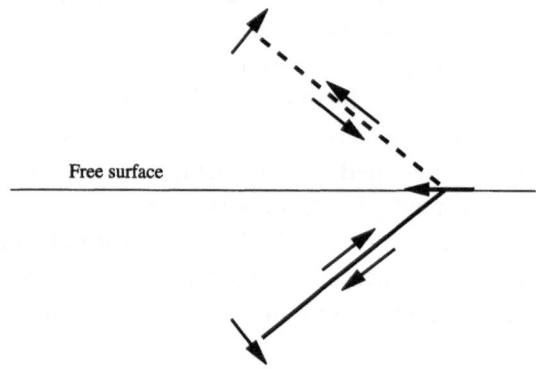

Free surface

Figure 5
Equivalent forces for a dip-slip fault that breaks the surface.

seismic moment of a surface-breaking fault. Brune (personal communication, 2000) noted that most authors who use equations like (23) to compute the seismic moment have not considered the equilibrium of forces. He was right, most authors did consider the deformation of the fault tip, assuming that since slip is discontinuous there is no force across the fault tip. Most centroid seismic moments are computed using displacement and stress eigenfunctions that satisfy the free surface boundary conditions by construction. Consequently, there is no error in the computation of centroid moment tensors. Errors may appear, however, in the estimation of seismic moments using body waves because in this case the free surface is simulated using reflection coefficients for surface reflected phases. These coefficients fail for very shallow faults; in particular the contribution of the surface-breaking phase is very likely to be in error.

7. The Rupture Pulse around a Propagating Dislocation Front

The effect of the rupture front in the radiation from a dip-slip fault is entirely included in the full solution (22). The rupture front is part of the solution and it occurs whenever the directivity factor $\xi(p(t), \theta) \to 0$. AKI and RICHARDS (1980, chapter 14) studied the rupture front, analyzing the complete solution. It is actually easier to compute the rupture front wave directly from the transformed solution (17).

We notice that (17) has complex poles located at the zeroes of the function $\chi^\ell(p, \theta)$ defined by (12), i.e., at

$$p\cos\theta + q_\ell\sin\theta = \Gamma \ .$$

Solving for p, we find that the the relevant pole is located at

$$p_p = \Gamma \cos \theta + i\sqrt{\Gamma^2 - c_\ell^2} \sin \theta \tag{28}$$

in the first quadrant of the complex p plane for $0 < \theta < \pi/2$. In the appropriate Riemmann sheet, we get

$$q_\ell = \Gamma \sin \theta - i\sqrt{\Gamma^2 - c_\ell^2} \cos \theta$$

and the derivative

$$\frac{\partial \chi(p)}{\partial p} = -\frac{i\sqrt{\Gamma^2 - c_\ell^2}}{q_\ell} .$$

The contribution of the rupture front is computed using the residue at the pole p_p in (17). After some computations we get

$$\bar{u}_i^\ell(x, 0, s) = \frac{\alpha D}{2\kappa^2} \Re \left[i \frac{\mathcal{T}_i^\ell(p_p)\mathcal{R}^\ell(p_p, \theta)}{R(p_p)q_\ell \partial \chi^\ell(p_p, \theta)/\partial p} e^{-\frac{s}{\alpha}[p_p x + q_\ell h]} \right] \tag{29}$$

We can transform this expression to the time domain since all the terms are independent of s except the exponential which can be rewritten as

$$e^{-\frac{s}{\alpha}[p_p x + q_\ell h]} = e^{-\frac{s}{\alpha}[\Gamma \xi - i\eta\sqrt{\Gamma^2 - c_\ell^2}]}$$

where ξ and η are the coordinates along the fault and perpendicular to it as shown in Figure 6. We find

$$\bar{u}_i^\ell(x, 0, t) = \frac{\alpha D}{4\pi\kappa^2} \Re \left[\frac{\mathcal{T}_i^\ell(p_p)\mathcal{R}^\ell(p_p, \theta)}{R(p_p)\sqrt{\Gamma^2 - c_\ell^2} \frac{1}{t - \xi/v + i\eta/\alpha\sqrt{\Gamma^2 - c_\ell^2}}} \right] . \tag{30}$$

This is very similar to the rupture wave due to strike-slip faults computed by BOORE and ZOBACK (1974), except that the amplitude is affected by the surface response factor $\mathcal{T}_i^\ell/R(p)$.

After some tedious computations we can show that the radiation patterns become very simple:

$$\begin{aligned} \mathcal{R}^p(p_p, \theta) &= -2i\Gamma\sqrt{\Gamma^2 - 1} \\ \mathcal{R}^s(p_p, \theta) &= (2\Gamma^2 - 1)\kappa^2 . \end{aligned} \tag{31}$$

Expression (30) is the near-field wave produced by a rupture front that starts running at constant speed v from the hypocenter at $(x = 0, z = h)$. In order to compute the radiation from a finite fault that stops at a depth $h_L \to 0$, we have to subtract to expressions like (30) another one that is appropriately retarded in time and space as in (22). It turns out that the rupture front wave and the stopping phase cancel exactly

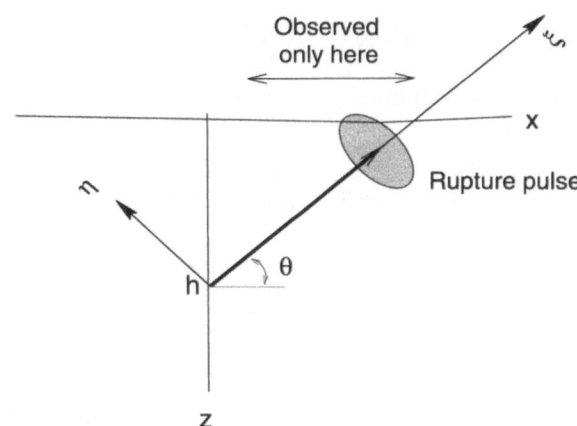

Figure 6
The rupture pulse and the area where it is observed around a dip-slip fault in a half space.

for $\xi > L$, so that the rupture waves can only be observed in the range $0 < \xi < L$. This is quite logical since the rupture wave should only occur in the areas of the fault where rupture propagates. Thus, rupture waves are only observable in the hanging wall; on the foot wall, as seen from Figure 6, $\xi > L$ and there is no rupture front wave.

8. Velocity Field on the Free Surface Due to a Finite Fault

We can now study the velocity field produced by a shallow reverse fault. We consider again the geometry shown in Figure 2. Since these computations are all self-similar, we assumed without loss of generality that depth $h = 1$, P-wave speed $\alpha = 1$ and S-wave speed $\beta = 1/\sqrt{3}$ so that $\kappa = \sqrt{3}$. We also assumed that $v = 0.4\alpha$ and $\theta = 30°$. We assume that rupture continued all the way to the free surface so that $L = h/\sin\theta$ and $h_L = 0$. The tip of the fault is at $x = h/\cos\theta$. Slip the rupture front is a step function.

Figure 7 depicts the horizontal component of particle velocities computed using equations (19) and (20). Figure 8 shows the corresponding vertical particle velocities. We observe the presence of both P and S waves radiated from the hypocenter; these are the weak waves that we identify as P0 and S0 in the figures. We also observe a weak surface converted phase SP0 in the forward direction. The seismic waves emitted when the rupture breaks the free surface are labeled P1, S1 and R1. We observe a clear Rayleigh wave in the forward direction just behind the S1 wave. Since the Rayleigh wave speed is very close to that of shear waves it is difficult to separate the two types of wave.

The most important event in both seismic sections is the large rupture front waves observed on the hanging wall from $0.5 < x < 1.73$. Initially at distances less than $x = 1$, the rupture front wave is just a subtle long wave oscillation, however it becomes sharper and stronger as it nears the break point at $x = 1.73$. This is the

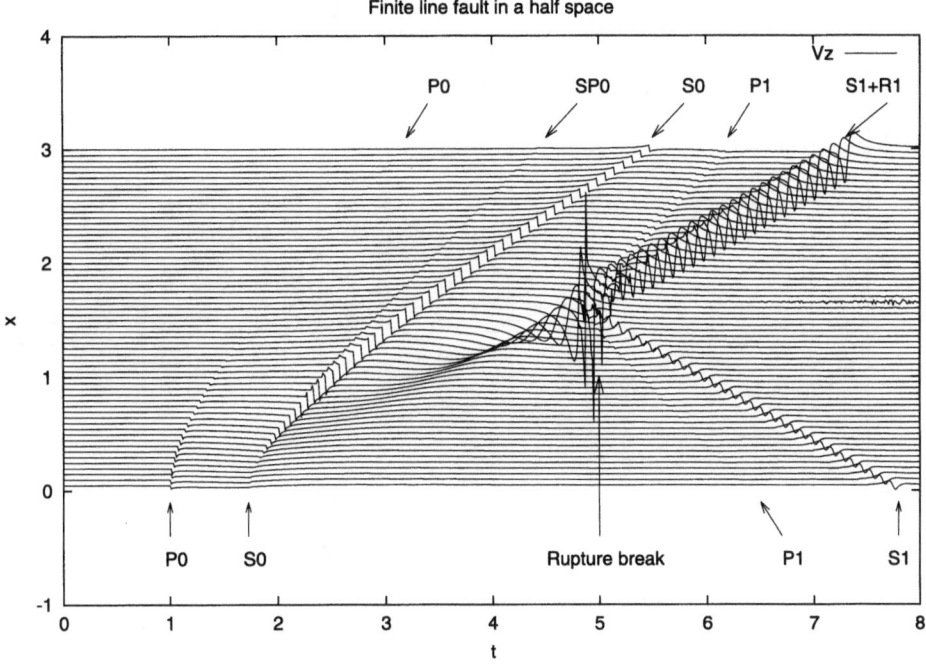

Figure 7

Vertical component of the particle velocity due to a finite fault that stops right below the surface. The arrows identify the phases radiated by the start of rupture (P0, SP0, S0) and from the stopping phase (P1, S1 and R1).

expected behavior of the rupture front waves that are inhomogeneous waves emitted by the rupture front. It is very clear from Figures 7 and 8 that the rupture front wave disappears suddenly for distances beyond the break out point at $x = 1.73$. The reason is that when the rupture front wave reaches the free surface it is exactly canceled by the stopping phase. Thus, there is no rupture front wave in the foot wall. As a consequence, rupture front waves can only be observed above the rupture front. This is the reason why particle velocities are markedly stronger on the hanging wall than in the foot wall. The motion on the foot wall is dominated by the waves radiated from the hypocenter and by the P, S and Rayleigh waves emitted when the rupture breaks the surface (P1, PS1, S1 and R1 in Figures 7 and 8). Of these waves, the most important are the Rayleigh waves radiated in the forward direction as is clearly visible in Figure 4.

It is interesting to compare the result we just obtained for a surface breaking shallow dip-slip fault with a similar solution for a finite dip-slip fault embedded in a homogeneous elastic medium. Particle velocities in this case are given by the same expression as those in a half space, except that the free surface responses $\mathscr{T}^{\ell}_i / R(p)$ are replaced by $\mathscr{D}^{\ell}_i$. Figure 9 displays the horizontal component of particle velocities for a finite fault buried in a full space and Figure 10 shows the corresponding vertical

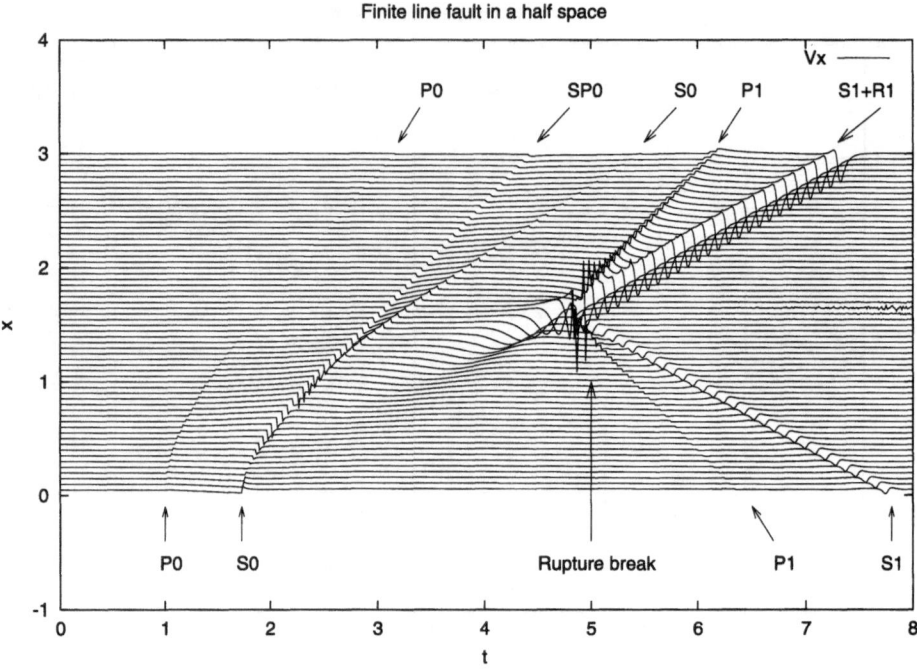

Figure 8
Horizontal component of the particle velocity due to a finite fault that stops right below the surface.
The arrows identify the phases radiated by the start of rupture (P0, SP0, S0) and from the stopping phase
(P1, S1 and R1).

particle velocities. Numerical computations for a full space are considerably more
stable than for a half space at long times. The most obvious difference between
figures for half space and full space is the absence of Rayleigh waves in the latter
case, as can be clearly observed comparing, for instance, Figures 8 and 10. Amplitude
scale in all these Figures is the same so that, as shown by BOUCHON (1978), the
amplitudes on the surface of a half space are about the double of those computed in a
full space. This is an excellent approximation for the P and S waves emitted by the
hypocenter (P0 and S0 in Figs. 7–10). On the other hand, the stopping phases
radiated from the surface-breaking event are completely different in a half and a full
space. In a half space the wavefield is dominated by the Rayleigh waves emitted in the
forward direction. There are also significant differences in the rupture front waves
observed clearly in the hanging wall of the fault for a half space, and which are much
weaker in the case of a full space. Although the amplitude ratio still approaches 2, the
frequency content and the waveforms are very different. Consequently the effect of
the free surface can only be approximated by doubling the amplitude far from the
hanging wall. Near the fault break out, on the other hand, there seems to be no short
cut to accurately computing the full seismograms, including Rayleigh and rupture
front waves.

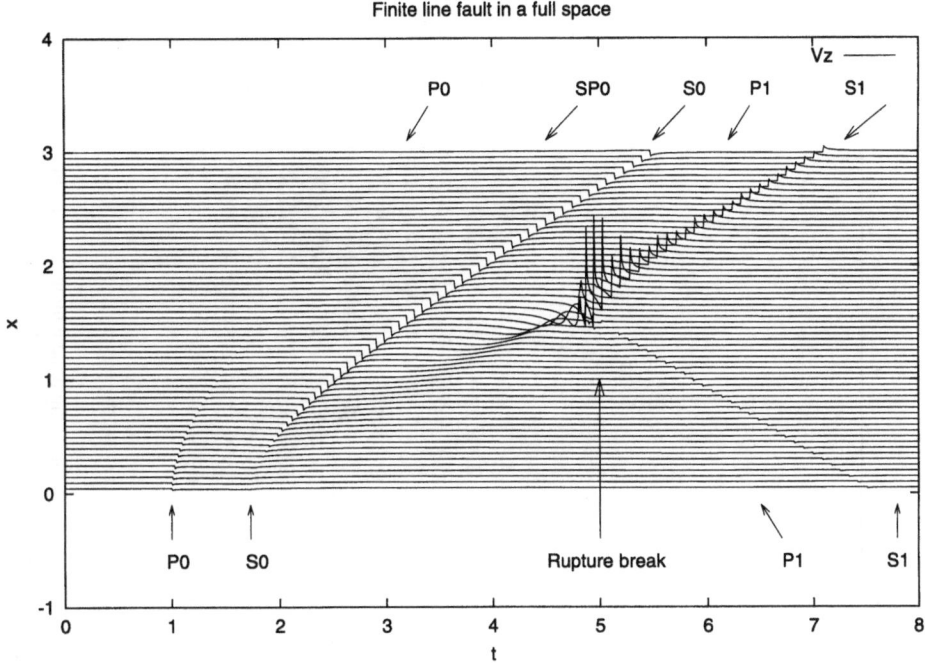

Figure 9

Vertical component of the particle velocity due to a finite fault that stops right below the surface. The arrows identify the phases radiated by the start of rupture (P0, SP0, S0) and from the stopping phase (P1, S1).

9. Conclusions

The effect of the free surface on the field created by a rupture front moving along a shallow dip-slip fault can be precisely computed using the Cagniard-de Hoop method. Numerical computations pose a number of problems that we showed were directly related to the singularities of the solution. It seems curious that not all the properties of this simple fault model that was solved in the late 1970s were fully understood. In part this is due to the fact that, although simple looking, the solution (19, 20) has several singularities that have to be carefully studied and smoothed in numerical computations. The most visible singularity is that due to the propagation of the rupture front that produces very large waves in the hanging wall. The other one is the so-called pole at infinity for P and S potentials: the synthetic seismograms for separate P and S waves diverge but the sum converges. This produces inaccuracies in the seismograms computed for long time delays. Careful numerical computations are required to remove these unwanted oscillations. Finally, the Rayleigh waves generated by the break-out of the rupture at the surface are singular on the free surface and, in the two-dimensional model studied here, they do not decay with distance. The only possible

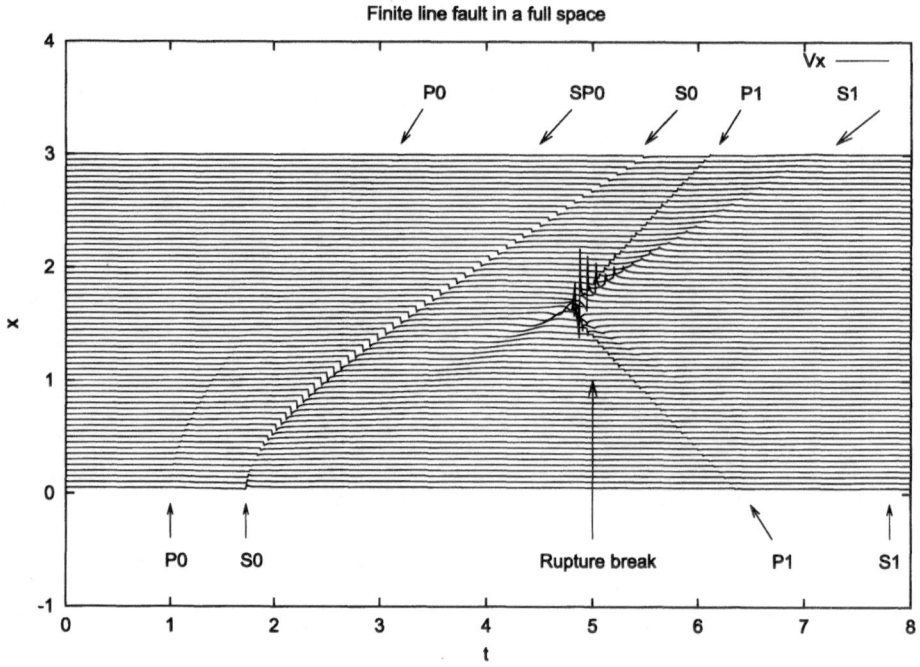

Figure 10
Horizontal component of the particle velocity due to a finite fault that stops right below the surface. The arrows identify the phases radiated by the start of rupture (P0, SP0, S0) and from the stopping phase (P1, S1).

way to write programs that can compute both static and dynamic solutions is to remove or smooth the effect of the singularities, otherwise the computations become unstable and inaccurate. We hope that the results presented in this paper can facilitate the regularization of spectral methods like that of BOUCHON and AKI (1977); they can also be used to test more sophisticated numerical computations by finite differences, finite elements or boundary integral equations. Our results demonstrate that when the rupture breaks the surface the solutions near the break-out point become singular so that the sum of forces and moments remain in equilibrium. Finally, we showed that the rupture front wave—the strong distur-bance observed in the vicinity of the rupture front—can only be observed in the hanging wall. On the foot wall it is exactly canceled by the stopping phase emitted when rupture stops at the free surface.

Acknowledgments

I thank K. Aki for his continuous support and encouragement spanning more than 30 years of research. Much of this work was inspired by a seminar given by J. Brune at the Institute of Crustal Studies of the University of California at Santa

Barbara, in July 2000. I thank Dr. Ralph Archuleta for calling my attention to the paper by FREUND and BARNETT (1976). I am most indebted to Dr. Taku Tada who checked the mathematics and found numerous errors. This work began when the author visited ISC during the summer of 2000. Dr. T. Yamashita and Y. Ben-Zion provided very helpful comments. A program to compute synthetic seismograms using the Cagniard-de Hoop method is available on request. I thank the support by CNRS (Centre National de la Recherche Scientifique) under contract 99PNRN13AS of the Programme National de Risques Naturels.

Appendix

Inversion of the Transforms

For a source located at depth $z = h$, we write the expression (17) in the general form

$$\dot{u}_i^{\ell}(p,z,s) = \frac{\alpha}{2s} \frac{\mathscr{A}_i^{\ell}(p,\theta)}{q_{\ell}} e^{-\frac{s}{\alpha}q_{\ell}h} \tag{32}$$

where $\mathscr{A}_i^{\ell}(p,\theta)$ is a function of the horizontal slowness p and the dip θ. The right-hand side of (32) is analytic in p, so that the inverse Laplace transform in the space domain (2) can be written as

$$\dot{u}_i^{\ell}(x,z,s) = \frac{1}{2\pi} \Re \left[\int_0^{i\infty} \frac{\mathscr{A}_i^{\ell}(p,\theta)}{iq_{\ell}} e^{-\frac{sr}{\alpha}(p\cos\phi + q_{\ell}\sin\phi)} dp \right] \tag{33}$$

where following Figure 1, we used $x = r\cos\phi$ and $h = r\sin\phi$, with $r = \sqrt{x^2 + h^2}$. Expression (33) is a homogeneous function of space and time that can be inverted using the Cagniard-de Hoop method, see, e.g., AKI and RICHARDS (1989) for more details.

As usual we deform the contour of integration from the Imaginary axis to the Cagniard contour defined by the complex valued solution of the equation

$$\tau = p\cos\phi + q_{\ell}\sin\phi \tag{34}$$

where τ is real parameter.

Solving (34) for p we find the Cagniard contour

$$\begin{aligned} p_{\ell} &= \tau\cos\phi + i\sqrt{\tau^2 - c_{\ell}^2}\sin\phi \\ q_{\ell} &= \tau\sin\phi - i\sqrt{\tau^2 - c_{\ell}^2}\cos\phi \ . \end{aligned} \tag{35}$$

For P waves we only take into account the part of the contour for which $\tau > c_p = 1$, while for S waves the contour contains a segment of the real axis in addition to the complex contour. Along the real axis we take $\sqrt{\tau^2 - \kappa^2} = i\sqrt{\kappa^2 - \tau^2}$ so that

$$
\begin{aligned}
p_\ell &= \tau \cos \phi - \sqrt{\kappa^2 - \tau^2} \sin \phi \\
q_\ell &= \tau \sin \phi + \sqrt{\kappa^2 - \tau^2} \cos \phi \ .
\end{aligned}
\tag{36}
$$

The Jacobian for the transformation of the integral from the $\Im(p)$ line to the Cagniard contour is

$$
\frac{dp}{d\tau} = i \frac{q_\ell}{\sqrt{\tau^2 - c_\ell^2}} \ .
$$

We discuss in the following the time-domain transformation for S waves. For P waves the transform is simpler because there is no headwave. Changing the integration path in (33) to the Cagniard contour we get

$$
\dot{u}_i^S(x, z, s) = \frac{1}{2\pi} \Re \left[\int_{\tau_{sp}}^{\infty} \frac{\mathscr{A}_i^S(p, \theta)}{\sqrt{\tau^2 - \kappa^2}} e^{-\frac{sr}{\alpha}\tau} \, d\tau \right]
\tag{37}
$$

where $\tau_{sp} = \min[\kappa, |\cos \phi| + \sqrt{\kappa^2 - 1} \sin \theta]$ is the arrival time of the PS wave when it precedes the S wave, or the arrival time of the S wave κ if not.

Computing the inverse Laplace transform in time we get

$$
\dot{u}_i^S(x, 0, t) = \frac{1}{2\pi} \int_1^{\infty} \Re \left[\frac{\mathscr{A}_i^S(p, \theta)}{\sqrt{\tau^2 - \kappa^2}} \right] \delta\left(t - \frac{r}{\alpha}\tau \right) d\tau
\tag{38}
$$

for $t > r/\alpha$ and 0 otherwise. So that, finally:

$$
\dot{u}_i^S(x, 0, t) = \frac{\alpha}{2\pi r} \Re \left[\frac{\mathscr{A}_i^S(p(t, \theta))}{\sqrt{\tau^2 - \kappa^2}} \right] H(t - t_{sp}) \ .
\tag{39}
$$

This is simply the 2-D Green's function for a double-couple source of any orientation. The same expression is valid for stresses. We must simply replace $\mathscr{A}_i^\ell$ by the appropriate expression extracted from (5).

For the simplicity of numerical computations it is convenient to change variables to $\tau = c_\ell \cosh u$, then $\sqrt{\tau^2 - c_\ell^2} = c_\ell \sinh u$ and all Cagniard contour calculations become very simple.

REFERENCES

AKI, K. (1968), *Seismic Displacements Near a Fault*, J. Geophys. Res. *73*, 5359–5376.

AKI, K. and RICHARDS, P. G., *Quantitative Seismology: Theory and Methods*, vol. II (W.H. Freeman and Co. San Francisco. 1980).

ARCHULETA, R. (1984), *A Faulting Model for the 1979 Imperial Valley Earthquake*, J. Geophys. Res. *89*, 4559–4585.

BOORE, D. M. and ZOBACK, M. D. (1974), *Near-field Motions from Kinematic Models of Propagating Faults*, Bull. Seismol. Soc. Am. *64*, 555–570.

BOUCHON, M. and AKI, K. (1977), *Discrete Wave Number Representation of Seismic-source Wavefields*, Bull. Seismol. Soc. Am. *67*, 259–277.

BOUCHON, M. (1978), *The Importance of the Surface or Interface P Wave in Near-Earthquake Studies*, Bull. Seismol. Soc. Am. *68*, 1293–1311.

COHEE, B. and BEROZA, G. (1994), *Slip Distribution of the 1992 Landers Earthquake and its Implications for Earthquake Source Mechanics*, Bull. Seismol. Soc. Am. *84*, 692–712.

COMMINOU, M. and DUNDURS, J. (1975), *The Angular Dislocation in a Half Space*, J. Elasticity *5*, 203–216.

COTTON, F. and CAMPILLO, M. (1995), *Frequency Domain Inversion of Strong Motions: Application to the 1992 Landers Earthquake*, J. Geophys. Res. *100*, 3961–3975.

CROUCH, S. L. (1976), *Solution of Plane Elasticity Problems by the Displacement Discontinuity Method*, Int. J. Numer. Meth. Engin. *10*, 301–343.

FREUND, L. B. and BARNETT, D. M. (1976), *A Two-dimensional Analysis of Surface Deformation due to Dip-slip Faulting*, Bull. Seismol. Soc. Am. *66*, 667–675.

MADARIAGA, R. (1980), *A Finite Two-dimensional Kinematic Fault in a Half Space*, Publ. Inst. Geophys. Pol. Acad. Sc. *A-10*, 33–47.

NIAZI, A. (1975), *An Exact Solution for a Finite, Two-dimensional Moving Dislocation in an Elastic Half Space with Applications to the San Fernando Earthquake of 1971*, Bull. Seismol. Soc. Am. *65*, 1797–1826.

NIELSEN, S. B. (1998), *Free Surface Effects on the Propagation of Dynamic Rupture*, Geophys. Res. Lett. *25*, 125–128.

OGLESBY, D. D., ARCHULETA, R. J., and NIELSEN, S. B. (1998), *Earthquakes on Dipping Faults; the Effects of Broken Symmetry*, Science *280*, 1055–1059.

OGLESBY, D. D., ARCHULETA, R. J., and NIELSEN, S. B. (2000a), *The Three-dimensional Dynamics of Dipping Faults*, Bull. Seismol. Soc. Am. *90*, 13,643–13,653.

OGLESBY, D. D., ARCHULETA, R. J., and NIELSEN, S. B. (2000b), *Dynamics of Dip-slip Faulting; Explorations in Two Dimensions*, J. Geophys. Res. *105*, 13,643–13,653.

OLSEN, K., MADARIAGA, R., and ARCHULETA, R. (1997), *Three-dimensional Dynamic Simulation of the 1992 Landers Earthquake*, Science *278*, 834–838.

PEYRAT, S., OLSEN, K., and MADARIAGA, R. (2002), *Dynamic Modelling of the 1992 Landers Earthquake*, J. Geophys. Res. *106*, 25467–25482.

KAO, H. and CHEN, W.-P. (2000), *The Chi-Chi Earthquake Sequence: Active Out-of-sequence Thrust Faulting in Taiwan*, Science *288*, 2346–2349.

SAVAGE, J. C. (1983), *A Dislocation Model of Strain Accumulation and Release Art a Subduction Zone*, J. Geophys. Res. *88*, 4984–4996.

SHI, B., ANOOSHEPOOR, A., BRUNE, J. N., and ZENG, Y. (1998), *Dynamics of Thrust Faulting: 2D Lattice Model*, Bull. Seismol. Soc. Am. *88*, 1484–1494.

SHIN, T. C., KUO, K. W., LEE, W. H. K., and TENG, T. L. (2000), *A Preliminary Report on the 1999 Chi-Chi (Taiwan) Earthquake*, Seismol. Res. Lett. *71*, 24–30.

WALD, D. and HEATON, T. (1994), *Spatial and Temporal Distribution of Slip for the 1992 Landers, California Earthquake*, Bull. Seismol. Soc. Am. *84*, 668–691.

YOFFE, E. H. (1960), *The Angular Dislocation*, Phil. Mag. *5*, 161–175.

(Received March 1, 2001, accepted May 15, 2001)

To access this journal online:
http://www.birkhauser.ch

Pure appl. geophys. 160 (2003) 579–602
0033–4553/03/040579–24

▌Pure and Applied Geophysics

Spontaneous Complex Earthquake Rupture Propagation

S. Das[1]

Abstract — The historical development of spontaneous rupture propagation, starting from the landmark paper of Griffith in 1920, through to the late 1980s is traced, with particular emphasis on the work carried out at MIT in the 1970s by K. Aki and his co-workers. Numerical applications of Kostrov's method for planar shear cracks were developed by Hamano, Das and Aki. Simultaneously at MIT, Madariaga considered the radiated field of a dynamic shear crack. The further development of these ideas, for example, three-dimensional spontaneous planar faulting models, continued through the 1980s. Major insight into the maximum possible rupture speeds for earthquakes developed, with the acceptance of the theoretical possibility of supersonic rupture speeds for faults with cohesion and friction, the theoretical developments spurring the search for such observations for earthquake ruptures. Possible mechanisms by which faults stop were elucidated. It was shown that a propagating rupture can jump over barriers for cracks with a cohesive zone at its tip. Complex faulting models, namely the barrier and asperity models, and their associated radiated field developed. In the late 1980s, it was shown that "dynamic" or transient asperities can develop during the complex rupturing process. Even seemingly relatively simple physical situations, can lead to such complex rupturing processes that the usual idea of "rupture velocity" needs to be abandoned in those cases. Some of the work initiated by Aki and his co-workers, such as the details of the transition from sub-Rayleigh to super-shear speeds in inplane shear mode, and the behavior of the cohesive zone size as the crack extends, still remains the subject of research today.

Key words: Earthquake rupture, complex faulting, spontaneous rupture propagation.

Introduction

Koto's (1893) study of the 1891 Mino-Owari earthquake finally confirmed the faulting origin of earthquakes, though a few seismologists still continued this debate into the 1960s! Before then, cause and effect were confused. Even a scientist as great as Darwin, in his description of the 1835 Chilean earthquake, wrote: "The most remarkable effect of this earthquake was the permanent elevation of the land; it would probably be far more correct to speak of it as the cause" (Darwin, 1889). In the history of seismology, "who first proposed it" (that earthquakes are due to faulting) "is not definitely known" (Howell, 1990). Clearly, Reid's (1910) brilliant study of the 1906 San Fransisco earthquake soon after Koto's paper helped in its acceptance. But though it was understood by Reid that shallow earthquakes were

[1] Department of Earth Sciences, University of Oxford, Parks Road, Oxford OX1 3PR, UK.
E-mail: das@earth.ox.ac.uk

due to rupture of the earth's crust in response to tectonic stresses, the physical basis for analyzing this phenomenon, namely, fracture mechanics, did not yet exist. It was only in 1920 that Griffith initiated the study of the mechanics of fracturing, and the subject really developed vigorously during and after the Second World War. GRIFFITH (1920) understood that for a pre-existing flaw in a material to extend, the energy required to create new crack surface (the fracture energy) can be at most equal to the strain energy available in the body. This criterion, now known as the "Griffith fracture criterion" is a global criterion. In 1957, Irwin first introduced the idea of stress intensity factor k at the crack tip, the stress at the crack tip being given by $k/(\sqrt{r})$, where r is the distance from the crack tip outside the crack, together with higher order terms in r, which can be neglected as one approaches the crack tip (that is, as $r \rightarrow 0$). Using this, IRWIN (1957, 1958, 1969) developed a local criterion which states that the crack tip extends when the stress-intensity factor exceeds some critical value, called the fracture toughness of the material. These ideas were applied at first to quasi-static rupture of tension cracks.

The static solution for displacement on a inplane shear crack was written down by STARR (1928). ESHELBY (1957) wrote down the closed form solution for static elliptical cracks, both in tensile and in shear modes. These solutions indicated that for constant stress drop on a simple earthquake fault, the displacement is variable, decreasing from zero at the edges to a maximum at the center. KOSTROV and DAS (1984) evaluated and plotted the stresses around circular and elliptical faults using this latter solution.

Development of Methods of Solution for the Dynamic Problem

In the 1960s, Kostrov pioneered the application of ideas developed in fracture mechanics to the study of shear fracture and hence set up the basis for analyzing earthquake ruptures. In KOSTROV (1964) he published the analytical solution for a self-similar shear crack extending at a prescribed velocity and showed that for a constant stress drop on the crack, the fault slip velocity varies at the crack edge as the inverse of the square root of the distance of any point on the crack to the crack edge. His solution also shows that the displacement in time at each point of the crack increases from zero to its final constant slope value, and that the time to reach this value increases as one moves away from the point of rupture initiation. This implies that the "rise time" of the source time function varies over the fault. In KOSTROV (1966), he considered the propagation of the semi-infinite antiplane shear crack which suddenly appears and starts extending at a prescribed (but not necessarily constant) velocity without stopping. He wrote down the complete closed form solution for this mixed boundary value problem, in which the stress changes are assumed known within the crack (the stress drop) and the displacements are known outside the crack (the slip is zero there). This solution gives the displacement on such a crack for any known stress drop distribution on it, as well as the stresses in the causal region outside the crack. This method has since been called the "Green function method" by FREUND (1990), and had been widely used in potential theory and in fluid mechanics studies of supersonic flow around aerofoil wings (developed by EVVARD (1950) and described in his textbook by WARD (1955)). Note that the equations for a crack tip moving through a solid is identical to those for fluid flowing past a solid object! BURRIDEGE and WILLIS (1969) developed the dynamic solution for a self-similar elliptical crack.

Kostrov's 1966 method formed the basis of the numerical boundary-integral equation (BIE) method later developed by HAMANO (1974) and DAS and AKI (1977a) for 2-D problems, and by DAS (1980) and DAS and KOSTROV (1987) for 3-D problems. In this method, the problem reduces to calculations of quantities on the fault surface only. The problem formulation in 3-D is briefly described below, mainly for completeness, and for definition of quantities to be used later in this paper. The earthquake source is modeled as a propagating plane shear crack in an infinite medium (Fig. 1) which is homogeneous and linearly elastic everywhere off the crack plane, the latter considered to have infinitesimal thickness. (Remember that earthquakes cannot occur in a medium that is truly homogeneous everywhere.) As the fault propagates on the planar surface $F: X_3 = 0$, waves are radiated out in three spatial dimensions. Initially, the infinite body is under a uniform state of stress σ_{ij}^0. The initial stress on the fault plane $X_3 = 0$ can be separated into the normal stress σ_{33}^0 and a shear stress $\sigma_{13}^0 = \sigma^0$, say. The component σ_{23}^0 can be taken as zero by taking the coordinate axis X_1 in the direction of the maximum initial shear (without loss of generality). The initial shear stress is increased sufficiently to initiate a fault at the origin, which then propagates on the $X_3 = 0$ plane. The normal stress σ_{33}^0 over the fault plane remains constant throughout the rupture process, for a planar fault. Let us take the origin of time $t = 0$ as the time when the fault initiates and starts

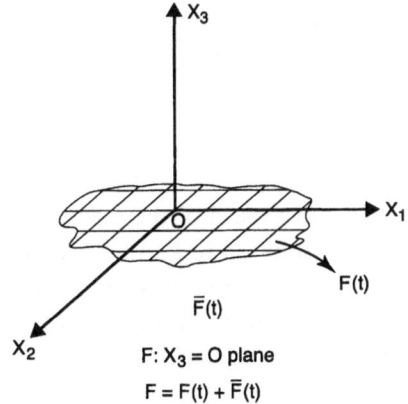

Figure 1
The geometry of the fault.

extending. We study the case when the fault propagation speed is rapid enough to generate elastic waves. The fault edges may move at some pre-assigned speed or the position of the fault edge may be found as a function of time, using some fracture criterion. (The latter is called "spontaneous" propagation in seismology.) Let us consider the former case only for the moment (that is, the fault propagation speed in all directions on the fault plane is known). As the fault propagates, there is relative motion between the two faces of the fault, that is between the regions $X_3 < 0$ and $X_3 > 0$, and a displacement discontinuity appears across the broken region of the fault plane. This discontinuity is a function only of the coordinates X_1 and X_2 and time t. The shear stress on the fault surface is zero if there is complete stress release; or, it can be equal to the the frictional stress σ on the fault faces, given by $\sigma = \mu\sigma_{33}^0$, where μ is the coefficient of friction. μ may be taken constant or a function of space, time, and any other desired parameter. Let the incremental stresses due to the displacement $\mathbf{u}$ from its initial configuration be τ_{ij}, so that $\sigma_{ij} = \sigma_{ij}^0 + \tau_{ij}$, that is, τ_{ij} is the stress change due to the motion, and all motions depend only on these stress changes on the fault. Exploiting the symmetries in the problem for planar shear cracks, the solution can be shown to be antisymmetric in X_3 that is, the displacement components u_1, u_2 and traction perturbation τ_{33} are odd in X_3 while u_3, τ_{13} and τ_{23} are even in X_3 (DAS and AKI, 1977a). Hence, it is sufficient to solve the problem for the upper half-space $X_3 \geq 0$. Further, from the continuity of tractions across $X_3 = 0$, it follows that τ_{33} vanishes everywhere on $X_3 = 0$. Then, the required representation relation is obtained as

$$u_k(\mathbf{X}, t) = \int\limits_{-\infty}^{\infty} dt' \iint\limits_F G_{ki}(\mathbf{X} - \mathbf{X}', t - t')\tau_{k3}(\mathbf{X}', t')dS \qquad (1)$$

where $\mathbf{X}$ and $\mathbf{X}'$ are two-dimensional vectors on F, u_k is the component of displacement in the k direction, G_{ki} is the displacement response of the medium in the k direction at $(\mathbf{X}, t)$ due to an impulse acting in the i direction at $(\mathbf{X}', t')$, $k = 1, 2, 3$, and F is the causal portion of the fault plane $X_3 = 0$, that is, the cone of dependence given by

$$v_P^2(t - t')^2 - (X_1 - X_1')^2 - (X_2 - X_2')^2 \geq 0, \quad t \geq t' \geq 0 \qquad (2)$$

where v_P is the compressional wave speed of the medium. The required components of the Green functions G are the solution to Lamb's problem and can be expressed in terms of elementary functions. The analytical expressions for G_{ki} for the two- and three-dimensional problems are given in Appendix I of KOSTROV and DAS (1988). The kernel G possesses only weak singularities and can be directly discretized for numerical computation.

   This mixed boundary value problem is solved numerically by discretizing the above equation. However, the stress changes τ_{k3} are known only on the broken part of the fault plane (the stress drop) but are unknown in the unbroken but causal portion, so these have to be determined before the integrations can be carried out in equation (1). This is done by using the fact that the slip is zero on the unbroken part so that the lhs of (1) is zero there for $k = 1, 2$. The solution then proceeds by a time-marching scheme. The region of integration in 3-D, the intersections of the cones of dependence and influence, is shown in Figure 2 (DAS, 1980). Note that even though the problem was solved by DAS and AKI (1977a) in 2-D only, the full set of equations for the 3-D problem were written down by DAS (1976) [in fact, it follows straightforwardly from BURRIDGE (1969)] and the required Green functions were already available, having been written down by CHAO (1960) for a Poisson solid, (the expressions for a general solid were given by RICHARDS (1979)), so that the development of the 3-D problem later by DAS (1980) followed naturally.

The normal component of displacement u_3 is non-zero during the dynamic process, though of course there is no discontinuity in this component across the fault for the shear crack problem. This property had already been used earlier by AKI (1968) to study the near-field transverse component of 1966 Parkfield earthquake, the only near-field seismogram of that earthquake that was recorded.

In 1969, Burridge had started working on the problem of dynamic crack propagation, first in 2-D, later extending his method, also a numerical boundary-integral method, to some simple (namely, the acoustic) 3-D problem (BURRIDGE and MOON, 1981). DAS and KOSTROV (1987) have discussed this form in detail, and shown that these two forms of writing the BIE are mutually inverse integral transforms of one another. In this second form of the BIE, the stress on the fault is written as a convolution between a kernel and the fault displacement. It has the advantage that the integrations extend only over the slipping portion of the fault. It

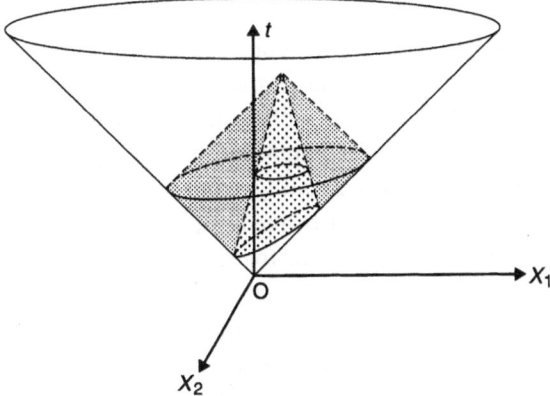

Figure 2

Volume of integration in BIE method of DAS (1980). The cone with vertex pointing down is the "cone of influence," and the cone with its vertex pointing up is the "cone of dependence." The inner cone (stippled) is the region where the Green function vanishes for this problem, so that the actual volume of integration for this problem formulation is the grey area.

has the disadvantage that the singularities in the kernel are strong and the kernel cannot be discretized as simply as in the previous form of the BIE. BURRIDGE (1969), mentioned above, did this. DAS and KOSTROV (1987) determined another numerical form of the kernel. Most recently, MADARIAGA and COCHARD (1992) and COCHARD and MADARIAGA (1994) have used this form of the BIE in their work, discretizing the kernel directly. This latter form of the BIE is particularly suited to "interior" crack problems such as an expanding fault, whereas that given in equation (1) is most efficient for "exterior" crack problems such as the rupture of an asperity on an infinite fault, where the zone of the unknown stress changes decreases with time.

Simultaneously, programs were developed to study the crack problem using finite-difference and finite-element methods, both in fracture mechanics for tension cracks in 2-D, and by seismologists for shear cracks, both in 2-D and 3-D. For the tension crack, the reader is referred to the very comprehensive bibliography given by FREUND (1990). For the shear problem, the 2-D work was carried out by ANDREWS (1976a,b; 1985) and by ARCHULETA (1976) and later by DAY (1982a,b) in 3-D.

Kostrov's *chef d'oeuvre* was probably his 1975 paper (KOSTROV, 1975) in which he wrote down closed form expressions for all three modes of semi-infinite and finite crack propagation for cracks with variable velocity, as long as this velocity did not exceed the Rayleigh wave speed. This paper is a landmark in the development of applied mathematical methods in fracture mechanics, as John Willis recently reminded us at the memorial meeting in honor of Kostrov, held at the EGS meeting in 1999 at The Hague. But probably due to its cumbersome nature, this solution has not been used widely by seismologists, though it is used by fracture mechanicists.

Spontaneous Propagation of Cracks

In his 1966 study, Kostrov also considered the problem of spontaneous propagation of a semi-infinite antiplane shear crack that suddenly appears and starts extending. Using a dynamic form of Griffith's criterion, he showed that if the material through which the crack is extending has constant fracture energy γ, the crack goes through a stage of accelerating from zero to its final speed and then continues to propagate at this speed. For the antiplane crack, this terminal speed was the shear-wave speed of the medium, but for stronger materials (i.e., with higher γ), the time to reach this speed was longer. Kostrov's 1966 paper was a very short paper, demonstrating that the length of a paper is unrelated to its impact on a field! In the introduction of his book "Cracks and Fracture," BROBERG (1990) singles out the work of Kostrov and Freund in dynamic fracture mechanics by saying "In the dynamic field, the significant and pioneering contributions of B. V. Kostrov and L. B. Freund deserve particular mention."

During his studies in the 1960s, Kostrov found that despite the sophisticated developments in fracture mechanics, for seismologists to apply the ideas to earthquake rupture was not straightforward. The reasons for this were twofold. First, engineering structures are in tensional stress regimes and all the theories were relevant to tensional fracture. In tensile cracks, the two faces of the crack are not in contact and hence there is no friction between them. For shear cracks, of course, this is not the case. In particular, when considering the total energy balance budget in a problem, the friction of the shear -fault surfaces becomes very large. Later, FREUND (1979) showed by considering the energy budget of dynamically propagating cracks that as a shear fault becomes larger and larger, the frictional term becomes more and more dominant over the term representing the energy needed to create new fracture surface. Studies of antiplane shear in fracture mechanics were considered in a similar way to the tension crack problem, and friction was always neglected. Secondly, at that time, most engineering studies related to quasi-static fracture, but earthquake ruptures are a dynamic phenomenon. KOSTROV and NIKITIN (1970) even had to redefine the idea of fracture by what they termed the "model of fracture." Without going into details here, we simply refer to his original paper as well as to a brief description in KOSTROV and DAS (1988). The crack tip energy flux for dynamic fracture was proposed by ATKINSON and ESHELBY (1968) simply by taking that for quasi-static fracture and guessing the result for dynamic propagation. KOSTROV and NIKITIN (1970) confirmed these expressions for dynamic fracture working directly from the field equations. KOSTROV and NIKITIN (1970) also extended the idea of the path-independent J-integral (ESHELBY, 1956; RICE, 1968) for quasi-static fracturing to dynamic fracturing (the J-integral can be obtained from their expression simply by setting the rupture speed to zero).

BURRIDGE (1973) demonstrated that for cohesionless cracks with friction, the maximum rupture speed could reach the compressional wave speed of the medium. His results were not taken seriously either in fracture mechanics or in seismology due to the cohesionless nature of the crack.

In parallel with developments of the theories which would lead to the study of spontaneous fault rupture, some important milestones occurred around this time in the study of the earthquake source. The body force equivalent in terms of the double-couple was developed by BURRIDGE and KNOPOFF (1964), and the scalar seismic moment was defined by AKI (1966). RANDALL (1971) understood that the seismic moment was actually a tensor. BRUNE (1970) wrote a simple relation between earthquake stress drops and fault radius, for a circular fault, and since this was a simple formula, it became very widely used, sometimes for faults that were far from being equidimensional.

The Cohesive Zone Model

In linear elastic brittle fracture, the transition from broken to unbroken material occurs over an infinitesimally small region. This sharp transition leads to infinite stresses at the crack tip in mathematical considerations. But since this cannot exist in reality, there is, in fact, a region between these two states where the material may be partially broken. Such a model of fracture is termed imperfectly or non-ideally brittle, and this intermediate region is called the "cohesive zone." BARENBLATT (1959) introduced the idea that the bonding force between atoms that end up on opposite faces of the crack after total separation, is proportional to the separation distance between them while the atoms are still in the transition or cohesive zone. Almost simultaneously, LEONOV and PANASYUK (1959) in the then USSR (the original paper was written in Ukrainian and hence is not easily accessible to most readers!) and DUGDALE (1960) in the United States developed the idea of the process zone further.

From 1972 to 1973, Ida, working beside Aki at MIT, developed a cohesive zone model in which the cohesive zone stress depends on the amount of relative slip between the two faces of the crack for a shear crack (IDA, 1972, 1973). PALMER and RICE (1973) introduced a similar idea to study the stability of a slope under gravitational sliding. In these models, now termed "slip-weakening" models, the stress just outside the crack is shown in Figure 3, and the work done at the crack tip, or the fracture energy, is the shaded area. FREUND (1990) explores such models more fully in his Chapters 5 and 6. ANDREWS (1976a) showed that for such a model the size of the cohesive zone decreases as the crack length increases, whereas for a "strain-weakening model," it remains constant. Based on laboratory results, OHNAKA (1996) stated that "the size of the breakdown zone is almost constant in the zone of dynamic, fast-speed rupture propagation." This is also seen from Figure 26 of OHNAKA and SHEN (1999).

BRACE and WALSH (1962) measured fracture energy in quartz in the laboratory under shear stress. IDA (1972) estimated the fracture energy for earthquakes from his

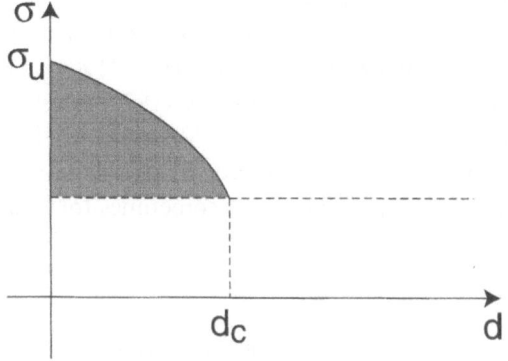

Figure 3
Cohesive zone model, showing the stress σ at the crack edge plotted against the slip weakening distance d.
The critical distance is d_c and the shaded area gives the fracture energy.

cohesive zone model, and found that it is several orders of magnitude greater than that obtained in laboratory experiments. All further estimations of fracture energy (DAS, 1976; AKI, 1979) found similar results.

The critical stress level fracture criterion. The ideas of Griffith, Irwin, Barenblatt, etc. all lead to seemingly different fracture criteria. KOSTROV (1966) showed that the Griffith and Barenblatt criteria were equivalent from energy arguments and WILLIS (1967) proved their equivalence by a direct stress analysis. For numerical applications however, these criteria are difficult to use. So first HAMANO (1974) and then DAS and AKI (1977a) introduced the "critical stress level fracture criterion" in which a grid ahead of the crack tip is allowed to break when the stress in that grid exceeds some critical stress level related to the resistance of the material to fracture. Hamano's preliminary results were never published, except as an AGU abstract. Work on this problem was continued by Das and Aki between 1974–1976. Hamano had started developing the 2-D numerical form of KOSTROV's (1966) Green function method for a semi-infinite antiplane shear-crack solution, and extended it to 2-D finite cracks for all three crack modes. He also implemented the critical stress level fracture criterion, but did not show its connection with the Irwin criterion, which was done later by DAS (1976) and DAS and AKI (1977a). They related this criterion to a discrete form of the Irwin criterion, but of course the relation is grid-size dependent. VIRIEUX and MADARIAGA (1982) studied this criterion further and by comparing the analytical and grid-size-dependent numerical solutions for the antiplane shear crack determined the range of normalized critical stress levels for which this criterion gives the same result as the analytical solution using the Irwin criterion.

The critical stress level criterion becomes nonproblematic from the point of view of grid size dependence if we consider the material to be imperfectly brittle and the stress at the crack tip to be nonsingular. Then for small enough grid sizes (but not so small as to be impracticable for computations) the average stress near the crack edge varies smoothly and becomes independent of the grid size. In seismological applications, no attempt has ever been made to relate this discretized resistance to fracture at the crack edge to actual laboratory measurements. This is partly because it is not yet technologically possible to conduct dynamic fracture experiments on large rock specimens at the temperatures and pressures that exist at depths in the Earth where earthquakes actually occur. So the numerical results have been discussed in the context of fracture on "relatively strong" or "relatively weak" faults or interfaces. The term "relative" is discussed next. For this, we define the dimensionless quantity $S = (\sigma_u - \sigma_0)/\sigma_e$, where σ_0 is the initial stress level, σ_u is the critical stress level ahead of the crack tip required for fracture, and σ_e is the stress drop on the fault. Then larger S implies "relatively" stronger material, that is, relative to the stress drop.

Relation between Different Fracture Criterion

DAS (1976) and DAS and AKI (1977a) extended Kostrov's analysis for the Griffith criterion to the Irwin criterion and compared both results with that for the critical stress level fracture criteria for a semi-infinite antiplane shear crack that suddenly appears and extends spontaneously. Using Griffith's criterion, the crack tip position x as a function of time t is

$$x = \beta t + \beta t_c(\pi/2 - 1 - 2 \ \arctan(t/t_c)) \ ,$$

where t_c is the time of onset of fracture (KOSTROV, 1966). Using Irwins's criterion, DAS and AKI (1977a) showed that the crack tip position in time is

$$x = \beta(t - t_c) - \beta t_c \ \log(t/t_c) \ .$$

Figure 4 shows the crack tip position as a function of time for this problem for the analytical forms of the Irwin and Griffith criteria and for the numerical criterion. It shows that for the same t_c, the Griffith locus always lies above the Irwin locus, i.e., the Griffith crack accelerates faster to the terminal velocity than the Irwin crack. For the numerical case, by finding the value of the (grid-size dependent) S for which a crack starts propagating at the same time as those for the Griffith and Irwin criteria, namely with the same t_c, it was shown that the critical stress level criterion could indeed be considered a numerical analog of the Irwin criterion. This criterion has since been used in numerical applications in seismology.

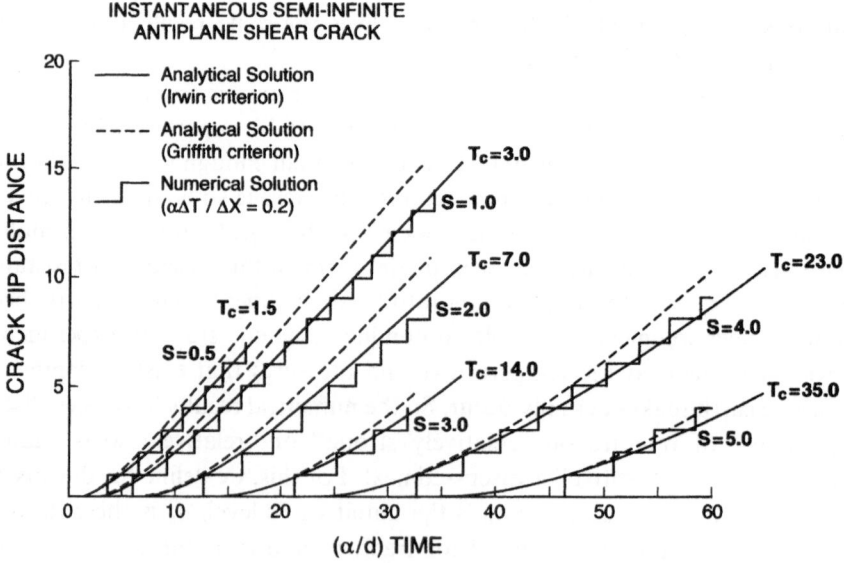

Figure 4
Comparison of the crack tip positions for the Griffith, the Irwin and the critical stress level fracture criteria
(after DAS and AKI, 1977a).

Maximum Permissible Rupture Speeds for Shear Faults

BARENBLATT and CHEREPANOV (1960) and BROBERG (1960) showed that the maximum speed for tension cracks in perfectly brittle materials (i.e., with infinite crack tip stresses) was the Rayleigh wave speed of the medium. (An incorrect solution had been published by MOTT (1948), in which he obtained the erroneous result that the maximum speed of a tensile crack is some fixed fraction, typically about half the shear-wave speed of the material.) CRAGGS (1960) showed that the maximum rupture speed was the Rayleigh wave speed both for the tensile as well as the inplane shear crack. The maximum rupture speed for antiplane shear cracks was shown to be the shear-wave speed.

The presence of the singularity at the crack tip in perfectly brittle material leads to the result that for tensile and inplane shear cracks the maximum speed cannot exceed the Rayleigh wave speed; for antiplane shear cracks the maximum permissible rupture speed is the shear-wave speed. In numerical problems where the stress singularity at the crack edge is replaced by a large but finite stress, application of the critical stress level fracture criterion to the antiplane shear crack and the tension crack gave the same terminal rupture speeds as obtained for the analytical problems discussed above. But for inplane shear cracks, terminal speeds as high as the P-wave speed were found for relatively weak materials (ANDREWS, 1976b; DAS, 1976; DAS and AKI, 1977a).

The transition from sub-Rayleigh to super-shear speeds for inplane shear cracks is reproduced from DAS and AKI (1977a) in Figure 5. Similar results were found by ANDREWS (1976b) using Ida's criterion and a finite-difference method. The maximum permissible rupture speed has since been confirmed in numerous numerical studies. There has been much discussion recently on whether the transition from sub-Rayleigh to super-shear is sudden or smooth. With the computing power available in the late 1970s, this was impossible to resolve, but it could be resolved today, if so desired.

BURRIDGE et al. (1979) showed that even in the case of a perfectly brittle solid, cracks can propagate at $\sqrt{2}$ times the shear-wave speed. Very recent laboratory measurements (ROSAKIS et al., 1999) confirm this.

Truly convincing observations of super-shear rupture speeds for earthquakes still remain elusive. In some reported cases it is not clear if the speed being measured is not the apparent rupture speed. Often such speeds are determined from one station close to the fault, or stations not close enough to the fault. Remembering the very unstable nature of the inverse problem of obtaining the fault rupture history from analysis of seismograms (KOSTROV and DAS, 1988; DAS and KOSTROV, 1990, 1994; DAS and SUHADOLC, 1996; DAS et al., 1996; SARAO et al., 1998), such velocities must not be unquestioningly accepted in situations without sufficient constraints (usually good station distribution and many three-component accelerograms). What is most important is that this theory has led to the search for such super-shear speeds, whereas previously such speeds would be considered impossible, and not considered at all.

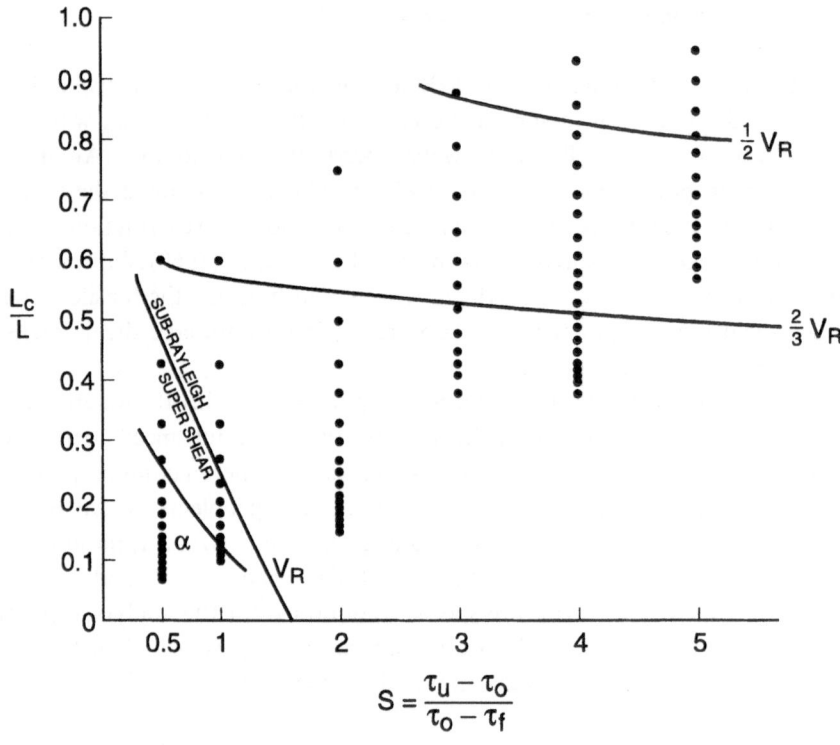

Figure 5

Transition from sub-Rayleigh to super-shear velocity for inplane cracks, (after DAS and AKI, 1977a), shown as contour plot of the crack tip velocity for different values of the fault strength parameter (S) against the dimensionless parameter L_c/L, where L_c is the initial crack length required for propagation to occur, i.e., the initial Griffith critical length, and L is the instantaneous crack length.

For the antiplane shear problem, the only wave speed in the problem is the shear-wave speed. But for the other two crack modes, P, S and Rayleigh waves all exist along the crack face. So why then does the tension crack not reach the P-wave speed even for the weakest materials? This can be explained by considering the Green functions for the two problems. The Green functions for the Mode II crack is such that the sign of the stress due to the body waves and the Rayleigh wave are the same, whereas for Mode I, the body wave stresses have opposite sign to the Rayleigh wave, and close the crack, inhibiting it from growing until the arrival of the Rayleigh wave. The tension crack, filled with fluid, was used by Aki and his co-workers, to model magma transport in volcanoes (AKI et al., 1977).

Radiation from Spontaneous Faults

MADARIAGA (1976) considered the radiated body-wave pulses from a dynamic circular crack propagating at a constant velocity. He solved the problem numerically to determine the slip rate on the fault. Then, summing these fault slip rates in appropriate directions around the source (HASKELL, 1964), he obtained the pulse shape in those directions. He demonstrated that the far-field pulse shapes in different directions in a plane perpendicular to the crack can be used to estimate the fault dimension, if the rupture speed is known. In earlier kinematic models (SAVAGE, 1966; SATO and HIRASAWA, 1973) fault slip was stopped artificially, to find the expected pulse shapes at different stations. Madariaga's study from propagating faults in which the slip was stopped using a physically realistic criterion, first gave us insight into how fault slip stops on the different parts of a fault by "healing" information from the fault edges, and the resulting pulse shapes. MADARIAGA (1976) also considered the corner frequencies for faults propagating at different constant speeds and showed that the rupture speed affects the corner frequency. This implies, as DAS and AKI (1977b) demonstrated, that corner frequency is not a measure of fault size but of rupture time, and the two can be related if and only if the rupture speed [which in reality is variable as shown, for example, by KOSTROV (1966)] is known.

Complex Faulting Models

The barrier model. Using their numerical BIE method, DAS and AKI (1977b) considered the propagation of spontaneous 2-D faults on planes with variable values of S. The critical stress level fracture criterion was used in these calculations to determine how the crack advanced. This led to the development of what has now become known as the "barrier model" for heterogeneous faults, the barriers being regions of large S. In addition to the maximum permissible rupture speed, discussed above, other unexpected results were found, which cannot occur for the previously studied cracks with singularities at their edges. For example, a fault can jump over very strong regions and continue to propagate, leaving behind some unbroken regions. If these regions had very large S they still remained unbroken at the end of the rupture process, but if they had some intermediate values, the region was unbroken when the fracture front first jumped over it, but the concentration of stress on it during the dynamic rupture process led to its rupture.

What is most important in seismology is that the different kinds of rupture processes lead to different pulse shapes and hence these differences can potentially be inferred from seismograms. Figure 6 shows the "far-field" pulse shapes obtained by DAS and AKI (1977b) for smooth and rough faults. Their major conclusions are summarized below:

(i) The smooth fault and the fault with barriers that break during the dynamic process result in single earthquakes whereas the heterogeneous faults with unbroken barriers result in multiple shocks.

(ii) The time history of slip on the fault and the resulting far-field radiation is most complicated in the case when the initially unbreakable barrier eventually breaks. In

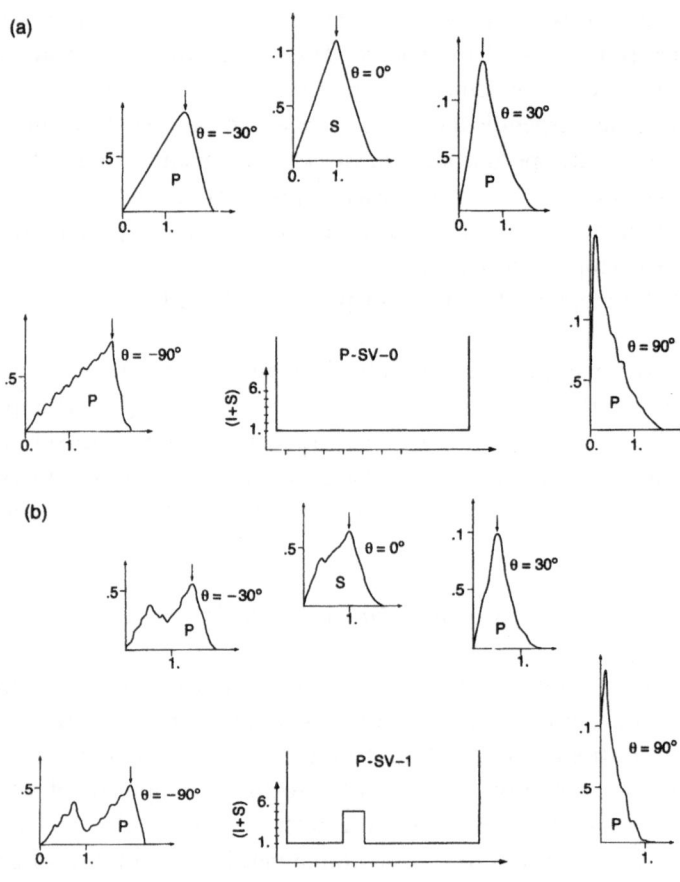

Figure 6

Pulse shapes for complex faulting models. (a), (b), (c) and (d) refer respectively to a fault without any complexity, a fault with one unbroken barrier, a fault with two unbroken barriers, and, a fault with two barriers which do not break when the rupture front passes it initially, but due to the increase of the stress on it caused by being surrounded by broken regions breaks while other parts of the fault are still fracturing dynamically.

this case the duration of the fracture and slipping process take longer than in the other cases for the same fault length.

(iii) The final slip on the fault and hence the seismic moment is largest for the smooth crack and smallest for the case of the fault with the most number of unbroken barriers. In the case of the barrier that eventually breaks, the final slip and moment are almost as large as that for the smooth fault. The slip for the fault with the largest number of unbreakable barriers has the most uniform value over the fault while the fault with no barriers at the end of the fracture process shows the largest amount of variation in slip distribution over the fault! This may explain why the uniform dislocation model (HASKELL, 1964) has often been able to explain observed

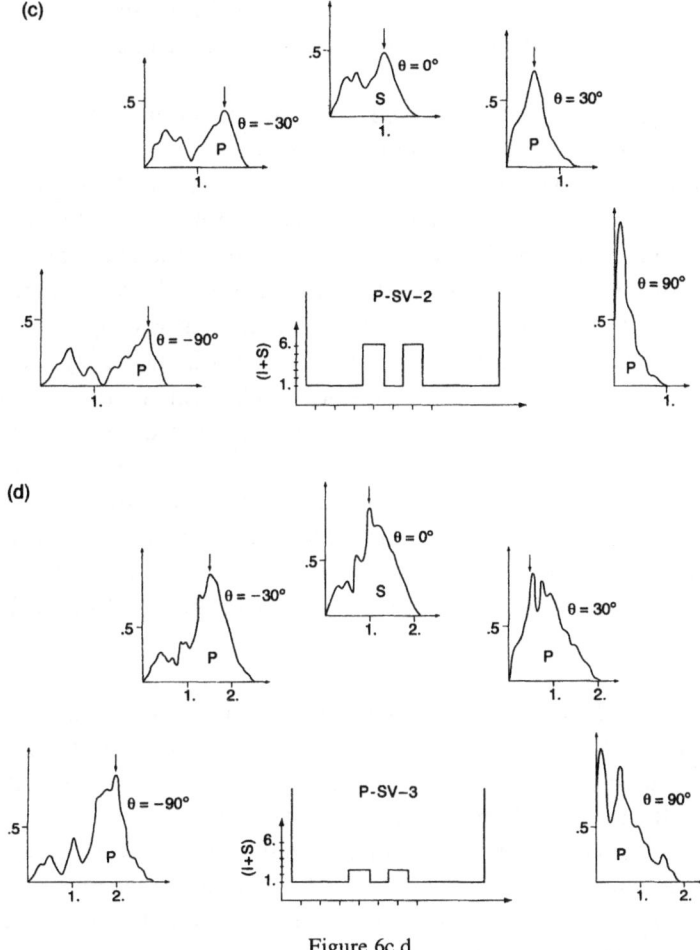

Figure 6c,d

overall features of seismograms satisfactorily, for example, BOUCHON's (1979) study of the 1966 Parkfield, California earthquake.

(iv) Clear directivity effects in the seismic radiation are seen in all cases, these effects being stronger for the fault with unbreakable barriers than for the smooth fault. However when the barriers eventually break the directivity effect is even weaker than that for the smooth fault.

(v) The time domain pulses are more sensitive to the complexity of the fracture process than the spectral shapes. In particular, when the barriers eventually break the pulses show complexity in all directions from the source but the amplitude spectra are not particularly revealing.

(vi) When the barriers remain unbroken, the spectra at the highest frequencies for which the numerical results are meaningful (this limit can be obtained by comparing the numerical solution for some simple case with an analytic solution, the spectra in all the cases plotted in this example being shown only up to the frequency where the numerical results are valid) have more energy than that for the smooth fault.

(vii) The corner frequency averaged over all directions from the source is unaffected by the presence of unbreakable barriers.

(viii) The stress drop averaged over the total fault length (including the barriers) is lower for the case with unbroken barriers than the other cases. In fact, there is a stress increase on these unbroken regions due to the earthquake. Thus, a complex earthquake with lower average stress drop can generate relatively higher frequency waves than a simple earthquake with relatively higher stress drop.

The idea that faults can jump over barriers was not immediately accepted, when first proposed by DAS and AKI (1977b), since in classical fracture mechanics with infinite crack tip stresses it cannot do so, as mentioned above. In the many observations since in which faults have been shown to jump across barriers, the barrier that is jumped over is relatively small, usually a few kilometers (AKI, 1979, 1980). A study of the great 1998 Antarctic plate earthquake shows that this earthquake jumped over a 70–100 km long barrier and kept propagating for another 60 km (HENRY et al., 2000). This is similar to the case P-SV-1 illustrated in Figure 6b.

The asperity model. The basic idea of this model was first suggested by MADARIAGA (1979) and then by RUDNICKI and KANAMORI (1981). According to this model, an earthquake is caused by the failure of isolated, highly stressed regions of the fault, the rest of the fault having little or no resistance to slip (being partially broken and preslipped, say) and contributing little or no stress drop to the earthquake process. This results in a nonuniform stress drop over the fault. Since the regions without slip are able to withstand the high stresses concentrated on it until the moment of commencement of the earthquake, the model implies that the parameter σ_u for these regions is higher than that for the rest of the fault.

The observational support for complex faulting models came both from seismology and geology. Observations of multiple shocks on seismograms, the measured surface slip after large earthquakes, direct evidence from fractures on mine faces showing that faults are usually very complex with side-steps and highly deformed but unbroken ligaments in the step-over regions (SPOTTISWOODE and McGARR, 1975; McGARR et al., 1979) all contributed to this. In spite of its idealizations, these models enhanced our understanding of the earthquake faulting process. It led to the characterization of barriers as being material (large S) or geometrical (when the fault plane deviated from planarity) by AKI (1979). It has led to the identification of barriers in the field by structural geologists and by seismologists in various locations world wide (LINDH and BOORE, 1981; KING and YIELDING, 1984; NABELEK and KING, 1985; SIBSON, 1986; BARKA and KADINSKY-

CADE, 1987; BRUHN et al.,1987; DAS, 1992, 1993; HENRY et al., 2000, to name only a few among many such examples). Finally, the most convincing evidence that faults are heterogeneous not only near the surface of the Earth but also at the depths where the main faulting in an earthquake occurs is that aftershocks do occur at these depths. Effort is under way to to identify barriers along faults and to try to understand the origin and geochemical characteristics of barriers. The primary reason for this general interest is that earthquakes often nucleate and terminate at barriers.

How Faults Stop

A problem that was considered in the mid-1970s is how faults stop. ESHELBY (1969) had shown that the crack tip has no inertia. For seismological applications, this implies that fractures can start and stop suddenly. HUSSEINI et al. (1975) considered two possible ways in which faults can stop, either by encountering a large high strength region or by running into a "seismic gap," i.e., running out of available strain energy for fault propagation to continue. DAS (1976) demonstrated that these two methods of fracture arrest lead to different far-field spectra. This part of the work was never published by the author except as part of a thesis, but is reviewed by DMOWSKA and RICE (1985).

Summary of Developments since the Late 1970s

In 1980, Das continued the work started with Aki at MIT and developed the fully 3-D numerical BIE method. DAS (1981) applied this method to truly 3-D shear cracks, and confirmed the maximum rupture speeds in the purely mode II and III directions as being the same as found for the 2-D case, implying that in relatively strong materials, a circular crack remains circular as it grows, but on weak planes, they become elongated. The problem was further continued by Das, in collaboration with Kostrov, until the late 1980s.

Since the unbroken barrier with its high residual stress concentration can become the "asperity" of a future earthquake on the same fault, the radiation due to the fracturing of such an unbroken barrier was considered by DAS and KOSTROV (1983), who studied the dynamic fracture of a single circular asperity and showed that the rupture process is so complex, that the idea of rupture velocity becomes almost meaningless. Of the different cases studied by DAS and KOSTROV (1983), one is shown in Figure 7. Interestingly, for an elliptical asperity the rupture propagates as a very simple straight front (DAS and KOSTROV, 1985). The rupture of a pre-existing circular crack was also shown to be complex, and is shown in Figure 8, which is redrawn from KOSTROV and DAS (1988). DAS and KOSTROV (1986) also studied the rupture of a single asperity on a finite pre-broken fault and showed that its spectrum has the properties of a "slow" or "weak" earthquake. DAS (1986) compared the radiated field generated by the rupture of a circular crack and a circular asperity. We refer the reader to KOSTROV and DAS (1988) for a full treatment of these problems.

DAS and KOSTROV (1987) increased the efficiency of the 3-D BIE method, enabling the use of fine grids in the numerical problem. DAS and KOSTROV (1988) used this to study complex 3-D rupture on faults, and showed that all peaks on source time functions are not due to rupture of pre-existing asperities, and that "dynamic asperities" can appear and be seen on the source time function. Such dynamic concentrations of stresses, involving persistent "crack front waves" have been shown numerically to exist for 3-D tensile crack problems by MORRISSEY and RICE (1998), and have been shown analytically to exist by RAMANATHAN and FISCHER (1997) based on the 3-D perturbation solutions by MOVCHAN and WILLIS (1995). Earlier 3-D analytical solutions by RICE, BEN-ZION and KIM (1994) for acoustic wave (scalar model) problem, later also studied by WOOLFRIES and WILLIS (1999), showed a non-persistent, but only slowly decaying (algebraically in time) type of dynamic stress concentration. The 3-D perturbation solution has also been given

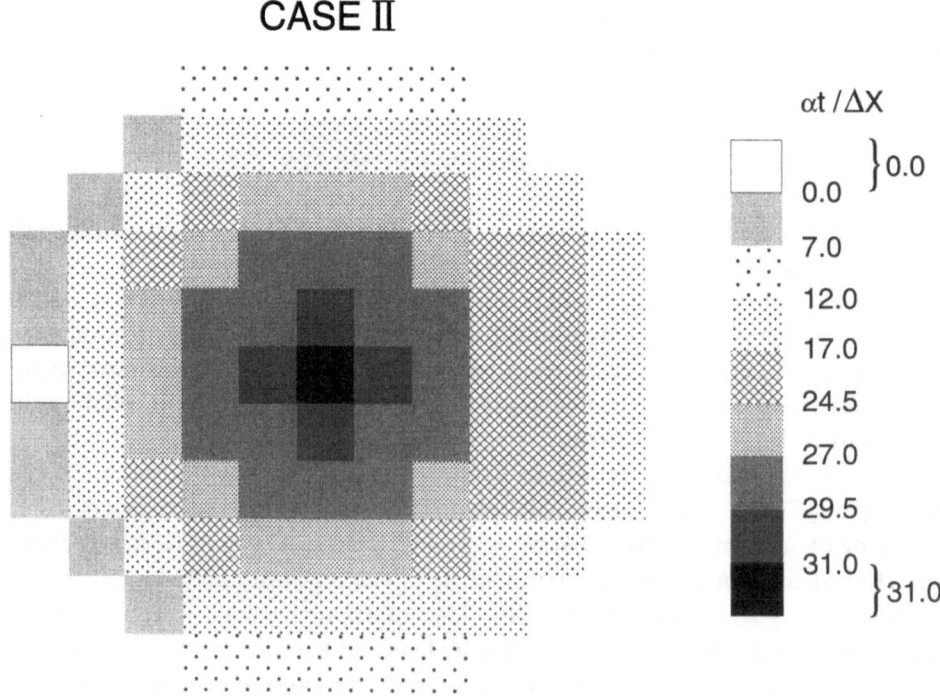

Figure 7
Rupture history for a pre-existing circular asperity on an infinite fault plane in shear, under a critical stress level fracture criterion.

**Spontaneous fracture of a pre-existing
circular shear crack**

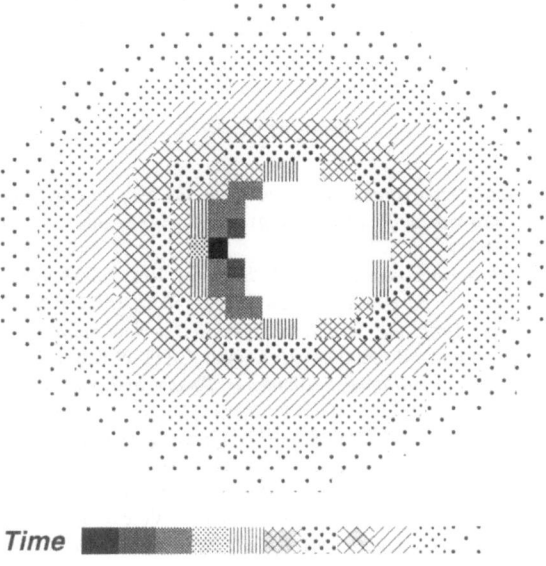

Figure 8
Rupture history for a pre-existing circular shear crack, under a critical stress level fracture criterion.

by MOVCHAN and WILLIS (1995) for shear cracks, and by WILLIS and MOVCHAN (1997) for cracks perturbed out of a plane, although it has not yet been established if and when these cases lead to persistent crack front waves.

Based on the work of DAS and KOSTROV (1988), DAS (1987) created movies which demonstrated that the existence of heterogeneities leads to narrow zones of slip propagating across faults, which have since been called the "Heaton pulse." Efforts are underway to show that such narrow pulses arise from complex friction laws, but the work of DAS and KOSTROV (1988) demonstrate that simple Coulomb friction combined with fault heterogeneities can also result in such narrow slip pulses on the fault.

Further developments in the late 1980s and the 1990s continued with rate- and state-dependent friction laws incorporated into the models (OKUBO, 1989). Some of this work is reviewed MADARIAGA and PEYRAT (2000).

Finally, we can ask what the impact of the work done in the 1970s by Aki and coworkers has been? One can answer this using the following famous statement:

"It is too soon to tell!"

Chou-en Lai, former prime minister of China, on being asked what the effect of the French Revolution was on history.

Acknowledgements

I would like to thank the Department of Earth Sciences, University of Southern California for financial support to participate in this conference.

REFERENCES

AKI, K. (1966), *Generation and Propagation of G–Waves from the Niigata Earthquake of June 16, 1964. 2. Estimation of Earthquake Movement, Released Energy and Stress–Strain Drop from G–Wave Spectrum*, Bull. Eq. Res. Inst. *44*, 23–88.

AKI, K. (1968), *Seismic Displacements near a Fault*, J. Geophys. Res. *73*, 5359–5376.

AKI, K. (1979), *Characterization of Barriers on an Earthquake Fault*, J. Geophys. Res. *84*, 6140–6148.

AKI, K. *Re-evaluation of Stress Drop and Seismic Energy using a New Model of Earthquake faulting*, In Source Mechanism and Earthquake Prediction (ed. Allègre, C.) (C.N.R.S. Publ., Paris 1980), pp. 23–50.

AKI, K. and RICHARDS, P. G., *Quantitative Seismology: Theory and Methods* (San Francisco: W. H. Freeman and Co., 1980).

AKI K., BOUCHON, M., CHOUET, B., and DAS, S. (1977), *Quantitative Prediction of Strong Motion for a Potential Earthquake Fault*, Annali de Geofisica *XXX*, 341–368.

AKI, K., FEHLER, M., and DAS, S. (1977), *Source Mechanism of Volcanic Tremor: Fluid-Driven Crack Models and their Application to the 1973 Kilaueau Eruption*, J. Volcanol. *2*, 259–287.

ANDREWS, D. J. (1976a), *Rupture Propagation with Finite Stress Antiplane Strain*, J. Geophys. Res. *81*, 3575–3582.

ANDREWS, D. J. (1976b), *Rupture Velocity of Plane Strain Shear Cracks*, J. Geophys. Res. *81*, 5679–5687.

ANDREWS, D. J. (1985), *Dynamic Plane-strain Shear Rupture with a Slip-weakening Friction Law Calculated by a Boundary Integral Method*, Bull. Seismol. Soc. Am. *75*, 1–22.

ARCHULETA, R. J. (1976), *Experimental and Numerical Three-dimensional Simulations of Strike-slip Earthquakes* (Ph.D. Dissertation, University of California, San Diego, 1976).

ATKINSON, C. and ESHELBY, J. D. (1968), *The Flow of Energy into the Tip of a Moving Crack*, Intl. J. Frac. *4*, 3–8.

BARENBLATT, G. I. (1959), *The Formation of Equilibrium Cracks During Brittle Fracture. General Ideas and Hypotheses*, J. Appl. Math. Mech. *23*, 434–444.

BARENBLATT, G. I. and CHEREPANOV, G. P. (1960), *On the Wedging of Brittle Bodies* (English translation) Phys. Math. Mech. *24*, 667–682.

BARKA, A. A. and KADINSKY-CADAE, K. (1988), *Strike-slip Fault Geometry in Turkey and its Influence on Earthquake Activity*, Tectonics 7, 663–684.

BOUCHON, M. (1979), *Predictability of Ground Displacement and Velocity near an Earthquake Fault: An Example: The Parkfield Earthquake of 1966*, J. Geophys. Res. *84*, 6149–6156.

BRACE, W. F. and WALSH, J. B. (1962), *Some Direct measurements of the Surface Energy of Quartz and Orthoclase*, Am. Mineral. *47*, 1111–1122.

BRUNE, J. N. (1970), *Tectonic Stress and the Spectra of Seismic Shear Waves from Earthquakes*, J. Geophys. Res. *75*, 4997–5009.

BROBERG, K. B. (1960), *The Propagation of a Brittle Crack*, Arkiv för Fysik *18*, 159–192.

BROBERG, K. B., *Cracks and Fracture* (Academic Press, New York, 1999).

BRUHN, R. L., GIBLER, P. R. and PARRY, W. T. (1987), *Rupture Characteristics of Normal Faults: An Example from the Wasatch Fault Zone, Utah*, Continental Extensional Tectonics, Geol. Soc. Special Publ. *29*, 337–353.

BURRIDGE, R. (1969), *The Numerical Solution of Certain Integral Equations with Non-Integrable Kernels Arising in the Theory of Crack Propagation and Elastic Wave Diffraction*, Phil. Trans. R. Soc. Lond. *A265*, 353–381.

BURRIDGE, R. (1973), *Admissible Speeds for Plane-strain Self-similar Shear Crack with Friction but Lacking Cohesion*, Geophys. J. R. Astron. Soc. *35*, 439–455.

BURRIDGE, R. and KNOPOFF, L. (1964), *Body Force Equivalents for Seismic Dislocations*, Bull. Seismol. Soc. Am. *54*, 1875–1888.

BURRIDGE, R. and MOON, R. (1981), *Slipping on a Frictional Fault Plane in Three Dimensions: A Numerical Simulation of a Scalar Analog*, Geophys. J. R. Astron. Soc. *67*(2), 325–342.

BURRIDGE, R. and WILLIS, J. R. (1969), *The Self-similar Problem of the Expanding Elliptical Crack in an Anisotropic Solid*, Proc. Camb. Phil. Soc. *66*, 443–468.

BURRIDGE, R., CONN, G., and FREUND, L. B. (1979), *The Stability of a Rapid Mode II Shear Crack with Finite Cohesive Traction*, J. Geophys. Res. *84*, 2210–2222.

CHAO, C. C. (1960), Dynamical Response of an Elastic Half-space to Tangential Surface Loadings, J. Appl. Mech. *27*, 559–567.

COCHARD, A. and MADARIAGA, R. (1994), *Dynamic Faulting Under Rate-dependent Friction*, Pure Appl. Geophys. *142*, 419–445.

COCHARD, A. and MADARIAGA, R. (1996), *Complexity of Seismicity due to Highly Rate-dependent Friction*, J. Geophys. Res. *101*, 25321–25336.

CRAGGS, J. W. (1960), *On the Propagation of a Crack in a Elastic-brittle Material*, J. Mech. Phys. Solids, *8*, 66–75.

DARWIN, C., *A Naturalist's Voyage*, (John Murray, New York, 1889).

DAS, S. (1976), *A Numerical Study of Rupture Propagation and Earthquake Source Mechanism* (Sc.D. Thesis, MIT, 1976).

DAS, S. (1980), *A Numerical Method for Determination of Source Time Functions for General Three-dimensional Rupture Propagation*, Geophys. J. R. Astron. Soc. *62*, 591–604.

DAS, S. (1981), *Three-dimensional Spontaneous Rupture Propagation and Implications for Earthquake Source Mechanism*, Geophys. J. R. Astron. Soc. *67*, 375–393.

DAS, S. (1985), *Application of Dynamic Shear Crack Models to the Study of the Earthquake Faulting Process*, Intl. J. Frac. *27*, 263–276.

DAS, S. (1986), *Comparison of the Radiated Fields Generated by the Fracture of a Circular Crack and a Circular Asperity*, Geophys. J. R. Astron. Soc. *85*, 601–615.

DAS, S. (1987), *Complex Earthquake Fault Dynamics: Color Movies*, Trans. Am. Geophys. Un. *68*, 1242.

DAS, S. (1992), *Reactivation of an Oceanic Fracture by the Macquarie Ridge Earthquake of 1989*, Nature *357*, 150–153.

DAS, S. (1993), *The Macquarie Ridge Earthquake of 1989*, Geophys. J. Intl. *115*, 778–798.

DAS, S. and AKI, K. (1977a), *A Numerical Study of Two-dimensional Rupture Propagation*, Geophys. J. R. Astron. Soc. *50*, 643–668.

DAS, S. and AKI, K. (1977b), *Fault Plane with Barriers: A Versatile Earthquake Model*, J. Geophys. Res. *82*, 5658–5670.

DAS, S. and KOSTROV, B. V. (1983), *Breaking of a Single Asperity: Rupture Process and Seismic Radiation*, J. Geophys. Res. *88*, 4277–4288.

DAS, S. and KOSTROV, B. V. (1985), *An Elliptical Asperity in Shear: Fracture Process and Seismic Radiation*, Geophys. J. R. Astron. Soc. *80*, 725–742.

DAS, S. and KOSTROV, B. V., *Fracture of a single asperity on a finite fault: A model for weak earthquakes?* In *Earthquake Source Mechanics* (eds. Das, S., Boatwright, J., and Scholz, C. H.) (AGU Monograph *37* 1986), pp. 91–96.

DAS, S. and KOSTROV, B. V. (1987), *On the Numerical Boundary Integral Equation Method for Three-dimensional Dynamic Shear Crack Problems*, J. App. Mech. *54*, 99–104.

DAS, S. and KOSTROV, B. V. (1988), *An Investigation of the Complexity of the Earthquake Source Time Function Using Dynamic Faulting Models*, J. Geophys. Res. *93*, 8035–8050.

DAS, S. and KOSTROV, B. V. (1990), *Inversion for Slip Rate History and Distribution on Fault with Stabilizing Constraints—The 1986 Andreanof Islands Earthquake*, J. Geophys. Res. *95*, 6899–6913.

DAS, S. and KOSTROV, B. V. (1994), *Diversity of Solutions of the Problem of Earthquake Faulting Inversion. Application to SH Waves for the Great 1989 Macquarie Ridge Earthquake*, Phys. Earth Planet. Int. *85*, 293–318.

DAS, S. and SUHADOLC, P. (1996), *On the Inverse Problem for Earthquake Rupture. The Haskell-Type Source Model*, J. Geophys. Res. *101*, 5725–5738.

DAS, S., SUHADOLC, P., and KOSTROV, B. V. (1996), *Realistic Inversions to Obtain Gross Properties of the Earthquake Faulting Process*, Tectonophysics, Special issue entitled *Seismic Source Parameters: from Microearthquakes to Large Events* (ed. C. Trifu), *261*, 165–177.

DAY, S. M. (1982a), *Three-dimensional Simulation of Spontaneous Rupture: The Effect of Non-uniform Prestress*, Bull. Seismol. Soc. Am. *72*, 1881–1902.

DAY, S. M. (1982b), *Three-dimensional Finite Difference Simulation of Fault Dynamics: Rectangular Faults with Fixed Rupture Velocity*, Bull. Seismol. Soc. Am. *72*, 705–727.

DMOWSKA, R. and RICE, J. R. *Fracture theory and its seismological applications.* In *Continuum Theories in Solid earth Physics* (ed. Teisseyre, R.) (Elsevier Publ. Co., Holland 1985) *III*, pp. 187–255.

DUGDALE, D. S. (1960), *Yielding of Steel Sheets Containing Slits*, J. Mech. Phys. Solids *8*, 100–110.

ESHELBY, J. D. *The continuum theory of lattice defects.* In *Progress in Solid State Physics* (eds. Seitz, F. and Turnbull, D.), (Academic Press, New York 1956) *3*, 79–144.

ESHELBY, J. D. (1957), *The Determination of the Elastic Field of an Ellipsoidal Inclusion, and Related Problems*, Proc. R. Soc. Lond., *A241*, 376–396.

ESHELBY, J. D. (1969), *The Elastic Field of a Crack Extending Nonuniformly under General Antiplane Loading*, J. Mech. Phys. Solids *17*, 177–199.

EVVARD, J. C. (1950), *Use of Source Distributions for Evaluating Theoretical Aerodynamics of Thin Finite Wings at Supersonic Speeds*, N. A. C. A. Report, *951*.

FREUND, L. B. (1979), *The Mechanics of Dynamic Shear Crack Propagation*, J. Geophys. Res. *84*, 2199–2209.

FREUND, L. B., *Dynamic Fracture Mechanics* (Appl. Math. Mech. Ser., Cambridge University Press, New York), 1990.

GRIFFITH, A. A. (1921), *The Phenomenon of Rupture and Flow in Solids*, Phil. Trans. R. Soc. Lond., Ser. A. *221*, 163–198.

HAMANO, Y. (1974), *Dependence of Rupture Time History on the Heterogeneous Distribution of Stress and Strength on the Fault* (abstract), Trans. Am. Geophys. Un. *55*, 352.

HASKELL, N. A. (1964), *Total Energy and Energy Spectral Density of Elastic Wave Radiation from Propagating Faults*, Bull. Seismol. Soc. Am. *54*, 1811–1841.

HENRY, C., DAS, S., and WOODHOUSE, J. H. (2000), *The Great March 25, 1998 Antarctic Plate Earthquake: Moment Tensor and Rupture History*, J. Geophys. Res. *105*, 16,097–16,118.

HOWELL, B. F., *An Introduction to Seismological Research. History and Development* (Cambridge, New York, 1990).

HUSSEINI, M. I., JOVANOVICH, D. B., RANDALL, M. J., and FREUND, L. B. (1975), *The Fracture Energy of Earthquakes*, Geophys. J. R. Astron. Soc. *43*, 367–385.

IDA, Y. (1972), *Cohesive Force across the Tip of a Longitudinal Shear Crack and Griffith' Specific Surface Energy*, J. Geophys. Res. *77*, 3796–3805.

IDA, Y. (1973), *Stress Concentrations and Unsteady Propagation of Longitudinal Shear Crack*, J. Geophys. Res. *78*, 3418–3429.

IRWIN, G. R. (1957), *Analysis of Stresses and Strains near the End of a Crack Traversing a Plate*, J. Appl. Mech. *24*, 361–364.

IRWIN, G. R. *Fracture dynamics.* In *Fracturing of Metals* (Cleveland: ASM 1948), pp. 147–166.

IRWIN, G. R. (1969), *Basic Concepts for Dynamic Fracture Testing*, Trans. ASME *91*, 519–524.

KING, G. and YIELDING, F. (1984), *The Evolution of a Thrust Fault System: Processes of Rupture Initiation, Propagation and Termination in the 1980 El Asnam (Algeria) Earthquake*, Geophys. J. R. Astron. Soc. *77*, 915–933.

KOSTROV, B. V. (1964), *Selfsimilar Problems of Propagation of Shear Cracks*, J. Appl. Math. Mech. *28*, 1077–1087.

KOSTROV, B. V. (1966), *Unsteady Propagation of Longitudinal Shear Cracks*, J. Appl. Math. Mech. *30*, 1241–1248.

KOSTROV, B. V. (1975), *On the Crack Propagation with Variable velocity*, Intl. J. Frac. *11*, 47–56.

KOSTROV, B. V. and DAS, S. (1984), *Evaluation of Stress and Displacement Fields due to an Elliptical Plane Shear Crack*, Geophys. J. R. Astron. Soc. *78*, 19–33.

KOSTROV, B. V. and DAS, S., *Principles of Earthquake Source Mechanics* (Appl. Math. Mech. Ser., Cambridge University Press, New York, 1988).

KOSTROV, B. V. and NIKITIN, L. V. (1970), *Some General Problems of Mechanics of Brittle Fracture*, Archiwum Mechaniki Stosowanej *22*, 749–775.

KOTO, B. (1893), *On the Cause of the Great Earthquake in Central Japan, 1891*, Tokyo Univ. Coll. Sci. J. *5*, 295–353.

LEONOV, M. YA. and PANASYUK, V. V. (1959), *Growth of the Minutest Cracks in a Brittle Body* (in Ukrainian), Prikladnaya Meckhanika *5*, 391–401.

LINDH, A. G. and BOORE, D. M. (1981), *Control of Rupture by Fault Geometry during the 1966 Parkfield Earthquake*, Bull. Seismol. Soc. Am. *71*, 95–116.

MADARIAGA, R. (1976), *Dynamics of an Expanding Circular Fault*, Bull. Seismol. Soc. Am. *66*, 639–666.

MADARIAGA, R. (1977), *High-Frequency Radiation from Crack (Stress Drop) Models of Earthquake Faulting*, Geophys. J. R. Astron. Soc. *51*, 625–651.

MADARIAGA, R. (1979), *On the Relation between Seismic Moment and Stress Drop in the Presence of Stress and Strength Heterogeneity*, J. Geophys. Res. *84*, 2243–2249.

MADARIAGA, R. (2000), *Earthquake Source Dynamics: Some Open Questions*, this volume.

MADARIAGA, R. and COCHARD, A. (1992), *Heterogeneous Faulting and Friction*, Intl. Symp. Earthquake Disaster, Mexico City.

MADARIAGA, R., PEYRAT, S., and OLSEN, K. B. (2000), Rupture Dynamics in 3D: A Review, In Problems in Geophysics for the New Millennium (Bologna, Italy: Editrice Composition), 89–110.

McGARR, A., SPOTTISWOODE, S. M., GAT, N. C., and ORTLEPP, W. D. (1979), *Observations Relevant to Seismic Driving Stress, Stress Drop and Efficiency*, J. Geophys. Res. *84*, 2251–2261.

MOTT, N. F. (1948), *Fracture of Metals: Theoretical Considerations*, Eng. *165*, 16–18.

MORRISSEY, J. W. and RICE, J. R. (1998), *Crack Front Waves*, J. Mech. Phys. Solids *46*, 467–487.

MOVCHAN, A. B. and WILLIS, J. R. (1995), *Dynamic Weight Functions for a Moving Crack. II. Shear Loading*, J. Mech. Phys. Solids *43*, 1369–1383.

NABELEK, J. and KING, G. (1985), *Role of Fault Bends in the Initiation and Termination of Earthquake Rupture*, Science *228*, 984–987.

OHNAKA, M. (1996), *Nonuniformity of the Constitutive Law Parameters for Shear Rupture and Quasistatic Nucleation to Dynamic Rupture: A Physical Model of Earthquake Generation Process*, Proc. Natl. Acad. Sci. U.S.A. *93*, 3795–3802.

OHNAKA, M. and SHEN, L. F. (1999), *Scaling of the Shear Rupture Process from Nucleation to Dynamic Propagation: Implications of Geometric Irregularity of the Rupturing Surfaces*, J. Geophys. Res. *104*, 817–844.

OKUBO, P. (1989), *Dynamic Rupture Modeling with Laboratory-derived Constitutive Relations*, J. Geophys. Res. *94*, 12,321–12,335.

PALMER, A. C. and RICE, J. R. (1973), *The Growth of Slip Surfaces in the Progressive Failure of Overconsolidated Clay*, Proc. R. Soc. (Lond.), *A332*, 527–548.

RAMANATHAN, S. and FISHER, D. S. (1997), *Dynamic Instabilities of Planar Tensile Cracks in Heterogeneous Media*, Phys. Rev. Lett. *79*, 877–880.

RANDALL, M. J. (1971), *Elastic Multipole Theory and Seismic Moment*, Bull. Seismol. Soc. Am. *61*, 1321–1326.

REID, H. F., *The Mechanics of the Earthquake*. In The California Earthquake of April 18, 1906, Report of the State Investigation Commission, (Washington, D. C.: Carnegie Institute of Washington, 1910), *2*.

RICE, J. R. (1968), *A Path Independent Integral and the Approximate Analysis of Strain Concentrations by Notches and Cracks*, J. Appl. Mech. *35*, 379–386.

RICE, J. R., BEN-ZION, Y. and KIM, K. S. (1994), *Three-dimensional Perturbation Solution for Dynamic Planar Crack Moving Unsteadily in a Model Elastic solid*, J. Mech. Phys. Solids *42*, 813–843.

RICHARDS, P. G. (1979), *Elementary Solutions to Lamb's Problem for a Point Source and their Relevance to Three-Dimensional Studies of Spontaneous Crack Propagation*, Bull. Seismol. Soc. Am. *69*, 947–956.

ROSAKIS, A. J., SAMUDRALA, O., and COKER, D. (1999), *Cracks Faster than the Shear Wave Speed*, Science *284*, 1337–1340.

RUDNICKI, J. W. and KANAMORI, H. (1981), *Effects of Fault Interaction on Moment, Stress-Drop and Strain Energy Release*, J. Geophys. Res. *86*, 1785–1793.

SARAO, A., DAS, S., and SUHADOLC, P. (1998), *Effect of Non-uniform Station Coverage on the Inversion for Seismic Moment Release History and Distribution for a Haskell-type Rupture Model*, J. Seismol. *2*, 1–25.

SATO, T. and HIRASAWA, T. (1973), *Body Wave Spectra from Propagating Shear Cracks*, J. Phys. Earth *21*, 415–431.

SAVAGE, J. C. (1966), *Radiation from a Realistic Model of Faulting*, Bull. Seismol. Soc. Am. *56*, 577–592.

SIBSON, R. *Rupture interaction with fault jogs*. In *Earthquake Source Mechanics* (eds. S. Das, J. Boatwright, and C. H. Scholz) (AGU Monograph *37* 1986), 157–167.

SPOTTISWOODE, S. M. and McGARR, A. (1975), *Source Parameters of Tremors in a Deep-level Gold Mine*, Bull. Seismol. Soc. Am. *65*, 93–112.

STARR, A. T. (1928), *Slip in a Crystal and Rupture in a Solid*, Proc. Camb. Phil. Soc. *24*, 489–500.

VIRIEUX, J. and MADARIAGA, R. (1982), *Dynamic Faulting Studied by a Finite Difference Method*, Bull. Seismol. Soc. Am. *72*, 345–369.

WARD, G. N., *Linearized Theory of Steady High-speed Flow* (Cambridge Monographs on Mechanics and Applied Mathematics, Cambridge University Press, 1955).

WILLIS, J. R. (1967), *A Comparison of the Fracture Criteria of Griffith and Barenblatt*, J. Mech. Phys. Solids *15*, 151–162.

WILLIS, J. R. and MOVCHAN, A. B. (1995), *Dynamic Weight Functions for a Moving Crack. I. Mode I Loading*, J. Mech. Phys. Solids *43*, 319–341.

WILLIS, J. R. and MOVCHAN, A. B. (1997), *Three-dimensional Dynamic Perturbation of a Propagating Crack*, J. Mech. Phys. Solids *45*, 591–610.

WOOLFRIES, S. and WILLIS, J. R. (1999), *Perturbation of a Dynamic Planar Crack Moving in a Model Elastic Solid*, J. Mech. Phys. Solids *47*, 1633–1661.

(Received July 23, 2000, accepted February 27, 2002)

To access this journal online:
http://www.birkhauser.ch

Pure appl. geophys. 160 (2003) 603–634
0033–4553/03/040603–32

Pure and Applied Geophysics

The Barrier Model and Strong Ground Motion

APOSTOLOS S. PAPAGEORGIOU[1]

Abstract—An overview of the most important developments in Engineering (or Strong Motion) Seismology is presented alongside Professor Keiiti Aki's contributions, who is one of the founders of this field. The mechanics of earthquake rupture are discussed with due emphasis on the various physical phenomena. The presentation is made in a tutorial manner, borrowing freely from Keiiti Aki's papers, and endeavoring to emulate his unique style of clarity, simplicity and synthetic ability.

Key words: Barrier, specific barrier model, earthquake source, strong ground motion.

Introduction

The work of Professor Keiiti Aki in the discipline of Seismology is unprecedented in its breadth, depth and originality. His contributions span virtually the entire frequency range (i.e., normal modes, surface waves, body waves, strong motion, seismic coda, harmonic tremor). In the present article, we aim to survey his contributions to the field of *Engineering* (or *Strong Motion*) *Seismology*. The presentation is made in a tutorial manner, borrowing freely from Keiiti Aki's papers, and trying to emulate his unique style of clarity, simplicity and synthetic ability.

Earthquake Seismology deals with the study of the generation, propagation, and recording of elastic waves in the earth, and of the physical processes occurring at the source of an earthquake. By the term *Engineering* (or *Strong-Motion*) *Seismology* we mean that part of seismology dealing with earthquakes close enough to the causative source where ground motion is strong enough to pose a threat to engineering structures. The principle problem of engineering seismology is the estimation of strength, frequency content, duration and spatial variability of the most destructive (in terms of its effects on a particular structure) ground-shaking that is likely to occur at a site. This estimation should be based on the physics of the generation and propagation of seismic waves. According to AKI (1980b), the ultimate objective of current research efforts is to "*compute seismic motion expected at a specific site of an*

[1] University at Buffalo, State University of New York, Department of Civil, Structural and Environmental Engineering, Ketter Hall, Rm. 212, Buffalo, New York 14260–4300.
E-mail: papaga@eng.buffalo.edu

engineering structure when the fault mapped by geologists breaks." In the early days of earthquake studies, before the development of sensitive seismographs, all seismology was of necessity "strong motion seismology," as this is evident in the work of ROBERT MALLET (1810–1881), who established the basis of observational field seismology in his detailed study of the destructive Neapolitan earthquake of 16 December 1857 in Italy (MALLET, 1862).

Tectonic Processes and the Mechanics of Earthquake Rupture

On the basis of overwhelming evidence it is now widely accepted that earthquakes are caused by the dynamic spreading of shear rupture on a fault plane (AKI, 1972a). This model of earthquake source is the *"fault model,"* initially proposed by REID (1910) in his *"elastic rebound theory."* On the basis of deformations observed on the surface or measured by geodetic methods and seismic data obtained at local and distant stations, Reid proposed that the San Francisco earthquake of 1906 was the release of strain energy stored in the vicinity of the San Andreas fault by a slip along the fault. This theory stirred considerable controversy. AKI (1979b, 1988) gives an historical account of the controversies which the fault model has survived from its early days until it was firmly established in the mid-60s, when a quantitative test of the model became possible with the use of the global network of calibrated stations, the advent of large-scale digital computers and the development of an appropriate mathematical framework, the so-called *"dislocation theory,"* which relates the observed seismogram with the slip motion across a fault plane. Furthermore, the success of the theory of plate tectonics provides the strongest support for the fault model. The *"theory of plate tectonics,"* which describes the kinematics of the upper layer of the earth, was implicit in Reid's elastic rebound theory. It is based upon the assumption that the upper part of the crust, called the lithosphere, is decidedly more rigid than the underlying asthenosphere. The lithosphere is composed of a number of plates which move relative to the mantle and to each other. Indeed, the consistency of plate motions with the direction and amount of slip during earthquakes everywhere is remarkable.

Kinematics of Fault Rupture

We have already pointed out above that earthquake ground motion results from unstable slip accompanying a sudden drop in shear stress on a geologic fault. Therefore, an earthquake is primarily a mechanical process. During the short span of this process, the earth, except in the earthquake source, behaves as an elastic body. Consequently, seismic waves are linear elastic waves propagating in a very complex, nonhomogeneous, dissipative, prestressed medium (the earth is in a prestressed state due to internal deformation and its own gravitational field). Therefore, the basic analytical tool for studying earthquakes is *classical elastodynamic theory* (e.g., GURTIN, 1972; ACHENBACH, 1973; ERINGEN and SUHUBI, 1975; MIKLOWITZ, 1978) supplemented with *fracture mechanics* (e.g., FREUND, 1990; KOSTROV and DAS, 1988; BROBERG, 1999).

In order to express mathematically the ground motion induced by an earthquake, we need a formula for the displacement—at a general point in space and time—in terms of the physical parameters that originated the motion. This formula is provided by the *elastodynamic representation theorem*. As noted by AKI and RICHARDS (1980), the representation theorem is a bookkeeping device by which the displacement from realistic source models is synthesized from the displacement produced by the simplest of sources—namely, the unidirectional unit impulse, which is localized precisely both in space and time. The displacement response due to such a singular source is referred to as *Green's function*.

A mathematical statement of the representation theorem for the *faulting source* (BURRIDGE and KNOPOFF, 1964)

$$u_n(\mathbf{x}, t) = \int\limits_{-\infty}^{+\infty} d\tau \iint\limits_{\Sigma} \Delta u_i(\boldsymbol{\xi}, \tau) c_{ijpq} v_j \frac{\partial}{\partial \xi_q} G_{np}(\mathbf{x}, t - \tau; \boldsymbol{\xi}, 0) \, d\Sigma(\boldsymbol{\xi}) \qquad (1)$$

where c_{ijpq} are the components of the elasticity tensor, $G_{np}(\mathbf{x}, t - \tau; \boldsymbol{\xi}, 0)$ is the Green's tensor which represents the n-th component of displacement at $(\mathbf{x}, t - \tau)$ due to an impulsive concentrated unit force acting in the p-th direction at $(\boldsymbol{\xi}, 0)$, $\Delta u_i(\boldsymbol{\xi}, \tau) = u_i(\boldsymbol{\xi}, \tau)\mid_{\Sigma^+} - u_i(\boldsymbol{\xi}, \tau)\mid_{\Sigma^-}$ is the displacement discontinuity across Σ (i.e., the slip on the fault plane) and $\mathbf{v}$ is a unit vector normal on surface Σ and pointing from Σ^- to Σ^+.

Starting with eq. (1) and making use of the properties of the Dirac delta function it may be demonstrated (e.g., AKI and RICHARDS, 1980) that a point shear dislocation is equivalent to a *double-couple*, i.e., the double-couple is the body force which would have to be applied in the absence of the fault to produce the same radiation as a given point dislocation. The first person to obtain the double-couple equivalence for an effective point source of slip was VVEDENSKAYA (1956).

In the far-field and for periods with wavelengths much larger than the source size, the fault appears as a point dislocation. The scalar value of the moment of one of the couples in the double-couple representation of the point dislocation is the *seismic moment* M_0. Assuming an *average slip* $\overline{\Delta u}$ over the fault plane then

$$M_0 \equiv \mu \overline{\Delta u} A = \mu \times \text{average slip} \times \text{fault area} \qquad (2)$$

where μ is the rigidity (i.e., shear modulus) of the lithosphere and A is the area of the fault plane which slipped.

The first precise determination of the seismic moment was accomplished by AKI (1966) for the 1964 Niigata earthquake using long-period Love waves observed by

WWSSN. It is the most important static parameter of the source; the value of which at the end of the rupture process measures the permanent inelastic strain produced by the event, and thus it is the simplest way to measure the strength or size of an earthquake. It can be reliably inferred from seismic observations such as displacement spectra of long-period surface waves, free oscillations of the earth or directly from field observations, as suggested by eq. (2) and thus serves as a direct link between seismological, geological and geodetic observations.

Kinematic Models

One of the fundamental problems of seismology is the inference of the earthquake faulting process from the analysis of seismic waves radiated from the source. In order to render such an *inverse problem* tractable, physically reasonable assumptions must be made about the rupture process and its evolution in time. Thus, as a first approximation to the solution of the above problem, dislocation models were introduced as simple kinematic descriptions of the evolution of faulting with time [for a discussion of the meaning of the terms "*kinematic*" and "*dynamic*" see AKI and RICHARDS (1980), Box 5.3]. In general, in kinematic models the faulting process is represented in terms of the *slip* (or *source*) *function*, the form of which usually is chosen intuitively, without rigorous analysis of the time-dependent stresses acting on the area. In particular, dislocation models represent simple geometrical idealizations of actual faulting in the earth, and as such they are extremely simplified, averaged out versions of the rupture process. Analyses and inversion studies using dislocation models have provided significant insights into the effects of *fault finiteness* and *fault geometry* on the radiation of elastic wave energy.

One of the first and most widely used dislocation models is *Haskell's model* (HASKELL, 1964, 1966, 1969) which represents faulting on a rectangular plane of length L and width W. According to this model, rupture initiates at one end of the fault with the appearance of a dislocation line segment spanning the width of the fault. This rupture front propagates along the length of the fault with velocity V. At each point of the fault plane, slippage is initiated when the rupture front reaches the point. The time that it takes for slip at a point to reach the final value Δu_0 is called the *rise time* τ. MADARIAGA (1978) calculated exactly the elastic waves radiated by Haskell's model in the far-and near-field.

When we take a closer look at the nucleation of the rupture process, we realize that the unidirectional propagation of rupture in Haskell's rectangular fault model is an oversimplification of physical reality. For this reason, other dislocation models were proposed that allow rupture to initiate at a point (rather than simultaneously everywhere along a line segment) and then spread out radially (rather than propagate in a single direction) at a uniform velocity until it covers an arbitrary two-dimensional surface on the fault plane (SAVAGE 1966, MOLNAR et al., 1973).

Focusing our attention to the near-field, it is fair to say that modern quantitative analysis of strong ground motion observations started with the now famous Station No. 2 record (HOUSNER and TRIFUNAC, 1967) obtained from the 1966 Parkfield earthquake at a distance of only 80 m from the fault break. AKI (1968) and HASKELL (1969) demonstrated that the observed transverse component of displacement of the above ground motion record, which exhibited a simple impulsive form, was precisely what is expected for a right-lateral strike-slip rupture propagating from northwest to southeast. They used the five parameter $(L, W, \Delta u_0, V, \tau)$ *Haskell model* described above. Within this model, the most important parameters affecting the level of strong ground motion are τ and D. Of these two parameters, the rise time τ is the one that is most difficult to determine. It is relevant to point out that the *slip velocity* $\Delta \dot{u}$, an estimate of which may be obtained by the ratio $(\Delta u_0 / \tau)$, was investigated by AKI (1983) and was found to exhibit a large variation. AKI (1983) attributed this significant variation of the value of the average slip velocity to the inability to resolve (i.e., infer accurate estimates) of the short rise times.

In the beginning, most simulation methods, including those presented in the pioneering studies of AKI (1968) and HASKELL (1969), used Green's function for an unbounded homogeneous medium. Subsequently, in order to make the propagation medium more realistic, the free surface effect and the effect of sedimentary layers were included in the simulations (e.g., BOUCHON, 1979; for a review of modeling studies extending to the beginning of the 1980s, see AKI, 1982, 1983).

It gradually became evident, though, that the assumption of uniform slip over the fault plane (as required by Haskell's model) was not adequate to simulate simultaneously the motions of an earthquake event recorded at more than one recording station. Therefore, the original source model had to be modified in the following two ways:

The first important modification was to allow the slip function to vary from place to place on the fault plane i.e., the fault had to be subdivided into subfaults, each with a different slip vector. The first attempt to allow non-uniform slip was made by TRIFUNAC and UDWADIA (1974) for the 1966 Parkfield earthquake. It was the study of the 1979 Imperial Valley earthquake, however, that marked the beginning of systematic waveform inversions (HARTZELL and HELMBERGER, 1982; OLSEN and APSEL, 1982; HARTZELL and HEATON, 1983; ARCHULETA, 1984). Similar inversion studies followed for other California earthquakes, and since then inversions of this type have become routine for well recorded earthquake events.

The above *deterministic kinematic modeling* approach has been limited to a rather smooth picture obtainable from the frequency range lower than about 1 Hz (i.e., spatial resolution ~5 km). Despite this limitation, the above inversion studies demonstrated that the spatial and temporal behavior of earthquake faults is rich in *complexity* and *heterogeneity*. The slip distributions inferred by all the above inversion studies exhibit two important features: (i) the existence of relatively localized areas of large static slip, and (ii) short dislocation rise time τ. The shortness of the dislocation rise time also has been confirmed recently by eyewitness observations of the coseismic fault movement during the 1990 Luzon, Phillipines, earthquake (M_s 7.8) as reported by YOMOGIDA and NAKATA (1994) (see also WALLACE, 1984, for a pertinent eyewitness account regarding the 1983 Borah Peak, Idaho, earthquake).

The short rise time was recognized earlier by AKI (1968, 1979) for the 1966 Parkfield earthquake; he introduced the *barrier model* to explain the observed shortness of rise time. According to the barrier model, the shortness of rise time is a consequence of the strong segmentation of the fault plane, so that rupture involves sequences of crack-like propagation over a small patch, arrest at its borders, renucleation on a neighboring patch, etc. The mathematical description of such a process is provided by the *"specific barrier model"* proposed by PAPAGEORGIOU and AKI (1982) and PAPAGEORGIOU and AKI (1983a) to be described below. An alternative explanation for the shortness of rise time has been proposed by HEATON (1990) who attributes it to a *"self-healing"* process due to a particular friction law with strong velocity dependence at high slip rate. However the "self-healing" hypothesis was found by BEROZA and MIKUMO (1996), IDE and TAKEO (1997) and DAY *et al.* (1998) unnecessary to explain earthquake kinematics. [The interested reader may find informative discussions related to the "self-healing" hypothesis in BIZZARI *et al.* (2001) and GUATTERI and SPUDICH (2000). In a related study ANDREWS and BEN-ZION (1997) investigated the conditions for and properties of dynamic ruptures consisting of narrow *"propagating slip pulses"* associated with variations of normal stress].

The other important modification in the source model originally used by AKI (1968) and HASKELL (1969) to simulate strong ground motion was the introduction of stochastic elements, thus reducing the number of parameters needed to describe the details of the slip function. We will elaborate on this in our discussion of the "specific barrier model."

Rupture Dynamics on a Heterogeneous Fault

In the previous section we pointed out the usefulness of dislocation models in earthquake seismology. However, such models are associated with very strong singularities which are physically unacceptable. Specifically, strong stress singularities of the type r^{-1} are found on the fault surface around the edges of the fault. These singularities, which are a consequence of the assumption of constant slip over the entire fault plane, are so strong that an infinite strain energy change is predicted independently of any source parameters. Since faulting is a failure along a fault plane, we expect that the fault plane, once ruptured, cannot sustain stress beyond the frictional stress.

Furthermore, from a purely continuum mechanical point of view, the constant slip is inadmissible because near the borders of the dislocation model there is interpenetration of matter. Therefore, the elastic solutions are not valid inside a core region around the edges of the dislocation model. The withdrawal from these physical inconsistencies is to eliminate these singularities by smoothing the slip near the rupture front, and thus rendering a slip function that is not only kinematically satisfactory for shear faults, but also compatible with a physically plausible stress distribution on the fault plane. This requirement forms the basis of crack models. The physics of dynamic crack growth is the domain of study of *dynamic fracture mechanics* (MADARIAGA, 1978; MADARIAGA, 1983a; AKI and RICHARDS, 1980; FREUND, 1990; DMOWSKA and RICE, 1986).

We consider a 2-D crack model in a homogeneous isotropic elastic half space as shown in Figure 1. A plane crack representing the fault lies on the plane (x, y) with its rupture front parallel to the y axis. The position of the rupture front as a function of time is described by the function $x = l(t)$. The material is assumed to be elastic everywhere (even at the crack tip) and the applied external stress (*tectonic stress*) $\sigma_{zj}^0(x)$ (where $j = x$ or y) is assumed to be uniform. Inside the crack, after the passage of the rupture the stress drops to the *dynamic friction* $\sigma_{zj}^f(x)$ and the difference $\Delta\sigma(x) = \sigma_{zj}^0(x) - \sigma_{zj}^f(x)$ is the *stress drop*.

Perhaps the most important results relevant to our discussion here have been obtained by KOSTROV and NITIKIN (1970) and FREUND (1972, 1979) who demonstrated that the solutions of these crack problems have a number of universal features which are independent of the details of the rupture front motion $l(t)$ and the

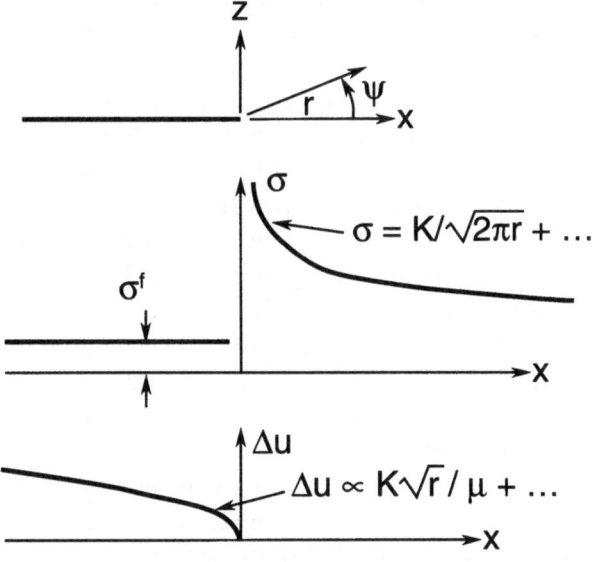

Figure 1
Two-dimensional elastic shear crack model.

stress drop distribution $\Delta\sigma(x)$. Specifically, the stress and velocity fields present characteristic inverse square-root singularities in the vicinity of the crack tip. The singularity of the stress field at the crack tip is a result of the assumption that the material remains elastic even in the immediate vicinity of the surface front. The inverse square-root singularity appears because this is the only way the elastic field can ensure a finite energy flow into the rupture front. This energy, referred to as *Griffith specific fracture energy G*, is used to create a new fault surface and is spent in the nonlinear processes taking place in the breakdown zone that exists in the vicinity of the crack tip. Thus, *linear elastic fracture mechanics* is only a large-scale approximation of the rupture process, according to which all the inelastic processes at the breakdown zone are characterized by one parameter, the *dynamic stress intensity factor K*.

To describe the fracture at an earthquake source using the crack model, two important pieces of information are necessary: (1) the initial distribution of stress on the fault surface before the earthquake, and (2) the constitutive law governing the fracture propagation, or as this law is referred to in linear elastic fracture mechanics, the *fracture criterion*.

Following ANDREWS (1978), the stress applied on the fault zone can be separated into two terms: (i) the *self-stress* that arises from irregular slip and whose sources are therefore local, and (ii) the *ambient tectonic stress* that has distant sources such as slip on distant faults, fault creep at depth and viscous drag at the base of the lithosphere. The tectonic stress arising from distant sources varies smoothly over the fault plane and has significant components at wavelengths of the order of the depth of the seismogenic region. On the contrary, the self-stress must vary strongly across the fault surface in order to explain the stationary occurrence of numerous small earthquakes.

In fracture mechanics, there are two different types of fracture criteria: (i) *Griffith's criterion* and (ii) *Irwin's criterion*, both of which have been used to describe crack growth in earthquake seismology (e.g., Box 15.2 in AKI and RICHARDS, 1980). Under both of the above criteria, we have the stress singularity in front of the tip of the advancing crack. In reality, however, no real material can sustain infinite stresses because as KNOPOFF (1981) points out, paraphrasing Spinoza, "*nature abhors an infinity.*" Thus, a zone of cohesive forces at the crack tip has been proposed to remove the stress singularity (BARENBLAT, 1959). This zone is used to model the breakdown process, small-scale yielding, microcrack formation, etc.; that takes place over a zone of finite area at the circumference of the crack (for a thorough review of the subject see e.g., RICE, 1980).

The most realistic cohesive zone model for a number of geophysical applications is the so-called *slip weakening model*. This model was used for the first time in a seismological context by IDA (1972, 1973) and its consequences on fault rupture evolution were investigated by IDA and AKI (1972) (see also ANDREWS, 1976a, 1976b, 1985, 1994). According to this model, in the simplest case slippage is modeled as rate insensitive. The strength of the fault zone reaches a peak value σ_u (also referred to as "*yield-stress*") which corresponds to the onset of slipping for fresh fractures or it is preceded by slip at lower stresses for pre-existing faults. The stress to maintain slippage reduces as the amount of slip increases up to a critical amount D (referred to as the "*characteristic weakening slip*"), above which the stress to maintain slippage remains constant, equal to the *dynamic friction* σ_f. Such a constitutive law of the fault gauge is depicted in Figure 2. The crosshatched area shown in Figure 2 represents the energy per unit area of crack absorbed at the crack tip by the breakdown process (as noted above $G = $ *Griffith's specific fracture energy*). The region of the crack near the tip where the applied stress is greater than the frictional stress is the *cohesive* (or *break-down*) *zone* d. The *average value* of the (*cohesive*) *stress*

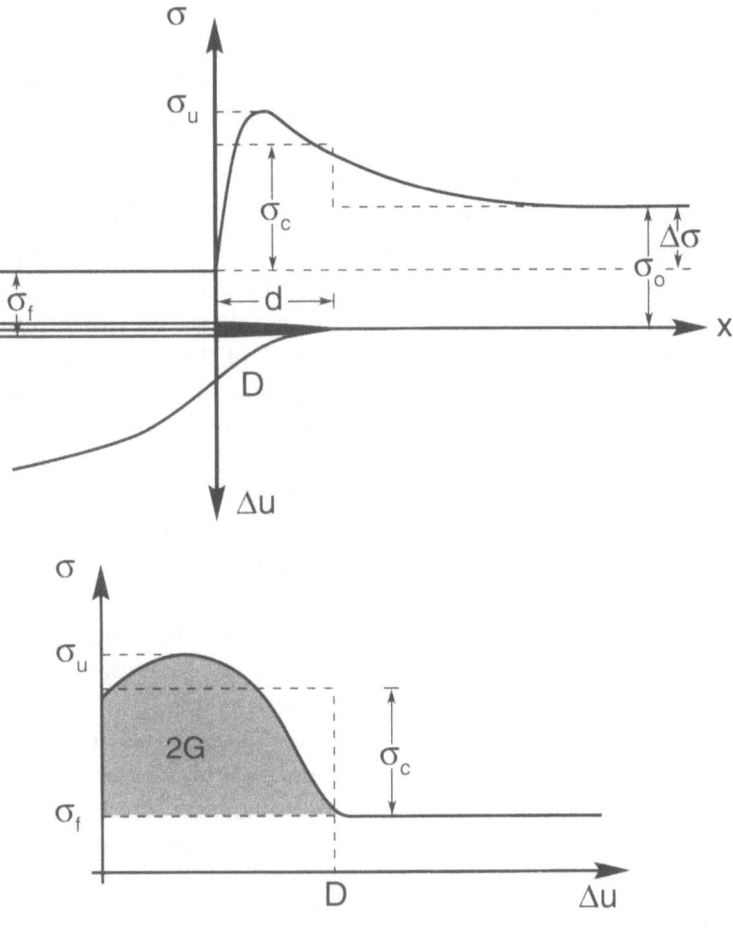

Figure 2
Constitutive law of the "slip-weakening" model.

distributed over the break-down zone is denoted by σ_c. The differences $(\sigma_u - \sigma_0)$ and $(\sigma_0 - \sigma_f)$ are referred to as the *"strength excess"* and *"stress drop"*, respectively, and the ratio $S = (\sigma_u - \sigma_0)/(\sigma_0 - \sigma_f)$ is referred to as the *"strength parameter."* Due to the finite strength of the material which is depicted by the constitutive law, the stress singularity and the slip distribution shown in Figure 1 have been replaced by a continuous stress distribution and a smooth slip distribution respectively, shown in Figure 2. [A competing constitutive law is the *"rate- and state-dependent friction law"* (DIETERICH, 1979, 1992; RUINA, 1983; PERRIN et al., 1995). In contrast to the "slip-weakening" constitutive law which assumes that friction (or total traction) is a function of the fault slip only, the "rate- and state-dependent" constitutive law implies that the friction is a function of slip velocity and state variables (BIZZARI et al., 2001)].

If the geometry and material properties of the fault zone are homogeneous and the tectonic stress uniform, then the crack, once it starts moving dynamically, will never stop. Stating this differently, without strength or stress heterogeneity, earthquake fault rupture would never stop. The only way that shear fracture may remain limited in space is that there be strong patches/ligaments on the fault surface to stop the rupture (e.g., HUSSEINI et al., 1975; MADARIAGA, 1979), or that the rupture would break into previously relaxed areas of the fault (i.e., the crack would *"run out of gas"* AKI, 1988; AKI and RICHARDS, 1980). Thus heterogeneity is a fundamental part of the earthquake process. The observed complexity of earthquake phenomena, which was extensively documented in the past two decades (for reviews see for example AKI et al., 1977; AKI, 1979a; AKI, 1980a; PAPAGEORGIOU and AKI, 1982), is a direct consequence of the heterogeneity of the physical properties of the fault zone. Once the earthquake starts, the growth and arrest of fracture is controlled in a very complex way by the distribution of stress and strength.

In order to describe heterogeneities on the fault plane and gain conceptual understanding of the complexity of rupture process, the terms *"asperities"* and *"barriers"* have been used in the published literature. *"Barrier"* is a strong patch on the fault plane which remains unbroken after the passage of the rupture front. The presence of barriers on the fault surface explains aftershocks as release of stress concentration through static fatigue. *"Asperity"* on the other hand, is a strong patch surrounded by a region where stress has been released by preslips and aftershocks (e.g., AKI, 1979a; AKI, 1984; PAPAGEORGIOU and AKI, 1982).

In the course of dynamic faulting, seismic radiation (and in particular high-frequency radiation) is controlled by the slip velocity field. In view of the fact that the most significant feature of slip velocity is the singularity at the rupture front, it follows naturally that the dominating part of seismic radiation is emitted by the rupture front. As the rupture front moves smoothly, it radiates continuously, generating the low-frequency part of the field. High frequency waves are produced by jumps in the rupture velocity and/or abrupt changes in the stress intensity factor. Accelerograms are dominated by these impulsive waves. Therefore, the radiation of high frequency waves is controlled by the motion of the rupture front. Because the rupture front is only a geometrical definition, it does not have *"inertia"* and hence its speed can change abruptly when the rupture reaches differing stress or frictional regimes which are controlled by the fault heterogeneities. As MADARIAGA (1983a) demonstrated there are two ways to produce jumps in the particle velocity radiation, and consequently strong acceleration pulses: (1) the rupture front stumbles on a barrier where the strength or rupture resistance increases suddenly, the rupture velocity changes abruptly and a strong wave (step change in particle velocity) is generated; (2) the rupture front encounters an asperity due to a previously unbroken ligament on the fault. Regardless of whether the rupture velocity changes or not, this generates a step of particle velocity. Therefore, barrier and asperities are the source of high frequency waves. The wave front discontinuities created in this fashion are evaluated by *asymptotic methods* and may be propagated away from the source by *ray theoretical methods* (MADARIAGA, 1977; MADARIAGA, 1983a; MADARIAGA, 1983b; MADARIAGA and BERNARD, 1985; ACHENBACH and HARRIS, 1978; ACHENBACH and HARRIS, 1987; BERNARD and MADARIAGA, 1984a; BERNARD and MADARIAGA, 1984b; SPUDICH and FRAZIER, 1984; ACHENBACH et al., 1982).

The above analytical results were confirmed observationally by SPUDICH and CRANSWICK (1984) who analyzed motions of the 1979 Imperial Valley earthquake recorded at the 5-element El Centro differential array, and by ZENG et al. (1993b), who by inversion mapped on the fault plane the sources of high frequency radiation of the 1989 Loma Prieta earthquake. Since both the state-of-stress on a fault and the strength of the fault material may be quite heterogeneous over a real fault surface, it is reasonable to expect that the advancement of the rupture front may be uneven, jumping around unyielding barriers—as was clearly demonstrated by the numerical studies of DAS and AKI (1977a, 1977b), MIKUMO and MIYATAKE (1978) and DAY (1982) and was verified by inversion studies such as that of BEROZA and SPUDICH (1988) and OLSEN et al. (1997)—and resulting in a pattern of broken and unbroken regions such as that of the 1966 Parkfield earthquake suggested by AKI (1979a) and shown in Figure 3. In particular, Figure 3 was obtained as follows: The hypocenters of the Parkfield aftershocks were projected on the fault plane. According to the barrier model few aftershocks are expected over a section of the fault that slipped smoothly. On the contrary, areas that act as barriers to the rupture experience little slip and are stress concentrators. This induced stress increase combined with static fatigue causes a sequence of aftershocks. With this reasoning, AKI (1979a) drew boundaries between regions with no aftershocks (slipped sections, indicated as white in Figure 3) and regions with aftershocks (unbroken barriers with little slip, indicated as gray in Figure 3). This *complementarity relation* between fault slip and aftershocks was further verified by MENDOZA and HARTZELL (1988) and TAKEO (1988).

However, the distribution of stress and strength on real faults is unknown, and thus it is impossible to describe deterministically the details of the rupture process which, as we argued above, are responsible for the generation of high frequency

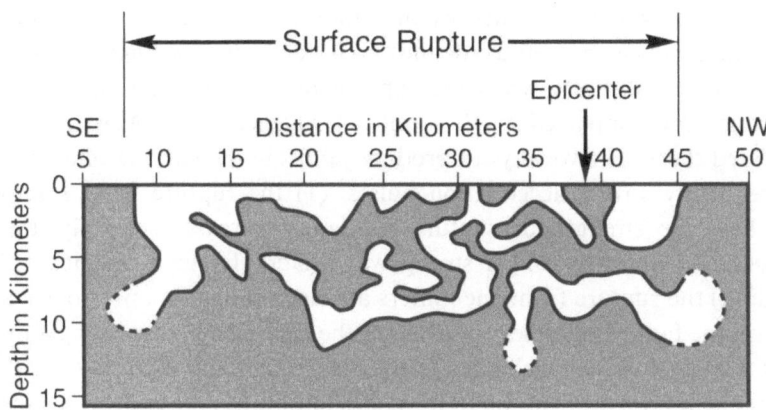

Figure 3
Heterogeneous rupture during the 1966 Parkfield, California, earthquake (modified from AKI, 1979a).

waves that dominate accelerograms. Hence, beginning with the works of HASKELL (1966) and AKI (1967), investigators have tried to introduce stochastic elements in the description of the source, and several attempts have been made to introduce a *hybrid* of deterministic and stochastic models, in which gross features of rupture propagation are specified deterministically while the details of the rupture process are described by a stochastic model specified by a small number of parameters (BOORE and JOYNER, 1978; HANKS, 1979; ANDREWS, 1981; IZUTANI, 1981; PAPAGEORGIOU and AKI, 1983a; BOATWRIGHT, 1982; BOATWRIGHT, 1988; KOYAMA, 1985).

In order to conceptualize rupture on a heterogeneous fault plane and provide the framework for its mathematical description, let us consider an idealized geometry consisting of a rectangular area containing small circles. As AKI (1982, 1983, 1984) points out, there are two opposing views of how slip can take place over this fault plane. In one of them, the circles represent strong ligaments resisting fracture, while the regions between circles have already slipped aseismically. Once the rupture starts, the ligaments will break in a more or less independent manner and will generate the high frequency waves that are observed in accelerograms. After the rupture, the entire area of the fault is broken, and the residual stress will be uniform over the fault plane, equal to the static friction. This viewpoint, that is referred to as the "*asperity model*," was adopted by KANAMORI and STEWART (1978) in interpreting the teleseismic *P* waveforms of the Guatemala earthquake of 1976, and was described by RUDNICKI and KANAMORI (1981) [For numerical studies of the rupture of an asperity see DAS and KOSTROV (1983, 1986) and FUKUYAMA and MADARIAGA (1998)].

In the other view, the circle represents a crack where a slip occurs during the fault rupture, but the region between cracks remains unbroken after the rupture. The possibility of such segmented ruptures was demonstrated by DAS and AKI (1977a) using numerical experiments. A rupture front may be stopped by a barrier, but elastic waves generated by the slip can break the fault plane ahead of the barrier in the case of shear crack. Thereafter, the rupture can propagate over the entire fault plane leaving unbroken barriers behind. The resultant irregular slip can explain observed accelerograms. This model is called the *"barrier model"* by Aki *et al.* (1977) and is supported by numerous examples of fault segmentation mapped by geologists (Aki, 1980a). In contrast to the asperity model, the residual stress over the fault plane is not uniform after the rupture. Excess stress will be induced at the unbroken barriers and may become the cause of aftershocks.

A real fault plane may contain a mixture of strong ligaments that during earthquake rupture may behave as asperities or barriers. In fact, in the numerical experiments of Das and Aki (1977a), the following three situations were found when a crack tip passes a barrier, depending on the initial stress: (1) The barrier is broken immediately; (2) the crack-tip proceeds beyond the barrier, leaving behind an unbroken barrier; and (3) the barrier is not broken at the initial passage of the crack tip, but eventually breaks due to a subsequent dynamic increase in stress, effectively behaving as an asperity.

Case (2) above was the basis of the "specific barrier model" proposed and developed by Papageorgiou and Aki (1983a, 1983b) to model and interpret strong motion acceleration spectra of major California earthquakes.

The *"specific barrier model"* consists of circular cracks of equal diameter $2\rho_0$, filling up a rectangular fault of length L and width W, as shown in Figure 4. As the rupture front sweeps the fault plane with the *"sweeping velocity"* V, a stress drop $\Delta\sigma$ (referred to as the *"local stress drop"*) takes place in each crack starting from its center and spreading with a *"spreading velocity"* v.

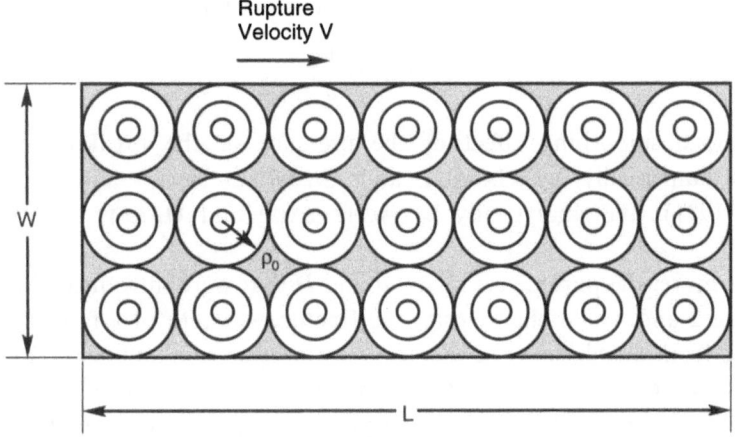

Figure 4
Specific Barrier Model (Papageorgiou and Aki, 1983a).

The slip stops abruptly when the crack radius reaches ρ_0. SATO and HIRASAWA (1973) proposed a kinematic (dislocation) model simulating the rupture of such a circular crack. The compact closed form expression of the far-field displacement waveform that they obtained has all the essential features of the waveform obtained from the more realistic numerical models studied by MADARIAGA (1976). Furthermore, PAPAGEORGIOU and AKI (1983a) demonstrated that the *stopping phases* (i.e., the phases radiated when the rupture front is arrested abruptly by the barrier that exists at the periphery of the crack) of the SATO and HIRASAWA (1973) model simulate to within a multiplicative factor the exact results obtained by MADARIAGA (1977). BERNARD and MADARIAGA (1984a) showed that these multiplicative factors may be calculated by approximate consideration of the healing waves on the fault. Finally, SATO (1994) investigated the effect that the finite deceleration time, at the final stage of rupture of the circular crack model, has on the stopping phases. Focusing again our attention on Figure 4, we point out that the region between circular cracks represents barriers left unbroken after the passage of the rupture front. The ruptures of individual cracks are statistically assumed to take place *independently*. Thus, the "specific barrier model" is a hybrid of a deterministic and stochastic one and is described by five parameters, namely L, W, $V(=v)$, $2\rho_0$ and $\Delta\sigma$.

PAPAGEORGIOU and AKI (1983b) chose major California earthquakes for which L, W, V and the maximum slip Δu_{max} were already known from observations other than strong ground acceleration data. Then, they estimated the barrier interval $2\rho_0$ and the local stress drop $\Delta\sigma$ by fitting the acceleration power spectra predicted by the model to the observed ones. In this process, they had to introduce a sixth parameter to define a cut-off frequency (called f_{max} by HANKS 1982) beyond which the acceleration spectrum decays sharply with increasing frequency.

PAPAGEORGIOU and AKI (1983a, 1983b) attributed f_{max} to the smoothing effect in the break-down zone at the crack tip and, following IDA (1973) and AKI (1979a), related it to the size d of the break-down zone using the following relation

$$f_{max} = \frac{v}{d}. \tag{3}$$

The above smoothing effect of the presence of the break-down zone was confirmed numerically by GABRIEL and CAMPILLO (1989) and FUJIWARA and IRIKURA (1991), and analytically by ACHENBACH and HARRIS (1978) and SATO (1994).

Furthermore, the parameters G, σ_c, d and D that were introduced in connection with Figure 2 and which characterize the fracture strength of a fault zone, may be determined from the observed acceleration power spectrum and eq. (3) as elaborated by PAPAGEORGIOU and AKI (1983a). The values of the above parameters thus inferred have the following physical interpretation: G represents the fracture energy of the barrier which is necessary to arrest the propagation of rupture of a subevent; d represents the length of the inelastic zone over which rupture is arrested; σ_c is the average cohesive force, distributed on the inelastic zone; D represents the slip that occurs in the break-down zone, which is required to break the bond completely. The above parameters determine the coseismic and long-term behavior of faults (e.g., CAO and AKI, 1984). For instance, it has been demonstrated theoretically that the critical slip displacement D plays a key role in determining the rupture nucleation dimension (DIETERICH, 1992), precursory deformation (YAMASHITA and OHNAKA, 1992), and high-frequency radiation of acceleration (IDA, 1973; PAPAGEORGIOU and AKI, 1983a; OHNAKA and YAMASHITA, 1989).

In the process of inferring the above parameters, PAPAGEORGIOU and AKI (1983b) corrected the observed acceleration spectra for propagation path effects, however they did not consider any correction for the local site effect of recording stations, inasmuch as they did not find obvious differences between rock and soil sites within their data set (PAPAGEORGIOU, 1988).

Later, however, a study of the site effect made by PHILLIPS and AKI (1986) at most of the stations of the USGS Central California network using the *coda method* (AKI, 1969; AKI and CHOUET, 1975) revealed a strong frequency-dependent site amplification for the average of all the network stations relative to the average of stations located on granitic rocks. Assuming that the latter site may be approximated by a homogeneous half space, the average of amplification factors for all stations was adopted in correcting the acceleration power spectra for the recording site effect (AKI and PAPAGEORGIOU, 1989). The revised source parameters (along with the corresponding ones of the Loma Prieta, 1989, earthquake which were inferred by CHIN and AKI (1991), using the "specific barrier model") are shown in Figure 5 as a function of Magnitude (M_s). It is evident that these source parameters show a remarkably systematic dependence on magnitude over the range 6.1 to 7.5. In particular, we point out the constancy (within a factor of 2) with magnitude of $\Delta\sigma$, and the linear variation of the sub-event size $2\rho_0$ with earthquake size. It is relevant also to point out that recently BERESNEV and ATKINSON (1999), in their simulations of strong ground motion using a model virtually identical to the "specific barrier model" (except that they use BRUNE's (1970) model to describe the radiation of the subevents), were compelled to use a subevent size that increases linearly with earthquake size in order to achieve best fits to the observed spectral amplitudes of the events that they analyzed. Finally, as pointed out originally by AKI *et al.* (1977) (see also PAPAGEORGIOU and AKI, 1982), the "barrier interval" $2\rho_0$ and the "rise time" τ are related as follows:

$$\text{(Barrier interval)} \sim \text{(Rise Time)} \cdot \text{(Rupture Velocity)}.$$

This relation suggests that knowledge of one of the two parameters i.e., the barrier interval or the rise time, permits the estimation of the other, given that the average rupture velocity ("sweeping velocity") V is a fairly stable parameter ($V = 0.6$ to $0.9\,\beta$, where $\beta =$ shear wave velocity).

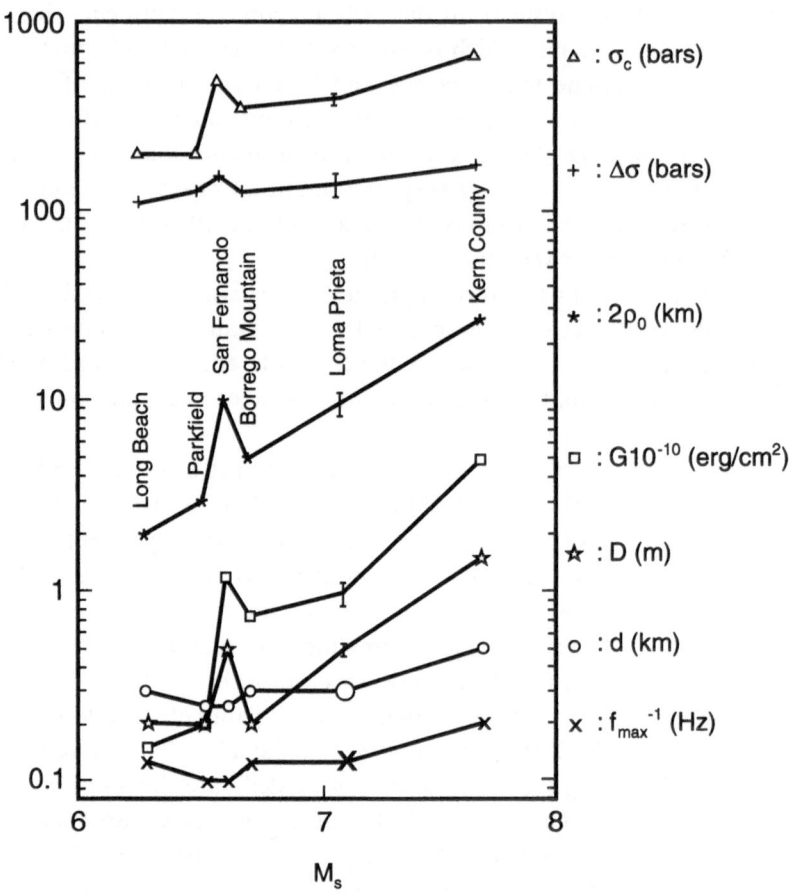

Figure 5
Source parameters of major California earthquakes.

The issue of f_{max}, whether it is due to source effects or recording site effects, has been controversial. HANKS (1982), ANDERSON and HOUGH (1984) and others found that f_{max} depends on the geologic condition of the recording site. On the other hand, AKI and PAPAGEORGIOU (1988) found that the f_{max} effect remained even after eliminating the site effect from the acceleration spectra. From an earthquake engineering point of view, f_{max} is an important parameter because it controls peak acceleration which is an important parameter for the seismic resistant design of structures. AKI and IRIKURA (1991) point out the work of KINOSHITA (1992) who found that f_{max} observed at the bottom of deep boreholes (about 3 km) in bedrock in Central Japan showed strong variation depending on the plate-tectonic setting of the seismic source. Furthermore, a weak but significant, in terms of its implications, increase of f_{max} with decreasing magnitude was observed by PAPAGEORGIOU and AKI

(1983b) for California earthquakes and by IRIKURA and YOKOI (1984) (see also AKI, 1988 and PAPAGEORGIOU, 1988) and UMEDA et al. (1984) for Japanese earthquakes, rendering another support for the source effect on f_{max}. The weak increase of the site of the break-down zone d with earthquake magnitude, which is evident in Figure 5, is a direct consequence of the above decrease of f_{max} with earthquake magnitude. As it was originally proposed by AKI (1988), (see also PAPAGEORGIOU, 1988, and references therein), the size of the break-down zone d is a measure of the width of the fault zone. With regard to this last statement, it is relevant to refer to the work of YAMASHITA and FUKUYAMA (1996). These investigators modeled that fault zone as a zone of densely distributed pre-existing cracks, consistent with seismological observations (LEARY et al., 1987; LI et al., 1994) and studied its behavior numerically. They found that the apparent critical slip displacement D is larger when the distribution density of the pre-existing cracks is larger and/or the fault zone width is greater. This is consistent with the observed variation of parameters D and d with earthquake magnitude shown in Figure 5. At any rate, we now have a more reliable estimator of the breakdown zone d, namely the low-velocity, low-Q zone measured by trapped modes in the fault zone (AKI, 2000, and references therein). It is relevant to point out that the estimates of d for the 1992 Landers earthquake zone (180 m; LI et al., 2000) and the 1966 Parkfield earthquake zone (160 m; LI, et al., 1997), based on the analysis of trapped modes, is in complete harmony with the estimates of d for the California earthquakes that PAPAGEORGIOU and AKI (1983b) analyzed (Fig. 5). Furthermore, for the 1992 Landers earthquake, a similar estimate of the fault zone width was made by an entirely different method at the same sites where the trapped modes were observed. From a detailed study of tension cracks on the surface, JOHNSON et al. (1994) concluded that the Landers fault rupture is *not* a distinct slip across a fault plane but rather a belt of localized shearing spread over a width of 50–100 m. AKI (2000) identifies this shear zone with the low-velocity, low-Q zone found from the trapped modes because their width is virtually the same at the same location on the fault. Since the trapped modes were observed from aftershocks with focal depths greater than 10 km, we conclude that the shear zone found by JOHNSON et al. (1994) extends to the same depth.

Regarding the issue of the origin of f_{max} of large earthquakes, we mention the work of AKI (1987) who proposed the hypothesis that f_{max} of large earthquakes is due to source effects and is causally related to the corner frequency of small earthquakes when the latter becomes constant below magnitude about 3 [for the tendency of the corner frequency to become constant for values of the seismic moment smaller than about 10^{21} dyn-cm (magnitude about 3 and corresponding source dimension of the order of 100 m) see CHOUET et al. (1978), ARCHULETA et al. (1982), among others]. AKI (1987) supported this hypothesis by analyzing borehole data (borehole located in the middle of the Newport-Inglewood fault) and demonstrating that there is a kink in the magnitude-frequency (of occurrence) relation at a magnitude around 3, reflecting a departure from self-similarity due to the effect of the fault width. ABERCROMBIE and LEARY (1993) and ABERCROMBIE (1995) investigated further Aki's above-mentioned hypothesis by analyzing also borehole data (borehole located at Cajon Pass, southern California) of small earthquakes. Based on the results of their analysis, the above authors concluded that there is no evidence of minimum source dimension at $\sim$100 m and that natural earthquakes are self-similar over a magnitude range $M \sim -2$ to $M \sim 8$. In correcting the analyzed data for attenuation, these authors assumed a constant Q. However, in a more recent study, ADAMS and ABERCROMBIE (1998), using data recorded in the same borehole and a robust method of analysis, found that Q is frequency-dependent (exhibiting strong frequency dependence for $f < 10$ Hz and weaker frequency dependence for $f > 10$ Hz). They conceed that the results (and therefore conclusions) of the previous study (that was based on the assumption of constant Q) may have been compromised. In fact, interpreting the variation of the observed frequency dependence of Q, ADAMS and ABERCROMBIE (1998) conclude that the earth's crust appears to be self-similar for length scales smaller than $\sim$100 m, and "smooth," possibly Gaussian, for longer length scales with correlation distances of a few kilometers. As a possible explanation of the above described apparent change, with length scale, of the crustal structure, these authors propose the presence of large crustal faults characterized by low velocity zones, about 100 m wide, in agreement with the findings of various other investigators (e.g., LI et al., 1994).

Comparing several major California earthquake events (including those analyzed by PAPAGEORGIOU and AKI 1983b) with other major events from different tectonic environments AKI (1992b) observed that the source parameters of the California events deviate systematically from *self-similarity*. [According to the assumption of self-similarity, all earthquake events may be specified by a single parameter, say seismic moment M_0, and that small events are similar to large ones. Self-similarity implies *geometric similarity* i.e., length L, width W and slip Δu_0 all scale as $\sim M_0^{1/3}$, and *physical similarity* i.e., all nondimensional products of source parameters are the same, while the rupture velocity is constant and all parameters with the dimension of time scale as $\sim M_0^{1/3}$ (AKI, 1967; KANAMORI and ANDERSON, 1975)]. In particular, AKI (1992a) noticed that while fault length and width were not much different among these major California events, the decrease in moment was primarily due to decreasing slip (i.e., the amount of slip varies almost as $\sim M_0$ rather than $\sim M_0^{1/3}$, as one would expect if self-similarity were valid).

AKI (1992b) interpreted this departure from self-similarity in terms of both the "specific barrier-model" and the asperity model and he found the asperity model to be inconsistent with the above peculiarity of major California earthquakes. Furthermore, he found the asperity model to be inconsistent with the following observations: (1) The "asperity model" *cannot* explain the observed sharp impulsive displacement perpendicular to the fault plane (AKI, 1968; HASKELL, 1969; BOUCHON, 1979; MENDEZ and LUCO, 1990; YOMOGIDA, 1988; CAMPILLO et al., 1989); (2) the observed *complementarity relation* between coseismic fault slip and aftershocks (AKI, 1979a; MENDOZA and HARTZELL, 1988; TAKEO, 1988; ZENG, 1991); (3) existence of barriers in the creeping segment of the San Andreas fault.

Also, in support of the "barrier model" is the work of ZENG (1991), ZENG *et al.* (1993a, 1993b), who by inversion found that the high-frequency energy sources of the 1989 Loma Prieta and the 1987 Whittier Narrows earthquakes are located along or near the boundaries of localized large slip zones, which is consistent with the theoretical consideration that high frequencies are primarily generated from the rupture stopping areas or places with large slip variation. These conclusions have been confirmed by other investigators based on the analysis of other earthquake events (KAKEHI and IRIKURA 1996, 1997; NAKAHARA *et al.*, 1998). There are also however notable exceptions. Specifically, from an analysis of the 1994 Northridge, California, earthquake, HARTZELL *et al.* (1996) found that of the two major sources of high-frequency radiation that they identified, the one located at the hypocenter (an area associated with a large final slip) is associated with the initiation of rupture and, apparently, the breaking of a high-stress-drop "asperity," while the second is associated with abrupt stopping of the rupture in a westerly direction, apparently by a "barrier."

Concluding, based on the above we may state that California earthquakes are of the "*barrier type*" family (AKI, 1984, 1988). On the other hand, asperities may be more important for great earthquakes along plate boundaries such as subduction zones (AKI, 1984, 1988; YOMOGIDA, 1988; CAMPILLO *et al.*, 1989).

The Barrier Model vis-a-vis Kinematic and Dynamic Source Models Inferred by Inversion of Strong Ground Motion

In the last two decades since the publication of the "specific barrier model" a sufficient number of earthquake slip models have been inferred by inversion of strong motion data so that certain systematic features emerged regarding the slip variation over the fault plane (MENDOZA and HARTZELL, 1988; SOMERVILLE *et al.*, 1999). Furthermore, the inferred kinematics of the earthquake source have been used to compute the dynamic features of the rupture process (i.e., "stress drop" $\Delta\sigma$, "strength excess" $(\sigma_u - \sigma_0)$ etc.) [e.g., BOUCHON (1997); GUATTERI and SPUDICH (2000) and numerous references therein]. In addition, by postulating a constitutive law (such as the slip-weakening model) investigators inferred parameters such as the "characteristic weakening slip" D and the "Griffith's specific fracture energy" G [for a critical and thorough review of the inference of constitutive law parameters such as D and G, see GUATTERI and SPUDICH (2000)]. Specifically, for the 1979 Imperial Valley, California, earthquake GUATTERI and SPUDICH (2000) estimate $G = 2 - 6 \cdot 10^9$ erg/cm^2 (consistent with an earlier estimate of $2 \cdot 10^9$ erg/cm^2 by BEROZA and SPUDICH, 1988) and they point out that there is a trade-off between D

and "strength excess" $(\sigma_u - \sigma_0)$ ($D = 1$ m with low "strength excess" or $D = 0.3$ m with high "stress excess," both estimates producing indistinguishable ground motion waveforms in the 0–1.6 Hz frequency band). For the 1992 Landers, California, earthquake OLSEN et al. (1997) estimate $D \sim 0.8$ m while for the same earthquake PULIDO and IRIKURA (2000) estimate $D \sim 1$ m for the Johnson Valley (southern) and Camp Rock/Emerson (northern) fault segments and $D \sim 3.5$ m for the Homestead Valley (central) fault segment. Finally, for the 1995 Hyogo-ken Nanbu (Kobe), Japan, earthquake IDE and TAKEO (1997) estimate $D \sim 0.5$ m for the deeper part of the fault and $D \sim 1$ m for the upper part of the fault. Comparing the above estimates of D with the corresponding values shown in Figure 5, we notice that they are remarkably close to—and in any case bound from above as expected (GUATTERI and SPUDICH 2000) — the estimates obtained using the parameters of the "specific barrier model" and high frequency waves (i.e., f_{max}). [Regarding the above comparison, we should keep in mind that the resolution of the kinematic inversion models is limited because they are based on the analysis of rather long period waves ($f < 1$ Hz) and due to the effects of spatial and temporal-smoothing constraints applied in such inverse-problem formulations].

Regarding the "local stress drop" $\Delta\sigma_L$ and "cohesive stress" σ_c parameters, we compare our estimates (Fig. 5) with those of BOUCHON (1997) [the latter properly averaged over the regions of high stress drop (which in most cases coincide with regions of high slip)] and of other investigators (e.g., GUATTERI and SPUDICH, 2000) and we find that they are in reasonable agreement.

Next, what caveats should one be aware of regarding the "specific barrier model"? Clearly, the model is an end member of a spectrum of models that represent a main earthquake event as a collection of subevents. For instance, the subevent is modeled as a crack, the rupture of which is arrested at the perimeter by a barrier. We have already seen above that heterogeneities on the fault plane may act also as "asperities," in which case a model such as that of KOSTROV and DAS (1988) of a *circular fault with a central asperity* (see also FUKUYAMA and MADARIAGA, 1998) may have to be considered in representing a subset of the subevents that compose the main event. For this we would need a closed-form mathematical expression of the far-field radiation of such a subevent model, analogous to the expression of SATO and HIRASAWA (1973) for the circular crack.

Another concern regarding the "specific barrier model" is that all subevents are assumed to be of the same size, contrary to the more complex picture that has emerged from the kinematic models that have been obtained by inversion. We performed preliminary calculations allowing for a distribution of crack size around a representative size (similar to the model proposed by BOATWRIGHT, 1982) and we have found that estimates of the "local stress drop" $\Delta\sigma_L$ are not affected significantly (less than 30% change). This should have been anticipated because, for an earthquake event of a given magnitude, there is an average/typical subevent size that contributes most of the radiation. [This is tacitly recognized by SOMERVILLE *et al.* (1999) when they plot "area of largest asperity" vs. "seismic moment".]

Then, in view of the above concerns, how does use of the "specific barrier model" provide estimates of various source parameters that are in general agreement with estimates of the same parameters obtained by other means? The answer to the above question, at least partially, lies in the following facts: In estimating source parameters using the "specific barrier model," we start with the "local stress drop" $\Delta\sigma_L$ which we estimate from the *power* spectrum of the "stationary" segment of the accelerogram. The geometric parameter that controls the power spectrum of the radiation emitted by the source is an *"effective width"* (see eq. (47) of PAPAGEORGIOU and AKI, 1983a) which, at least for the strike slip California events that were analyzed using the model, is well approximated by the *"nominal"* width of the fault. Next, securing a reliable estimate of the "local stress drop" $\Delta\sigma_L$ we proceed to estimate the barrier interval from the ratio $(\Delta\bar{u}/2\rho_0)$ which represents the "local strain drop" which in turn is proportional to the "local stress drop." The uncertainty here lies with estimates of $\Delta\bar{u}$ in view of the fact that it appears that only a fraction (say 50%) of the nominal fault plane slips significantly (and radiates seismic energy) while in the "specific barrier model" we assume a uniform distribution of cracks covering the entire nominal rupture area of the fault plane. This would involve an uncertainty of a factor of not more than 2. However, usually this is the uncertainty associated with estimates of the seismic moment M_0. [It is evident that as more data accumulate regarding the percentage of the nominal rupture area that slips significantly radiating seismic energy, the information could readily be incorporated in the procedure of estimating parameters of earthquake sources using the "specific barrier model".]

Summarizing, the "specific barrier model" provides the most complete, yet *parsimonious*, self-consistent description of the faulting processes that are responsible for the generation of high-frequency waves. The model, in spite of its simplicity, is robust enough to provide reasonable estimates of various important source parameters and provides an effective tool to simulate/model strong ground motion for engineering applications. Until the kinematic inversion studies of the earthquake source are based on considerably higher frequency waves than the present ones, the "specific barrier model" contributes to earthquake source studies.

Numerical Simulation of Strong Ground Motion – Forward Modeling

Considerable progress has been made in recent years in understanding strong ground motion in terms of source, propagation path and recording site effect. As a demonstration of the level of the achieved understanding one may refer to the plethora of successful numerical simulations of observed strong ground motions using various mathematical models of the earthquake source and earth medium.

The simplest and least expensive method for simulating strong ground motion is based on the assumption that accelerograms are realizations of a stochastic process with time varying intensity and, possibly, frequency content. HOUSNER (1947, 1955) interpreted the erratic appearance of strong-motion accelerograms by reasoning that seismic waves are initiated by irregular slippage along faults followed by numerous random reflections, refractions and attenuations along the propagation path. Following Housner's *paradigm* many investigators (e.g., HUDSON, 1956; BYCROFT, 1960; HOUSNER and JENNINGS, 1964; JENNINGS *et al.*, 1968; JOYNER and BOORE, 1988; SHINOZUKA, 1988, see last one for a recent review) developed stochastic models for the analysis of recorded accelerograms or the computation of synthetic ones.

At the time Housner formed the above hypothesis, earthquake source theory and methods for evaluating Green's functions for realistic earth media were not well developed. Thus, Housner proceeded by considering simple, yet effective for earthquake engineering purposes, functional forms for radiated waves. As it was elaborated in the previous section, now we know that high frequency waves emanate from the rupture front as it interacts with heterogeneities (i.e., barriers and asperities) of the fault plane, and that ground motion may be computed by convolving the slip function with the Green's tensor of the earth (see eq. (1)). By now it should be apparent to the reader that Housner's original idea of modeling high-frequency seismic radiation has obtained a more concrete expression with the "specific barrier model" of PAPAGEORGIOU and AKI (1983a) that was presented in the previous section.

The developments related to the stochastic modeling of accelerograms came about thanks primarily to the efforts of the engineering community. Not having a physical model to describe the frequency content of the elastic waves radiated by the earthquake source, earthquake engineers adopted simple and/or empirical spectral models (e.g., white noise spectrum, Kanai-Tajimi spectrum; see for example CLOUGH and PENZIEN, 1975).

Recently seismologists, recognizing the stochastic character of high-frequency waves, adopted the engineering approach in simulating strong motion accelerograms, based on the assumption that they are realizations of "*band limited gaussian white noise*" with time varying intensity (HANKS, 1982; HANKS and McGUIRE, 1981; BOORE, 1983). The contribution that the seismologists made to this development consists of the fact that they used a physical model (instead of an empirical one) to describe the spectral content of the simulated motions. In particular, they adopted BRUNE'S (1970) "ω^2-*model*" to describe the *source spectrum* (i.e., the spectrum of the elastic waves radiated by the source before these has been modified by the propagation path and site effects) and assumed self-similarity to establish the *scaling law* of the source spectrum (i.e., how the source spectrum scales/varies with earthquake size; AKI, 1967, 1972b).

The stochastic modeling of accelerograms, as was originally used by earthquake engineers or even in its most refined form proposed by seismologists (e.g., BOORE,

1983), has the following limitations: (1) The model is based on a point source; (2) the model is not adequate for simulating the long-period *near-field effect* expected in the *near-source region* from the slip on the fault; and (3) the model provides a description of the *temporal* variation of ground motion of a single point of the ground but cannot provide a description of the *spatial* variability of ground motion which is necessary for the analysis of extended structures (e.g., pipelines, tunnels, bridges, dams).

The above three limitations apply irrespective of the spectral model that one may adopt. In addition, the "ω^2-model" fails to explain the observed $M_s - M_0$ relation for large earthquakes (BOORE, 1986) and there appears to be a consensus that a single corner frequency model (such as the "ω^2-model") cannot explain observations for the entire frequency range of large events (e.g., GUSEV, 1983; PAPAGEORGIOU and AKI, 1985; PAPAGEORGIOU, 1988; ATKINSON, 1993; BOATWRIGHT, 1994; HADDON, 1995, 1996a). Furthermore, a fundamental problem with the "ω^2-model" is the ambiguous nature of the key parameter called "*stress parameter*" (BOORE and ATKINSON, 1987). [Parenthetically we point out that the stress parameter that appears in the "specific barrier model" and is referred to as the "local stress drop," has a clear physical meaning—being the stress drop of the subevents that compose the earthquake event—has been found to be a very stable parameter for a given tectonic region (see Fig. 5) and may be estimated even by geological exploration methods (*paleoseismology*) thus rendering the model potentially useful in predicting strong ground motion even for tectonic areas for which there are no recordings (AKI, 1984).]

The limitations of the engineering approach for simulating strong motion may be eliminated by using the deterministic kinematic modeling approach based on the representation theorem (eq. (1)). It was pointed out earlier that in order to apply eq. (1) one needs to know: (1) the slip history of the fault rupture (i.e., when, how much, for how long and in what direction each point of the fault slipped, or will slip, during an event), and (2) the Green's function of the earth with enough accuracy for the frequency range of interest.

Regarding the first requirement above, if one is interested in simulating ground motion to be generated by a fault which may potentially rupture, it is evident that the slip history involves many parameters that must be specified. One possible way to proceed is to adopt the slip history, inferred by inversion, of an event of similar size, with the same *source mechanism* (i.e., strike-slip, dip-slip, etc.) and preferably from the same tectonic region (e.g., SAIKIA, 1993; HEATON *et al.*, 1995). Alternatively, one may adopt the "specific barrier model" to parameterize the slip history and obtain reliable estimates of the size of the subevents.

Regarding the evaluation of the Green's function, the chief factor limiting its accuracy is ignorance of the earth structure at the source-site region. For example, in order to simulate deterministically ground motion reaching a maximum frequency of 5 Hz, it is necessary to know the 3-D structure of the earth on a scale of a few hundred meters. In view of the fact that both engineers and seismologists recognize the stochastic character of high frequency strong motion, the following approach for earthquake motion simulation may be proposed: Use the *Empirical Green's Function Method* (e.g., HARTZELL, 1978, 1989; WU, 1978; HUTCHINGS and WU, 1990; HADDON, 1996b) or stochastic modeling (e.g., ZENG *et al.*, 1993a, 1995) to simulate ground motions in the high frequency range (say above 1 Hz) and combine these results with those obtained using deterministic kinematic modeling in the low frequency range (say below 1 Hz) (e.g., ZENG *et al.*, 1993a; HEATON *et al.*, 1995). The above recommended procedure tacitly recognizes the fact that ground motions at periods longer than 3 sec of past events have not been reliably recorded by the analog strong motion instruments.

Finally, it is well recognized that ground shaking and associated damage to engineered structures are strongly influenced by the geology in their vicinities. The *Coda Method*, exploiting the well established separability of source, path and site effects on coda waves of local seismic events originally proposed in the seminal work of AKI (1969), offers a cost-effective way to empirically determine the *site amplification factor* for regional microzonation (PHILLIPS and AKI, 1986; SU and AKI, 1995; AKI, 1993, see last one for review). Complications that are caused by the nonlinear behavior that unconsolidated sediments exhibit when subjected to large strains, were first detected seismologically by CHIN and AKI (1991) for the 1989 Loma Prieta earthquake. In order to correct the coda (i.e., weak motion) site amplification factor to be applicable to strong motions (i.e., motions exceeding a threshold peak acceleration), AKI and CHIN (1994) have proposed a very simple method that they tested in connection with 1992 Landers earthquake data. The method appears promising and requires further testing with strong motion data that have been recorded by accelerographs collocated with high frequency (i.e., weak motion) instruments or broad-band instruments (KATO *et al.*, 1995). Since the publication of the work of CHIN and AKI (1991), various investigators have observed seismologically the effect of soil nonlinearities for other earthquakes (e.g., SU *et al.*, 1998).

Ultimately, it should be evident that the intent of the engineering approach to strong motion simulation is to capture the essential characteristics of high-frequency motion at an average site from an average earthquake of specified size. Phrasing this differently, the accelerograms artificially generated by engineers do not duplicate any specific earthquake but rather embody certain average properties of past earthquakes of a given magnitude (SHINOZUKA, 1988). Contrastingly, the kinematic modeling approach adopted by seismologists involves the prediction of motions from a fault that was identified by geologists and which has specific dimensions and orientation in a specified geologic setting. This latter approach is useful for *site-specific* simulations.

Conclusions

The last three decades have witnessed remarkable advances in the field of Engineering Seismology. It is fair to say that earth scientists have developed the capabilities to synthesize ground motions generated by realistic sources embedded in realistic propagation media to such a degree, that they are currently capable of assessing the range of plausible ground motions at the site of an engineering structure. Every aspect of Strong Motion Seismology has been influenced by the seminal works of Keiiti Aki. His contributions include the first modeling of near-source ground motion, introduction in seismology of the concept of earthquake source spectrum and its scaling with magnitude, study of fault rupture using models of Fracture Mechanics, documentation and characterization of fault heterogeneity responsible for the generation of the short-period and high-frequency waves, introduction of the "barrier model" (and a mathematical expression of it referred to as the "specific barrier model"), inversion study to identify the sources of high-frequency radiation on the fault plane, study and modeling of coda waves, analysis and numerical modeling of site effects including a cost-effective method for regional microzonation using the coda method. All of Keiiti Aki's contributions plowed new ground and opened new vistas of research. Like Galileo, he focused on fundamental seismological phenomena and quantified them. He left an indelible mark both as a superb scientist and as a great teacher. Let all of us follow in his footsteps.

REFERENCES

ABERCROMBIE, R. (1995), *Earthquake Source Scaling Relationship from* −1 *to* 5 M_L *Using Seismograms Recorded at 2.5 km Depth*, J. Geophys. Res. *100*, 24,015–24,036.

ABERCROMBIE, R. and LEARY, P. (1993), *Source Parameters of Small Earthquakes Recorded at 2.5 km Depth, Cajon Pass, Southern California: Implications for Earthquake Scaling*, Geophys. Res. Lett. *20*, 1511–1514.

ACHENBACH, J., *Wave Propagation in Elastic Solids* (North-Holland Publisher, Amsterdam 1973).

ACHENBACH, J., GAUTESEN, A., and MCMAKEN, H., *Ray Methods for Waves in Elastic Solids* (Pitman, Boston, Massachusetts 1982).

ACHENBACH, J. and HARRIS, J. (1978), *Ray Method for Elastodynamic Radiation from a Slip Zone of Arbitrary Shape*, J. Geophys. Res. *83*, 2283–2291.

ACHENBACH, J. and HARRIS, J., *Asymptotic Modeling of Strong Ground Motion Excited by Subsurface Sliding Events* (Academic Press, Inc., Orlando, Florida 1987), pp. 1–52.

ADAMS, D. A. and ABERCROMBIE, R. E. (1998), *Seismic Attenuation Above 10 Hz in Southern California from Coda Waves Recorded in the Cajon Pass Borehole*, J. Geophys. Res. *103*, 24,257–24,270.

AKI, K. (1966), *Generation and Propagation of G Waves from the Niigata Earthquake of June 16, 1964. 2. Estimation of Earthquake Movement Released Energy, and Stress-Strain Drop from G Wave Spectrum*, Bulletin of the Earthquake Research Institute *44*, 23–88.

AKI, K. (1967), *Scaling Law of Seismic Spectrum*, J. Geophys. Res. *72*, 1217–1231.

AKI, K. (1968), *Seismic Displacement Near a Fault*, J. Geophys. Res. *73*, 5359–5376.

AKI, K. (1969), *Analysis of the Seismic Coda of Local Earthquakes as Scattered Waves*, J. Geophys. Res. *74*, 615–631.

AKI, K. (1972a), *Earthquake Mechanism*, Tectonophysics *13*, 423–446.

AKI, K. (1972b), *Scaling Law of Earthquake Source-time Function*, Geophys. J. R. Astron. Soc. *31*, 3–25.

AKI, K. (1979a), *Characterization of Barriers on an Earthquake Fault*, J. Geophys. Res. *84*, 6140–6148.

AKI, K., *Evolution of Quantitative Models of Earthquakes*, In *SIAM-AMS Proceedings* (1979b), vol. 12, pp. 43–58.

AKI, K. (1980a), *Attenuation of Shear-waves in the Lithosphere for Frequencies from 0.05 to 25 Hz*, Phys. Earth Planet. Inter. *21*, 50–60.

AKI, K. (1980b), *Possibilities of Seismology in the 1980's, Presidential Address*, Bull. Seismol. Soc. Am. *70*, 1969–1976.

AKI, K. (1982), *Scattering and Attenuation*, Bull. Seismol. Soc. Am. *72*, S319–S330.

AKI, K., *Strong-motion Seismology*, In *Earthquakes: Observation, Theory and Interpretation* (eds. Kanamori, H. and Boschi, E.) (North-Holland Publ., Amsterdam 1983), vol. Course LXXXV, pp. 223–250.

AKI, K. (1984), *Asperities, Barriers, Characteristic Earthquakes and Strong Motion Prediction*, J. Geophys. Res. *89*, 5867–5872.

AKI, K. (1987), *Magnitude-frequency Relation for Small Earthquakes: A Clue to the Origin of f_{max} of Large Earthquakes*, J. Geophys. Res. *92*, 1349–1355.

AKI, K. (1988), *Impact of Earthquake Seismology on the Geological Community Since the Benioff Zone*, Geolog. Soc. Am. Bull. *100*, 625–629.

AKI, K., *Earthquake Sources and Strong Motion Prediction*, In *Proc. IDNDR International Symposium on Earthquake Disaster Technology* (Sponsored by the International Institute of Seismology and Earthquake Engineering, Tsukuba, Science City, Japan 1992a).

AKI, K. (1992b), *Higher Order Interpretations Between Seismogenic Structures and Earthquake Processes*, Technophysics *211*, 1–12.

AKI, K. (1993), *Local Site Effects on Weak and Strong Ground Motion*, Tectonophysics *218*, 93–111.

AKI, K. (2000), *Scale-dependence in Earthquake Processes and Seismogenic Structures*, submitted for publication.

AKI, K., BOUCHON, M., CHOUET, B., and DAS, S. (1977), *Quantitative Prediction of Strong Motion for a Potential Earthquake Fault*, Ann. Geofis. *XXX*, 341–368.

AKI, K. and CHIN, B.-H., *The Use of Coda S Waves for Characterizing the Site Effect on Strong Ground Motion*. In *Proceedings, Structures Congress XII* (eds. Baker, N. and Goodno, B.) (ASCE, Atlanta, GA 1994), vol. 1, pp. 579–584.

AKI, K. and CHOUET, B. (1975), *Origin of Coda-Waves: Source, Attenuation, and Scattering Effects*, J. Geophys. Res. *80*, 3322–3342.

AKI, K. and IRIKURA, K., *Characterization and Mapping of Earthquake Shaking for Seismic Zonation*. In *Proc. 4th International Conference on Seismic Zonation* (Stanford, Calif. 1991), vol. I, pp. 61–110.

AKI, K. and PAPAGEORGIOU, A., *Separation of Source and Site Effects in Acceleration Power Spectra of Major California Earthquakes*. In *Ninth World Conference on Earthquake Engineering* (Tokyo-Kyoto, Japan 1988), vol. VIII, SB-8, pp. 163–167.

AKI, K. and RICHARDS, P. G., *Quantitative Seismology. Theory and Methods*, vols. I, II (W. H. Freeman and Company, San Francisco, USA 1980), 948 pp.

ANDERSON, J. and HOUGH, S. (1984), *A Model for the Shape of the Fourier Amplitude Spectrum of Acceleration at High Frequencies*, Bull. Seismol. Soc. Am *74*, 1969–1993.

ANDREWS, D. (1978), *Coupling of Energy Between Tectonic Processes and Earthquakes*, J. Geophys. Res. *83*, 2259–2264.

ANDREWS, D. (1981), *A Stochastic Fault Model, 2, Time-Dependent Case*, J. Geophys. Res. *86*, 10,821–10,834.

ANDREWS, D. J. (1976a), *Rupture Propagation with Finite Stress in Antiplane Strain*, J. Geophys. Res. *81*, 3575–3582.

ANDREWS, D. J. (1976b), *Rupture Velocity of Plane Strain Shear Cracks*, J. Geophys. Res. *81*, 5679–5687.

ANDREWS, D. J. (1985), *Dynamic Plane-strain Shear Rupture with a Slip-weakening Friction Law Calculated by a Boundary Integral Method*, Bull. Seismol. Soc. Am. *75*, 1–21.

ANDREWS, D. J. (1994), *Dynamic Growth of Mixed-mode Shear Cracks*, Bull. Seismol. Soc. Am. *84*, 1184–1198.

ANDREWS, D. J. and BEN-ZION, Y. (1997), *Wrinkle-like Slip Pulse on a Fault Between Different Materials*, J. Geophys. Res. *102*, 553–571.

ARCHULETA, R. (1984), *A Faulting Model for the 1979 Imperial Valley, California Earthquake*, J. Geophys. Res. *89*, 4559–4585.

ARCHULETA, R., CRANSWICK, J. E., MUELLER, C., and SPUDICH, P. (1982), *Source Parameters of the 1980 Mammoth Lakes, California, Earthquake Sequence*, J. Geophys. Res. *87*, 4595–4608.

ATKINSON, G. M. (1993), *Earthquake Source Spectra in Eastern North America*, Bull. Seismol. Soc. Am. *83*, 1778–1798.

BARENBLAT, G. (1959), *The Formation of Equilibrium Cracks During Brittle Fracture: General Ideas and Hypothesis, Axially Symmetric Cracks*, J. Appl. Math. Mech. *23*, 434–444.

BERESNEV, I. A. and ATKINSON, G. M. (1999), *Generic Finite-fault Model for Ground-motion Prediction in Eastern North America*, Bull. Seismol. Soc. Am. *89*, 608–625.

BERNARD, P. and MADARIAGA, R. (1984a), *High-frequency Seismic Radiation from a Buried Circular Fault*, Geophys. J. R. Astr. Soc. *78*, 1–17.

BERNARD, P. and MADARIAGA, R. (1984b), *A New Asymptotic Method for the Modeling of Near-field Accelerograms*, Bull. Seismol. Soc. Am. *74*, 539–557.

BEROZA, G. and SPUDICH, P. (1988), *Linearized Inversion for Fault Rupture Behavior: Application to the 1984 Morgan Hill, California Earthquake*, J. Geophys. Res. *93*, 6275–6296.

BEROZA, G. C. and MIKUMO, T. (1996), *Short Slip Duration in Dynamic Rupture in the Presence of Heterogeneous Fault Properties*, J. Geophys. Res. *101*, 22,449–22,460.

BIZZARI, A., COCCO, M., ANDREWS, D. J., and BOSCHI, E. (2001), *Solving the Dynamic Rupture Problem with Different Numerical Approaches and Constitutive Laws*, Geophys. J. Int. *144*, 656–678.

BOATWRIGHT, J. (1982), *A Dynamic Model for Far-field Acceleration*, Bull. Seismol. Soc. Am. *72*, 1049–1068.

BOATWRIGHT, J. (1988), *The Seismic Radiation from Composite Models of Faulting*, Bull. Seismol. Soc. Am. *78*, 489–508.

BOATWRIGHT, J. (1994), *Regional Propagation Characteristics and Source Parameters of Earthquakes in Northeastern North America*, Bull. Seismol. Soc. Am. *84*(1), 1–15.

BOORE, D. and JOYNER, W. (1978), *The Influence of Rupture Incoherence on Seismic Directivity*, Bull. Seismol. Soc. Am. *68*, 283–300.

BOORE, D. M. (1983), *Stochastic Simulation of High-frequency Ground Motions Based on Seismological Models of the Radiated Spectra*, Bull. Seismol. Soc. Am. *73*, 1865–1894.

BOORE, D. M. (1986), *Short-period P- and S-wave Radiation from Large Earthquakes: Implications for Spectral Scaling Relations*, Bull. Seismol. Soc. Am. *76*, 43–64.

BOORE, D. M. and ATKINSON, G. M. (1987), *Stochastic Prediction of Ground Motion and Spectral Response Parameters at Hard-rock Sites in Eastern North America*, Bull. Seismol. Soc. Am. *77*, 440–467.

BOUCHON, M. (1979), *Predictability of Ground Displacement and Velocity Near an Earthquake Fault. An Example: The Parkfield Earthquake of 1966*, J. Geophys. Res. *84*, 6149–6156.

BOUCHON, M. (1997), *The State of Stress on some Faults of the San Andreas System as Inferred from Near-field Strong Motion Data*, J. Geophys. Res. *102*, 11,731–11,744.

BROBERG, B. K., *Cracks and Fracture* (Academic Press, San Diego, CA 1999).

BRUNE, J. N. (1970), *Tectonic Stress and the Spectra of Seismic Shear Waves from Earthquakes*, J. Geophys. Res. *75*, 4997–5009.

BURRIDGE, R. and KNOPOFF, L. (1964), *Body Force Equivalents for Seismic Dislocations*, Bull. Seismol. Soc. Am. *54*, 1874–1888.

BYCROFT, G. (1960), *White Noise Representation of Earthquakes*, ASCE J. Eng. Mech. Div. *86*, 1–16.

CAMPILLO, M., GARIEL, J.-C., AKI, K., and SANCHEZ-SESMA, F. (1989), *Destructive Strong Ground Motion in Mexico City: Source, Path and Site Effects During the Great 1985 Michoacan Earthquake*, Bull. Seismol. Soc. Am. *79*, 1718–1735.

CAO, T. and AKI, K. (1984), *Seismicity Simulation with a Mass-spring Model and a Displacement Hardening-softening Friction Law*, Pure Appl. Geophys. *122*, 10–24.

CHIN, B. and AKI, K. (1991), *Simultaneous Study of the Source Path and Site Effects on Strong Ground Motion During the 1989 Loma Prieta Earthquake: A Preliminary Result on Pervasive Nonlinear Site Effects*, Bull. Seismol. Soc. Am. *81*(5), 1859–1884.

CHOUET, B., AKI, K., and TSUJIURA, M. (1978), *Regional Variation of the Scaling Law of Earthquake Source Spectra*, Bull. Seismol. Soc. Am. *68*, 49–79.

CLOUGH, R. and PENZIEN, J., *Dynamics of Structures* (McGraw-Hill Book Company, New York 1975).

DAS, S. and AKI, K. (1977a), *Fault Plane with Barriers: A Versatile Earthquake Model*, J. Geophys. Res. *82*, 5648–5670.

DAS, S. and AKI, K. (1977b), *A Numerical Study of Two-dimensional Spontaneous Rupture Propagation*, Geophys. J. R. Astron. Soc. *50*, 643–668.

DAS, S. and KOSTROV, B. V. (1983), *Breaking of a Single Asperity: Rupture Process and Seismic Radiation*, J. Geophys. Res. *88*, 4277–4288.

DAS, S. and KOSTROV, B. V., *Fracture of a Single Asperity on a Finite Fault: A Model for Weak Earthquakes?* In *Earthquake Source Mechanics* (eds. Das, S., Boatwright, J., and Scholz, C.) (American Geophysical Union, Washington 1986), pp. 91–96.

DAY, S. (1982), *Three-dimensional Simulation of Spontaneous Rupture: The Effect of Nonuniform Prestress*, Bull. Seismol. Soc. Am. *72*, 1881–1902.

DAY, S. M., YU, G., and WALD, D. J. (1998), *Dynamic Stress Changes During Earthquake Rupture*, Bull. Seismol. Soc. Am. *88*, 512–522.

DIETERICH, J. H. (1979), *Modeling of Rock Friction. I. Experimental Results and Constitutive Equations*, J. Geophys. Res. *84*, 2169–2175.

DIETERICH, J. H. (1992), *Earthquake Nucleation on Faults with Rate- and State-dependent Strength*, Tectonophysics *211*, 115–134.

DMOWSKA, R. and RICE, J. R., *Fracture Theory and its Seismological Applications*. In *Continuum Theories in Solid Earth Physics* (ed. Teisseyre, R.) (Elsevier, New York, NY 1986), chap. 3.

ERINGEN, A. and SUHUBI, E. S., *Elastodynamics*, vol. II, Linear Theory (Academic Press 1975).

FREUND, L. B. (1972), *Energy Flux Into the Tip of an Extending Crack in an Elastic Solid*, J. Elasticity *2*, 341–349.

FREUND, L. B. (1979), *The Mechanics of Dynamic Shear Crack Propagation*, J. Geophys. Res. *84*, 2199–2209.

FREUND, L. B., *Dynamic Fracture Mechanics* (Cambridge University Press, Cambridge 1990).

FUJIWARA, H. and IRIKURA, K. (1991), *High-frequency Seismic Wave Radiation from Antiplane Cohesive Zone Model and f_{max} as Source Effect*, Bull. Seismol. Soc. Am. *81*, 1115–1128.

FUKUYAMA, E. and MADARIAGA, R. (1998), *Rupture Dynamics of a Planar Fault in a 3D Elastic Medium: Rate- and Slip-weakening Friction*, Bull. Seismol. Soc. Am. *88*, 1–17.

GABRIEL, J.-C. and CAMPILLO, M. (1989), *The Influence of the Source on the High-frequency Behavior of the Near-field Acceleration Spectrum: A Numerical Study*, Geophys. Res. Lett. *16*, 279–282.

GUATTERI, M. and SPUDICH, P. (2000), *What Can Strong-motion Data Tell Us About Slip-weakening Fault-friction Laws?* Bull. Seismol. Soc. Am. *90*, 98–116.

GURTIN, M. E., *The Linear Theory of Elasticity*. In *Handbuch der Physik* (ed. Flügge, S.) (Berlin 1972).

GUSEV, A. A. (1983), *Descriptive Statistical Model of Earthquake Source Radiation and its Application to an Estimation of Short-period Strong Motion*, Geophys. J. R. Astr. Soc. *74*, 787–808.

HADDON, R. A. W. (1995), *Modeling of Source Rupture Characteristics for the Saguenay Earthquake of November 1988*, Bull. Seismol. Soc. Am. *85*, 525–551.

HADDON, R. A. W. (1996a), *Earthquake Source Spectra in Eastern North America*, Bull. Seismol. Soc. Am. *86*, 1300–1313.

HADDON, R. A. W. (1996b), *Use of Empirical Green's Functions, Spectral Ratios and Kinematic Source Models for Simulating Strong Ground Motion*, Bull. Seismol. Soc. Am. *86*, 597–615.

HANKS, T. (1979), *b Values and $\omega^{-\gamma}$ Seismic Source Models: Implications for Tectonic Stress Variations Along Active Crustal Fault Zones and the Estimation of High-frequency Strong Ground Motion*, J. Geophys. Res. *84*, 2235–2242.

HANKS, T. C. (1982), *f_{max}*, Bull. Seismol. Soc. Am. *72*, 1867–1880.

HANKS, T. C. and McGUIRE, R. K. (1981), *The Character of High-frequency Strong Ground Motion*, Bull. Seismol. Soc. Am. *71*, 2071–2095.

HARTZELL, S., LIU, P., and MENDOZA, C. (1996), *The 1994 Northridge, California, Earthquake: Investigation of Rupture Velocity, Risetime and High-frequency Radiation*, J. Geophys. Res. *101*, 20,091–20,108.

HARTZELL, S. H. (1978), *Earthquake Aftershocks as Green's Functions*, Geophys. Res. Lett. *5*, 1–4.

HARTZELL, S. H. (1989), *Comparison of Seismic Waveform Inversion Techniques for the Rupture History of a Finite Fault: Application to the 1986 North Palm Springs, California, Earthquake*, J. Geophys. Res. *94*, 7515–7534.

HARTZELL, S. H. and HEATON, T. (1983), *Inversion of Strong Ground Motion and Teleseismic Waveform Data for the Fault Rupture History of the 1979 Imperial Valley, California Earthquake*, Bull. Seismol. Soc. Am. *73*, 1553–1583.

HARTZELL, S. H. and HELMBERGER, D. V. (1982), *Strong-motion Modeling of the Imperial Valley Earthquake of 1979*, Bull. Seismol. Soc. Am. *72*, 571–596.

HASKELL, N. A. (1964), *Radiation Pattern of Surface Waves from Point Sources in a Multi-layered Medium*, Bull. Seismol. Soc. Am. *54*, 377–394.

HASKELL, N. A. (1966), *Total Energy and Energy Spectral Density of Elastic Wave Radiation from Propagating Faults*, Bull. Seismol. Soc. Am. *56*, 125–140.

HASKELL, N. A. (1969), *Elastic Displacements in the Near-field of a Propagating Fault*, Bull. Seismol. Soc. Am. *59*, 865–908.

HEATON, T. H. (1990), *Evidence for and Implications of Self-healing Pulses of Slip in Earthquake Rupture*, Phys. Earth Planet. Int. *64*, 1–20.

HEATON, T. H., HALL, J. F., WALD, D. J., and HALLING, M. H. (1995), *Response of High-rise and Base-isolated Buildings to a Hypothetical M_w 7.0 Blind Thrust Earthquake*, Science *267*, 206–211.

HOUSNER, G. and JENNINGS, P. C. (1964), *Generation of Artificial Earthquakes*, ASCE J. Eng. Mech. Div. *90*, 113–150.

HOUSNER, G. W. (1947), *Characteristics of Strong-motion Earthquakes*, Bull. Seismol. Soc. Am. *37*, 19–31.

HOUSNER, G. W. (1955), *Properties of Strong Ground Motion Earthquakes*, Bull. Seismol. Soc. Am. *45*, 197–218.

HOUSNER, G. W. and TRIFUNAC, M. D. (1967), *Analysis of Accelerograms—Parkfield Earthquake*, Bull. Seismol. Soc. Am. *57*, 1193–1220.

HUDSON, D., *Response Spectrum Techniques in Engineering Seismology*. In *Proc. 1st World Conference on Earthquake Engineering* (Berkeley, California 1956).

HUSSEINI, M. I., JOVANOVICH, D. B., RANDALL, M. J., and FREUND, L. B. (1975), *The Fracture Energy of Earthquakes*, Geophys. J. Royal Astr. Soc. *43*, 367–385.

HUTCHINGS, L. and WU, F. (1990), *Empirical Green's Functions from Small Earthquakes — A Waveform Study of Locally Recorded Aftershocks of the San Fernando Earthquake*, J. Geophys. Res. *95*, 1187–1214.

IDA, Y. (1972), *Cohesive Force Across the Tip of a Longitudinal Shear Crack and Griffith's Specific Surface Energy*, J. Geophys. Res. *77*, 3796–3805.

IDA, Y. (1973), *The Maximum Acceleration of Seismic Ground Motion*, Bull. Seismol. Soc. Am. *63*, 959–968.

IDA, Y. and AKI, K. (1972), *Seismic Source Time Function of Propagating Longitudinal Shear Cracks*, J. Geophys. Res. *77*, 2034–2044.

IDE, S. and TAKEO, M. (1997), *Determination of Constitutive Relations of Fault-slip Based on Seismic Wave Analysis*, J. Geophys. Res. *102*, 27,379–27,391.

IRIKURA, K. and YOKOI, T. (1984), *Scaling Law of Seismic Source Spectra for the Aftershock of the 1983 Central-Japan-Sea Earthquake*, Abstracts of the Seismological Society of Japan.

IZUTANI, Y. (1981), *A Statistical Model for Prediction of Quasi-realistic Ground Motion*, J. Phys. Earth *29*, 537–558.

JENNINGS, P., HOUSNER, G., and TSAI, N., *Simulated Earthquake Motions* (Earthquake Engineering Research Laboratory, California Institute of Technology, Pasadena, California 1968).

JOHNSON, A. M., FLEMING, R. W., and CRUIKSHANK, K. M. (1994), *Analysis of Structures Formed During the 28 June 1992 Landers-Big Bear, California, Earthquake Sequence*, Bull. Seismol. Soc. Am. *84*, 499–510.

JOYNER, W. B. and BOORE, D. M., *Measurement, Characterization and Prediction of Strong Ground Motion*, In *Proc. Earthq. Eng. Soil Dynamics II* (Geotechnical Division/ASCE, Park City, Utah 1988).

KAKEHI, Y. and IRIKURA, K. (1996), *Estimation of High-frequency Wave Radiation Areas on the Fault Plane by the Envelope Inversion of Acceleration Seismograms*, Geophys. J. *125*, 892–900.

KAKEHI, Y. and IRIKURA, K. (1997), *High-frequency Radiation Process During Earthquake Faulting — Envelope Inversion of Acceleration Seismograms from the 1993 Hokkaido-Nansei-Oki, Japan, Earthquake,* Bull. Seismol. Soc. Am. *87*, 904–917.

KANAMORI, H. and ANDERSON, D. (1975), *Theoretical Basis of Some Empirical Relations in Seismology,* Bull. Seismol. Soc. Am. *65*, 1073–1095.

KANAMORI, H. and STEWART, G. S. (1978), *Seismological Aspects of the Guatemala Earthquake of February 4, 1976,* J. Geophys. Res. *83*, 3427–3434.

KATO, K., AKI, K., and TAKEMURA, M. (1995), *Site Amplification from Coda waves: Validation and Application to S-wave Site Response,* Bull. Seismol. Soc. Am. *85*, 467–677.

KINOSHITA, S. (1992), *Local Characteristics of the f_{max} of Bedrock Motion in the Tokyo Metropolitan Area, Japan,* J. Phys. Earth *40*, 487–515.

KNOPOFF, L., *The Nature of the Earthquake Source. In Identification of Seismic Sources—Earthquake or Underground Explosion* (eds. Husebye, E. S. and Mykkeltveit, S.) (Reidel Publishing Company 1981), pp. 49–69.

KOSTROV, B. and NITIKIN, L. (1970), *Some General Problems of Mechanics of Brittle Fracture,* Arch. Mech. Stosow *22*, 749–776.

KOSTROV, B. V. and DAS, S., *Principles of Earthquake Source Mechanics* (Cambridge University Press 1988).

KOYAMA, J. (1985), *Earthquake Source Time Function from Coherent and Incoherent Rupture,* Tectonophysics *118*, 227–242.

LEARY, P., LI, Y.-G., and AKI, K. (1987), *Observation and Modelling of Fault-zone Fracture Seismic Anisotropy – I. P, SV and SH Travel Times,* Geophys. J. R. Astron. Soc. *91*, 461–484.

LI, Y.-G., AKI, K., ADAMS, D., HASEMI, A., and LEE, W. (1994), *Seismic Guided Waves Trapped in the Zone of the Landers, California, Earthquake of 1992,* J. Geophys. Res. *99*, 11,705–11,722.

LI, Y.-G., ELLSWORTH, W. L., THURBER, C. H., MALIN, P. E., and AKI, K. (1997), *Fault-zone Guided Waves from Explosions in the San Andreas Fault Strands Near Anza, California,* Bull. Seismol. Soc. Am. *87*, 210–221.

LI, Y.-G., VIDALE, J. E., AKI, K., and XU, F. (2000), *Depth-dependent Structure of the Landers Fault Zone from Trapped Waves Generated by Aftershocks,* J. Geophys. Res. *105*, 6237–6254.

MADARIAGA, R. (1976), *Dynamics of an Expanding Circular Fault,* Bull. Seismol. Soc. Am. *66*, 639–666.

MADARIAGA, R. (1977), *High-frequency Radiation from Crack (Stress Drop) Models of Earthquake Faulting,* Geophys. J. R. Astr. Soc. *51*, 625–651.

MADARIAGA, R. (1978), *The Dynamic Field of Haskell's Rectangular Dislocation Fault Model,* Bull. Seismol. Soc. Am. *68*, 869–888.

MADARIAGA, R. (1979), *On the Relation Between Seismic Moment and Stress Drop in the Presence of Stress and Strength Heterogeneity,* J. Geophys. Res. *84*, 2243–2250.

MADARIAGA, R., *Earthquake Source Theory: A Review. In Earthquakes: Observation, Theory and Interpretation* (eds. Kanamori, H. and Boschi, E.), Proc. Int. Sch. Phys. Enrico Fermi, Course LXXXV (North-Holland, Amsterdam 1983a), pp. 1–44.

MADARIAGA, R. (1983b), *High-frequency Radiation from Dynamic Earthquake Fault Models,* Annales Geophysicae *1*, 17–23.

MADARIAGA, R. and BERNARD, P. (1985), *Ray Theoretical Strong Motion Synthesis,* J. Geophys. *58*, 73–81.

MALLET, R., *Great Neapolitan Earthquake of 1857: The First Principles of Observational Seismology* (Chapman and Hall, London 1862).

MENDEZ, A. J. and LUCO, J. E. (1990), *Steady State, Near-source Models of the Parkfield, Imperial Valley, and Mexicali Valley Earthquakes,* J. Geophys. Res. *95*, 327–340.

MENDOZA, C. and HARTZELL, S. (1988), *Inversion for Slip Distribution Using GDSN P Waves: North Palm Springs, Borah Peak, and Michoacan Earthquakes,* Bull. Seismol. Soc. Am. *78*, 1092–1111.

MIKLOWITZ, J., *The Theory of Elastic Waves and Waveguides,* vol. 22 of *North-Holland Series in Applied Mathematics and Mechanics* (North-Holland, Amsterdam 1978).

MIKUMO, T. and MIYATAKE, T. (1978), *Dynamic Rupture Process on a Three-dimensional Fault with Non-uniform Frictions and Near-field Seismic Waves,* Geophys. J. R. Astron. Soc. *54*, 417–438.

MOLNAR, P., TUCKER, B. E., and BRUNE, J. N. (1973), *Corner Frequencies of P and S Waves and Models of Earthquake Sources,* Bull. Seismol. Soc. Am. *63+*, 2091–2104.

NAKAHARA, H., NISHIMURA, T., SATO, H., and OHTAKE, M. (1998), *Seismogram Envelope Inversion for the Spatial Distribution of High-frequency Energy Radiation from the Earthquake Fault: Application to the 1994 Far East of Sanriku Earthquake, Japan*, J. Geophys. Res. *103*, 855–867.

OHNAKA, M. and YAMASHITA, T. (1989), *A Cohesive Zone Model for Dynamic Shear Faulting Based on Experimentally Inferred Constitutive Relations and Strong Motion Source Parameters*, J. Geophys. Res. *94*, 4089–4104.

OLSEN, A. and APSEL, R. (1982), *Finite Faults and Inverse Theory with Applications to the 1979 Imperial Valley Earthquake*, Bull. Seismol. Soc. Am. *72*, 1969–2001.

OLSEN, K., MADARIAGA, R., and ARCHULETA, R. (1997), *Three-dimensional Dynamic Simulation of the 1992 Landers Earthquake*, Science *278*, 834–838.

PAPAGEORGIOU, A. S. (1988), *On Two Characteristic Frequencies of Acceleration Spectra: Patch Corner Frequency and f_{max}*, Bull. Seismol. Soc. Am. *78*, 509–529.

PAPAGEORGIOU, A. S. and AKI, K. (1982), *Aspects of the Mechanics of Earthquake Rupture Related to the Generation of High-frequency Waves and the Prediction of Strong Ground Motion.*, Soil Dyn. Eq. Eng. *1*, 67–74.

PAPAGEORGIOU, A. S. and AKI, K. (1983a), *A Specific Barrier Model for the Quantitative Description of Inhomogeneous Faulting and the Prediction of Strong Ground Motion. I. Description of the Model*, Bull. Seismol. Soc. Am. *73*, 693–722.

PAPAGEORGIOU, A. S. and AKI, K. (1983b), *A Specific Barrier Model for the Quantitative Description of Inhomogeneous Faulting and the Prediction of Strong Ground Motion. Part II. Applications of the Model*, Bull. Seismol. Soc. Am. *73*, 953–978.

PAPAGEORGIOU, A. S. and AKI, K. (1985), *Scaling Law of Far-field Spectra Based on Observed Parameters of the Specific Barrier Model*, Pure Appl. Geophys. *123*, 354–374.

PERRIN, G., RICE, J. R., and ZHENG, G. (1995), *Self-healing Slip Pulse on a Frictional Surface*, J. Mech. Phys. Solids *43*, 1461–1495.

PHILLIPS, W. S. and AKI, K. (1986), *Site Amplification of Coda Waves from Local Earthquakes in Central California*, Bull. Seismol. Soc. Am. *76*, 627–648.

PULIDO, N. and IRIKURA, K. (2000), *Estimation of Dynamic Rupture Parameters from the Radiated Seismic Energy and Apparent Stress*, Geophys. Res. Lett. *27*, 3945–3948.

REID, H., *The Mechanics of the Earthquake. In The California Earthquake of April 18, 1906* (Report of the State Investigation Commission, Carnegie Institute of Washington, Washington, D.C. 1910).

RICE, J., *The Mechanics of Earthquake Rupture. In Physics of the Earth's Interior* (ed. Boschi, E.) (North-Holland, Amsterdam 1980), pp. 555–649.

RUDNICKI, J. and KANAMORI, H. (1981), *Effects of Fault Interaction on Moment, Stress Drop, and Strain Energy Release*, J. Geophys. Res. *86*, 1785–1793.

RUINA, A. L. (1983), *Slip Instability and State Variable Friction Laws*, J. Geophys. Res. *88*, 10,359–10,370.

SAIKIA, C. (1993), *Estimated Ground Motions in Los Angeles Due to $M_w = 7$ Earthquake on the Elysian Thrust Fault*, Bull. Seismol. Soc. Am. *83*, 780–810.

SATO, T. (1994), *Seismic Radiation from Circular Cracks Growing at Variable Rupture Velocity*, Bull. Seismol. Soc. Am. *84*, 1199–1215.

SATO, T. and HIRASAWA, T. (1973), *Body Wave Spectra from Propagating Shear Cracks*, J. Phys. Earth *21*, 415–431.

SAVAGE, J. C. (1966), *Radiation rom a Realistic Model of Faulting*, Bull. Seismol. Soc. Am. *56*, 577–592.

SHINOZUKA, M., *Engineering Modeling of Ground Motion: State-of-the-art Report. In Proceedings of the Ninth World Conference on Earthquake Engineering* (Tokyo-Kyoto, Japan 1988), vol. VIII.

SOMERVILLE, P., GRAVES, R., SAWADA, S., WALD, D., ABRAHAMSON, N., IWASAKI, Y., KAGAWA, T., SMITH, N., and KOWADA, A. (1999), *Characterizing Crustal Earthquake Slip Models for the Prediction of Strong Ground Motion*, Seism. Res. Lett. *70*, 59–80.

SPUDICH, P. and CRANSWICK, E. (1984), *Direct Observation of Rupture Propagation During the 1979 Imperial Valley Earthquake Using a Short Baseline Accelerometer Array*, Bull. Seismol. Soc. Am. *74*, 2038–2114.

SPUDICH, P. and FRAZIER, L. (1984), *Use of Ray Theory to Calculate High-frequency Radiation from Earthquake Sources Having Spatially Variable Rupture Velocity and Stress Drop*, Bull. Seismol. Soc. Am. *74*, 2061–2082.

Su, F. and Aki, K. (1995), *Site Amplification Factors in Central and Southern California Determined from Coda Waves*, Bull. Seismol. Soc. Am. *85*, 452–466.

Su, F., Anderson, J. G., and Zeng, Y. (1998), *Study of Weak and Strong Ground Motion Including Nonlinearity from the Northridge, California, Earthquake Sequence*, Bull. Seismol. Soc. Am. *88*, 1411–1425.

Takeo, M. (1988), *Rupture Process of the 1980 Izu-Hanto-Toho-Oki Earthquake Deduced from Strong Motion Seismograms*, Bull. Seismol. Soc. Am. *78*, 1074–1091.

Trifunac, M. and Udwadia, F. E. (1974), *Parkfield, California, Earthquake of June 27, 1966: A Three-dimensional Moving Dislocation Model*, Bull. Seismol. Soc. Am. *64*, 511–533.

Umeda, Y., Lio, Y., Luroiso, A., Ito, K., and Murakami, H. (1984), *Scaling of Observed Seismic Spectra*, Zisin *37*, 559–567.

Vvedenskaya, A. (1956), *The Determination of Displacement Fields by Means of Dislocation Theory*, Isvestiya Akad. Nauk. S.S.S.R., Ser. Geofiz., 227–284.

Wallace, R. (1984), *Eyewitness Account of Surface Faulting During the Earthquake of 28 October 1983, Borah Peak, Idaho*, Bull. Seismol. Soc. Am. *74*, 1091–1094.

Wu, F., *Prediction of Strong Ground Motion Using Small Earthquakes*. In *Proceedings of the 2nd International Conference on Microzonation* (San Francisco 1978), pp. 701–704.

Yamashita, T. and Fukuyama, E. (1996), *Apparent Critical Slip Displacement Caused by the Existence of a Fault Zone*, Geophys. J. Int. *125*, 459–472.

Yamashita, T. and Ohnaka, M. (1992), *Precursory Surface Deformation Expected from a Strike-slip Fault Model Into Which Rheological Properties of the Lithosphere are Incorporated*, Tectonophysics *211*, 179–199.

Yomogida, K. (1988), *Crack-like Rupture Process Observed in Near-fault Strong Motion Data*, Geophys. Res. Lett. *15*, 1223–1226.

Yomogida, K. and Nakata, T. (1994), *Large Slip Velocity of the Surface Rupture Associated with the 1990 Luzon Earthquake*, Geophys. Res. Lett. *21*, 1799–1802.

Zeng, Y. (1991), *Deterministic and Stochastic Modeling of the High Frequency Seismic Wave Generation and Propagation in the Lithosphere*, Ph.D. Thesis, University of Southern California, Los Angeles, California.

Zeng, Y., Aki, K., and Teng, L. (1993a), *Source Inversion of the 1987 Whittier Narrows Earthquake, California, Using the Isochrone Method*, Bull. Seismol. Soc. Am. *83*, 358–377.

Zeng, Y., Aki, K., and Teng, T.-L. (1993b), *Mapping of the High-frequency Source Radiation for the Loma Prieta Earthquake, California*, J. Geophys. Res. *98*, 11,981–11,933.

Zeng, Y., Anderson, J. G., and Su, F. (1995), *Subevent Rake and Random Scattering Effects in Realistic Strong Ground Motion Simulation*, Geophys. Res. Lett. *22*, 17–20.

(Received August 31, 2000, accepted May 17, 2001)

 To access this journal online:
http://www.birkhauser.ch

Pure appl. geophys. 160 (2003) 635–676
0033–4553/03/040635–42

Pure and Applied Geophysics

Simulation of Ground Motion Using the Stochastic Method

DAVID M. BOORE[1]

Abstract — A simple and powerful method for simulating ground motions is to combine parametric or functional descriptions of the ground motion's amplitude spectrum with a random phase spectrum modified such that the motion is distributed over a duration related to the earthquake magnitude and to the distance from the source. This method of simulating ground motions often goes by the name "the stochastic method." It is particularly useful for simulating the higher-frequency ground motions of most interest to engineers (generally, $f > 0.1$ Hz), and it is widely used to predict ground motions for regions of the world in which recordings of motion from potentially damaging earthquakes are not available. This simple method has been successful in matching a variety of ground-motion measures for earthquakes with seismic moments spanning more than 12 orders of magnitude and in diverse tectonic environments. One of the essential characteristics of the method is that it distills what is known about the various factors affecting ground motions (source, path, and site) into simple functional forms. This provides a means by which the results of the rigorous studies reported in other papers in this volume can be incorporated into practical predictions of ground motion.

Key words: Stochastic, simulation, ground motion, random vibration, earthquake, strong motion, site amplification.

Introduction

Keiiti Aki was one of the first seismologists to derive an expression for the spectrum of seismic waves radiated from complex faulting. In a 1967 paper (AKI, 1967) he used assumptions about the form of the autocorrelation function of slip as a function of space and time to derive an ω-square model of the spectrum (and he coined the term "ω-square model" in that paper). He then used the assumption of similarity to derive a source-scaling law, showing that the spectral amplitude at the corner frequency goes as the inverse-cube power of the corner frequency. He explicitly recognized that this is a constant-stress-drop model. His work has been used knowingly and unknowingly by several generations of seismologists to predict ground motions for earthquakes, particularly at high frequencies where the space- and time-distribution of fault slip is complicated enough to warrant a stochastic description of the source. Usually these predictions are for a specified seismic

[1] U.S. Geological Survey, Mail Stop 977, 345 Middlefield Road, Menlo Park, California 94025, U.S.A. E-mail: boore@usgs.gov

moment, and this is another place in which Kei's work had a long-term impact: in 1966 (AKI, 1966) he determined the seismic moment of an earthquake for the first time and also explicitly related the seismic moment to the product of rigidity, slip, and fault area. His research on the shape and scaling of source spectra and on seismic moment form the basis for the method for simulating ground motions discussed in this paper. In recognition of its use of a partially stochastic, rather than a completely deterministic, description of the source and path, this method is often referred to as "the stochastic model" or "the stochastic method." A word about terminology may be in order here: I refer to the means of simulating ground motions as the "stochastic method," whereas a particular application of the method results in a "stochastic model" of the ground motion (often associated with a particular study, such as the FRANKEL et al. (1996) model). The terminology is not standardized, however, and more usually (and loosely) people refer to any application of the stochastic method as the stochastic model; the distinction between the two is important, because the ground motions for different applications of the method (different models) might be very different.

There are several methods that use stochastic representations of some or all of the physical processes responsible for ground shaking (e.g., PAPAGEORGIOU and AKI, 1983a; ZENG et al., 1994). In this paper I review the particular stochastic method that I and a number of others developed in the last several decades. The paper includes a few new figures and an improvement in the calculation of random vibration results that previously appeared only in an USGS open-file report (BOORE, 1996), Other authors have published papers applying the stochastic method and extending the method in various ways. Table 1 contains a partial list of papers primarily concerned with development of the method; a table of references applying the method is given later.

Most of the discussion assumes that the motions to be simulated are S waves—these are the most important motions for seismic hazard. The method can be modified to simulate P-wave motions, as was done in BOORE (1986).

The Essence of the Method

The stochastic method described in this paper has its basis in the work of Hanks and McGuire, who combined seismological models of the spectral amplitude of ground motion with the engineering notion that high-frequency motions are basically random (HANKS, 1979; MCGUIRE and HANKS, 1980; HANKS and MCGUIRE, 1981). Assuming that the far-field accelerations on an elastic half space are band-limited, finite-duration, white Gaussian noise, and that the source spectra are described by a single corner-frequency model whose corner frequency depends on earthquake size according to the BRUNE (1970, 1971) scaling, they derived a remarkably simple relationship for peak acceleration that was in good agreement with data from 16

Table 1

Some references on methodology

BERESNEV and ATKINSON (1997, 1998a), BOORE (1983, 1984, 1989b, 1996, 2000),
BOORE and JOYNER (1984), CAMPBELL (1999), CARTWRIGHT and LONGUET-HIGGINS (1956),
CORREIG (1996), ERDIK and DURUKAL (2001), GHOSH (1992), HANKS and McGUIRE (1981),
HERRMANN (1985), JOYNER (1984, 1995), JOYNER and BOORE (1988), KAMAE and IRIKURA (1992), KAMAE
et al. (1998), KOYAMA (1997), LAM *et al.* (2000), LIAO and JIN (1995), LIU and PEZESHK (1998, 1999), LOH
and YEH (1988), MILES and HO (1999), ÓLAFSSON and SIGBJÖRNSSON (1999), OU and HERRMANN
(1990a, 1990b), PAPAGEORGIOU and AKI (1983a), PEZESHK *et al.* (2001), RATHJE *et al.* (1998), SABETTA and
PUGLIESE (1996), ŞAFAK and BOORE (1988), SCHNEIDER *et al.* (1991), SHAPIRA and VAN ECK (1993),
SILVA (1992), SILVA and LEE (1987), SILVA *et al.* (1988, 1990, 1997), TAMURA *et al.* (1991),
WENNERBERG (1990), YU *et al.*

earthquakes. I generalized their work to allow for arbitrarily complex models, extended it to the simulation of time series, and considered many measures of ground motions, the most important of which are response spectra (BOORE, 1983). The underlying simplicity of the method, however, remains unchanged. The essence of the method is shown in Figure 1: The top of the figure shows the spectrum of the ground motion at a particular distance and site condition for magnitude 5 and 7 earthquakes, based on a standard seismological model; by assuming that this motion is distributed with random phase over a time duration related to earthquake size and propagation distance, the time series shown in the bottom of the figure are produced.

The essential ingredient for the stochastic method is the spectrum of the ground motion—this is where the physics of the earthquake process and wave propagation are contained, usually encapsulated and put into the form of simple equations. Most of the effort in developing a model is in describing the spectrum of ground motion. As is traditional, I find it convenient to break the total spectrum of the motion at a site ($Y(M_0, R, f)$) into contributions from earthquake source (E), path (P), site (G), and instrument or type of motion (I), so that

$$Y(M_0, R, f) = E(M_0, f)P(R, f)G(f)I(f) \ , \tag{1}$$

where M_0 is the seismic moment, introduced into seismology in 1966 by K. AKI (AKI, 1966). I usually use moment magnitude **M** rather than seismic moment as a more familiar measure of earthquake size; there is a unique mapping between the two:

$$\mathbf{M} = \frac{2}{3}\log M_0 - 10.7 \tag{2}$$

(HANKS and KANAMORI, 1979).

Seismic moment has a number of advantages as the predictor variable for earthquake size in applications:

- It is the best single measure of overall size of an earthquake and is not subject to saturation.
- It can be determined from ground deformation or from seismic waves.

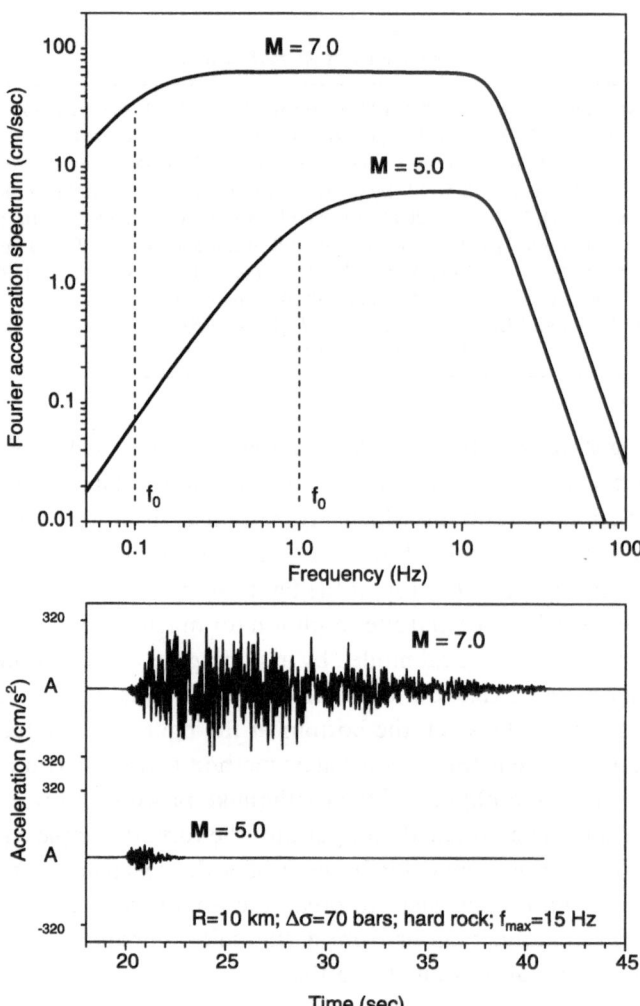

Figure 1

Basis for stochastic method. Radiated energy described by the spectra in the upper part of the figure is assumed to be distributed randomly over a duration equal to the inverse of the lower corner frequency (f_0). Each time series is one realization of the random process for the actual spectrum shown. When plotted on a log scale, the levels of the low-frequency part of the spectra are directly proportional to the logarithm of the seismic moment and thus to the moment magnitude. Various peak ground-motion parameters (such as response spectra, instrument response, and velocity and acceleration) can be obtained by averaging the parameters computed from each member of a suite of acceleration time series or more simply by using random vibration theory, working directly with the spectra. The examples in this figure came from an actual simulation and are not sketched in by hand.

- It can be estimated from paleoseismological studies.
- It can be related to slip rates on faults.
- It is the variable of choice for empirically and theoretically based equations for the prediction of ground motions.

By separating the spectrum of ground motion into source, path, and site components, the models based on the stochastic method can be easily modified to account for specific situations or to account for improved information about particular aspects of the model.

The Source $(E(M_0, f))$

Both the shape and the amplitude of the source spectrum must be specified as a function of earthquake size. This is the most critical part of any application of the method. References given later should be consulted to see how various authors have approached this issue. The most commonly used model of the earthquake source spectrum is the ω-square model, a term coined by AKI (1967). Figure 2 shows this spectrum for earthquakes of moment magnitude 6.5 and 7.5. The scaling of the spectra from one magnitude to another is determined by specifying the dependence of the corner frequency f_0 on seismic moment. AKI (1967) recognized that assuming similarity in the earthquake source implies that

$$M_0 f_0^3 = \text{constant} , \qquad (3)$$

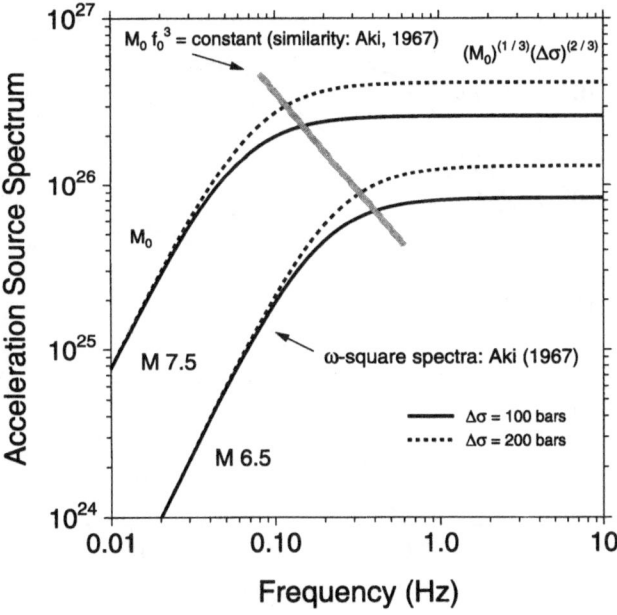

Figure 2

Source scaling for single-corner-frequency ω-square spectral shape. For constant stress drop $M_0 f_0^3$ is a constant (AKI, 1967), and this dependence of the corner frequency f_0 on the moment M_0 (given by the shaded line) determines the scaling of the spectral shapes.

where the constant can be related to the stress drop ($\Delta\sigma$). Following BRUNE (1970, 1971), the corner frequency is given by the following equation:

$$f_0 = 4.9 \times 10^6 \beta_s (\Delta\sigma/M_0)^{1/3} \; , \qquad (4)$$

where f_0 is in Hz, β_s (the shear-wave velocity in the vicinity of the source) in km/s, $\Delta\sigma$ in bars, and M_0 in dyne-cm.

Although the ω-square model is widely used, in practice a variety of other models have been used with the stochastic method. Figure 3 shows a number of those that have been used to predict ground motions in eastern North America. It turns out that the source spectra for all of the models can be given by the following equation:

$$E(M_0, f) = C M_0 S(M_0, f) \; , \qquad (5)$$

where C is a constant, given below, and $S(M_0, f)$ is the displacement source spectrum, given by the equation

$$S(M_0, f) = S_a(M_0, f) \times S_b(M_0, f) \; , \qquad (6)$$

and S_a, S_b for the various models shown in Figure 3 are given in Table 2. The moment dependence of the two factors S_a and S_b is given by the relations between the

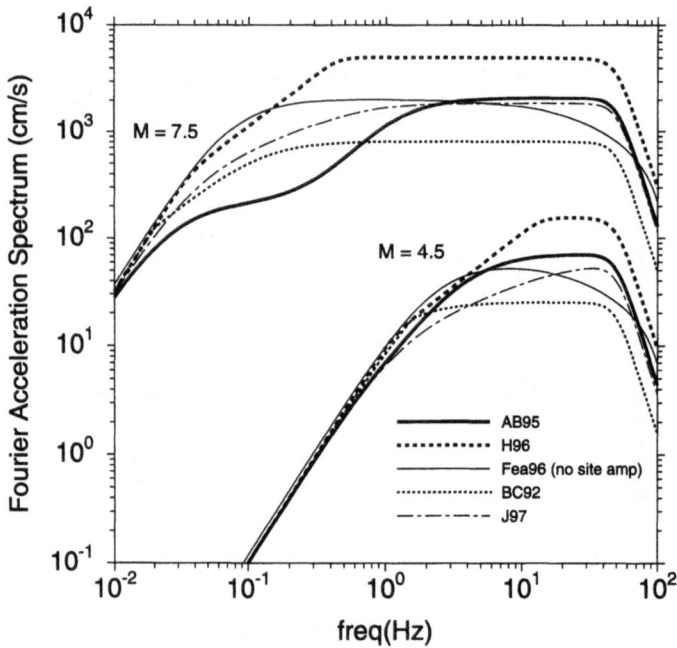

Figure 3

Fourier spectrum of acceleration at R = 1 km, according to the source spectral models given in Tables 2 and 3 (from ATKINSON and BOORE, 1998). (The roll-off at high frequencies is produced by using equation (19) with $f_{max} = 50$ Hz).

Table 2

Shape of source spectral $(S(f) = S_a(f) * S_b(f))$

Model[†]	S_a	S_b
BC92	$f < f_a : 1$ $f \geq f_a : f_a/f$	$\dfrac{1}{(1+(f/f_b)^2)^{1/2}}$
AB95	$\dfrac{1-\epsilon}{1+(f/f_a)^2} + \dfrac{\epsilon}{1+(f/f_b)^2}$	1
Fea96*	$\dfrac{1}{1+(f/f_a)^2}$	1
H96	$\dfrac{1}{(1+(f/f_a)^8)^{1/8}}$	$\dfrac{1}{(1+(f/f_b)^8)^{1/8}}$
J97	$\dfrac{1}{(1+(f/f_a)^2)^{3/4}}$	$\dfrac{1}{(1+(f/f_b)^2)^{1/4}}$
AS00	$\dfrac{1-\epsilon}{1+(f/f_a)^2} + \dfrac{\epsilon}{1+(f/f_b)^2}$	1

[†] The references to the models are as follows: BC92 = BOATWRIGHT and CHOY (1992); AB95 = ATKINSON and BOORE (1995); Fea96 = FRANKEL *et al.* (1996); H96 = HADDON (1996); J97 = JOYNER (1997), as modified in a written communication to D. Boore; AS00 = ATKINSON and SILVA (2000) for California.
* This is the ω-square model.

corner frequencies f_a and f_b appearing in the factors and the seismic moment, as shown in Table 3 (which also contains the scaling for the ATKINSON and SILVA (2000) model for California; a number of illustrations later in the paper use their model). The constant C in equation (5) is given by

$$C = \langle R_{\Theta\Phi}\rangle VF/(4\pi\rho_s\beta_s^3 R_0) \ , \tag{7}$$

where $\langle R_{\Theta\Phi}\rangle$ is the radiation pattern, usually averaged over a suitable range of azimuths and take-off angles (BOORE and BOATWRIGHT, 1984), V represents the partition of total shear-wave energy into horizontal components $(= 1/\sqrt{2})$, F is the effect of the free surface (taken as 2 in almost all applications, which strictly speaking

Table 3

Corner frequencies and moment ratios

Model	$\log f_a$	$\log f_b$	$\log \epsilon$
BC92	M ≥ 5.3:[†] 3.409 − 0.681M M < 5.3: 2.452 − 0.5M	1.495 − 0.319M 2.452 − 0.5M	− −
AB95	M ≥ 4.0:[‡] 2.41 − 0.533M M < 4.0: 2.678 − 0.5M	1.43 − 0.188M 2.678 − 0.5M	2.52 − 0.637M 0.0
Fea96*	2.623 − 0.5M	−	−
H96	2.3 − 0.5M	3.4 − 0.5M	−
J97	2.312 − 0.5M	3.609 − 0.5M	−
AS00	M ≥ 2.4:[†] 2.181 − 0.496M M < 2.4: 1.431 − 0.5(M − 2.4)	2.41 − 0.408M 1.431 − 0.5(M − 2.4)	0.605 − 0.255M 0.0

[†] The specified magnitude corresponds to the point at which $f_a = f_b$.
[‡] The specified magnitude corresponds to the point at which $\epsilon = 1.0$.
* This is the ω-square model, for which $\log f_0 = 1.341 + \log(\beta(\Delta\sigma)^{1/3}) - 0.5\,M$, with $\beta = 3.6$ km/s and $\Delta\sigma = 150$ bars.

is only correct for SH waves), ρ_s and β_s are the density and shear-wave velocity in the vicinity of the source, and R_0 is a reference distance, usually set equal to 1 km. In applications, care must be taken if mixed units are used. For example, if ground motion is to be in cm and ρ_s, β_s, and R_0 are in units of gm/cc, km/s and km, respectively, then C in equation (7) should be multiplied by the factor 10^{-20}. It is probably safer to convert all quantities into common units.

The Path $(P(R, f),$ duration$)$

Now that the source has been specified, it remains to discuss the other components of the process that affect the spectrum of motion at a particular site. The next component is the path effect. For some applications involving a specific path from source to site it might be desirable to convolve the radiation from the source with theoretically calculated path effects. An example of calculated path response is shown in Figure 4 for a four-layer model of the crust in the central United States. The response is complicated because of the critical-angle arrivals and reverberations of the waves. Even though complicated, however, the response is probably simpler than reality because the crust may not be laterally uniform and because scattering has not been included. For most applications it is advisable to represent the effects of the path by simple functions that account for geometrical spreading, attenuation (combining intrinsic and scattering attenuation), and the general increase of duration with distance due to wave propagation and scattering.

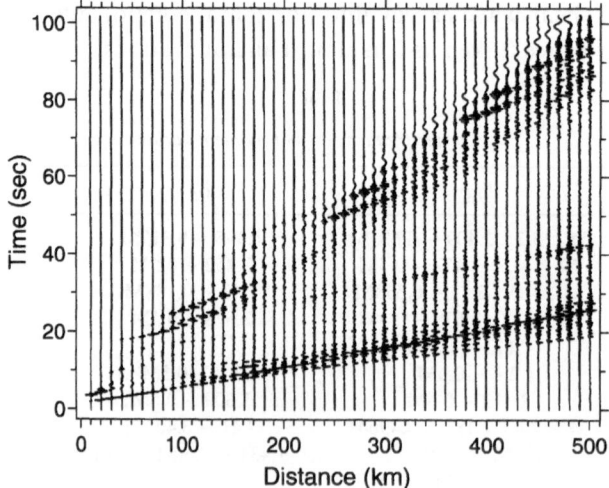

Figure 4
Synthetic seismograms for a 4-layer model of the crust in the central United States, showing the complexity of the waveforms and duration due to reverberations within the crust (written commun., R. HERRMANN, 2000).

The simplified path effect P is given by the multiplication of the geometrical spreading and Q functions

$$P(R,f) = Z(R)\exp\left[-\pi f R / Q(f)c_Q\right],\qquad(8)$$

where c_Q is the seismic velocity used in the determination of $Q(f)$, and the geometrical spreading function $Z(R)$ is given by a piecewise continuous series of straight lines:

$$Z(R) = \begin{cases} \frac{R_0}{R} & R \le R_1 \\ Z(R_1)\left(\frac{R_1}{R}\right)^{p_1} & R_1 \le R \le R_2 \\ \vdots \\ Z(R_n)\left(\frac{R_n}{R}\right)^{p_n} & R_n \le R \end{cases}\qquad(9)$$

In applications, R is usually taken as the closest distance to the rupture surface, rather than the hypocentral distance. In some applications it may be appropriate to include a period and/or moment dependent "pseudo-depth" h in a manner consistent with the effectively point-source models used in fitting empirical strong-motion data. For example, following BOORE et al. (1997) R would be given by $R = \sqrt{D^2 + h^2}$, where D is the closest distance to the vertical projection of the rupture surface onto the ground surface, and h is taken from the empirical results in BOORE et al. (1997). Other empirically-based prediction equations use different relations to determine the distance—see the review by CAMPBELL (2002)—but the idea is the same. By defining R in this way rather than as hypocentral distance, the method is more applicable to extended ruptures. As an example of $Z(R)$, Figure 5 shows the three-segment geometrical spreading operator used in ATKINSON and BOORE's (1995) predictions of ground motions in eastern North America. For this example, $R_0 = 1$, $R_1 = 70$, $p_1 = 0.0$, $R_2 = 130$, and $p_2 = 0.5$.

The form of the attenuation operator is motivated by K. Aki's compilation of seismic attenuation Q shown in Figure 6. As a simple way of capturing the variation of Q, the attenuation operator is made up of three piecewise-continuous line segments (Fig. 7). The outer lines are specified by slopes and intercepts at specified reference frequencies, and the middle line joins the outer lines between frequencies $ft1$ and $ft2$. In applications the various parameters describing the attenuation operator can be obtained from analysis of weak-motion data, if available. If determined from data, it is important to keep in mind the tradeoffs between geometrical spreading and attenuation. Both functions are needed in fitting data, and for consistency, the same functions must be used in applications. An example of the combined path effect is shown in Figure 8, which compares observed spectral amplitudes as a function of distance with geometrical spreading and attenuation operators fit to the data. In this case the geometrical spreading function is that shown in Figure 5, and the Q function is given by $Q = 680f^{0.38}$, which is the s2 branch in Figure 6 (the data were not sufficient to determine the longer-period s1 branch).

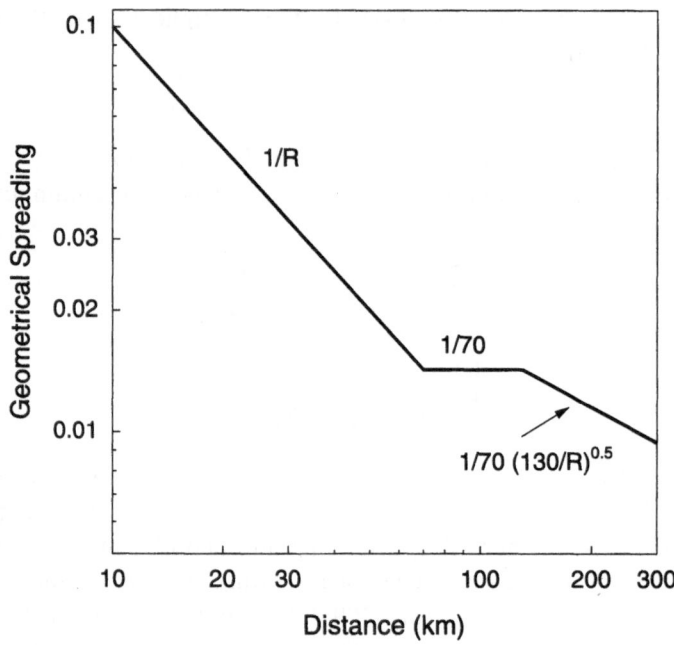

Figure 5
The geometrical spreading function used in applications in central and eastern North America by
ATKINSON and BOORE (1995) and FRANKEL *et al.* (1996).

The distance-dependent duration is an important function, for the peak motions decrease with increasing duration, all other things being equal. Although the Fourier amplitude spectrum of the ground motion (equation (1)) is not dependent on the duration, I include a discussion of duration here because it is a function of the path, as well as the source; the way it is used in the calculations of ground motion is given later. The ground-motion duration (T_{gm}) is the sum of the source duration, which is related to the inverse of a corner frequency (e.g., $0.5/f_a$ for the AB95 and $1/f_a$ for the Fea96 models in Tables 2 and 3) and a path-dependent duration. Empirical observations and theoretical simulations suggest that the path-dependent part of the duration can be represented by a connected series of straight-line segments. The function used in ATKINSON and BOORE (1995) is shown in Figure 9, along with the data from which the function was determined.

The Site ($G(f)$)

In the strictest sense, the modification of seismic waves by local site conditions is part of the path effect. Because local site effects, however, are largely independent of the distance traveled from the source (except for nonlinear effects for which the amplitudes of motion are important), it is convenient to separate site and path

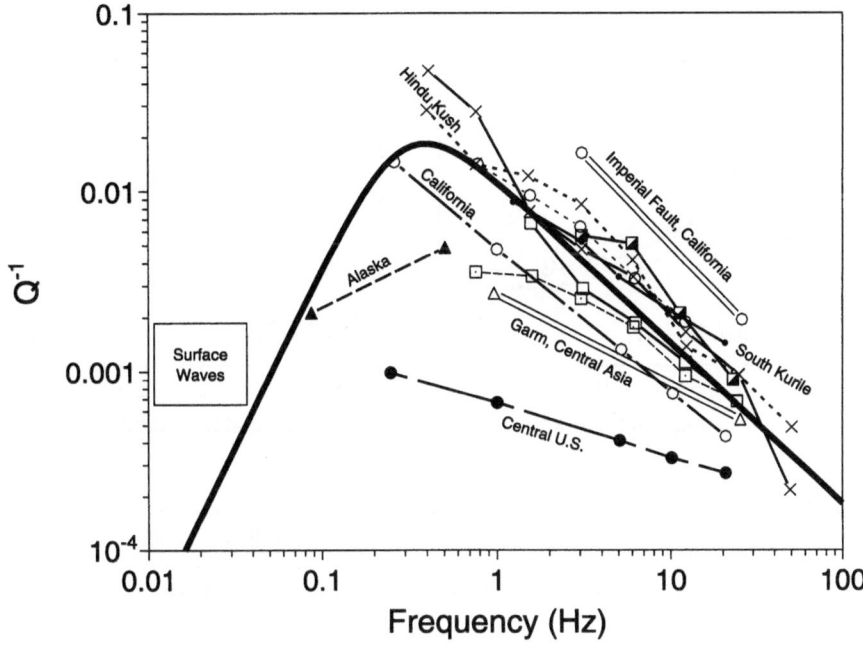

Figure 6
Observed inverse shear-wave Q from Aki (AKI, 1980, summarized by CORMIER, 1982); the heavy solid line is an "eyeball" average of the observations. (Figure modified from BOORE, 1984.)

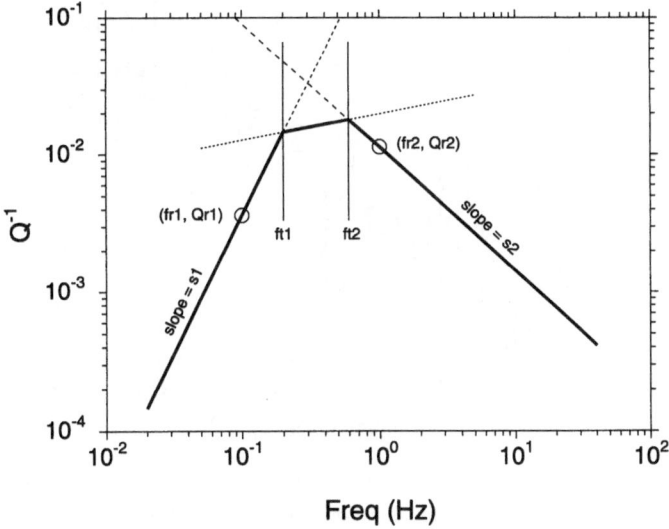

Figure 7
Illustration of the specification of $Q(f)$: it is made up of three lines in log–log space. The lines shown are an approximation of the $Q(f)$ function shown in the previous figure.

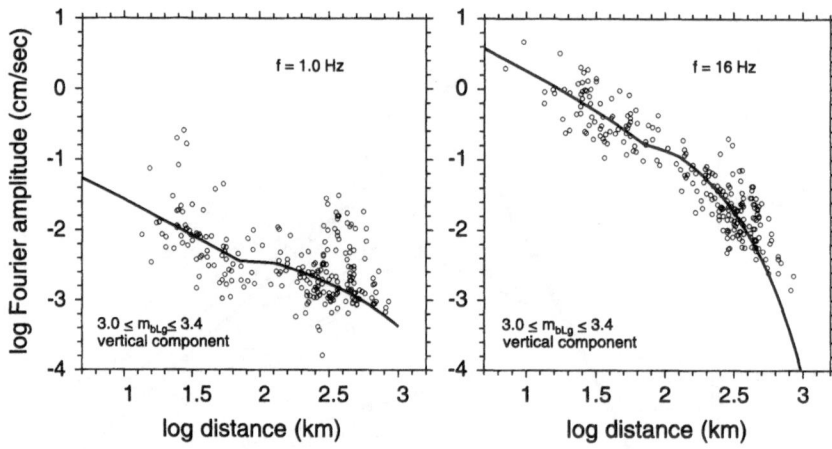

Figure 8
Observed attenuation of motions with distance in eastern North America for a narrow range of magnitudes (data: written commun. from G. ATKINSON, 2000), along with the combination of geometrical spreading and whole path attenuation used by ATKINSON and BOORE (1995) and FRANKEL *et al.* (1996) in simulating ground motions in central and eastern North America.

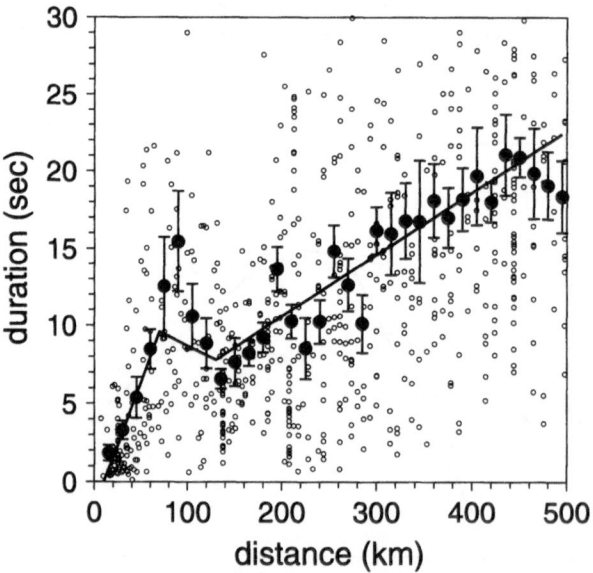

Figure 9
Observed duration (after subtracting source duration) from earthquakes in eastern North America. The data were used by ATKINSON (1993) and ATKINSON and BOORE (1995) to define path-dependent durations for use in stochastic method simulations. The solid circles are averages within 15-km-wide bins, and the error bars are plus and minus one standard error of the mean. The three-part solid line is the duration function used by ATKINSON and BOORE (1995) in simulations of ground motions in eastern North America.

effects. Much effort can go into accounting for the modifications of the ground motion due to local site geology. This is a situation where site-specific effects might best be used. On the other hand, in many cases the simulations from the stochastic method are intended to be used for the prediction of motion at a generic site—such as a generic rock or a generic soil site. In such cases, a simplified function can be used to describe the frequency-dependent modifications of the seismic spectrum. I find it convenient to separate the amplification ($A(f)$) and attenuation ($D(f)$), as follows:

$$G(f) = A(f)D(f) \ . \tag{10}$$

The amplification function $A(f)$ is usually relative to the source unless amplitude variations due to wave propagation, separate from the geometrical spreading, have been accounted for. In contrast, the diminution function $D(f)$ is used to model the path-independent loss of energy (the path-dependent part is modeled by the exponential function in equation (8)). It is important in applications to be specific about the reference conditions for the A and D functions. In general, G can be a function of the amplitude of shaking, but I do not account for nonlinear effects in my method, preferring to compute rock motions using a linear model and account for nonlinear effects as part of an additional site-response calculation. W. Silva, however, has incorporated nonlinear effects into his version of the stochastic method (SILVA et al., 1991).

The starting point for deriving the amplification $A(f)$ is a function of shear-wave velocity vs. depth. Figure 10 shows such a function for a generic rock site appropriate for coastal California. The top 100 m is based on averaging of travel times measured in boreholes, while the deeper parts of the curve are based on judgment and a few data (BOORE and JOYNER, 1997). The amplification $A(f)$ can be given by wave-calculation solutions that account for reverberations, or approximately and more simply by assuming that the amplification of the waves is given by the square root of the impedance ratio between the source and the surface. The algorithm is the following:

$$A(f(z)) = \sqrt{Z_s/\bar{Z}(f)} \ , \tag{11}$$

where the seismic impedance near the source (Z_s) is given by

$$Z_s = \rho_s \beta_s \ , \tag{12}$$

and ρ_s and β_s are the density and shear-wave velocity near the source. $\bar{Z}(f)$ is an average of near-surface seismic impedance; it is a function of frequency because it is a time-weighted average from the surface to a depth equivalent to a quarter wavelength:

$$\bar{Z}(f) = \int\limits_{0}^{t(z(f))} \rho(z)\beta(z) \, dt \bigg/ \int\limits_{0}^{t(z(f))} dt \ , \tag{13}$$

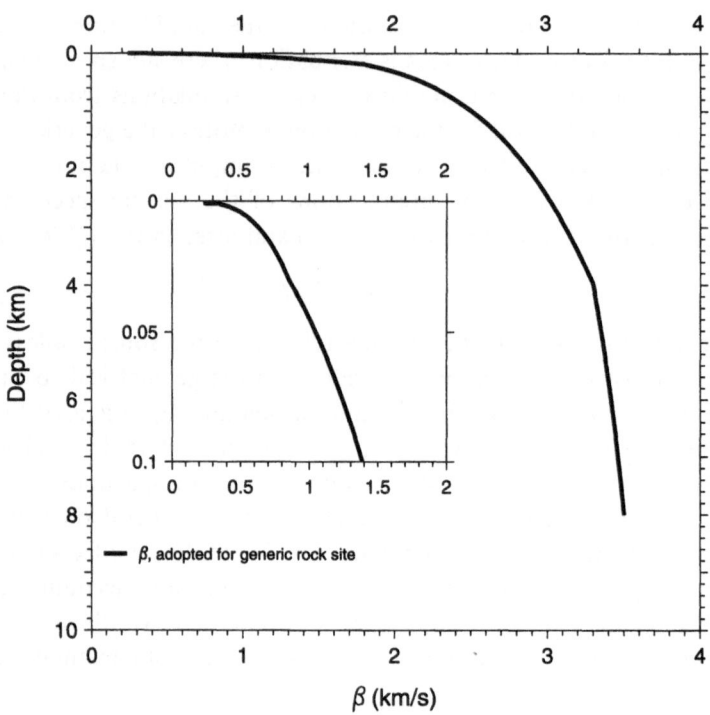

Figure 10
S-wave velocity versus depth used by BOORE and JOYNER (1997) for computing amplifications on generic
"soft" rock sites (adapted from BOORE and JOYNER, 1997.)

in which the upper limit of the integral is the time for shear waves to travel from depth $z(f)$ to the surface. The depth is a function of frequency and is chosen such that z is a quarter-wavelength for waves traveling at an average velocity given by $\bar{\beta} = z(f)/\int_0^{z(f)}[1/\beta(z)]\ dz$. The condition of a quarter-wavelength $z = (1/4)\bar{\beta}/f$ then yields the following implicit equation for $z(f)$:

$$f(z) = 1\left/\left[4\int\limits_0^{z(f)}\frac{1}{\beta(z)}\ dz\right]\right. . \tag{14}$$

In practice, it is easiest to compute f and $\bar{Z}$ for a given z. By changing variables from time to depth, equation (13) becomes

$$\bar{Z}(f) = \int\limits_0^{z(f)}\rho(z)\ dz\left/\int\limits_0^{z(f)}\frac{1}{\beta(z)}\ dz\right. . \tag{15}$$

Equation (15) can be simplified to

$$\bar{Z}(f) = \bar{\rho}\bar{\beta} \ , \tag{16}$$

where

$$\bar{\rho} = \frac{1}{z(f)} \int_0^{z(f)} \rho(z) \ dz \ , \tag{17}$$

and

$$\bar{\beta} = z(f) \left[\int_0^{z(f)} \frac{1}{\beta(z)} \ dz \right]^{-1} . \tag{18}$$

Figure 11 compares the amplification computed using equation (11) for the generic rock velocity profile in Figure 10 and wave propagation for two angles of incidence.

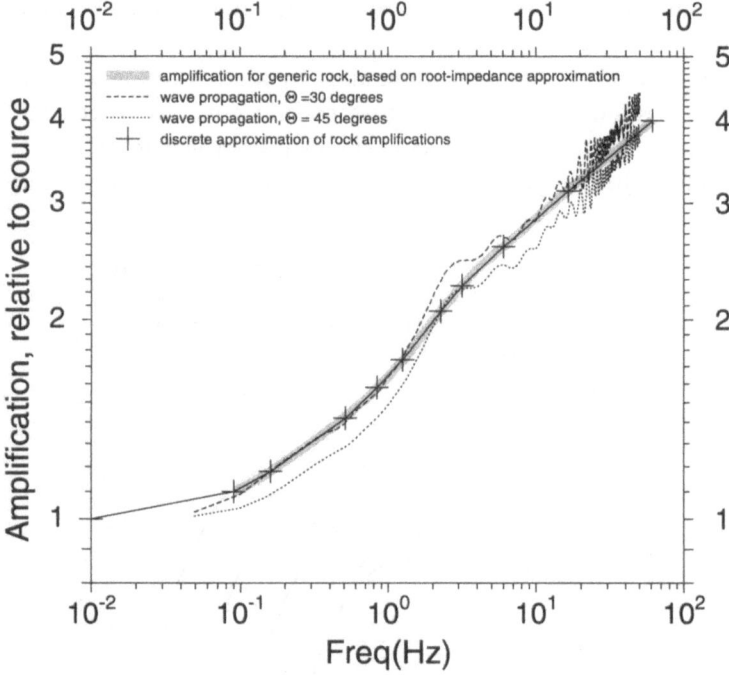

Figure 11

Amplification vs. frequency. The wide shaded line is computed using the root-impedance approximation and the velocity profile shown in the preceeding figure. The results from plane *SH* waves incident at the base of a 8-km thick stack of constant-velocity layers (with $Q = 10000$) closely approximating the continuous shear-wave velocity in the previous figure are shown by the light lines for angles of incidence of 30 and 45 degrees; the results were computed from the Haskell matrix method, as implemented by program *Rattle* by C. Mueller. The segmented-line function used in the stochastic method is given by lines joining the plus symbols. (Adapted from BOORE and JOYNER, 1997.)

For application, it is convenient to approximate the amplification by a series of connected line segments; these are also shown in Figure 11.

The attenuation, or diminution, operator $D(f)$ in equation (10) accounts for the path-independent loss of high-frequency in the ground motions. This loss may be due to a source effect, as suggested by PAPAGEORGIOU and AKI (1983b) or a site effect, as suggested by a number of authors, including HANKS (1982), or by a combination of these effects. If a source effect, D may also depend on the size of the earthquake. It is not my intention to argue for a particular cause, but only to point out that a simple multiplicative filter can account for the diminution of the high-frequency motions. Two filters are in common use: the f_{max} filter

$$D(f) = \left[1 + (f/f_{max})^8\right]^{-1/2} , \tag{19}$$

(HANKS, 1982; BOORE, 1983), and the κ_0 filter

$$D(f) = \exp(-\pi\kappa_0 f) , \tag{20}$$

(ANDERSON and HOUGH, 1984). Of course, both filters can be combined in an application.

The combined effect of amplification and attenuation for a series of diminution parameters κ_0 is shown in Figure 12 for a generic rock site in coastal California. Comparisons with data suggest that κ_0 near 0.04 is appropriate (BOORE and JOYNER, 1997). Filters for other types of site geology can be obtained by combining the results

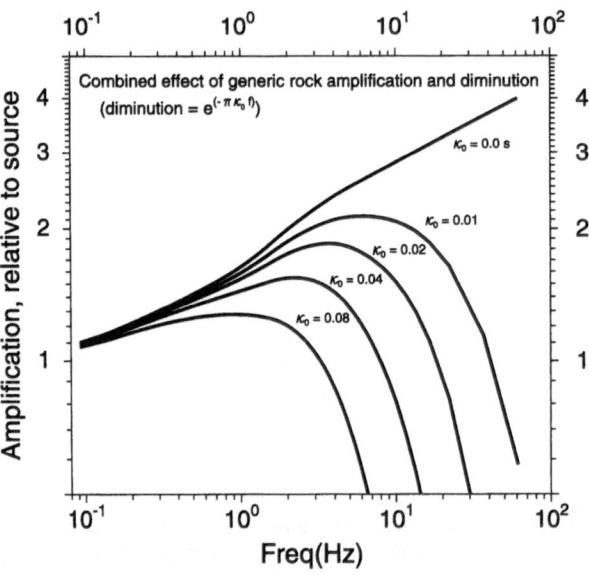

Figure 12

Combined effect of the site amplification in the previous figure and path-independent diminution. (Adapted from BOORE and JOYNER, 1997.)

in Figure 12 with the site effects from empirical attenuation curves. The results are shown in Figure 13 (for more detail, see BOORE and JOYNER, 1997).

Accounting for Type of Ground Motion $(I(f))$

The particular type of ground motion resulting from the simulation is controlled by the filter $I(f)$. If ground motion is desired, then

$$I(f) = (2\pi fi)^n \ , \tag{21}$$

where $i = \sqrt{-1}$ and $n = 0, 1,$ or 2 for ground displacement, velocity, or acceleration, respectively. For the response of an oscillator, from which response spectra or Wood-Anderson magnitudes can be derived,

$$I(f) = \frac{-Vf^2}{(f^2 - f_r^2) - 2ff_r\zeta i} \ , \tag{22}$$

for an oscillator with undamped natural frequency f_r, damping ζ, and gain V (for computation of response spectra, $V = 1$).

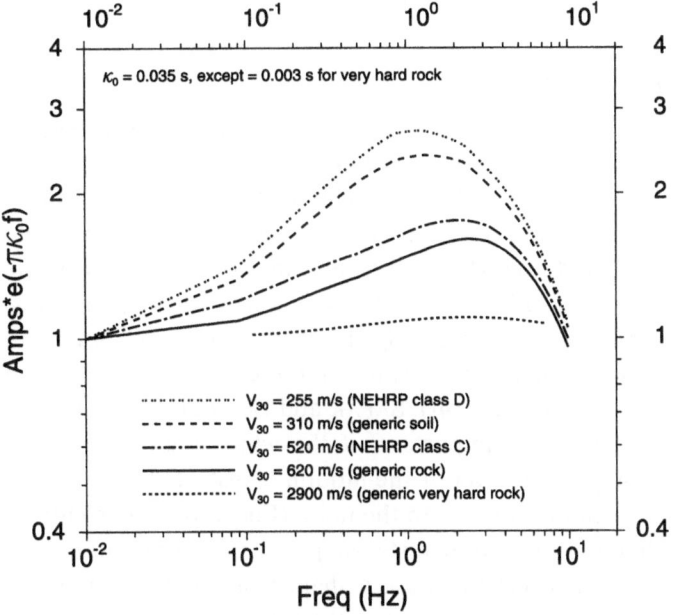

Figure 13

The product of Fourier spectral amplifications and the diminution factor $\exp(-\pi\kappa_0 f)$ for various site conditions, as measured by the average shear-wave velocity in the upper 30 m. (From BOORE and JOYNER, 1997.)

Integral Measures of Ground Motion

Measures of ground motion based on some average of the motion over time or of the spectrum over frequency are sometimes used in seismic hazard (e.g., JIBSON, 1993; JIBSON *et al.*, 1998; WILSON, 1993). The most common of these may be the Arias intensity (I_{xx}), defined as

$$I_{xx} \equiv \frac{\pi}{2g} \int_0^{t_d} a(t)^2 \, dt \ , \tag{23}$$

where g is the acceleration of gravity, a is the ground acceleration, and t_d is the duration of the motion (ARIAS, 1970). This intensity measure can be easily computed within the context of the stochastic method, as shown below.

Obtaining Ground Motions

Given the spectrum of motion at a site, there are two ways of obtaining ground motions: 1) time-domain simulation and 2) estimates of peak motions using random vibration theory.

Simulations of Time Series

Time-domain simulations are easy to obtain. This is illustrated in Figure 14 for an actual application, using the AS00 model as given in Tables 2, 3, and 4 (this model is used for all but the last of the remaining figures). White noise (Gaussian or uniform) is generated for a duration given by the duration of the motion (Fig. 14a); this noise is then windowed (Fig. 14b); the windowed noise is transformed into the frequency domain (Fig. 14c); the spectrum is normalized by the square-root of the mean square amplitude spectrum (Fig. 14d); the normalized spectrum is multiplied by the ground motion spectrum Y (Fig. 14e); the resulting spectrum is transformed back to the time domain (Fig. 14f). SAFAK and BOORE (1988) show that the order of windowing and filtering is important; if the white noise is first filtered and then windowed the long-period level of the motion is distorted.

The shaping window applied to the noise (Fig. 14b) can be either a box window or a window that gives a more realistic shape for the acceleration time series (as will be shown shortly, the decision to use a shaped rather than a box window is based more on aesthetics than on differences in the derived ground-motion parameters). By studying a number of recorded motions, SARAGONI and HART (1974) found that the following function is a good representation of the envelope of acceleration time series:

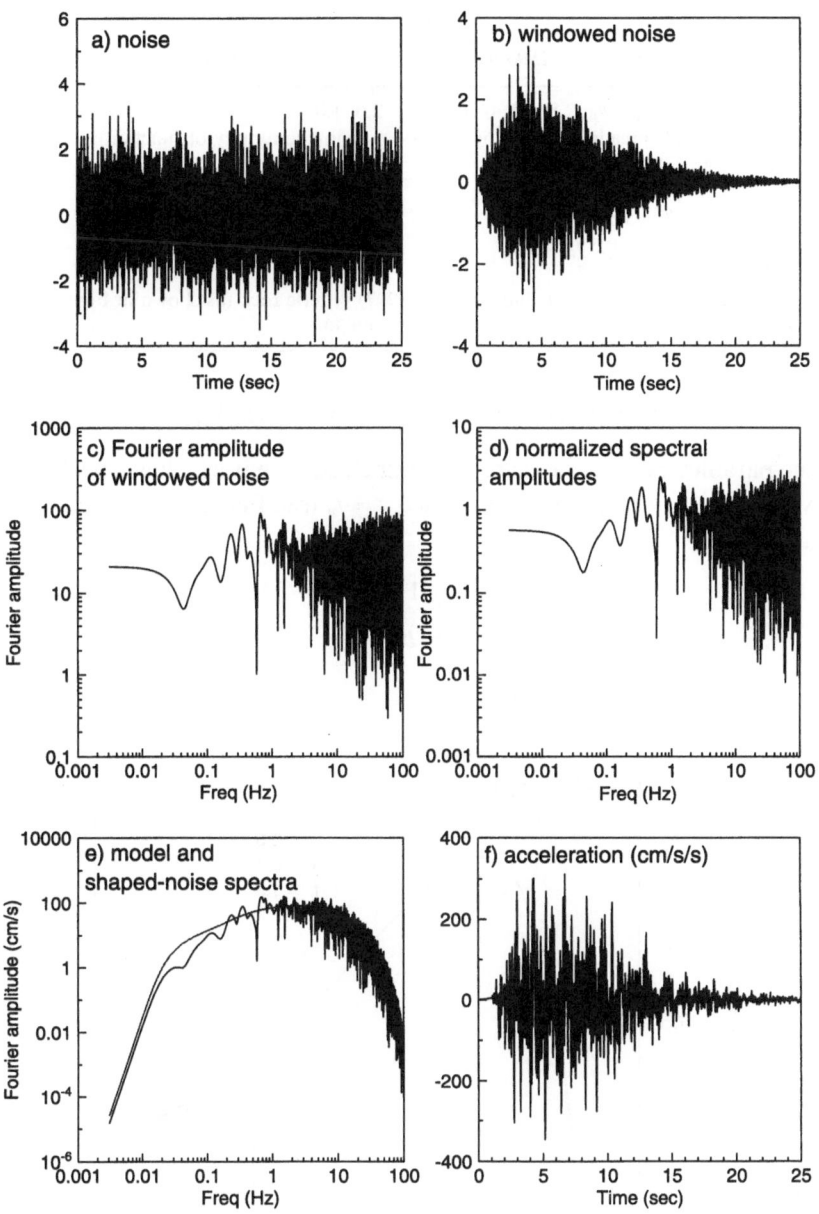

Figure 14

Basis of the time-domain procedure for simulating ground motions using the stochastic method. These are from an actual simulation, using the AS00 model as specified in Tables 2, 3, and 4. An acausal low-cut filter with a cut-off frequency of 0.02 Hz was applied to the acceleration time series. Various other measures of ground motion, such as peak velocity, peak displacement, Arias intensity, and response spectral amplitudes, can be computed from the simulated acceleration.

Table 4

Parameters for AS00 model (from ATKINSON and SILVA, 2000)

- $\rho_s, \beta_s, V, \langle R_{\Theta\Phi}\rangle, F, R_0$: 2.8, 3.5, 0.707, 0.55, 2.0, 1.0
- Geometrical spreading (including factors to insure continuity of function):
  $r < 40\,\text{km} : 1/r$
  $40\,\text{km} \le r : (1/40)(40/r)^{0.5}$
- Q, c_Q: $180f^{0.45}, 3.5\,\text{km/s}$
- Source duration: $0.5/f_a$
- Path duration: $0.05\,R$
- Site amplification: BOORE and JOYNER (1997) generic rock (as shown in Figure 11).
- Site diminution parameters $(f_{\max}, \kappa)$: 100.0, 0.030.

$$w(t; \epsilon, \eta, t_\eta,) = a(t/t_\eta)^b \exp(-c(t/t_\eta)) \ , \tag{24}$$

where the parameters a, b, and c are determined such that $w(t)$ has a peak with value of unity when $t = \epsilon \times t_\eta$ and $w(t) = \eta$ when $t = t_\eta$ (see Fig. 15). The equations for a, b, and c follow:

$$b = -(\epsilon \ln \eta)/[1 + \epsilon(\ln \epsilon - 1)] \ , \tag{25}$$

$$c = b/\epsilon \ , \tag{26}$$

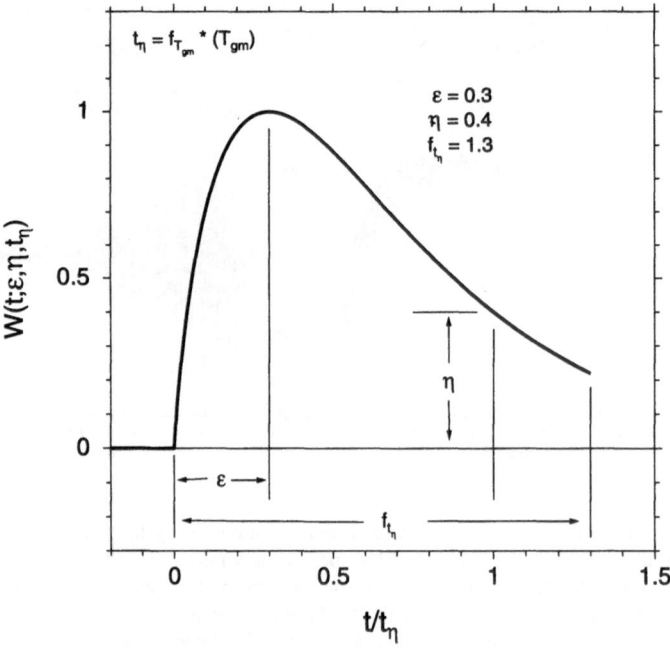

Figure 15
Exponential window and the variables controlling its shape.

and

$$a = (\exp(1)/\epsilon)^b \ . \tag{27}$$

As discussed in BOORE (1983), a can also be chosen such that the integral of the square of $w(t)$ equals unity; this is appropriate if the spectrum of the windowed noise is not normalized so that it has a mean square amplitude of unity. The time t_η is given by

$$t_\eta = f_{T_{gm}} \times T_{gm} \ , \tag{28}$$

where T_{gm} is the duration of ground motion. Based on SARAGONI and HART (1974), I use $\epsilon = 0.2$ and $\eta = 0.05$ in applications. I find a good comparison between response spectra computed using the box and exponential windows if $f_{T_{gm}} = 2.0$. A comparison of accelerations derived from the box and the exponential windows is given in Figure 16. Also shown are the 5%-damped pseudo-velocity response spectra obtained from averaging the response spectra computed from 640 simulated accelerations. It is clear that the response spectra obtained from the two windows are similar.

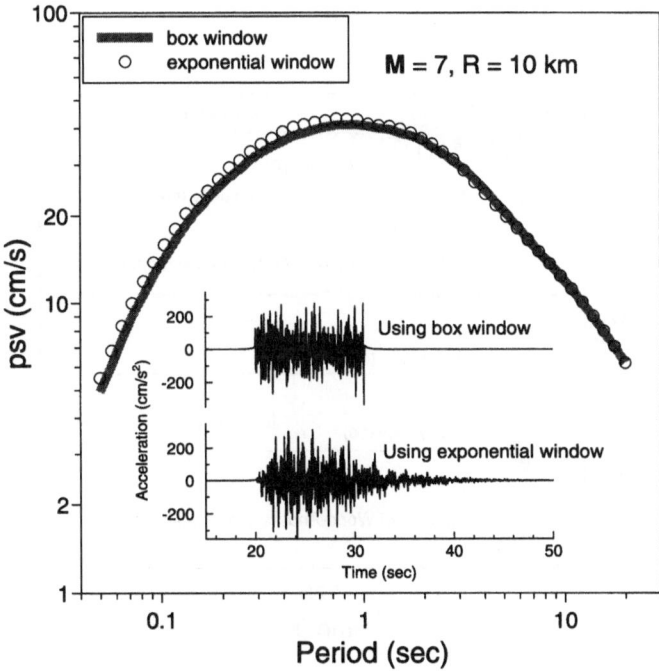

Figure 16

Comparison of waveforms and response spectra for time-domain simulations using the box and the exponential windows to shape the noise. The response spectra are averages from a suite of 640 simulations, whereas the time series are for a single realization. The simulations are for the AS00 model, as specified in Table 2, 3, and 4.

In applications, it is most common to compute the ground acceleration $(I(f) = (2\pi f \sqrt{-1})^2$ in equation (1)) and then derive other measures of ground motion from the time series of ground acceleration. Figure 17 shows examples of various types of motion for magnitude 4 and 7 earthquakes; magnitude was the only thing that changed in the program input. Individual time series should be used with caution, however, for there is no guarantee that the spectrum of each realization will be close to the "target" spectrum $Y(M_0, R, f)$; it is only the mean of the individual spectra for a number of simulations that will match the target spectrum. An example of this is shown in Figure 18, in which the mean of the spectra from 640 realizations is almost indistinguishable from the target spectrum, although the spectrum of a randomly chosen individual realization deviates significantly from the mean at some frequencies.

It is important to note that the variability of ground-motion parameters obtained from a suite of simulations does not represent the variability observed in real ground-motion parameters. Simulating the observed variability requires running the simulations for model parameters chosen from distribution functions for those parameters (see, e.g., EPRI, 1993).

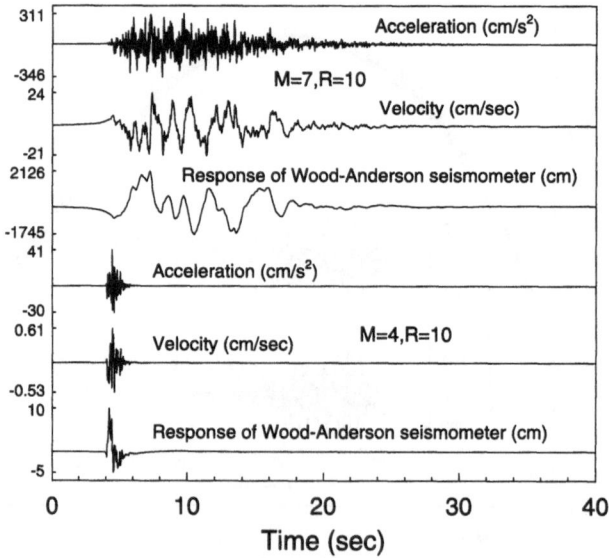

Figure 17

Time series for magnitude 4 and 7 earthquakes. The acceleration was computed using the stochastic method and the AS00 model, as specified in Tables 2, 3, and 4, and the velocity and response of a Wood-Anderson seismometer were obtained from the simulated accelerations; an acausal low-cut filter with a cut-off frequency of 0.02 Hz was applied to the acceleration time series before the velocity and Wood-Anderson response were computed.

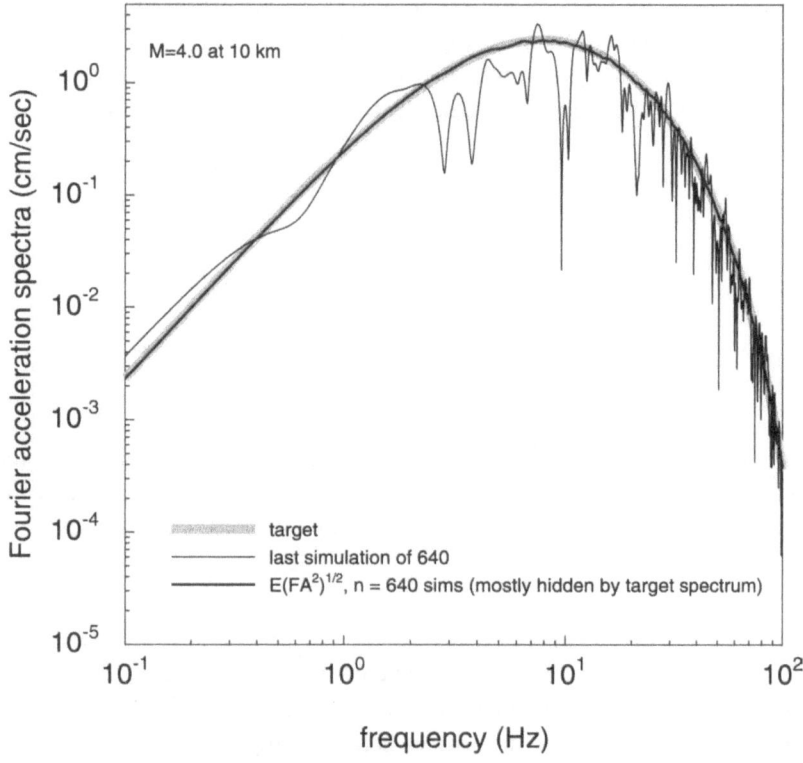

Figure 18

The model (target) spectrum, the spectrum from a single realization, and the spectrum from an average of 640 realizations. Any one realization can differ markedly from the model spectrum, but on average the simulations match the model spectrum. The simulations are for the AS00 model, as specified in Table 2, 3, and 4.

Peak Motions from Random Vibration Theory

A very rapid way of obtaining measures of peak motion (response spectra, peak acceleration, peak velocity, peak displacement, peak response of instruments for magnitude determination, Arias intensity, etc.) is to use random-vibration theory. In essence, random-vibration theory provides an estimate of the ratio of peak motion (y_{max}) to rms motion (y_{rms}), and Parseval's theorem is used to obtain y_{rms} in terms of an integral of the squared amplitude spectrum $|Y|^2$, where $|Y|^2$ contains the response of the particular measure of ground motion (e.g., equation (21) or (22)) for which peak values are desired.

The ratio of peak to rms motion is given by equations from CARTWRIGHT and LONGUET-HIGGINS (1956), who used the analysis in RICE (1954) to develop a method for predicting extrema of ocean waves from spectral characteristics of a continuous record of sea heights. In order to use their results for the extrema of transient earthquake ground motions, I had to pay special attention to the definition of duration used in the equations, as described below.

After a change of variable to remove an integrable singularity, CARTWRIGHT and LONGUET-HIGGINS' (1956) equation (their equation (6.8)) for the ratio of peak to rms motion is

$$\frac{y_{\max}}{y_{\mathrm{rms}}} = 2 \int_{0}^{\infty} \{1 - [1 - \xi \exp(-z^2)]^{N_e}\}\, dz \ , \tag{29}$$

where

$$\xi = \frac{N_z}{N_e} \ , \tag{30}$$

and N_z, N_e are the number of zero crossings and extrema, respectively (extrema correspond to all places where the first derivative of the time series equals zero; for a broadband function, there can be numerous local extrema). For large N

$$\frac{y_{\max}}{y_{\mathrm{rms}}} = [2 \ln(N_z)]^{1/2} + \frac{0.5772}{[2 \ln(N_z)]^{1/2}} \ . \tag{31}$$

The integral in equation (29) is well-behaved numerically, and therefore in my applications it, rather than the asymptotic equations in equation (31), is used.

In the equations above, the number of zero crossings and extrema are related to the frequencies of zero crossings (f_z) and extrema (f_e) and to duration (T) by the equation

$$N_{z,e} = 2\tilde{f}_{z,e}T \ , \tag{32}$$

where the frequencies are given by

$$\tilde{f}_z = \frac{1}{2\pi}(m_2/m_0)^{1/2} \ , \tag{33}$$

and

$$\tilde{f}_e = \frac{1}{2\pi}(m_4/m_2)^{1/2} \ . \tag{34}$$

In these equations, m_k, $k = 0, 2, 4$ are moments of the squared spectral amplitude. These play a fundamental role in random vibration theory and are defined for any integer k as

$$m_k = 2 \int_{0}^{\infty} (2\pi f)^k |Y(f)|^2\, df \ , \tag{35}$$

where the spectrum Y is given by equation (1) and includes the specific type of ground motions, as specified by equations (21) or (22). y_{rms} is simply

$$y_{rms} = (m_0/T)^{1/2} .$$ (36)

Being an integral of the squared acceleration, the Arias intensity is closely related to the 0-th spectral moment:

$$I_{xx} = \frac{\pi}{2g} m_0 .$$ (37)

Seismic waves from earthquakes are inherently nonstationary, and the response of resonant systems (local site layering or mechanical oscillators) to those waves will have significant correlation between adjacent peaks. Both of these characteristics violate basic assumptions of the random vibration theory just discussed. Despite this, the theory works very well in predicting ground motions, although some simple refinements are needed for oscillator response when the oscillator period is longer than the duration of ground motion or for lightly damped oscillators, for which the response continues well past the random ground-motion excitation. Examples of these cases, computed using time-domain simulations, are shown in Figure 19. For the small earthquake, the 10-sec oscillator response is almost equal to the ground displacement and has a short duration. On the other hand, the response of the oscillator to the larger earthquake rings on for a duration significantly in excess of the ground motion duration. The problem is in defining durations to use in determining rms and in determining the number of cycles of quasi-stationary motion to be used in the relation between y_{max} and y_{rms}. BOORE and JOYNER (1984) found that good results could be obtained if two durations were used: one duration (T_{rms}) for the computation of the rms in equation (36), and the other, smaller, duration for the determination in equation (32) of the number of zero crossings (N_z) or extrema (N_e) used in evaluating y_{max}/y_{rms}. For the latter BOORE and JOYNER (1984) use the duration of ground motion (T_{gm}), such as that shown in Figure 9. From considerations of oscillator response and numerical experiments with time-domain simulations, they proposed the following equation for the time T_{rms} to be used in the computation of rms:

$$T_{rms} = T_{gm} + T_o \left(\frac{\gamma^n}{\gamma^n + \alpha} \right) ,$$ (38)

where $\gamma = T_{gm}/T_o$ and the oscillator duration is given by $T_o = 1/(2\pi f_r \zeta)$. For small and large earthquakes T_{rms} approaches T_{gm} and $T_{gm} + T_o$, respectively, which is consistent with the oscillator responses shown in Figure 19. The constants n and α were determined from numerical experimentation, with values $n = 3$ and $\alpha = 1/3$. Recently, LIU and PEZESHK (1999) have found somewhat better comparisons between time domain and random vibration theory results by setting $n = 2$ and

$$\alpha = \left[2\pi \left(1 - \frac{m_1^2}{m_0 m_2} \right) \right]^{1/2} ,$$ (39)

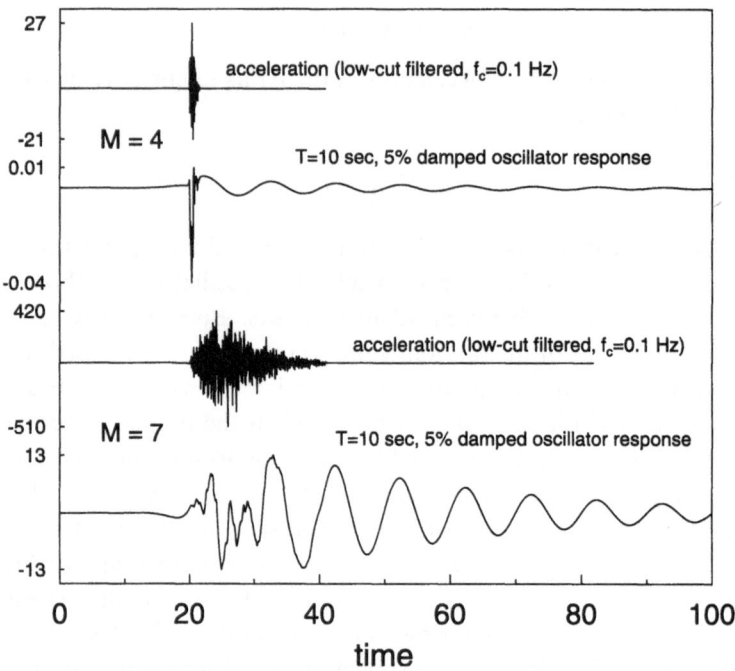

Figure 19

Simulated acceleration time series and computed response of 10.0-sec, 5-percent-damped oscillator for magnitude 4 and 7 earthquakes at 10 km. Because the relative shape is important, each trace has been scaled individually (the actual amplitudes are given to the left of the y-axis—acceleration in cm/s² and oscillator response in cm), The simulations are for the AS00 model, as specified in Table 2, 3, and 4. The accelerations differ from those in Figure 17 because the seeds used in generating the random numbers needed in the simultations were not the same.

where $m_i, i = 0, 1, 2$ are given by equation (35). According to Liu and Pezeshk, equation (39) accounts for the bandwidth of the ground motion. Comparisons of response spectra computed using time-domain calculations (for 10, 40, 160, and 640 simulations) and random-vibration calculations with both the Boore-Joyner and the Liu-Pezeshk oscillator corrections are shown in Figures 20 and 21. The figures show good agreement between the time-domain and the random-vibration theory calculations, with the Liu-Pezeshk correction giving somewhat better answers for **M** 7 earthquake at periods between 5 and about 12 secs (Fig. 21). The comparisons between the different ways of doing the oscillator correction, however, is model- and period-dependent. For example, the comparison in Liu and Pezeshk's paper indicates that their correction is significantly better than the Boore-Joyner correction for small earthquakes, which is a different conclusion than obtained from the comparisons shown in Figure 20. Figures 20 and 21 also indicate that more than 40 simulations may be required adequately capture the mean of the ground motions.

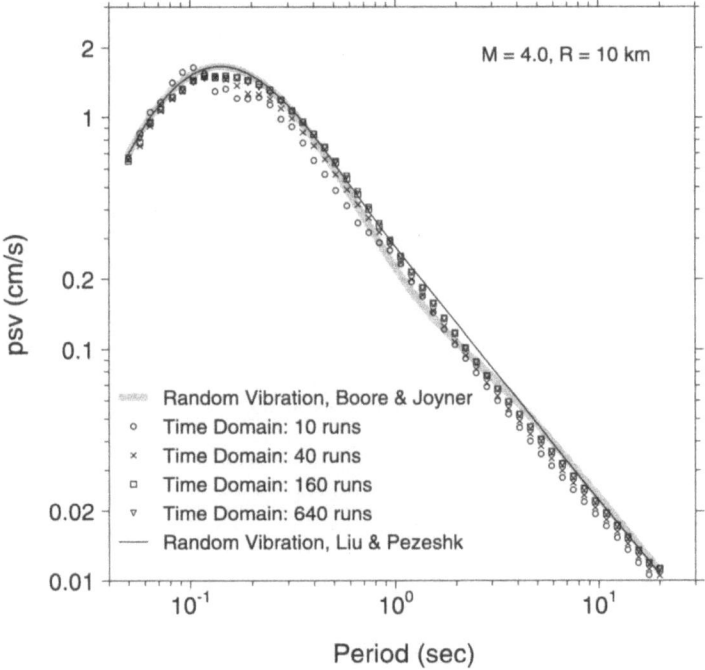

Figure 20

Comparison of simulations using the time-domain calculations with various values of the number of simulations, with a different seed for the random-number generator for each set of simulations. The random-vibration results are shown for comparison, using both the BOORE and JOYNER (1984) and LIU and PEZESHK (1999) modification of random-vibration theory for oscillator response. The calculations are for magnitude 4 at 10 km. The simulations are for the AS00 model, as specified in Table 2, 3, and 4.

Applications

The stochastic method has been widely applied, but rather than attempt to discuss a number of specific applications, I have included in Table 5 a fairly comprehensive list of applications, separated by primary geographic region. There are many ways in which the stochastic method has been used, and the effectiveness of the method has been demonstrated by fitting observations ranging from negative-magnitude rockbursts to great earthquakes at teleseismic distances. Calibrations of the method, which may involve finding the parameters so as to fit empirically-derived equations for predicting ground motions, and validations of the method, which consist of checking predictions against data (but not the same data used in deriving the necessary parameters) are included in a number of the references. The method can be used in absolute or relative senses. For example, predicting ground-shaking going from the source to the site is an absolute prediction, whereas predicting the ratio of ground-shaking for two source models is a relative prediction. Examples of both of these uses are given in Figure 22—an admittedly complicated figure, but one

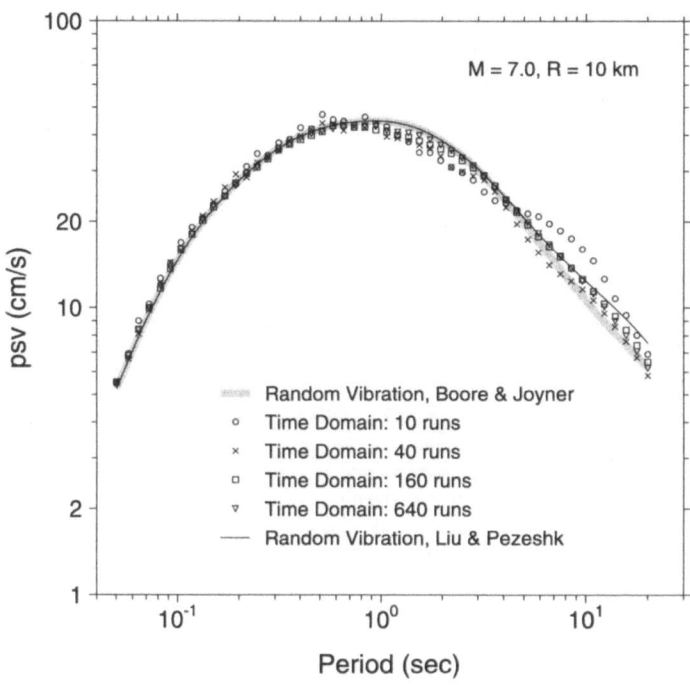

Figure 21

Comparison of simulations using the time-domain calculations with various values of the number of simulations, with a different seed for the random-number generator for each set of simulations. The random-vibration results are shown for comparison, using both the BOORE and JOYNER (1984) and LIU and PEZESHK (1999) modification of random-vibration theory for oscillator response. The calculations are for magnitude 7 at 10 km. The simulations are for the AS00 model, as specified in Table 2, 3, and 4.

which makes a number of points; see BOORE (1999) for details. In Figure 22 the circles are absolute predictions of response spectra for magnitude 5.6 and 7.6 earthquakes using the stochastic method. For comparison, the dashed lines are response spectra from empirical analyses of data, and the heavy solid line for **M** = 5.6 is the observed spectrum for a recording of the 1990 Upland, California, earthquake; the light solid line is the response spectrum computed for just the *S*-wave portion of the record, and excludes the longer-period surface waves. The absolute predictions for **M** = 5.6 are in reasonable agreement with the observations for shorter periods and for longer periods when the surface waves are excluded; the mismatch for longer-period response spectra obtained from the whole record is due to the lack of the surface waves in the stochastic method, which is a limitation of the method as usually applied. The heavy solid line for **M** = 7.5 is based on the observed spectrum of the smaller earthquake, corrected by the relative difference of motions for magnitude 7.5 and 5.6 earthquakes, as predicted by the stochastic method. The relative prediction of ground motions has also been used by CAMPBELL (1999) in the hybrid prediction of ground motions in eastern North America, in which he uses

Table 5

Some references for applications of the stochastic method

Western North America

ANDERSON and LEI (1994), ATKINSON (1995, 1997), ATKINSON and BOORE (1997b), ATKINSON and CASSIDY (2000), ATKINSON and SILVA (1997, 2000), AVILES and PEREZ-ROCHA (1998), BEN-ZION and ZHU (2002), BERESNEV (2002), BERESNEV and ATKINSON (1998b), BOORE (1986a, 1995, 1999), BOORE and JOYNER (1997), BOORE *et al.* (1992), CHIN and AKI (1991, 1996), HANKS and BOORE (1984), HARTZELL *et al.* (1999, 2002), IGLESIAS *et al.* (2002), LUCO (1985), MAHDYIAR (2002), McGUIRE and HANKS (1980), McGUIRE *et al.* (1984), PAPAGEORGIOU and AKI (1983b), SCHNEIDER and SILVA (2000), SCHNEIDER *et al.* (1993), SILVA and WONG (1992), SILVA et al. (1991), SINGH *et al.* (1989), VETTER *et al.* (1996), WENNERBERG (1996), WONG and SILVA (1990, 1993, 1994), WONG *et al.* (1993), YOUNGS and SILVA (1992)

Central and Eastern North America

ATKINSON (1984, 1989, 1990), ATKINSON and BERESNEV (1998, 2002), ATKINSON and BOORE (1987, 1990, 1995, 1997a, 1998), ATKINSON and HANKS (1995), ATKINSON and SOMERVILLE (1994), BERESNEV and ATKINSON (1999), BOLLINGER *et al.* (1993), BOORE (1989a), BOORE and ATKINSON (1987), BOORE and JOYNER (1991), CAMPBELL (2002), CHAPMAN *et al.* (1990), EPRI (1993), FRANKEL *et al.* (1996), GREIG and ATKINSON (1993), HANKS and JOHNSTON (1992), HARIK *et al.* (1997), HERRMANN and AKINCI (2000), HWANG (2001), HWANG and HUO (1994, 1997), HWANG *et al.* (1997, 2001a, 2001b), KUMAR (2000), SILVA *et al.* (1989), TORO (1985), TORO and McGUIRE (1987), TORO *et al.* (1988, 1992, 1997), WEN and WU (2001)

Other Parts of the World or Several Regions Combined

AKINCI *et al.* (2001), ASCE (2000), ATKINSON and GREIG (1994), BERARDI *et al.* (1999), BERESNEV and ATKINSON (2002), BOORE (1986b), CASTRO *et al.* (2001), CHEN and ATKINSON (2002), CHERNOV and SOKOLOV (1999), DE NATALE *et al.* (1988), FACCIOLI (1986), HARTZELL and HEATON (1988), HARMSEN (2002), HLATYWAYO (1997), MALAGNINI and HERRMANN (2000), MALAGNINI *et al.* (2000), MARGARIS and BOORE (1998), MARGARIS and HATZIDIMITRIOU (2002), MARGARIS and PAPAZACHOS (1999), McGUIRE (1984), ÓLAFSSON *et al.* (1998), MIYAKE *et al.* (2001), PITARKA *et al.* (2000, 2002), ROUMELIOTI and KIRATZI (2002), ROVELLI *et al.* (1988, 1991, 1994a, 1994b), SATOH (2002), SATOH *et al.* (1997), SCHERBAUM (1994), SCHERBAUM *et al.* (1994), SILVA (1997), SILVA and COSTANTINO (1999), SILVA and DARRAGH (1995), SILVA and GREEN (1989), SILVA *et al.* (2000a, 2000b, 2002), SINGH *et al.* (1999, 2002), SOKOLOV (1997, 1998, 2000a, 2000b), SOKOLOV *et al.* (2000, 2001), SUZUKI *et al.* (1998), TREMBLAY and ATKINSON (2001), TSAI (1997, 1998a, 1998b), WONG *et al.* (1991)

---

the stochastic method to modify empirically derived western U.S. ground motions for differences in source, propagation, and site.

Other general areas in which the stochastic method has been applied include:

- Generate suites of ground motions for many magnitudes and distance, and use these to derive ground-motion prediction equations and tables of motion. This is the basis for the CEUS motions used in the U.S. National Hazard Maps.
- Use as a basis for design-motion specification of critical structures.
- Find parameters controlling spectral content (e.g., $\Delta\sigma$, κ).
- Use in parameter sensitivity studies.
- Relate time-domain measures of ground motion to frequency-domain descriptions.
- Generate time series for use in nonlinear analyses (structural, site response, landslides, liquefaction).
- Use to compute subfault motions in simulations of extended ruptures.

One topic I have not discussed is that of uncertainties in the predictions; this has been a major focus of a study by EPRI (1993) (see also SILVA, 1992; TORO *et al.*,

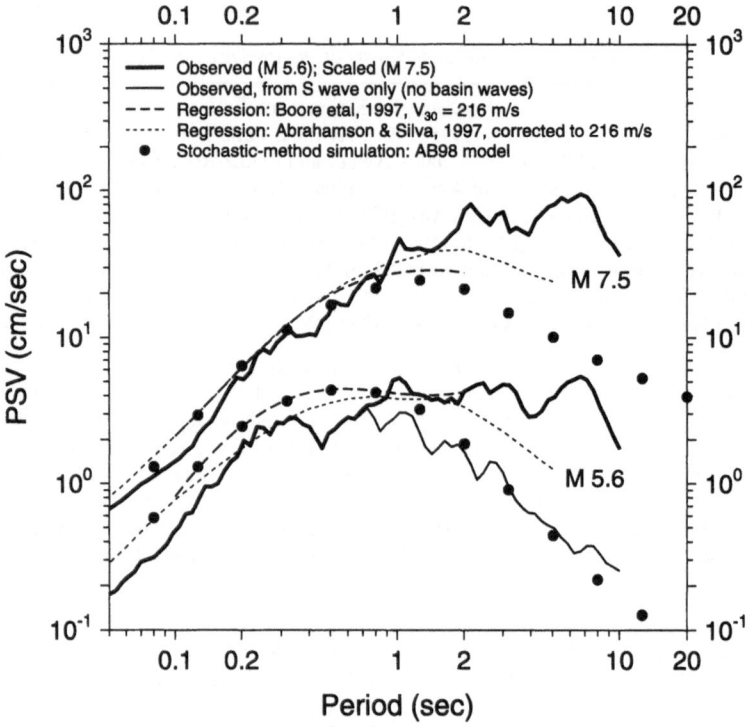

Figure 22

5%-damped, pseudo-velocity response spectra (*PSV*) for a small earthquake (**M** = 5.6) and a large earthquake (**M** = 7.5) (heavy solid lines). The *PSV* for the large event has been derived from the small event assuming ATKINSON and BOORE (1998) (AB98) source models. Also shown are the predictions from two regression analyses (dashed lines) and from stochastic-method simulations (solid circles). The light solid line for the **M** = 5.6 event was computed from the *S*-wave portion of the event (the first 35 sec of the recorded motion). (Modified from BOORE, 1999.)

1997). An example of this uncertainty was shown in Figure 3, which displays the range of Fourier spectra for predictions of ground motions in eastern North America (the variations in predicted ground motions are similar to the variations in Fourier spectra). ATKINSON and BOORE (1998) showed that the ATKINSON and BOORE (1995) model best fits response spectra computed from earthquake records in eastern North America and from other tectonically comparable areas. More than half of the events providing data used in the comparisons, however, had magnitudes less than or equal to 5. For this reason uncertainty exists in how applicable any one of the proposed source models would be in predicting ground motions from earthquakes in eastern North America large enough to constitute significant seismic hazard. A sensible way of dealing with this uncertainty is to base hazard calculations on a weighted average of ground motions from a number of the proposed source models (but all calculations still employ the stochastic-method simulations). The choice of weights then becomes the issue; this topic has been dealt with for eastern North America by expert elicitations (SAVY *et al.*, 1998) using the concepts discussed in BUDNITZ *et al.* (1997, 1998) and for the development of the National Hazard Maps by informed subjective opinion, based on a series of regional workshops (e.g., FRANKEL *et al.*, 1996 (currently being updated—see http://geohazards.cr.usgs.gov/eq/)).

Limitations and Improvements

Comparisons of stochastic-method predictions with empirically-determined ground motions indicates that the stochastic method is useful for simulating *mean* ground motions expected for a suite of earthquakes having a specified magnitude and fault–station distance. Care must be used, however, when the method is used to simulate site-specific and earthquake-specific ground motions. As described in this paper, the method does not include any phase effects due to the propagating rupture and to the wave propagation enroute to the site (including local site response). In addition, the differences between the various components of motion and different wave types are ignored. For these reasons, fault-normal effects, phase differences over horizontal distances, spatial correlations, directivity, etc. are not captured by the simulated motions. It should be possible to include some of these effects in the method, and I am aware of some efforts along these lines (e.g., LOH, 1985; LOH and YEH, 1988; TAMURA *et al.*, 1991; TAMURA and AIZAWA, 1992).

As noted before (Figure 18), the Fourier spectrum of each time series realization may diverge from the "target" Fourier spectrum $Y(M_0, R, f)$. For this reason, when the method is used to simulate a suite of time series for use in engineering design, it is important to check the Fourier or response spectrum of each simulation to be sure that it does not deviate too far from the desired spectrum. In practice, this will mean choosing the best subset from a number of simulations. This approach has been used by WEN and WU (2001) and by HARMSEN (2002), both of whom used the similarity to a specified response spectrum as the basis for choosing the time series (but the two papers used different "goodness-of-fit" criteria).

The method also assumes stationarity of the frequency content with time. As the example in Figure 22 shows, this is a poor assumption for situations, such as deep sedimentary basins (e.g., JOYNER, 2000), where long-period surface waves occur. It should be possible to incorporate these waves into the method.

The duration in applications of the method is independent of frequency. RAOOF *et al.* (1999), however, find that duration is frequency-dependent. Modifications of the method to account for frequency-dependent duration would be relatively easy for time-domain simulations (simulating the motions for a series of narrow-band filters), but might be more difficult for simulations of motions using random-vibration theory.

An apparent limitation often expressed is that most of the models based on the stochastic method are fundamentally point-source models. This may not be as important a limitation as might at first be thought. Although it is true that near- and intermediate-field terms are lacking, in most applications the frequencies are high enough that the far-field terms dominate, even if the site is near the fault. Furthermore, the effects of a finite-fault averaged over a number of sites distributed around the fault (to average over radiation pattern and directivity effects) can be captured in several ways: 1) using the closest distance to faulting (as is done in empirically derived ground-motion prediction equations) as the source-to-site distance; 2) using a two-corner source spectrum (ATKINSON and SILVA, 2000); 3) allowing the geometrical spreading to be magnitude dependent (SILVA et al., 2002). In addition, it should be possible to extend the method to account for specific fault-station geometries in a simple way, perhaps combining the simple computation of envelopes of acceleration (MIDORIKAWA and KOBAYASHI, 1978; COCCO and BOATWRIGHT, 1993) with statistical descriptions of the source (e.g., LOMNITZ-ADLER and LUND, 1992; HERRERO and BERNARD, 1994; JOYNER, 1995; BERNARD et al., 1996; HISADA, 2000). The overriding philosophy of such an effort would be to capture the essence of motions from an extended rupture without sacrificing the conceptual simplicity of the stochastic method.

Conclusions

The stochastic method is a simple, yet powerful, means for simulating ground motions. It is particularly useful for obtaining ground motions at frequencies of interest to earthquake engineers, and it has been widely applied in this context.

My source codes, written in FORTRAN, and executables that can be used on a PC can be obtained from my web site (http://quake.usgs.gov/~boore) or via anonymous ftp on *samoa.wr.usgs.gov* in directory *get*. Programs are included both for time-domain and for random-vibration simulations. The user should download the files README.TXT, SMSIMxxx.ZIP, SITEAxxx.ZIP, and SMSIM_MANUAL.PDF where "xxx" is the current version number, and follow the instructions in README.TXT to extract and to use the programs. SMSIM_MANUAL.PDF contains the manual for the program (BOORE, 2000). The manual is also available online at http://geopubs.wr.usgs.gov/open-file/of00-509/.

Acknowledgments

I wish to thank Bob Herrmann for the record section shown in Figure 4, Gail Atkinson for sending various data files, and Nancy Blair for help in finding papers that have used the stochastic method. In addition, the comments John Anderson, Tom Hanks, Bob Herrmann, Bill Joyner, Chuck Mueller, and Walt Silva were very helpful.

Most importantly, I want to thank Keiiti Aki for the guidance and inspiration that started me on my career path over 30 years ago.

REFERENCES

ABRAHAMSON, N. A. and SILVA, W. J. (1997), *Empirical Response Spectral Attenuation Relations for Shallow Crustal Earthquakes*, Seism. Res. Lett. *68*, 94–127.

AKI, K. (1966), *Generation and Propagation of G Waves from the Niigata Earthquake of June 16, 1964. Part 2. Estimation of Earthquake Moment, Released Energy, and Stress-strain Drop from the G-Wave Spectrum*, Bull. Earthq. Res. Inst. *44*, 73–88.

AKI, K. (1967), *Scaling Law of Seismic Spectrum*, J. Geophys. Res. *72*, 1217–1231.

AKI, K. (1980), *Attenuation of Shear-waves in the Lithosphere for Frequencies from 0.05 to 25 Hz*, Phys. Earth Planet. Inter. *21*, 50–60.

AKINCI, A., MALAGNINI L., HERRMANN R. B., PINO N. A., SCOGNAMIGLIO L., and EYIDOGAN, H. (2001), *High-frequency ground motion in the Erzincan region, Turkey: Inferences from small earthquakes*, Bull. Seism. Soc. Am. *91*, 1446–1455

ANDERSON, J. G. and HOUGH, S. E. (1984), *A Model for the Shape of the Fourier Amplitude Spectrum of Acceleration at High Frequencies*, Bull. Seismol. Soc. Am. *74*, 1969–1993.

ANDERSON, J. G. and LEI, Y. (1994), *Nonparametric Description of Peak Acceleration as a Function of Magnitude, Distance, and Site in Guerero, Mexico*, Bull. Seismol. Soc. Am. *84*, 1003–1017.

ARIAS, A. *A Measure of Earthquake Intensity*. In *Seismic Design for Nuclear Power Plants* (Robert J. Hansen, ed.) (The M.I.T. Press, Cambridge, Mass. 1970) pp. 438–483.

ASCE (2000), *Seismic Analysis of Safety-Related Nuclear Structures*, American Society of Civil Engineers, ASCE-98.

ATKINSON, G. M. (1984), *Attenuation of Strong Ground Motion in Canada from a Random Vibrations Approach*, Bull. Seismol. Soc. Am. *74*, 2629–2653.

ATKINSON, G. M. (1989), *Implications of Eastern Ground-Motion Characteristics for Seismic Hazard Assessment in Eastern North America*. In *Earthquake Hazards and the Design of Constructed Facilities in the Eastern United States* (K. H. Jacob and C. J. Turkstra, eds.) Annals of the New York Academy of Sciences *558*, 128–135.

ATKINSON, G. M. (1990), *A Comparison of Eastern North American Ground Motion Observations with Theoretical Predictions*, Seism. Res. Lett. *61*, 171–180.

ATKINSON, G. M. (1993a), *Notes on Ground Motion Parameters for Eastern North America: Duration and H/V Ratio*, Bull. Seismol. Soc. Am. *83*, 587–596.

ATKINSON, G. M. (1993b), *Earthquake Source Spectra in Eastern North America*, Bull. Seismol. Soc. Am. *83*, 1778–1798.

ATKINSON, G. M. (1995), *Attenuation and Source Parameters of Earthquakes in the Cascadia Region*, Bull. Seismol. Soc. Am. *85*, 1327–1342.

ATKINSON, G. M. (1996), *The High-frequency Shape of the Source Spectrum for Earthquakes in Eastern and Western Canada*, Bull. Seismol. Soc. Am. *86*, 106–112.

ATKINSON, G. M. (1997), *Empirical Ground Motion Relations for Earthquakes in the Cascadia Region*, Canadian J. Civil Eng. *24*, 64–77.

ATKINSON, G. M. and BERESNEV, I. A. (1998), *Compatible Ground-motion Time Histories for New National Seismic Hazard Maps*, Canadian J. Civil Eng. *25*, 305–318.

ATKINSON, G. M. and BERESNEV, I. A. (2002), *Ground motions at Memphis and St. Louis from M 7.5–8.0 earthquakes in the New Madrid seismic zone*, Bull Seism. Soc. Am. *92*, 1015–1024.

ATKINSON, G. M. and BOORE, D. M. (1987), *On the m_N, M Relation for Eastern North American Earthquakes*, Seism. Res. Lett. *58*, 119–124.

ATKINSON, G. M. and BOORE , D. M. (1990), *Recent Trends in Ground Motion and Spectral Response Relations for North America*, Earthquake Spectra *6*, 15–35.

ATKINSON, G. M. and BOORE, D. M. (1995), *Ground Motion Relations for Eastern North America*, Bull. Seismol. Soc. Am. *85*, 17–30.

ATKINSON, G. M. and BOORE, D. M. (1997a), *Some Comparisons between Recent Ground-motion Relations*, Seism. Res. Lett. *68*, 24–40.

ATKINSON, G. M. and BOORE, D. M. (1997b), *Stochastic Point-source Modeling of Ground Motions in the Cascadia Region*, Seism. Res. Lett. *68*, 74–85.

ATKINSON, G. M. and BOORE, D. M. (1998), *Evaluation of Models for Earthquake Source Spectra in Eastern North America*, Bull. Seismol. Soc. Am. *88*, 917–934.

ATKINSON, G. M. and CASSIDY J. F. (2000), *Integrated use of seismograph and strong-motion data to determine soil amplification: Response of the Fraser River Delta to the Duvall and Geoprgia Strait earthquakes*, Bull. Seism. Soc. Am. *90*, 1028–1040.

ATKINSON, G., and GREIG, G. (1994), *On the relationship between linear and nonlinear response parameters*. Proc. Fifth U.S. National Conference on Earthquake Engineering, Chicago, July 10–14, 1994, *4*, 561–570.

ATKINSON, G. M. and HANKS, T. C. (1995), *A High-frequency Magnitude Scale*, Bull. Seismol. Soc. Am. *85*, 825–833.

ATKINSON, G. M. and MEREU, R. F. (1992), *The Shape of Ground Motion Attenuation Curves in Southeastern Canada*, Bull. Seismol. Soc. Am. *82*, 2014–2031.

ATKINSON, G. M. and SILVA, W. (1997), *An Empirical Study of Earthquake Source Spectra for California Earthquakes*, Bull. Seismol. Soc. Am. *87*, 97–113.

ATKINSON, G. M. and SILVA, W. (2000), *Stochastic Modeling of California Ground Motions*, Bull. Seismol. Soc. Am. *90*, 255–274.

ATKINSON, G. M. and SOMERVILLE, P. G. (1994), *Calibration of Time History Simulation Methods*, Bull. Seismol. Soc. Am. *84*, 400–414.

AVILES, J. and PEREZ-ROCHA, L. E. (1998), *Site Effects and Soil-structure Interaction in the Valley of Mexico*, Soil Dyn. Earthq. Eng. *17*, 29–39.

BEN-ZION, Y. and ZHU, L. (2002), *Potency-magnitude scaling relations for southern California earthquakes with $1.0 < M_L < 7.0$*, Geophys. J. Int. *148*, F1–F5.

BERARDI, R., JIMENEZ, M., ZONNO, G., and GARCIA-FERNANDEZ, M. (1999), *Calibration of Stochastic Ground Motion Simulations for the 1997 Umbria-Marche, Central Italy, Earthquake Sequence.* Proc. 9th Intl. Conf. *On Soil Dynamics and Earthquake Engineering*, SDEE 99, Aug. 1999, Bergen, Norway.

BERESNEV, I.A. (2002). Nonlinearity at California generic soil sites from modeling recent strong-motion data, Bull. Seism. Soc. Am. *92*, 863–870

BERESNEV, I. A. and ATKINSON, G. M. (1997), *Modeling Finite-fault Radiation from the ω^n Spectrum*, Bull. Seismol. Soc. Am. *87*, 67–84.

BERESNEV, I. A. and ATKINSON, G. M. (1998a), *FINSIM—a FORTRAN Program for Simulating Stochastic Acceleration Time Histories from Finite Faults*, Seism. Res. Lett. *69*, 27–32.

BERESNEV, I. A. and ATKINSON, G. M. (1998b), *Stochastic Finite-fault Modeling of Ground Motions from the 1994 Northridge, California Earthquake. I. Validation on Rock Sites*, Bull. Seismol. Soc. Am. *88*, 1392–1401.

BERESNEV, I. A. and ATKINSON, G. M. (1999), *Generic Finite-fault Model for Ground-motion Prediction in Eastern North America*, Bull. Seismol. Soc. Am. *89*, 608–625.

BERESNEV, I. A. and ATKINSON, G. M. (2002), *Source parameters of earthquakes in eastern and western North America based on finite-fault modeling*, Bull. Seism. Soc. Am. *92*, 695–710.

BERNARD, P., HERRERO, A., and BERGE, C. (1996), *Modeling Directivity of Heterogeneous Earthquake Ruptures*, Bull. Seismol. Soc. Am. *86*, 1149–1160.

BOATWRIGHT, J. and CHOY, G. (1992), *Acceleration Source Spectra Anticipated for Large Earthquakes in Northeastern North America*, Bull. Seismol. Soc. Am. *82*, 660–682.

BOLLINGER, G. A., CHAPMAN, M. C., and SIBOL, M. S. (1993), *A Comparison of Earthquake Damage Areas as a Function of Magnitude across the United States*, Bull. Seismol. Soc. Am. *83*, 1064–1080.

BOORE, D. M. (1983), *Stochastic Simulation of High-frequency Ground Motions Based on Seismological Models of the Radiated Spectra*, Bull. Seismol. Soc. Am. *73*, 1865–1894.

BOORE, D. M. (1984), *Use of Seismoscope Records to Determine M_L and Peak Velocities*, Bull. Seismol. Soc. Am. *74*, 315–324.

BOORE, D. M. (1986a), *The Effect of Finite Bandwidth on Seismic Scaling Rrelationships.* In *Earthquake Source Mechanics* (S. Das, J. Boatwright, and C. Scholz, eds.) Geophysical Monograph 37 (American Geophysical Union, Washington, D. C. 1986a) pp. 275–283.

BOORE, D. M. (1986b), *Short-period P- and S-wave Radiation from Large Earthquakes: Implications for Spectral Scaling Relations*, Bull. Seismol. Soc. Am. *76*, 43–64.

BOORE, D. M. (1989a), *Quantitative Ground-Motion Estimates*. In *Earthquake Hazards and the Design of Constructed Facilities in the Eastern United States* (K. H. Jacob and C. J. Turkstra, eds.) *Annals of the New York Academy of Sciences 558*, 81–94.

BOORE, D. M. (1989b), *The Richter Scale: Its Development and Use for Determining Earthquake Source Parameters*, Tectonophysics *166*, 1–14.

BOORE, D. M. (1995), *Prediction of Response Spectra for the Saguenay earthquake*. In *Proceedings: Modeling Earthquake Ground Motion at Close Distances*, Electric Power Research Institute Report EPRI TR-104975, 6-1-6-14.

BOORE, D. M. (1996), *SMSIM – Fortran Programs for Simulating Ground Motions from Earthquakes: Version 1.0*, U.S. Geol. Surv. Open-File Rept. *96-80-A, 96-80-B*, 73 pp.

BOORE, D. M. (1999), *Basin Waves on a Seafloor Recording of the 1990 Upland, California, Earthquake: Implications for Ground Motions from a Larger Earthquake*, Bull. Seismol. Soc. Am. *89*, 317–324.

BOORE, D. M. (2000), *SMSIM – Fortran Programs for Simulating Ground Motions from Earthquakes: Version 2.0—A Revision of OFR 96-80-A*, U.S. Geol. Surv. Open-File Rept. *OF 00-509*, 55 pp. (available at http://geopubs.wr.usgs.gov/open-file/of00-509/).

BOORE, D. M. (2002), *SMSIM: Stochastic Method SIMulation of Ground Motion from Earthquakes*. In *IASPEI Centennial International Handbook of Earthquake and Engineering Seismology* (W. Lee, K. Kanamori, P. Jennings, and C. Kisslinger, eds), Academic press, Chapter 85.13, (in press).

BOORE, D. M. and ATKINSON, G. M. (1987), *Stochastic Prediction of Ground Motion and Spectral Response Parameters at Hard-rock Sites in Eastern North America*, Bull. Seismol. Soc. Am. *77*, 440–467.

BOORE, D. M. and BOATWRIGHT, J. (1984), *Average Body-wave Radiation Coefficients*, Bull. Seismol. Soc. Am. *74*, 1615–1621.

BOORE, D. M. and JOYNER, W. B. (1984), *A Note on the Use of Random Vibration Theory to Predict Peak Amplitudes of Transient Signals*, Bull. Seismol. Soc. Am. *74*, 2035–2039.

BOORE, D. M. and JOYNER, W. B. (1991), *Estimation of Ground Motion at Deep-soil Sites in Eastern North America*, Bull. Seismol. Soc. Am. *81*, 2167–2185.

BOORE, D. M. and JOYNER, W. B. (1997), *Site amplifications for Generic Rock Sites*, Bull. Seismol. Soc. Am. *87*, 327–341.

BOORE, D. M., JOYNER, W. B., and WENNERBERG, L. (1992), *Fitting the Stochastic ω^{-2} Source Model to Observed Response Spectra in Western North America: Trade-offs between $\Delta\sigma$ and κ*, Bull. Seismol. Soc. Am. *82*, 1956–1963.

BOORE, D. M., JOYNER, W. B., and FUMAL, T. E. (1997), *Equations for Estimating Horizontal Response Spectra and Peak Acceleration from Western North American Earthquakes: A Summary of Recent Work*, Seism. Res. Lett. *68*, 128–153.

BRUNE, J. N. (1970), *Tectonic Stress and the Spectra of Seismic Shear Waves from Earthquakes*, J. Geophys. Res. *75*, 4997–5009.

BRUNE, J. N. (1971), *Correction*, J. geophys. Res. *76*, 5002.

BUDNITZ, R. J., APOSTOLAKIS, G., BOORE, D. M., CLUFF, L. S., COPPERSMITH, K. J., CORNELL, C. A., and MORRIS, P. A. (1997), *Recommendations for Probabilistic Seismic Hazard Analysis: Guidance on Uncertainty and Use of Experts*, NUREG/CR-6372, Washington, D.C.: U.S. Nuclear Regulatory Commission.

BUDNITZ, R. J., APOSTOLAKIS, G., BOORE, D. M., CLUFF, L. S., COPPERSMITH, K. J., CORNELL, C. A., and MORRIS, P. A. (1998), *Use of Technical Expert Panels: Applications to Probabilistic Seismic Hazard Analysis*, Risk Analysis *18*, 463–469.

CAMPBELL, K. W. (1999), *Hybrid Empirical Model for Estimating Strong Ground Motion in Regions of Limited Strong-motion Recordings*, presented at OECD-NEA Workshop on Engineering Characterization of Seismic Input, Brookhaven National Laboratory, Upton, NY, Nov. 15–17, 1999.

CAMPBELL, K. W. (2002), *Strong Motion Attenuation Relations: Commentary and Discussion of Selected Relations*. In *IASPEI Centennial International Handbook of Earthquake and Engineering Seismology* (W. Lee, K. Kanamori, P. Jennings, and C. Kisslinger, eds.), Academic press, Chapter 60, (in press).

CAMPBELL, K. W. (2002), *Prediction of strong ground motion using the hybrid empirical method: Example application to eastern North America*, Bull. Seism. Soc. Am. *92* (in press).

CARTWRIGHT, D. E. and LONGUET-HIGGINS, M. S. (1956), *The Statistical Distribution of the Maxima of a Random Function*, Proc. R. Soc. London *237*, 212–232.

CASTRO, R. R., ROVELLI, A., COCCO, M., DI BONA, M., and PACOR, F. (2001), *Stochastic simulation of strong-motion records from the 26 September 1997 (M_W6), Umbria-Marche (central Italy) earthquake*, Bull. Seism. Soc. Am. *91*, 27–39.

CHAPMAN, M. C., BOLLINGER, G. A., SIBOL, M. S., and STEPHENSON, D. E. (1990), *The Influence of the Coastal Plain Sedimentary Wedge on Strong Ground Motions from the 1886 Charleston, South Carolina, Earthquake*, Earthquake Spectra *6*, 617–640.

CHEN, S.-Z. and ATKINSON, G. M. (2002), *Global comparisons of earthquake source spectra*, Bull. Seism. Soc. Am. *92*, 885–895.

CHERNOV, Y. K. and SOKOLOV, V. Y. (1999), *Correlation of Seismic Intensity with Fourier Acceleration Spectra*, Phys. Chem. Earth *24*, 523–528.

CHIN, B.-H. and AKI, K. (1991), *Simultaneous Study of the Source, Path, and Site Effects on Strong Ground Motion during the 1989 Loma Prieta Earthquake: A Preliminary Result on Pervasive Nonlinear Site Effects*, Bull. Seismol. Soc. Am. *81*, 1859–1884.

CHIN, B.-H. and AKI, K. (1996), *Reply to Leif Wennerberg's Comment on "Simultaneous Study of the Source, Path, and Site Effects on Strong Ground Motion during the 1989 Loma Prieta earthquake: A Preliminary Result on Pervasive Nonlinear Site Effects"*, Bull. Seismol. Soc. Am. *86*, 268–273.

COCCO, M. and BOATWRIGHT, J. (1993), *The Envelopes of Acceleration Time Histories*, Bull. Seismol. Soc. Am. *83*, 1095–1114.

CORMIER, V. F. (1982), *The Effect of Attenuation on Seismic Body Waves*, Bull. Seismol. Soc. Am. *72*, S169–S200.

CORREIG, A. M. (1996), *On the Measurement of the Predominant and Resonant Frequencies*, Bull. Seismol. Soc. Am. *86*, 416–427.

DE NATALE, G., FACCIOLI, E., and ZOLLO, A. (1988), *Scaling of Peak Ground Motions from Digital Recordings of Small Earthquakes at Campi Flegrei, Southern Italy*, Pure Appl. Geophys. *126*, 37–53.

EPRI (1993), *Guidelines for Determining Design Basis Ground Motions*, Electric Power Research Institute, Palo Alto, Calif., Rept. No. EPRI TR-102293, vols. 1–5.

Erdik, M. and Durukal, E. (2001), *A Hybrid Procedure for the Assessment of Design Basis Earthquake Ground Motions for Near-fault Conditions*, Soil Dyn. Earthq. Eng. *21*, 431–443.

FACCIOLI, E. (1986), *A Study of Strong Motions from Italy and Yugoslavia in terms of Gross Source Properties*. In *Earthquake Source Mechanics* (S. Das, J. Boatwright, and C. Scholz, eds.), Amer. Geophys. Union Geophysical Monograph *37*, 297–309.

FRANKEL, A., MUELLER, C., BARNHARD, T., PERKINS, D., LEYENDECKER, E., DICKMAN, N., HANSON, S. and HOPPER, M. (1996), *National Seismic Hazard Maps: Documentation June 1996*. U.S. Geol. Surv. Open-File Rept. 96-532, 69 pp.

GHOSH, A. K. (1992), *A Semianalytical Model for Fourier Amplitude Spectrum of Earthquake Ground Motion*, Nuclear Engin. and Design *133*, 199–208.

GREIG, G. L. and ATKINSON, G. M. (1993), *The damage potential of eastern North American earthquakes*, Seism. Res. Lett. *64*, 119–137.

GUSEV, A. A., GORDEEV, E. I., GUSEVA, E. M., PETUKHIN, A. G., and CHEBROV, V. N. (1997), *The First Version of the $A_{(max)}(M(W),R)$ Relationship for Kamchatka*, Pure Appl. Geophys. *149*, 299–312.

HADDON, R. (1996), *Earthquake Source Spectra in Eastern North America*, Bull. Seismol. Soc. Am. *86*, 1300–1313.

HANKS, T. C. (1979), *b Values and $\omega^{-\gamma}$ Seismic Source Models: Implications for Tectonic Stress Variations along Active Crustal Fault Zones and the Estimation of High-frequency Strong Ground Motion*, J. Geophys. Res. *84*, 2235–2242.

HANKS, T. C. (1982), *f_{max}*, Bull. Seismol. Soc. Am. *72*, 1867–1879.

HANKS, T. C. and BOORE, D. M. (1984), *Moment-Magnitude Relations in Theory and Practice*, J. Geophys. Res. *89*, 6229–6235.

HANKS, T. C. and JOHNSTON, A. C. (1992), *Common Features of the Excitation and Propagation of Strong Ground Motion for North American Earthquakes*, Bull. Seismol. Soc. Am. *82*, 1–23.

HANKS, T. C. and KANAMORI, H. (1979), *A Moment Magnitude Scale*, J. Geophys. Res. *84*, 2348–2350.

HANKS, T. C. and McGUIRE, R. K. (1981), *The Character of High-frequency Strong Ground Motion*, Bull. Seismol. Soc. Am. *71*, 2071–2095.

HARIK, I. E., ALLEN, D. L., STREET, R. L., GUO, M., GRAVES, R. C., HARISON, R. C., and GAWRY, M. J. (1997), *Seismic Evaluation of Brent-Spence Bridge*, J. Struct. Eng. *123*, 1269–1275.

HARMSEN, S. (2002), SEISMOGRAPHS FROM THE INTERACTIVE DEAGGREGATION WEB PAGE, HTTP://EQ-INT1.CR.USGS.GOV/EQ/HTML/STOCHASTIC_SEISMOGRAM_THEORY.HTML.

HARTZELL, S. H. and HEATON, T. H. (1988), *Failure of Self-similarity for Large ($M_w > 8\,1/4$) Earthquakes*, Bull. Seismol. Soc. Am. *78*, 478–488.

HARTZELL, S., HARMSEN, S., FRANKEL, A., and LARSEN, S. (1999), *Calculation of Broadband Time Histories of Ground Motion: Comparison of Methods and Validation Using Strong-ground Motion from the 1994 Northridge Earthquake*, Bull. Seismol. Soc. Am. *89*, 1484–1504.

HARTZELL, S., LEEDS, A., FRANKEL, A., WILLIAMS, R. A., ODUM, J., STEPHENSON, W., and SILVA, W. (2002), *Simulation of broadband ground motion including nonlinear soil effects for a magnitude 6.5 earthquake on the Seattle fault, Seattle, Washington*, Bull. Seism. Soc. Am. *92*, 831–853.

HERRERO, A. and BERNARD, P. (1994), *A Kinematic Self-similar Rupture Process for Earthquakes*, J. Geophys. Res. *84*, 1216–1228.

HERRMANN, R. B. (1985), *An Extension of Random Vibration Theory Estimates of Strong Ground Motion to Large Distances*, Bull. Seismol. Soc. Am. *75*, 1447–1453.

HERRMANN, R. B. and AKINCI, A. (2000), *Mid-America Ground Motion Models*, http://www.eas.slu.edu/People/RBHerrmann/MAEC/maecgnd.html.

HISADA, Y. (2000), *A Theoretical Omega-square Model Considering the Spatial Variation in Slip and Rupture Velocity*, Bull. Seismol. Soc. Am. *90*, 387–400.

HLATYWAYO, D. J. (1997), *Seismic Hazard in Central Southern Africa*, Geophys. J. Int. *130*, 737–745.

HWANG, H. (2001), *Simulation of Earthquake ground motion*. In *Monte Carlo Simulation* (Schuelle & Spanos, eds.) (Balkema, Rotterdam 2001) pp. 467–473.

HWANG, H. and HUO, J.-R. (1994), *Generation of Hazard-consistent Ground Motion*, Soil Dyn. Earthq. Eng. *13*, 377–386.

HWANG, H. and HUO, J.-R. (1997), *Attenuation Relations of Ground Motion for Rock and Soil Sites in Eastern United States*, Soil Dyn. Earthq. Eng. *13*, 363–372.

HWANG, H., LIN, H., and HUO, J.-R. (1997), *Site Coefficients for Design of Buildings in Eastern United States*, Soil Dyn. Earthq. Eng. *16*, 29–40.

HWANG, H., PEZESHK, S., LIN, Y. W., HE, J., and CHIU, J. M. (2001a), *Generation of Synthetic Ground Motion*, CD Release 01-02, Mid-America Earthquake Center, University of Illinois at Urbana-Champaign, Urbana, IL.

HWANG, H., LIU, J. B., and CHIU, Y. H. (2001b), *Seismic Fragility Analysis of Highway Bridges*, CD Release 01-06, Mid-America Earthquake Center, University of Illinois at Urbana-Champaign, Urbana, IL.

IGLESIAS, A., SINGH, S. K., PACHECO, J. F., and ORDAZ, M. (2002), *A source and wave propagation study of the Copalillo, Mexico, earthquake of 21 July 2000 ($M_w5.9$): Implications for seismic hazard in Mexico City from inslab earthquakes*, Bull. Seism. Soc. Am. *92*, 1060–1071.

JIBSON, R. W. (1993), *Predicting Earthquake-induced Landslide Displacements Using Newmark's Sliding Block Analysis*, Transportation Research Record 1411, National Research Council, Washington, D.C., 9–17.

JIBSON, R. W., HARP, E. L., and MICHAEL, J. A. (1998), *A Method for Producing Digital Probabilistic Seismic Landslide Hazard Maps: An Example from the Los Angeles, California, Area*, U.S. Geol. Surv. Open-File Rept. *98-113*, 17 pp.

JOYNER, W. B. (1984), *A Scaling Law for the Spectra of Large Earthquakes*, Bull. Seismol. Soc. Am. *74*, 1167–1188.

JOYNER, W. B. (1995), *Stochastic Simulation of Near-source Earthquake Ground Motion*. In *Proceedings: Modeling Earthquake Ground Motion at Close Distances*, Electric Power Research Institute report EPRI TR-104975, 8-1–8-24.

JOYNER, W. B. (1997), *Ground Motion Estimates for the Northeastern U.S. or Southeastern Canada*. In *Recommendations for Probabilistic Seismic Hazard Analysis: Guidance on Uncertainty and Use of Experts*, Senior Seismic Hazard Analysis Committee (R. Budnitz, G. Apostolakis, D. Boore, L. Cluff, K. Coppersmith, A. Cornell, and P. Morris eds.), U.S. Nuclear Reg. Comm. Rept. NUREG/CR-6372, Washington, D.C.

JOYNER, W. B. (2000), *Strong Motion from Surface Waves in Deep Sedimentary Basins*, Bull. Seismol. Soc. Am. *90*, S95–S112.

JOYNER, W. B. and BOORE, D. M. (1988), *Measurement, Characterization, and Prediction of Strong Ground Motion*, In *Earthquake Engineering and Soil Dynamics II, Proc. Am. Soc. Civil Eng. Geotech. Eng. Div. Specialty Conf.*, June 27–30, 1988, Park City, Utah, 43–102.

KAMAE, K. and IRIKURA, K. (1992), *Prediction of site-specific strong ground motion using semiempirical methods, in* Proc. of the Tenth World Conference on Earthquake Engineering, Madrid, Spain, 19–24 July 1992, 801–806.

KAMAE, K., IRIKURA, K., and PITARKA, A. A. (1998), *A Technique for Simulating Strong Ground Motion Using Hybrid Green's Function*, Bull. Seismol. Soc. Am. *88*, 357–367.

KOYAMA, J. *The Complex Faulting Process of Earthquakes* (Kluwer Academic Publishers, Dordrecht, The Netherlands (1997)), 194 pp.

KUMAR, S. (2000), *Evaluation and Reduction of Liquefaction Potential at a Site in St. Louis, Missouri*, Earthquake Spectra *16*, 455–472.

LAM, N., WILSON, J., and HUTCHINSON, G. (2000), *Generation of Synthetic Earthquake Accelerograms Using Seismological Modeling: A Review*, J. Earthq. Eng. *4*, 321–354.

LIAO, Z.-P. and JIN, X. (1995), *A Stochastic Model of the Fourier Phase of Strong Ground Motion*, Acta Seismologica Sinica *8*, 435–446.

LIU, L. and PEZESHK, S. (1998), *A stochastic approach in estimating the pseudo-relative spectral velocity*, Earthquake Spectra *14*, 301–317.

LIU, L. and PEZESHK, S. (1999), *An Improvement on the Estimation of Pseudoresponse Spectral Velocity Using RVT Method*, Bull. Seismol. Soc. Am. *89*, 1384–1389.

LOH, C.-H. (1985), *Analysis of the Spatial Variation of Seismic Waves and Ground Movements from SMART-1 Array Data*, Earthq. Eng. Struct. Dyn. *13*, 561–581.

LOH, C.-H. and YEH, Y.-T. (1988), *Spatial Variation and Stochastic Modeling of Seismic Differential Ground Movement*, Earthq. Eng. Struct. Dyn. *16*, 583–596.

LOMNITZ-ADLER, J. and LUND, F. (1992), *The Generation of Quasi-dynamical Accelerograms from Large and Complex Seismic Fractures*, Bull. Seismol. Soc. Am. *82*, 61–80.

LUCO, J. E. (1985), *On Strong Ground Motion Estimates Based on Models of the Radiated Spectrum*, Bull. Seismol. Soc. Am. *75*, 641–649.

MAHDYIAR, M. (2002), *Are NEHRP and Earthquake-based Site Effects in Greater Los Angeles Compatible?*, Seism. Res. Lett. *73*, 39–45.

MALAGNINI, L. and HERRMANN, R. B. (2000), *Ground-motion scaling in the region of the 1997 Umbria-Marche earthquake* (Italy), Bull. Seism. Soc. Am. *90*, 1041–1051.

MALAGNINI, L., HERRMANN, R. B. and DI BONA, M. (2000), *Ground-motion scaling in the Appennines (Italy)*, Bull. Seism. Soc. Am. *90*, 1062–1081.

MARGARIS, B. N. and BOORE, D. M. (1998), *Determination of $\Delta\sigma$ and κ_0 from Response Spectra of Large Earthquakes in Greece*, Bull. Seismol. Soc. Am. *88*, 170–182.

MARGARIS, B. N. and HATZIDIMITRIOU, P. M. (2002), *Source spectral scaling and stress release estimates using strong-motion records in Greece*, Bull. Seism. Soc. Am. *92*, 1040–1059.

MARGARIS, B. N. and PAPAZACHOS, C. B. (1999), *Moment-magnitude Relations Based on Strong-motion Records in Greece*, Bull. Seismol. Soc. Am. *89*, 442–455.

MCGUIRE, R. K. *Ground Motion Estimation in Regions with Few Data, Proc. 8th World Conf. Earthquake Engineering* (Prentice-Hall, Inc., Englewood Cliffs, New Jersey, (1984)) *II*, pp. 327–334.

MCGUIRE, R. K. and Hanks, T. C. (1980), *RMS Accelerations and Spectral Amplitudes of Strong Ground Motion during the San Fernando, California, Earthquake*, Bull. Seismol. Soc. Am. *70*, 1907–1919.

MCGUIRE, R. K., BECKER, A. M., and DONOVAN, N. C. (1984), *Spectral Estimates of Seismic Shear Waves*, Bull. Seismol. Soc. Am. *74*, 1427–1440.

MIDORIKAWA, S. and KOBAYASHI, H. (1978), *On Estimation of Strong Earthquake Motions with Regard to Fault*, Proceedings 2nd Intern. Conf. Microzonation *2*, 825–836.

MILES, S. B. and HO, C. L. (1999), *Rigorous Landslide Hazard Zonation Using Newmark's Method and Stochastic Ground Motion Simulation*, Soil Dyn. and Earthq. Eng. *18*, 305–323.

MIYAKE, H., IWATA, T., and IRIKURA, K. (2001), *Estimation of Rupture Propagation Direction and Strong Motion Generation area from Azimuth and Distance dependence of Source Amplitude Spectra*, Geophys. Res. Lett. *28*, 2727–2730.

ÓLAFSSON, S. and SIGBJÖRNSSON, R. (1999), *A Theoretical Attenuation Model for Earthquake-induced Ground Motion*, J. Earthq. Eng. *3*, 287–315.

ÓLAFSSON, S., SIGBJÖRNSSON, R., and EINARSSON, P. (1998), *Estimation of Source Parameters and Q from Acceleration Recorded in the Vatnafjoll Earthquake in South Iceland*, Bull. Seismol. Soc. Am. *88*, 556–563.

OU, G.-B. and HERRMANN, R. B. (1990a), *A Statistical Model for Ground Motion Produced by Earthquakes at Local and Regional Distances*, Bull. Seismol. Soc. Am. *80*, 1397–1417.

OU, G.-B. and HERRMANN, R. B. (1990b), *Estimation Theory for Strong Ground Motion*, Seism. Res. Lett. *61*, 99–107.

PAPAGEORGIOU, A. S. and AKI, K. (1983a), *A Specific Barrier Model for the Quantitative Description of Inhomogeneous Faulting and the Prediction of Strong Ground Motion. I. Description of the Model*, Bull. Seismol. Soc. Am. *73*, 693–722.

PAPAGEORGIOU, A. S. and AKI, K. (1983b), *A Specific Barrier Model for the Quantitative Description of Inhomogeneous Faulting and the Prediction of Strong Ground Motion. Part II. Applications of the Model*, Bull. Seismol. Soc. Am. *73*, 953–978.

PEZESHK, S., CAMP, C. V., LIU, L., and GREVE, W. M. (2001), Site Specific Analysis Program (SSAP), version 1.06, Dept. of Civil Eng., U. of Memphis, Memphis, TN.

PITARKA, A., SOMERVILLE, P., FUKUSHIMA, Y., UETAKE, T., and IRIKURA, K. (2000), *Simulation of near-fault strong-ground motion using hybrid Green's functions*, Bull. Seism. Soc. Am. *90*, 566–586.

PITARKA, A., SOMERVILLE, P. G., FUKUSHIMA, Y., and UETAKE, T. (2002), *Ground-motion attenuation from the 1995 Kobe earthquake based on simlations using the hybrid Green's function method*, Bull. Seism. Soc. Am. *92*, 1025–1031

RAOOF, M., HERRMANN, R. B., and MALAGNINI, L. (1999), *Attenuation and Excitation of Three-component Ground Motion in Southern California*, Bull. Seismol. Soc. Am. *89*, 888–902.

RATHJE, E. M., ABRAHAMSON, N. A., and BRAY, J. D. (1998), *Simplified Frequency Content Estimates of Earthquake Ground Motions*, J. Geotech. Geoenviron. Eng. *124*, 150–159.

RICE, S. O. *Mathematical Analysis of Random Noise*, reprinted in *Selected Papers on Noise and Stochastic Processes* (N. Wax, ed.) (Dover Publications, New York, 1954), pp. 133–294.

ROUMELIOTI, Z. and KIRATZI, A. (2002), *Stochastic simulation of strong-motion records from the 15 April 1979 (M7.1) Montenegro earthquake*, Bull. Seism. Soc. Am. *92*, 1095–1101.

ROVELLI, A., BONAMASSA, O., COCCO, M., DI BONA, M., and MAZZA, S. (1988), *Scaling Laws and Spectral Parameters of the Ground Motion in Active Extensional Areas in Italy*, Bull. Seismol. Soc. Am. *78*, 530–560.

ROVELLI, A., COCCO, M., CONSOLE, R., ALESSANDRINI, B., and MAZZA, S. (1991), *Ground Motion Waveforms and Source Spectral Scaling from Close-distance Accelerograms in a Compressional Regime Area (Friuli, Northeastern Italy)*, Bull. Seismol. Soc. Am. *81*, 57–80.

ROVELLI, A., CASERTA, A., MALIGNINI, L., and MARRA, F. (1994a), *Assessment of Potential Strong Motions in the City of Rome*, Annali di Geofisica *37*, 1745–1769.

ROVELLI, A., MALAGNINI, L., CASERTA, A., and MARRA, F. (1994b), *Using 1-D and 2-D Modeling of Ground Motions for Seismic Zonation Criteria: Results for the City of Rome*. Annali di Geofisica *38*, 591–605.

SABETTA, F. and PUGLIESE, A. (1996), *Estimation of Response Spectra and Simulation of Nonstationary Earthquake Ground Motions*, Bull. Seismol. Soc. Am. *86*, 337–352.

ŞAFAK, E. and BOORE, D. M. (1988), *On Low-frequency Errors of Uniformly Modulated Filtered White-noise Models for Ground Motions*, Earthq. Eng. Struct. Dyn. *16*, 381–388.

SARAGONI, G. R. and HART, G. C. (1974), *Simulation of Artificial Earthquakes*, Earthq. Eng. Struct. Dyn. *2*, 249–267.

SATOH, T. (2002), *Empirical frequency-dependent radiation pattern of the 1998 Miyagiken-nanbu earthquake in Japan*, Bull. Seism. Soc. Am. *92*, 1032–1039.

SATOH, T., KAWASE, H., and SATO, T. (1997), *Statistical Spectral Model of Earthquakes in the Eastern Tohoku District, Japan, Based on the Surface and Borehole Records Observed in Sendai*, Bull. Seismol. Soc. Am. *87*, 446–462.

SAVY, J. B., FOXALL, W., and ABRAHAMSON, N. (1998), *Guidance for Performing Probabilistic Seismic Hazard Analysis for a Nuclear Plant Site: Example Application to the Southeastern Unted States*, NUREG/CR 6607, UCRL-ID-133494.

SCHERBAUM, F. (1994), *Modeling the Roermond Earthquake of 1992 April 13 by Stochastic Simulation of its High-frequency Strong Ground Motion*, Geophys. J. Int. *119*, 31–43.

SCHERBAUM, F., PALME, C. and LANGER, H. (1994), *Model parameter optimization for site-dependent simulation of ground motion by simulated annealing - reevaluation of the Ashigara valley prediction experiment*, Natural Hazards *10*, 275–296.

SCHNEIDER, J. and SILVA, W. J. (2000), *Earthquake Scenario Ground Motion Hazard Maps for the San Francisco Bay Region*, Final report, USGS Grant award #98-HQ-GR-1004.

SCHNEIDER, J. F., SILVA, W. J., CHIOU, S.-J., and STEPP, J. C. (1991), *Estimation of Ground Motion at Close Distances Using the Band-limited-white-noise Model*, Proc. Fourth International Microzonation Conf. **II**, 187–194.

SCHNEIDER, J. F., SILVA, W. J., and STARK, C. L. (1993), *Ground Motion Model for the 1989 M 6.9 Loma Prieta Earthquake Including Effects of Source, Path and Site*, Earthquake Spectra *9*, 251–287.

SHAPIRA, A. and VAN ECK, T. (1993), *Synthetic Uniform-Hazard Site Specific Response Spectrum*, Natural Hazards *8*, 201–215.

SILVA, W. J. (1992), *Factors Controlling Strong Ground Motions and their Associated Uncertainties*, Proc. *Dynamic Analysis and Design Considerations for High Level Nuclear Waste Repositories*, Structures Div./ Am. Soc. Civil Eng., 132–161.

SILVA, W. J. (1997), *Characteristics of Vertical Strong Ground Motions for Applications to Engineering Design*, Proc. Of the FHWA/NCEER Workshop on the National Representation of Seismic Ground Motion for New and Existing Highway Facilities (I.M. Friedland, M.S. Power, and R.L. Mayes, eds.), Technical Report NCEER-97-0010.

SILVA, W. J. and COSTANTINO, C. (1999), *Assessment of Liquefaction Potential for the 1995 Kobe, Japan Earthquake Including Finite-source Effects*, Final Report, U.S Army Engineer Waterways Experiment Station, Corps of Engineers Contract #DACW39-97-K-0015.

SILVA, W. J. and DARRAGH, R. B. (1995), *Engineering Characterization of Strong Ground Motion Recorded at Rock Sites*, Electric Power Research Institute, Palo Alto, Calif., Report No. TR-102262.

SILVA, W. J. and GREEN, R. K. (1989), *Magnitude and Distance Scaling of Response Spectral Shapes for Rock Sites with Applications to North American Tectonic Environment*, Earthquake Spectra *5*, 591–624.

SILVA, W. J. and LEE, K. (1987), *WES RASCAL code for Synthesizing Earthquake Ground Motions*, State-of-the-Art for Assessing Earthquake Hazards in the United States, Report 24, U.S. Army Engineers Waterways Experiment Station, Misc. Paper *S-73-1*.

SILVA, W. J. and WONG, I. G. (1992), *Assessment of Strong Near-field Earthquake Ground Shaking Adjacent to the Hayward fault, California*. In *Proc. Second Conf. on Earthq. Hazards in Eastern San Francisco Bay Area* (Glenn Borchardt and others, eds.), Calif. Dept. of Conservation, Div. of Mines and Geology Special Publication 113, 503–510.

SILVA, W. J., TURCOTTE, T., and MORIWAKI, Y. (1988), *Soil Response to Earthquake Ground Motion*, Electric Power Research Institute, Palo Alto, California, Report No. NP-5747.

SILVA, W. J., DARRAGH, R. B., GREEN, R. K., and TURCOTTE, F. T. (1989), *Estimated Ground Motions for a New Madrid Event*, U.S. Army Engineers Waterways Experiment Station, Misc. Paper *GL-89-17*.

SILVA, W. J., DARRAGH, R., STARK, C., WONG, I., STEPP, J. C., SCHNEIDER, J., and CHIOU, S.-J. (1990), *A Methodology to Estimate Design Response Spectra in the Near-source Region of Large Earthquakes Using the Band-limited-white-noise Ground Motion Model*, Proc. Fourth U.S. Conf. on Earthq. Eng. *1*, 487–494.

SILVA, W. J., WONG, I. G., and DARRAGH, R. B. (1991), *Engineering Characterization of Earthquake Strong Ground Motions with Applications to the Pacific northwest*, U.S. Geol. Surv. Open-File Rept. 91-441-H.

SILVA, W. J., ABRAHAMSON, N., TORO, G., and COSTANTINO, C. (1997), *Description and Validation of the Stochastic Ground Motion Model*, Final Report, Brookhaven National Laboratory, Associated Universties, Inc. Upton, New York.

SILVA, W. J., McGUIRE, R., and COSTANTINO, C. (1999), *Comparison of Site Specific Soil UHS to Soil Motions Computed with Rock UHS*, Proc. of the OECE-NEA Workshop on Engineering Characterization of Seismic Input, Nov. 15–17, 1999, NEA/CSNI/R(2000)2.

SILVA, W. J., DARRAGH, R., GREGOR, N., MARTIN, G., KIRCHER, C., and ABRAHAMSON, N. (2000a), *Reassessment of Site Coefficients and Near-fault Factors for Building Code Provisions*, Final Report, USGS Grant award #98-HQ-GR-1010.

SILVA, W. J., YOUNGS, R. R., and IDRISS, I. M. (2000b), *Development of Design Response Spectral Shapes for Central and Eastern U.S. (CEUS) and Western U.S. (WUS) Rock Site Conditions.* Proc. of the OECE-NEA Workshop on Engineering Characterization of Seismic Input Nov. 15–17, 1999 NEA/CSNI/R(2000)2.

SILVA, W., GREGOR, N., and DARRAGH, R. (2002), *Department of Regional Hard Rock Attenuation Relations for Central and Eastern North America*, ftp:// ftp.pacificengineering.org/CEUS/

SINGH, S. K., ORDAZ, M., ANDERSON, J. G., RODRIGUEZ, M., QUAAS, R., MENA, E., OTTAVIANI, M., and ALMORA, D. (1989), *Analysis of Near-source Strong-motion Recordings along the Mexican Subduction Zone*, Bull. Seismol. Soc. Am. *79*, 1697–1717.

SINGH, S. K., ORDAZ, M., DATTATRAYAM, R. S., and GUPTA, H. K. (1999), *A Spectral Analysis of the 21 May 1997, Jabalpur, India, Earthquake (M_w = 5.8) and Estimation of Ground Motion from Future Earthquakes in the Indian shield region*, Bull. Seismol. Soc. Am. *89*, 1620–1630.

SINGH, S. K., MOHANTY, W. K., BANSAL, B. K., and ROONWAL, G. S. (2002), *Ground motion in Delhi from future large/great earthquakes in the central seismic gap of the Himalayan arc*, Bull. Seism. Soc. Am. *92*, 555–569.

SOKOLOV, V. (1997), *Empirical Models for Estimating Fourier-amplitude Spectra of Ground Acceleration in the Northern Caucasus (Racha Seismogenic Zone)*, Bull. Seismol. Soc. Am. *87*, 1401–1412.

SOKOLOV, V. Y. (1998), *Spectral Parameters of the Ground Motions in Caucasian Seismogenic Zones*, Bull. seismol. Soc. Am. *88*, 1438–1444.

SOKOLOV, V. (2000a), *Spectral Parameters of Ground Motion in Different Regions: Comparison of Empirical Models*, Soil Dyn. Earthq. Eng. *19*, 173–181.

SOKOLOV, V. Y. (2000b), *Hazard-consistent ground motions: Generation on the basis of the uniform hazard Fourier spectra*, Bull. Seism. Soc. Am. *90*, 1010–1027.

SOKOLOV, V., LOH, C. H., and WEN, K. L (2000), *Empirical Model for Estimating Fourier Amplitude Spectra of Ground Acceleration in Taiwan region*, Earthq. Eng. Struct. Dyn. *29*, 339–357.

SOKOLOV, V., Loh, C. H. and Wen, K. L. (2001), *Empirical models for site- and region-dependent ground-motion parameters in the Taipei area: A unified approach*, Earthquake Spectra *17*, 313–331.

SUZUKI, S., HADA, K., and ASANO, K. (1998), *Simulation of Strong Ground Motions Based on Recorded Accelerograms and the Stochastic Method*, Soil Dyn. Earthq. Eng. *17*, 551–556.

TAMURA, K. and AIZAWA, K. (1992), *Differential Ground Motion Estimation Using a Time-space Stochastic Process Model*, Proc. Japan Soc. Civil Eng. *8*, 217–223.

TAMURA, K., WINTERSTEIN, S. R., and SHAH, H. C. (1991), *Spatially Varying Ground Motion Models and their Application to the Estimation of Differential Ground Motion*, Proc. Japan Soc. Civil Eng. *8*, 153–161.

TORO, G. R. (1985), *Stochastic Model Estimates of Strong Ground Motion*, Section 3 of *Seismic Hazard Methodology for Nuclear Facilities in the Eastern United States*, Report Prepared for EPRI, Project Number P101-29.

TORO, G. R., and MCGUIRE, R. K. (1987), *An Investigation into Earthquake Ground Motion Characteristics in Eastern North America*, Bull. Seismol. Soc. Am. *77*, 468–489.

TORO, G. R., MCGUIRE, R. K., and SILVA, W. J. (1988), *Engineering Model of Earthquake Ground Motion for Eastern North America*, Electric Power Research Institute, Palo Alto, Calif., Rept. No. RP-6074.

TORO, G. R., SILVA, W. J., MCGUIRE, R. K., and HERRMANN, R. B.(1992), *Probabilistic Seismic Hazard Mapping of the Mississippi Embayment*, Seism. Res. Lett. *63*, 449–475.

TORO, G. R., ABRAHAMSON, N. A., and SCHNEIDER, J. F. (1997), *Model of Strong Ground Motions from Earthquakes in Central and Eastern North America: Best Estimates and Uncertainties*, Seism. Res. Lett. *68*, 41–57.

TREMBLAY, R. and ATKINSON, G. M. (2001), *Comparative study of the inelastic seismic demand of eastern and western Canadian sites*, Earthquake Spectra *17*, 333–358.

TSAI, C. C. P. (1997), *Ground Motion Modeling for Seismic Hazard Analysis in the Near-source Regime: An Asperity Model*, Pure Appl. Geophys. *149*, 265–297.

TSAI, C. C. P. (1998a), *Ground Motion Modeling in the Near-source Regime: A Barrier Model*, Terrestrial Atmosph. Oceanic Sci. *9*, 15–30.

TSAI, C. C. P. (1998b), *Engineering Ground Motion Modeling in the Near-source Regime Using the Specific Barrier Model for Probabilistic Seismic Hazard Analysis*, Pure Appl. Geophy. *152*, 107–123.

TUMARKIN, A. G. and ARCHULETA, R. J. (1994), *Empirical Ground Motion Prediction*, Annali di Geofisica *37*, 1691–1720.

VETTER, U. R., AKE, J. P., and LAFORGE, R. C. (1996), *Seismic Hazard Evaluation for Dams in Northern Colorado, USA*, Natural Hazards *14*, 227–240.

WEN, Y. K. and WU, C. L. (2001), *Generation of Ground Motions for Mid-America Cities*, Earthquake Spectra *17*, 359–384.

WENNERBERG, L. (1990), *Stochastic Summation of Empirical Greens Functions*, Bull. Seismol. Soc. Am. *80*, 1418–1432.

WENNERBERG, L. (1996), *Comment on "Simultaneous Study of the Source, Path, and Site Effects on Strong Ground Motion During the 1989 Loma Prieta Earthquake: A Preliminary Result on Pervasive Nonlinear Site Effects" by Byau-Heng Chin and Keiiti Aki*, Bull. Seismol. Soc. Am. *86*, 259–267.

WILSON, R. C. (1993), *Relation of Arias Intensity to Magnitude and Distance in California*, U. S. Geol. Surv. Open-File Rept, 93-556 42 pp.

WONG, I. G. and SILVA, W. J. (1990), *Preliminary Assessment of Potential Strong Earthquake Ground Shaking in the Portland, Oregon, Metropolitan Area*, Oregon Geology *52*, 131–134.

WONG, I. G. and SILVA, W. J. (1993), *Site-specific Strong Ground Motion Estimates for the Salt Lake Valley, Utah*, Utah Geological Survey Misc. Publ. 93-9.

WONG, I. G. and SILVA, W. J. (1994), *Near-field Strong Ground Motions on Soil Sites: Augmenting the Empirical Data Base through Stochastic Modeling*, Proc. Fifth U.S. National Conference on Earthquake Engineering, Chicago, July 10–14, 1994, **III**, 55–65.

WONG, I. G., SILVA, W. J., DARRAGH, R. B., STARK, C., and WRIGHT, D. H. (1991), *Applications of the Band-limited-white-noise Source Model for Predicting Site-specific Strong Ground Motions*, Proc. Second Int. Conf. on Recent Advances in Geotech. Earthq. Eng. and Soil Dynamics, **Paper** *9.13*, 1323–1331.

WONG, I. G., SILVA, W. J., and MADIN, I. P. (1993), *Strong Ground Shaking in the Portland, Oregon, Metropolitan Area: Evaluating the Effects of Local crustal and Cascadia Subduction Zone Earthquakes and Near-surface Geology*, Oregon Geology *55*, 137–143.

YOUNGS, R. R. and SILVA, W. J. (1992), *Fitting the ω^{-2} Brune Source Model to California Empirical Strong Motion Data* (abs.), Seism. Res. Lett. *63*, 34.

YU, G., ANDERSON, J. G. and SIDDHARTHAN, R. (1993), *On the Characteristics of Nonlinear Soil Response*, Bull. Seismol. Soc. Am. *83*, 218–244.

ZENG, Y. H., ANDERSON, J. G., and YU, G. A. (1994), *Composite Source Model for Computing Realistic Synthetic Strong Ground Motions*, Geophys. Res. Lett. *21*, 725–728.

(Received July 2, 2000, accepted February 21, 2001)

 To access this journal online:
http://www.birkhauser.ch

Pure appl. geophys. 160 (2003) 677–715
0033–4553/03/040677–39

❘ Pure and Applied Geophysics

Characterization of Fault Zones

Yehuda Ben-Zion[1] and Charles G. Sammis[1]

Abstract — There are currently three major competing views on the essential geometrical, mechanical, and mathematical nature of faults. The standard view is that faults are (possibly segmented and heterogeneous) Euclidean zones in a continuum solid. The *continuum-Euclidean* view is supported by seismic, gravity, and electromagnetic imaging studies; by successful modeling of observed seismic radiation, geodetic data, and changes in seismicity patterns; by detailed field studies of earthquake rupture zones and exhumed faults; and by recent high resolution hypocenter distributions along several faults. The second view focuses on granular aspects of fault structures and deformation fields. The *granular* view is supported by observations of rock particles in fault zone gouge; by studies of block rotations and the mosaic structure of the lithosphere (which includes the overall geometry of plate tectonics); by concentration of deformation signals along block boundaries; by correlation of seismicity patterns on scales several times larger than those compatible with a continuum framework; and by strongly heterogeneous wave propagation effects on the earth's surface. The third view is that faults are fractal objects with rough surfaces and branching geometry. The *fractal* view is supported by some statistical analysis of regional hypocenter locations; by long-range correlation of various measurements in geophysical boreholes; by the fact that observed power-law statistics of earthquakes are compatible with an underlying scale-invariant geometrical structure; by geometrical analysis of fault traces at the earth's surface; and by measurements of joint and fault surfaces topography.

There are several overlaps between expected phenomenology in continuum-Euclidean, granular, and fractal frameworks of crustal deformation. As examples, highly heterogeneous seismic wavefields can be generated by granular media, by fractal structures, and by ground motion amplification around and scattering from an ensemble of Euclidean fault zones. A hierarchical granular structure may have fractal geometry. Power-law statistics of earthquakes can be generated by slip on one or more heterogeneous planar faults, by a fractal collection of faults, and by deformation of granular material. Each of the three frameworks can produce complex spatio-temporal patterns of earthquakes and faults. At present the existing data cannot distinguish unequivocally between the three different views on the nature of fault zones or determine their scale of relevance. However, in each observational category, the highest resolution results associated with mature large-displacement faults are compatible with the standard continuum-Euclidean framework. This can be explained by a positive feedback mechanism associated with strain weakening rheology and localization, which attracts the long-term evolution of faults toward progressive regularization and Euclidean geometry. A negative feedback mechanism associated with strain hardening during initial deformation phases and around persisting geometrical irregularities and conjugate sets of faults generates new fractures and granularity at different scales. We conclude that long-term deformation in the crust, including many aspects of the observed spatio-temporal complexity of earthquakes and faults, may be explained to first order within the continuum-Euclidean framework.

Key words: Fault zone evolution, continuum-Euclidean, fractal, granular, complexity, simplicity.

[1] Department of Earth Sciences, University of Southern California, Los Angeles, CA 90089-0740, U.S.A. E-mails: benzion@terra.usc.edu, sammis@earth.usc.edu

1. Introduction

Earthquakes are *processes* associated with *objects* that are called *fault zones.*
Earthquake physics is thus dictated to a very large extent by fault zone properties. In
this paper we review conceptual frameworks and data on the character and properties
of earthquake fault zones, guided by the following three questions: 1) What are the
geometrical properties of a fault zone? 2) What is the best theoretical framework to
describe fault mechanics? 3) How do the answers depend on scale? Figure 1 illustrates
the complexity involved in addressing these questions. On the global plate-tectonics
scale, all of western California may be considered as a fault zone between the Pacific and
North American plates. It appears granular in the sense of having crustal blocks that
translate and rotate to accommodate the deformation. It contains several strands of
localized deformation that form the major sub-parallel faults in the San Andreas
system, bordered by a network of subsidiary faults with complex geometry. Focusing
down in scale to the main trace of the San Andreas fault reveals a core of crushed rock
containing multiple shear localizations, bordered by zones of intense fracturing and
damaged rock. Numerous bends and jogs along the main strands tend to be sites of
additional structural complexity. The California plate boundary may thus be viewed as
a nested hierarchy of shear localizations within shear localizations, each surrounded by
a granular or a continuum matrix. A fundamental question is whether this complex
structure is self-similar geometrically and mechanically, or whether different frameworks are required at different scales. A related key issue is whether all components of
the visible complex structure, or perhaps just a few or even one, play a dominant role in
accommodating the long-term tectonic deformation.

As suggested by the foregoing description, the three major competing views on
the essential geometrical, mechanical, and mathematical (GMM) nature of faults are
continuum-Euclidean, granular, and *fractal.* Each of these frameworks carries a very
different set of implications. In the first standard view, faults are collections of planar
or tabular Euclidean zones in a continuum solid. In the *continuum-Euclidea*n
framework, the underlying "macroscopic" GMM structure is fundamentally smooth
and continuous (e.g., FUNG, 1977). This implies the possibility of stable or
convergent averaging of abrupt fluctuations over smaller space-time scales that are
referred to as "microscopic," and clear separation between the microscopic and
macroscopic scales. The suitably averaged macroscopic description has gradual
variations of all fields. Slip on Euclidean faults in a continuum solid can be analyzed
in terms of fracture mechanics, friction, and other constitutive laws measured in
laboratory rock-mechanics experiments. The constitutive laws, like all other
functions, vary smoothly with the ongoing deformation. Stress transfer from a slip
region falls in the far field like $1/r^3$ with r being the distance from the source. This
provides an estimate for the size of expected correlation of stress and other dynamic
variables in a continuum solid. In a medium governed by a *strain weakening*
rheology, deformation processes and structures are expected to evolve toward the

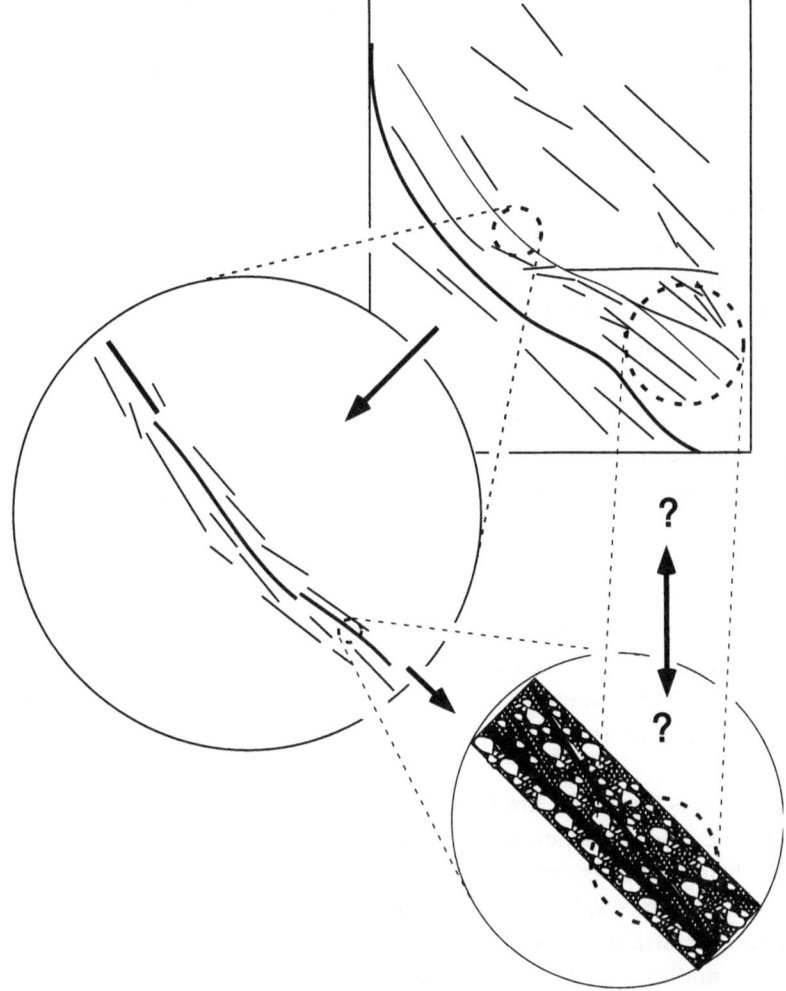

Figure 1
Schematic map views of fault structures at different scales, each with possible Euclidean, granular, and fractal geometrical features. At the largest scale containing boundary regions of different plates, shear is distributed over a network of faults. At a plate boundary scale, individual major faults in the network consist of a quasi-linear array of subparallel strands. At internal fault-zone scale, each strand consists of a distributed band of intense fracturing containing one or more tabular zones of strain localization and intense fragmentation. The arrows between the different scales suggest the possibility that the key structural elements repeat at different scales.

continuum-Euclidean framework. This is because strain weakening produces zones of localized deformation and strength reduction, leading to further strain localization and strength reduction. In an ideal homogeneous quasi-static case, this positive feedback mechanism cascades into deformation that is concentrated on a planar fault in a surrounding elastic continuum. In actual cases, heterogeneities of material properties and applied fields, dynamic branching, etc. produce complications that prevent complete localization. Nevertheless, the long-term deformation in a brittle solid governed by a strain weakening rheology will still be *dominated* by Euclidean structures of size comparable to that of the overall medium dimensions (e.g., depth of seismogenic zone), surrounded by a more-or-less continuum matrix that contains a variety of lesser structures.

The second view focuses on granular aspects of fault structures and deformation fields. In the *granular* framework, the fundamental GMM structure is discrete and strongly heterogeneous. Abrupt fluctuations of fields are present and cannot be averaged out. Load is supported mostly along a few "connectivity chains" rather than the whole medium, producing strong macroscopic anisotropy of all fields (e.g., JAEGER *et al.*, 1996). As a consequence, correlation lengths of dynamic variables exhibit strong directivity effects. For example, stress transfer along the connectivity chains can decay much slower than $1/r^3$, at the expense of much faster decay in other directions. While deformation of granular media includes strong fluctuations, it is still possible to use concepts from fracture mechanics and friction with appropriate modifications. In contrast to a continuum description, however, constitutive laws of granular material may vary abruptly at places. The granular framework is expected to hold in a medium governed by a *strain hardening* rheology that creates a negative feedback mechanism opposite to that associated with strain weakening. This leads to ongoing creation of new fractures, overall distributed or diffused deformation, and structures of relatively small size compared to the overall dimensions of the deforming domain.

The third view is that faults are fractal objects with rough surfaces and branching geometry. In the *fractal* framework, the fundamental GMM structure is irregular, discrete, and heterogeneous on all scales (e.g., MANDELBROT, 1983). If we take the fractal framework at face value, differential calculus and associated continuum-based concepts like stress, strain, fracture, and friction are not valid. At present there are no corresponding mathematical and mechanical quantities, or effective constitutive laws, to describe deformation in a solid with a truly fractal geometry. The fractal framework implies a *dynamic balance* between strain weakening and strain hardening processes that is *perfectly* or *critically tuned* to produce neither positive nor negative overall feedback during deformation. In such a case, the long-term deformation is accommodated by a collection of structures that have no preferred size scale, i.e., structures following a power-law frequency-size distribution. Fractal geometry has been reported to characterize brittle deformation structures in the crust over several bands of length scales, from regional fault networks through main traces of individual faults to the internal structure of fault zones. In this study we examine critically these observations and the mechanical significance of the fractal structures for long-term accommodation of slip on faults.

In the following sections we review observational evidence on the GMM character of faults from a number of different categories of imaging methods and data sources. For each category we discuss several classes of observations, distinguished by their imaging resolution and by whether they apply to regional or fault specific studies. The existing data cannot distinguish unequivocally between the three major views on the nature of faults. However, in each observational category the highest resolution results associated with large-displacement *mature faults* are compatible, in general, with the standard *continuum-Euclidean* framework. Here the term mature designates an evolutionary stage for which the GMM properties of a fault zone achieve asymptotically stable values. The term mature fault also implies a through-going structure at the largest available scale of a fault network that dominates the tectonic slip accommodation. The observational data, as well as modeling results, indicate that rock deformation has a relatively short initial transient phase involving strain hardening and the creation of granularity and band-limited fractal structures at several hierarchies. With small increasing deformation under the same applied loads, and high enough strain rate compared to the rate of healing, this process is replaced by strain weakening and localization to tabular zones that become the main carriers of subsequent deformation. At that stage, most of the complex initial structure becomes mechanically passive and the dominant localized fault zones evolve with continuing deformation toward Euclidean geometry and progressive simplicity and regularization. Fault offsets, kinks, and bends, end regions of earthquake ruptures and faults, and transition zones between different tectonic regimes with several active sets of faults, continue to produce local complexity at different scales. However, the overall structural evolution at different hierarchies is toward progressive regularization. Global transient phases of renewed generation of complexity at different scales occur when a mature fault zone rotates away from a favored orientation compatible with the remote loading.

2. Surface Fault Traces and Structure of Exhumed Fault Zones

The category of studies discussed in this section is based on direct observations of structural elements ranging from the geometry of surface traces and topography of the wall rocks to the internal geometry of fault zones and particle size distribution of the gouge.

2.1. Detailed Geological, Geophysical, and Geochemical Measurements on Exhumed Fault Zones

The data from these works provide multi-disciplinary information on specific fault segments over a range of length scales varying from sub-mm to several km. These studies have broadband widths of both resolution and fault slip and they form the core of the discussion in this section. CHESTER and LOGAN (1986, 1987), CHESTER *et al.* (1993), EVANS and CHESTER (1995), CHESTER and CHESTER (1998) and SCHULZ and EVANS (2000) give extensive field and lab results on the structure of the San Gabriel and Punchbowl faults in southern California. These faults are large abandoned branches of the San Andreas fault system with total slips of several tens of km, and the studied sites have exhumation depths of 3–5 km. Figure 2 shows a mapping example at an intermediate resolution level from the Punchbowl fault zone. At first sight, the figure exhibits all the elements associated with the three different frameworks. The map contains a complex network of faults that divides the medium into blocks of different sizes forming a hierarchy of granularity and shear localizations. However, detailed mapping and a variety of related quantitative analyses have established that most of the 40 km or so of slip on the Punchbowl fault occurred in a contiguous narrow layer of ultracataclasite that is only 10–20 cm wide. This extremely narrow Euclidean layer is parallel to the macroscopic slip vector and is referred to as the "core" of the fault zone (CHESTER and CHESTER, 1998; SCHULZ and EVANS, 2000). Moreover, much of the slip appears to have been accommodated along a single surface within the narrow core layer that is nearly planar in all mapped sections and is referred to as the "principal fracture surface." The ultracataclasite core layer is surrounded by a few m thick cataclasite layer, which in turn is surrounded by a tabular zone of damaged rock, 100–200 m thick at the Punchbowl

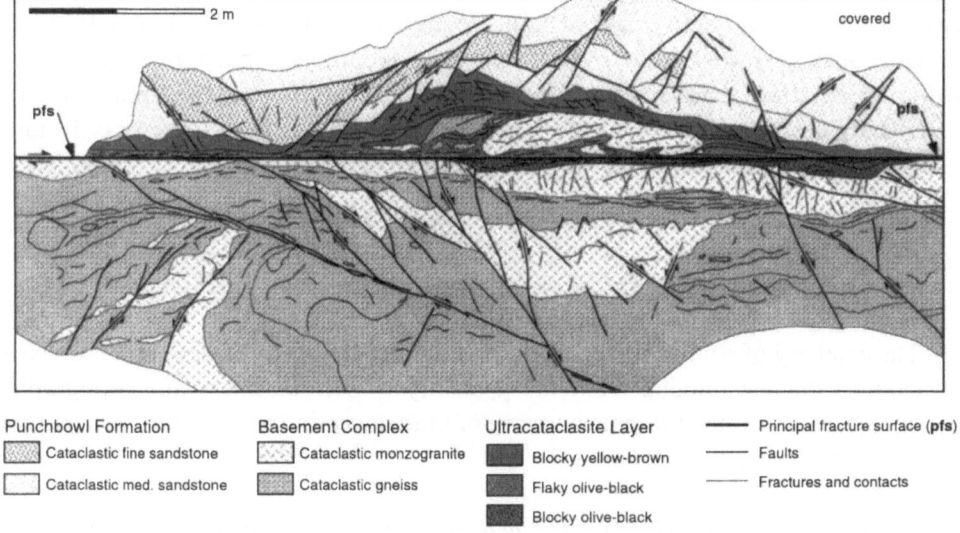

Figure 2

Structure of the Punchbowl fault. The core of the fault consists of an ultracataclasite layer of shear localization with a thickness of cm to tens of cm containing a single planar slip surface labeled "principal fracture surface" that accommodated several tens of km of slip. The ultracataclasite layer and principal fracture surface preserve their simple Euclidean character in all exposures in the field area of CHESTER and CHESTER (1998) and SCHULZ and EVANS (2000). The core structure is surrounded by a tabular zone of cataclastic material that is few m thick and a broader tabular zone of highly fractured rock with various fault branches that is several hundred m wide. Modified from CHESTER and KIRSCHNER (2000).

fault zone, with higher fracture density than the regional background level. The granulation process and shear localizations in the 100–200 m thick damage zone do not appear to have accommodated significant shear strain beyond that associated with their initial formation.

Similar observations characterize the structure of the San Gabriel fault (e.g., CHESTER et al., 1993; EVANS and CHESTER, 1995), the structure of the Kern Canyon fault at sites with total slip of about 15 km and exhumation depth over 5 km (NEAL et al., 2000), and the structure of various faults in the Sierra Nevada at sites with slip up to 150 m and exhumation depth in the range 4–15 km (EVANS et al., 2000). In all those cases, the fault zone structures have a narrow Euclidean core layer (2–3 mm thick in small faults with 10 cm of slip in the Sierra Nevada; 10–20 cm thick in the Punchbowl fault with 40 km of slip) that accommodates most of the deformation, surrounded by a tabular damage zone that is several orders of magnitude wider and is largely passive in slip accommodation. The great similarity in the detailed structures of these faults, having slips and lengths that range over several orders of magnitude, indicates that localization to a dominant narrow core fault zone layer occurs at a very early stage of the shear deformation.

The faults at the Sierra Nevada sites in the study of EVANS et al. (2000) underwent an earlier development phase associated with reactivation in shear of a pre-existing set of cooling joints (e.g., SEGALL and POLLARD, 1983; MARTEL, 1990; MARTEL et al., 1988). In this earlier phase, reactivated joints formed small strike-slip faults that linked to create a larger through-going shear structure. Further deformation within the network of linked faults then localized to primary slip surfaces at one or both edges of the newly formed fault zone. The earlier phase of fault zone formation by linkage of pre-existing joints may be common to rock domains with a suitably oriented set of joints (e.g., WILLEMSE et al., 1997). However, the progression from initial activation of a complex network to localization onto a few planar structures is essentially the same as that observed in the other cases described in this section. AYDIN and JOHNSON (1978, 1983) documented another style of early development of fault zones in porous sandstone, where deformation bands combine initially to form zones of deformation bands which are then spanned by through-going slip surfaces that become the main carriers of continuing slip. Here too, distributed shear in relatively complex initial structures collapses with slight slip onto simple Euclidean surfaces.

An early transition from an initial complex deformation with a regional network of joints and faults to a narrow localized dominant tabular zone was also pointed out by TCHALENKO (1970), based on several scales of observations, including the rupture zone of the 1968 Dasht-e Bayaz earthquake in Iran and "Riedel" and "shear box" laboratory experiments with clays. TCHALENKO (1970) described the deformation structures in all these and other cases cited in his paper as progressing along three main stages associated with peak strength, post peak strength, and residual strength of the material. The structures in the three stages are formed by creation and different arrangements of Riedel, conjugate Riedel, P, and Y shear zones. In the first stage, around the peak shear resistance, there is a transition from relatively homogeneous bulk shear to deformation dominated by localized Riedel zones, at angles of about 10°–15° to the direction of loading, and a conjugate set of Riedel shears. With increasing displacement of about half the slip-weakening distance in the laboratory experiments, the Riedel shears rotate to shallower angles and become connected by P and Y shear zones, while the conjugate Riedel zones are left abandoned. At this stage, estimated by TCHALENKO (1970) to occur in the rupture zone of the Dasht-e Bayaz earthquake at about 250 cm of slip, the Riedel, P, and Y shear zones are concentrated in a relatively narrow and contiguous tabular zone. With additional slip for which the strength in the laboratory experiments achieves a stable residual value, estimated by TCHALENKO to be about 300 cm for the Dasht-e Bayaz earthquake, the structure reached a mature stage and slip becomes concentrated in one or sometime two "principal displacement zones" that are parallel to the direction of loading (Figure 3).

It is interesting to note that the work of TCHALENKO (1970) is often referred to as showing structural complexity at different scales, although he emphasized the *similarity at different scales of early structural evolution toward simple narrow zones that accommodate the continuing deformation*. A similar evolution from a distributed complex initial deformation toward a simple localized shear is seen clearly in the high resolution acoustic emission experiments of LOCKNER *et al.* (1992) with different rocks, and in additional types of data discussed below.

2.2. Analysis of Rupture Surface Topography

These observations involve specific rupture sites and they have a resolution of sub-mm. In most cases the study sites in this class are joints or fault surfaces with very little slip. Thus the rupture topography data typically describe properties of faults at very initial stages. BROWN (1995) summarized field and lab observations of rough fracture surfaces having self-affine distributions of topography. POWER and TULLIS (1991) found that roughness of natural fault surfaces in the direction parallel to that of the slip *deviates significantly from self-similarity for length scales larger than about 1 mm, and has a considerably smaller amplitude* than the roughness in the direction normal to the slip. The latter is approximately self-similar in the data of POWER and TULLIS (1991) from about 10^{-2} mm to about 10 m. A likely explanation for the differences between the amplitude and character of the roughness in the different directions is that cumulative slip tends to smooth the fault surface and make it increasingly more Euclidean-like.

2.3. Evolution of Fault Trace Complexity with Cumulative Slip

These works have a resolution on the order of a km and they fall between those analyzing regional fault networks and those examining specific fault structures.

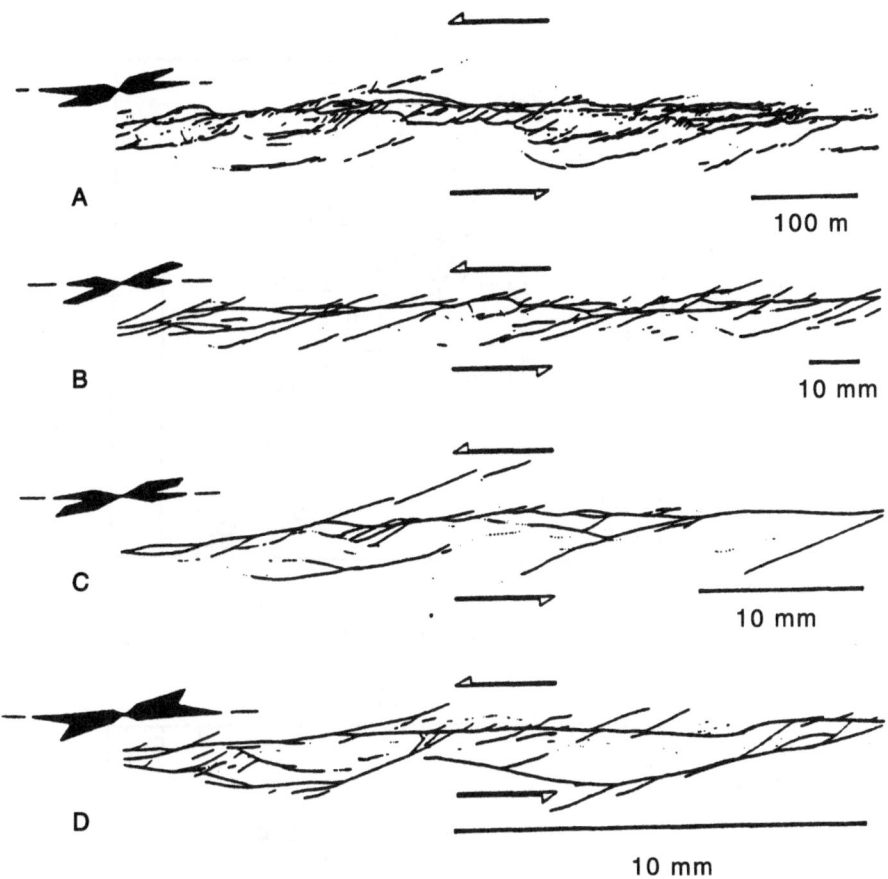

Figure 3

Mature residual structures illustrating localization of deformation at different scales into narrow shear zones that accommodate most of the displacement. (A) Dasht-e Bayaz earthquake in Iran. (B) Riedel experiment with clay. (C) Shear-box experiment with clay. (D) Details of shear-box sample. The rose diagrams show Riedel, P, and principal shear directions at the various scales. From TCHALENKO (1970).

WESNOUSKY (1988, 1994) and STIRLING *et al.* (1996) measured the density of fault steps larger than 1 km per unit distance along strike for some 30 strike-slip fault zones, and assembled frequency-size statistics of earthquakes on these faults. They found that the step density of the examined fault traces decreases as a function of cumulative slip (Fig. 4a), showing an evolution with continuing deformation from a disordered network of linked fault segments to simpler dominant localized fault zones. These results provide examples of structural regularization at larger scales of fault length and slip than the examples of sections 2.1 and 2.2. WESNOUSKY (1988) documented an increase in the average segment length of a fault with increasing total slip, consistent with the idea that increasing slip progressively destroys the small-scale structure and produces a more linear fault. AN and SAMMIS (1996) simulated fault

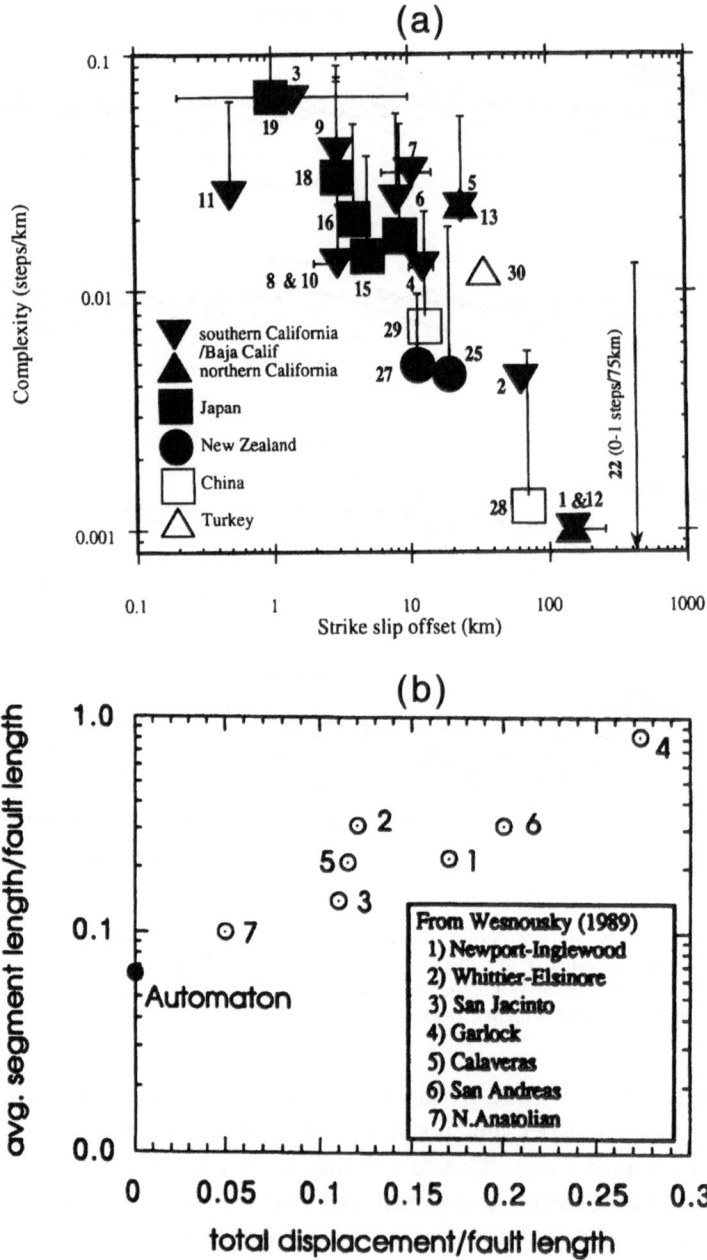

Figure 4

(A) Density of fault steps per unit distance along strike as a function of total strike-slip displacement from
STIRLING *et al.* (1996). (B) Similar results from WESNOUSKY (1989) converted to average segment length per
total fault length as a function of total strike-slip displacement. The data are compared with a computer
simulation (automaton) by AN and SAMMIS (1996) of crack growth and coalescence at zero slip.

traces using an automaton model that incorporated physically based rules for the nucleation, growth and interaction of fractures (Fig. 4b). The number of segments per unit fault length for the initial through-going structure was consistent with the scaling found by WESNOUSKY (1988) in the limit of zero slip.

The compiled frequency-size earthquake statistics of WESNOUSKY (1994) and STIRLING et al. (1996) show a transition from the Gutenberg-Richter power-law distribution on relatively disordered immature faults with small slip, to the characteristic earthquake distribution on relatively regular mature faults with large cumulative slip. These different types of observed statistics as a function of disorder in fault zone properties are compatible with computer simulations and analytical results for models of seismicity patterns along fault systems with different levels of heterogeneity (BEN-ZION and RICE, 1993, 1995; BEN-ZION, 1996; FISHER et al., 1997; DAHMEN et al., 1998). We note that observed power-law earthquake statistics are often assumed to imply an underlying fractal network of faults (e.g., KAGAN, 1992, 1994). However, the above observational and theoretical studies reveal that it is possible to obtain power-law earthquake statistics on individual heterogeneous fault systems occupying narrow or even planar regions of space.

A structural evolution toward increasing localization and simplicity is an expected outcome for deformation in a medium governed by any strain weakening rheology. Figure 5 from LYAKHOVSKY et al. (2001) presents examples of structural evolution in computer simulations of coupled evolution of earthquakes and faults in a model consisting of a seismogenic zone governed by damage rheology over a viscoelastic half space. The employed damage rheology is a generalization of Hookean elasticity to a nonlinear continuum mechanics framework accounting for large strain and irreversible deformation (LYAKHOVSKY et al., 1997). The evolving damage in the seismogenic layer simulates the creation and healing of fault systems as a function of the deformation history. The upper crust is coupled viscoelastically to the substrate where steady plate motion drives the deformation. A parameter space study for this model (BEN-ZION et al., 1999; LYAKHOVSKY et al., 2001) indicates that the types of generated fault structures and earthquake statistics are governed by the ratio of the time scale for material healing (τ_H) to the time scale for loading (τ_L). In general, each brittle failure is associated locally with both strength degradation and stress drop. The value of τ_H depends on the rheology and it controls the time for strength recovering after the occurrence of a brittle event. The value of τ_L depends on the boundary conditions and large-scale parameters (which also determine the average time of a large earthquake cycle) and it controls the time for re-accumulation of stress at a failed location. Relatively high ratios of τ_H/τ_L (left panels) lead to the development of geometrically regular fault zones and frequency-size statistics of earthquakes compatible with the characteristic earthquake distribution. Relatively low ratios of τ_H/τ_L (right panels) lead to the development of highly disordered fault zones and frequency-size event statistics compatible with the Gutenberg-Richter distribution. Interestingly, intermediate cases of τ_H/τ_L produce a "mode-switching"

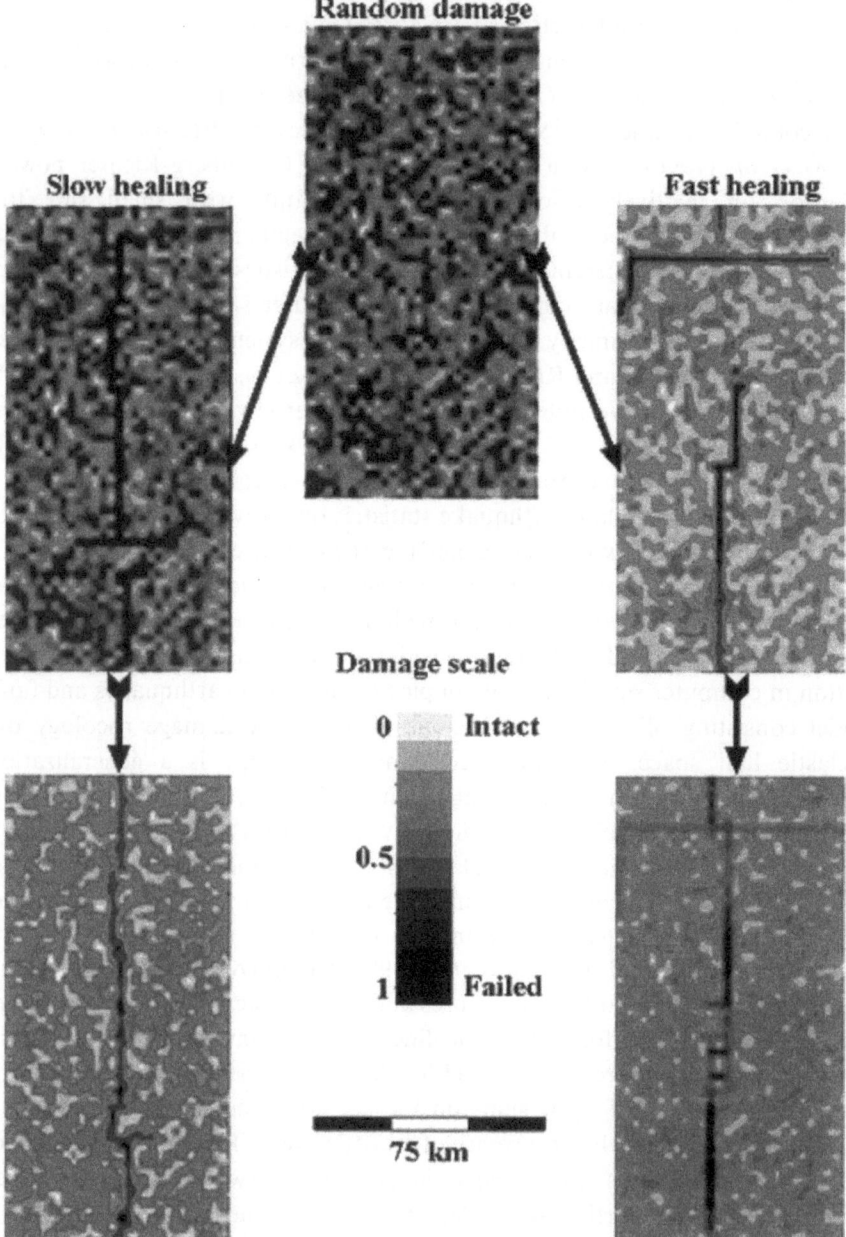

Figure 5

Map views of evolving fault structures in a model consisting of a seismogenic upper crust governed by damage rheology over a viscoelastic substrate. The top central panel shows an initial damage distribution with uncorrelated random heterogeneities. The other panels show evolving damage patterns (fault zones) for cases with relatively slow (left) and fast (right) material healing compared with loading rate. From LYAKHOVSKY *et al.* (2001).

behavior in which the fault zone structures and seismicity patterns alternate between time intervals associated with relatively regular structures and characteristic earthquake distribution and intervals associated with relatively disordered structures and the Gutenberg-Richter statistics. The results are compatible with the observations of WESNOUSKY (1994), STIRLING et al. (1996), MARCO et al. (1996), and ROCKWELL et al. (2000), as well as with the theoretical results of BEN-ZION and RICE (1993, 1995), BEN-ZION (1996), FISHER et al. (1997), and DAHMEN et al. (1998) for various cases of a heterogeneous planar fault in a continuum solid.

2.4. Scaling Analysis of Surface Traces of Faults, Internal Fault Zone Structures, and Fault Networks

These studies may be considered "regional at different hierarchies" and they have various nominal resolutions from sub-mm to several km. However, at each hierarchical level the statistical analyses mix structural units from different deformation phases that may have played significantly different roles in long-term strain accommodation. Thus it is important to try to separate results dominated by structural units reflecting short-lived initial deformation processes from results reflecting properties of long-lived mature surviving structures. Various studies have argued that traces of individual faults have a fractal structure. AVILES et al. (1987) used a ruler method to measure the dimension of the main trace of the San Andreas at several locations. They found dimensions in the range $D_0 = 1 \pm 0.004$. OKUBO and AKI (1987) used a variant of the standard box counting algorithm to measure the capacity dimension of the main trace plus parallel strands and splays of the San Andreas at several locations, and found a slightly higher dimension $D_0 = 1.2 \pm 0.2$. These low fractal dimensions close to 1.0 are consistent with shear localization to an essentially Euclidean structure on the active trace of the San Andreas. It is not clear at this point whether the parallel strands and splays analyzed by OKUBO and AKI (1987) are active participants in slip accommodation or are relict structures from an earlier phase of deformation.

Several ideas for the possible mechanical significance of fractal fault traces have been proposed. OKUBO and AKI (1987) suggested that fault segments with an anomalously high fractal dimension may act as barriers to slip propagation by delocalizing shear onto a more complex set of sub-faults. POWER et al. (1988) proposed that a fractal fault roughness could provide a mechanism by which fault zones grow wider with increased displacement. While these suggestions are interesting, it is important to note that they are non-unique. For example, BEN-ZION and ANDREWS (1998) proposed another explanation, based on the continuum-Euclidean framework, in which fault zone width increases with slip due to wear caused by successive localization of dynamic rupture at the interface between a relatively compliant fault zone layer and a stiffer surrounding host rock. As another example, ANDREWS (1989) and HARRIS and DAY (1993) analyzed quantitatively the capacity of fault bends and offsets to act as barriers using planar fault surfaces and continuum mechanics.

MARONE and KILGORE (1993) suggested that the "critical slip distance" (D_C) in rate- and state-dependent friction may reflect a characteristic strain in granular layers, and thus be proportional to the width of fault zone gouge. This can have important implications since the size of the minimum unstable slip patch (and hence the size of the maximum stable nucleation zone) is proportional to D_C multiplied by rigidity and divided by strength drop (e.g., DIETERICH, 1986; RICE, 1993; BEN-ZION, 2001). However, the existence of narrow core layers and principal slip surfaces in the large-displacement fault zones of section 2.1, and the observation of very small earthquakes on the San Andreas fault (down to about magnitude M = -1 (e.g., NADEAU et al., 1994) with the available recording resolution), indicate that values of D_C on mature faults are not very different than those measured in the lab (and hence are not associated with wide granular fault zone layers or other significantly scaled-up feature).

Other studies have found evidence for fractal-like features in the internal structure of fault zones. Three structures that are observed to scale are: 1) the topography of the wall rock, 2) the particle distribution of the breccia and gouge, and 3) hierarchical shear localizations. Item 1 was discussed in section 2.2. SAMMIS et al. (1987) and SAMMIS and BIEGEL (1989) found a power-law particle size distribution over a range of particle diameters from 10 µm to 1 cm with a fractal dimension in planar cross section of $D_0 = 1.6 \pm 0.1$. AN and SAMMIS (1994) measured the particle distributions of a suite of natural fault gouges by direct counting. They used sieves for particles in the range 62.5 µm $< d <$ 16 mm and a Coulter-Counter for those in the range 1 µm $< d <$ 62.5 µm. The volumetric fractal dimension of the Lopez Canyon gouge was found to be 2.7 $\pm$ 0.2, in agreement with $D_0 = 1.6 \pm 0.1$ measured in 2-D section by SAMMIS et al. (1987). They also examined particles from gouges of the San Andreas and San Gabriel fault zones in southern California and found a correlation between the peak particle size (by weight) and the fractal dimension. They observed that finer gouges tended to have a higher fractal dimension which they interpreted as being the consequence of the existence of a "grinding limit" at about 1 µm (see, e.g., PRASHER, 1987). CHESTER et al. (1993) analyzed the particle size distributions of cataclasites from the North Branch of the San Gabriel fault. They also found power-law particle distributions with a fractal dimension near $D_0 = 2.6$. Particles from the Euclidean ultracataclasite layer have a very small upper fractal limit of about 0.1 mm consistent with strain concentration in this layer.

SAMMIS et al. (1987) proposed a mechanism, which they dubbed "constrained comminution," for generating power-law (fractal) particle size distribution in fault gouge based on the geometry of nearest-neighbor inter-grain loading. This term was chosen to distinguish fragmentation under high confining pressures where particles are not able to change neighbors, from the more common crushing and ball milling processes where the particles are free to move relative to each other. These latter processes are known to produce Rosin-Ramler and other exponential particle distributions (PRASHER, 1987). However, when the particles are constrained to a given neighborhood during the fragmentation process, their fragility is controlled by the relative sizes of their immediate neighbors. More specifically, a particle is most fragile when it is loaded by neighbors of its size since this configuration produces a bipolar load and maximum tensile stress within the particle. Starting with the largest (weakest) grains in the distribution, constrained comminution selectively eliminates nearest neighbors of same size at all scales. This process leads to a distribution that has no neighbors of the same size at any scale. Sierpinski gaskets and carpets have exactly this property. Further, they are characterized by a fractal dimension of $D_0 = 1.58$ in 2-D or $D_0 = 2.58$ in 3-D, similar to that observed in natural gouges. As discussed below, however, the fragmentation process becomes saturated at a relatively small amount of strain and is replaced after an initial transient phase by slip along a dominant localized surface.

BIEGEL et al. (1989) produced power-law distribution of particle sizes in the laboratory, which they characterized over the range 0.02–0.6 mm, by deforming a 3 mm thick layer of 750 μm rock fragments between the sliding rock surfaces of a double-shear friction apparatus. After a shear strain of only about 1, the distribution had evolved into one that looked very similar to the cross sections from natural fault zones (Fig. 6 top). Measurements of the fragment size distribution yielded $D_0 = 1.6$ in 2-D section, consistent with measurements in natural gouges. MARONE and SCHOLZ (1989) also produced fractal gouge in the laboratory using a triaxial saw-cut testing configuration. STEACY and SAMMIS (1991) used a computer automaton to show that the systematic elimination of same-sized nearest neighbors does indeed lead to a random fractal distribution with a dimension $D_0 = 2.6$ in 3-D. BIEGEL et al. (1989) demonstrated that the initial evolution to a fractal particle size distribution is closely related to the transition from a velocity strengthening to a velocity weakening friction commonly observed in granular layers (Fig. 6 bottom). Initially, shear strain is accommodated mostly by grain fracturing, which is inherently a strengthening process. However, with the emergence of a fractal distribution, an increasing proportion of shear is accommodated by slip between the grains that is associated with time-dependent contact strength leading to velocity weakening. The elimination of same-sized neighbors in the initial transient deformation phase minimizes the fragility of grains, thereby suppressing fracture. The same process also minimizes dilatancy, thereby enhancing slip. SAMMIS and STEACY (1994) used the friction data from BIEGEL et al. (1989) in a "grain-bridge" model to quantify this change in friction behavior.

Deformation within fault zones is usually localized on a set of Riedel shears (e.g., Fig. 3) as noted earlier on by TCHALENKO (1970). Occasionally one finds a nested hierarchy of Riedel shears within Riedel shears. A well documented example of such a hierarchy was discussed by ARBOLEYA and ENGELDER (1995) who observed three

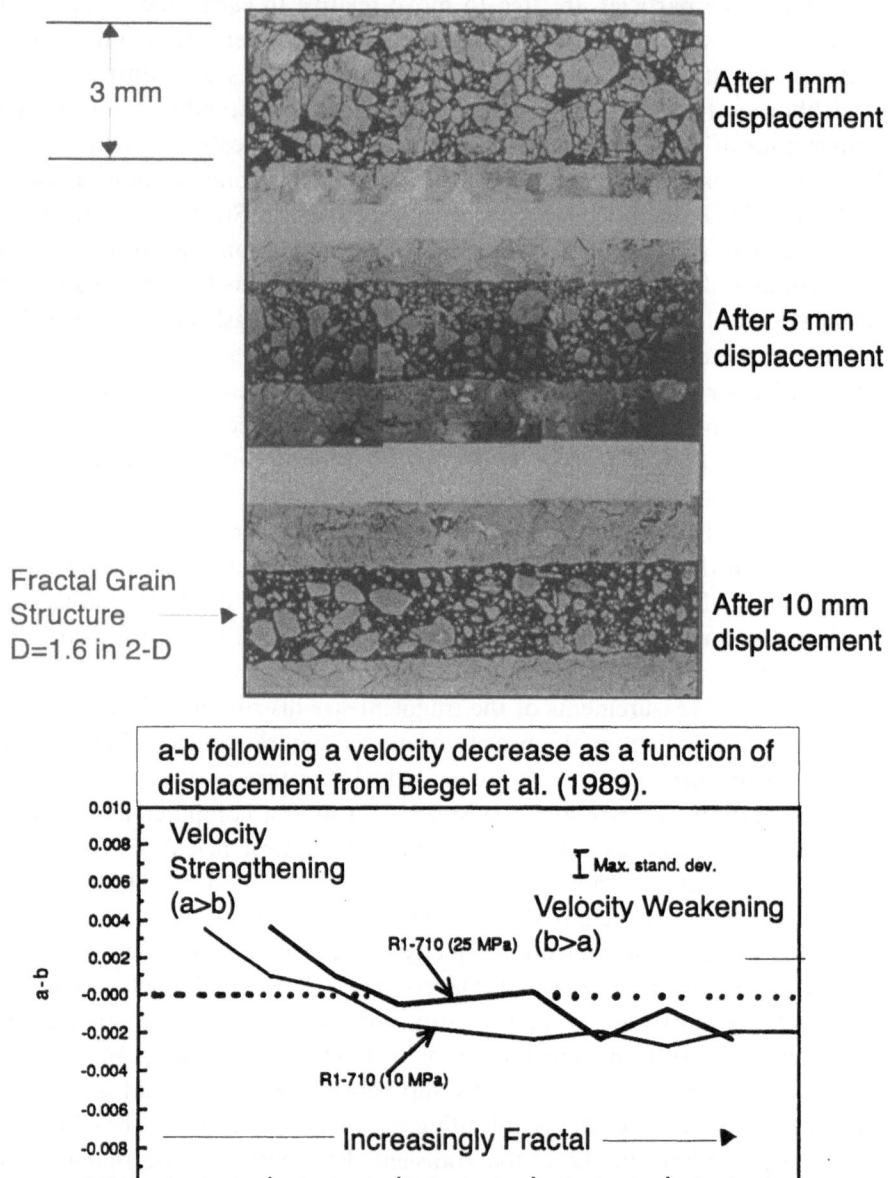

Figure 6
Transition from grain crushing (hardening) to shear localization (weakening) in a granular layer deformed in simple shear. The upper panel shows cross sections of a portion of the layer after 1, 5 and 10 mm of shear displacement. Note the isolation of particles of all sizes leading to the formation of a fractal grain structure as proposed by SAMMIS et al. (1987). The lower panel shows the evolution with slip of the combination of parameters $a - b$ of the rate and state-dependent friction law. The friction has a transition from velocity strengthening ($a > b$) and stable slip to velocity weakening ($b > a$) and potentially unstable slip. The figures are from BIEGEL et al. (1989) who argue that the change in mechanical behavior is due to the establishment of a fractal grain distribution.

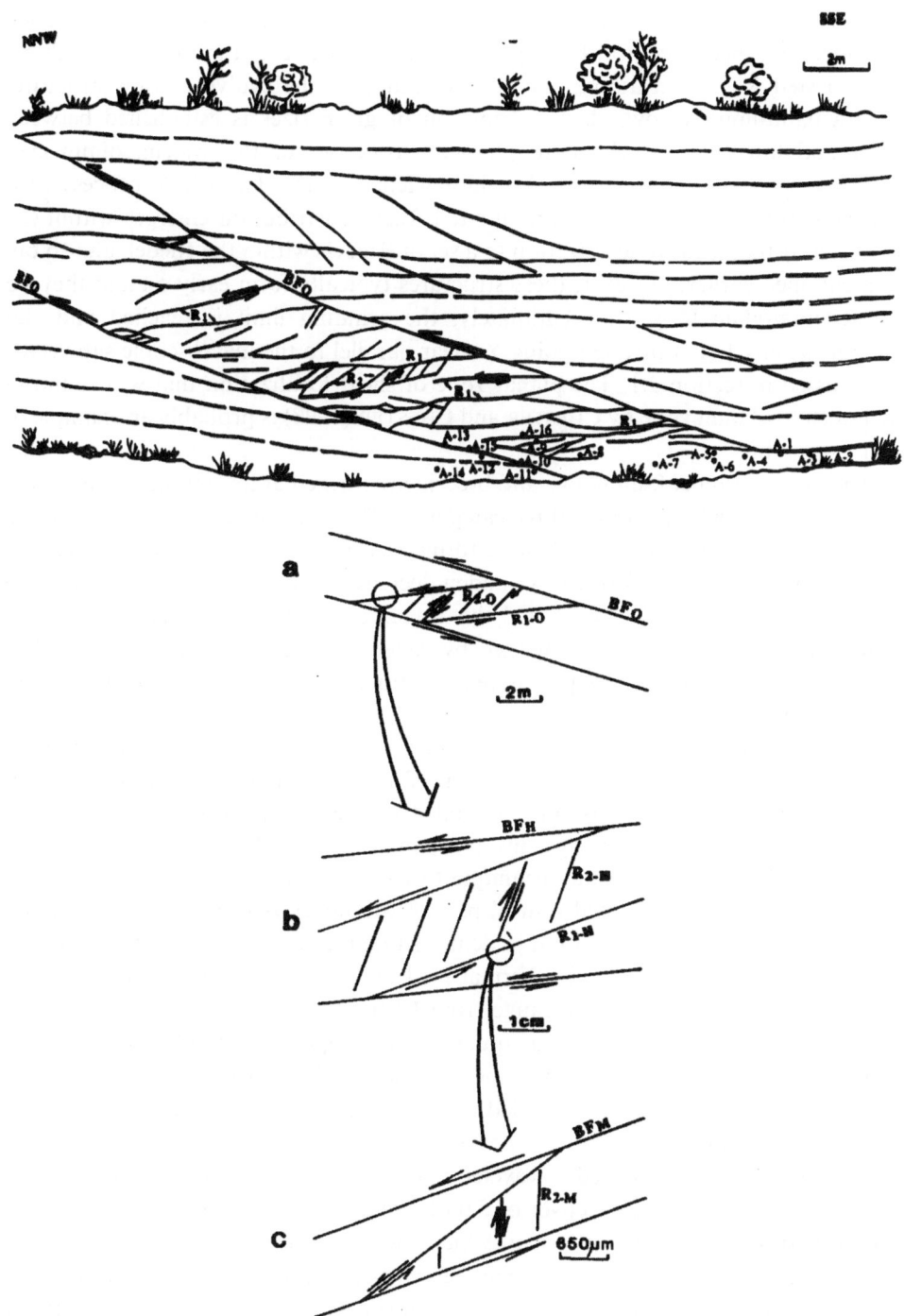

Figure. 7

A nested hierarchy of shear localizations from ARBOLEYA and ENGELDER (1995). The mapped section is shown on top and the interpretation as a nested hierarchy of Riedel shears is given at the bottom.

nested sets of Riedel shears in the Cerro Brass fault zone in the Appalachian valley and ridge province in central Pennsylvania (Fig. 7). The evolution of a hierarchical set of nested Riedel shears may be closely related to the evolving grain structure. Once a band-limited power-law distribution of grain sizes is established between upper and lower scales related to physical properties of the system, diminished dilatancy will cause strain to localize into Riedel shears. The whole process then repeats at the smaller scale. A band-limited fractal distribution emerges within the Riedel shear that leads to localization to a set of Riedels within the Riedel, and so on. As mentioned earlier, however, these structures typically form early on and they do not accommodate large slip. Ultimately, the grinding limit is reached and slip remains localized on a through-going Y shear parallel to the macroscopic slip vector as discussed in section 2.1. The planar zone of shear localization that was found in the Punchbowl fault zone by CHESTER and CHESTER (1998) is probably an example of such a structure.

The idea that most fault gouge and breccia are formed at low strains is supported by SIBSON (1986) who identified three categories of natural gouge: a) attrition breccia due to wear, b) distributed crush breccia formed by the destruction of local asperities, and c) implosion breccia formed by sudden decreases in fault zone pressure (usually at dilatational jogs and bends). Of the three, only the first is associated with significant fault slip. It is characterized by rolled clasts and is relatively rare. The other two are characterized by pervasive microfractures and a jigsaw texture and are far more common.

At the plate boundary scale, deformation is accommodated on a network of faults often having a width on the order of tens of kilometers. Various studies claimed that fault networks also have fractal structure. HIRATA (1989) found that the mapped surface traces of fault systems in Japan are self-similar over scales from 2 to 20 km. He used a box-counting technique to measure a fractal dimension of 1.5 to 1.6 in the more heavily faulted central part of the Japan Arc. SAMMIS et al. (1992) also used a box-counting algorithm to demonstrate that the surface expression of the fault network in the Geysers geothermal field in northern California has a fractal dimension of $D_0 = 1.9$ over the range from 0.8 to 10 km. LEARY (1991) discussed results based on analysis of various quantities in geophysical boreholes (e.g., sonic velocity, mass density, porosity) and seismic coda waves. From long-range correlations in these data he concluded that the fracture density in the crust follows a fractal distribution.

OUILLON et al. (1996) used a wavelet analysis to measure the structure of fault networks on the sedimentary cover of Saudi Arabia at scales from 1 cm to 100 km. They found a multifractal pattern in which the fractal dimension and anisotropic fabric were different for distinct ranges of scale-length separated by characteristic dimensions that correspond to known rheological and lithological units in the Arabian crust. SCHOLZ (1991) noted a transition in geometrical properties of fault segments in the San Andreas system and the rupture zone of the Dasht-e Bayaz earthquake in Iran at a size scale corresponding to the width of the seismogenic zone. The fault networks at these locations appear self-similar at length scales smaller than the seismogenic depth. However, at larger scales the network is replaced by a single through-going structure. As with other items in this section it is not clear whether all the fractures and faults in the regional networks have long-term mechanical significance, or whether most of the displacement is carried by a few dominant structures. The last two references and fractal analyses of regional seismicity discussed below suggest the latter.

Why such structures exhibit fractal characteristics is still an open question. KING (1983) pointed out that 3-D brittle deformation is accommodated in general by several faults, and that fault bends, junctions, and other geometrical irregularities lead to generation of additional smaller scale faults. He suggested that the attractor of this process is a fractal network of faults and showed that such a network can accommodate complex deformation patterns. The process envisioned by KING (1983) is based essentially on purely geometrical considerations in a uniform solid and it involves a cascade of deformation structures from large faults to small ones. However, evolution of material properties that accompanies faulting and slip, not considered by KING (1983), can lead to structural evolution in the opposite direction. The multi-disciplinary observations reviewed in this paper suggest that "large-to-small" cascades of structures characterize transient phases of deformation at different scales that are superposed on "small-to-large" longer-term evolution involving smoothing of geometrical irregularities, coalescence of small segments, and the weakening-localization positive feedback mechanism discussed earlier.

ROBERTSON et al. (1995) suggested that faulting is an example of a percolation process. In this explanation, the fractal structure observed at the surface corresponds to the percolation threshold. However, they suggested that once the percolation threshold is achieved, deformation localizes to the percolation backbone since this is the through-going structure that can accommodate the plate boundary motion. We return to this issue in section 5. Another possibility is that brittle deformation at plate boundaries granulates the entire crust. As discussed for internal fault zone structures, fragmentation in shear with strong constraints on dilatation, as expected to exist at seismogenic depths with normal pore pressure conditions, produces a band-limited fractal distribution of block sizes having a fractal capacity dimension of $D_0 = 1.6$ in planar section. In this interpretation, faults are the boundaries of the blocks. Major faults such as the San Andreas represent major shear localizations within the block structure. GALLAGHER (1981) proposed such a block model for crustal deformation in China based on regional satellite images. FREUND (1974) and NUR et al. (1989) made similar suggestions for the crust in Israel and California. A blocky crustal model is also supported by concentration of deformation signals along block bounding faults (e.g., BEN-ZION et al., 1990), long-range correlation of precursory seismic activity (e.g., KEILIS-BOROK and KOSSOBOKOV, 1990), and rotated coherent terrains (e.g., LUYENDYK, 1991). However, the strong anisotropy of fields implied by a truly blocky (granular) structure is not observed in general. This is indicated, for example, by successful routine modeling of regional seismic (e.g., WALD, 1992; ZHU and HELMBERGER, 1996) and geodetic (e.g., SEGALL and DAVIS, 1997; BURGMANN et al., 2000) fields generated by earthquakes in terms of dislocation surfaces in elastic solid. As discussed earlier, the complex regional fault patterns often visible at the surface are probably relic structures of earlier deformation phases that do not contribute significantly to the long-term slip accommodation.

Most studies discussed in this section are based on direct observations of fault zone properties although they are limited to structures that are presently at the surface. One possible objection to these observations is that they may not represent structures that are currently active at seismogenic depths. In general, fault structures at seismogenic depth are expected to be *simpler* than shallow structures because increasing pressure and temperature tend to suppress brittle branching and other sources of structural complexity. In the next section we summarize observations of *in situ* fault structures at seismogenic depths, based on various indirect geophysical imaging methods.

3. Inversions of Geophysical Data

This category of works involves inferences relative to the geometry and material properties of faults using standard geophysical techniques including gravity, electromagnetic surveys, and seismology.

3.1. Gravity Anomalies

These are regional studies, although they can center on specific faults, with a resolution on the order of a few km. STIERMAN (1984) found a Bouguer gravity low along the San Jacinto fault in southern California and concluded from modeling the data that the fault at seismogenic depth is surrounded by a several km wide tabular zone of damaged rock with reduced mass density. WANG et al. (1986) obtained a similar conclusion from a gravity study across the Bear Valley section of the central San Andreas fault.

3.2. Electromagnetic Signals

These are also regional studies that may center on specific faults. In general, the resolution of these and seismic studies depends on the spatial distribution of sources and receivers, ranges of frequencies generated by the sources and detectable by the (surface or sometimes shallow borehole) receivers, the portion of the source-receiver path spent in the fault zone itself, and whether travel time or whole waveform observations are used. EBERHART-PHILLIPS *et al.* (1995) reviewed various electromagnetic and seismic techniques and concluded based on synthetic data tests that typical electromagnetic studies have a resolution on the order of a few km. A dense instrument spacing may increase the resolution of electromagnetic surveys to 500 m or so (UNSWORTH *et al.*, 1999). The electromagnetic methods provide an image of structural resistivity (or conductivity) that may be interpreted in terms of various manifestations of damaged fault zone rock (high fluid pressure, clay minerals, and/or deposited conductive minerals). EBERHART-PHILLIPS and MICHAEL (1993), MACKIE *et al.* (1997), and UNSWORTH *et al.* (1999) found that various sections of the San Andreas fault are associated with prominent tabular resistivity structures.

3.3. Seismic Reflection/Refraction and Travel-time Tomography

These methods are lumped together here because they all use seismic travel time (and sometime also amplitude) information, as opposed to waveform modeling. Studies in this class have typically a resolution of up to about 500 m, and like the previous two classes they are regional surveys that may center on specific faults. EBERHART-PHILLIPS *et al.* (1995) give an excellent review of various seismic (and electromagnetic) methods, and their ability to image fault zone properties. Additional reviews of reflection/refraction seismology and travel-time tomography are given by MOONEY (1989) and THURBER (this volume). Reflection/refraction surveys are not well suited for imaging narrow vertical structures. They have been used widely to image sub-horizontal structures and results typically show, within the imaging resolution, Euclidean fault structures (e.g., MOONEY and BROCHER, 1987; FUIS and MOONEY, 1990). Travel-time tomography has a better ability to image vertical structures; in the first application of body-wave tomography to fault imaging, AKI and LEE (1976) found a tabular low velocity fault zone layer along the San Andreas fault south of Hollister.

One general shortcoming of regular seismic tomography is that P and S body waves tend to avoid low velocity media. This, together with the fact that each travel time is a function of the entire source-receiver path, limits considerably the ability of body-waves tomography to image narrow low-velocity structures. An improvement in the imaging resolution may be obtained by adding travel-time information of phases that spend much of their travel path along the fault zone structure. BEN-ZION *et al.* (1992) developed a joint travel-time tomography of body waves and fault zone head waves that propagate along material interfaces in the fault zone structure. Application of the method to a small data set from the Parkfield segment of the San Andreas fault led to separate depth profiles of seismic velocities for the different sides of the fault, consistent with the existence of a sharp Euclidean contrast across the fault. Regular tomographic inversions of large data sets of body-wave arrivals at Parkfield show a several km wide transition zone of P and S wave velocity across the fault with a possible internal low velocity layer (e.g., LEES and MALIN, 1990; MICHELINI and MCEVILLY, 1991; EBERHART-PHILLIPS and MICHAEL, 1993). Similar tomographic images are found along other sections of the San Andreas and other faults (e.g., Feng and MCEVILLY, 1983; MICHAEL and EBERHART-PHILLIPS, 1991; EBERHART-PHILLIPS and MICHAEL, 1998).

3.4. Seismic Fault Zone Guided (Head and Trapped) Waves

The highest imaging resolution of fault zone structure at depth with surface or shallow observations is probably provided by waveform modeling of seismic fault zone head and trapped waves. These studies can have a resolution on the order of meters to tens of meters and they involve specific fault segments. Fault zone head and trapped waves can exist *only* in structures that are sufficiently Euclidean and uniform to act as a waveguide. As mentioned above, fault zone head waves propagate along material interfaces in the structure. Fault zone trapped waves are generated by constructive interference of critically reflected phases within low-velocity fault zone layers.

Fault zone head waves have been observed along the subduction zone of the Philippine Sea plate underneath Japan (FUKAO et al., 1983), the Parkfield segment of the San Andreas fault (BEN-ZION and MALIN, 1991; BEN-ZION et al., 1992) and small fault segments in the aftershock zone of the 1992 Joshua-Tree California earthquake (HOUGH et al., 1994). Fault zone trapped waves were observed along the Philippine Sea plate underneath Japan (FUKAO et al., 1983), the Middle America Trench near Mexico (SHAPIRO et al., 1998), a small normal fault in Oroville, California (LEARY et al., 1987), several segments of the San Andreas (LI et al., 1990, 1997; MICHAEL and BEN-ZION, 2002) and San Jacinto (LI et al., 1998) faults, and the rupture zones of the 1992 Landers, California (LI et al., 1994; PENG et al., 2000), 1995 Kobe Japan (LI et al., 1998; NISHIGAMI et al., 2000; KUWAHARA and ITO, 2000), and 1999 Izmit Turkey (BEN-ZION et al., 2000) earthquakes.

At present there are still considerable uncertainties in the interpretation of these observations due to the nonuniqueness of modeling and the limited scope of the analysis to date (BEN-ZION, 1998; MICHAEL and BEN-ZION, 2002). Collectively, however, the observations of fault zone head and trapped waves along a number of fault and rupture segments where detailed data are available indicate that fault structures at depth have coherent Euclidean interfaces and/or low-velocity layers that are tens to hundreds of meters wide. The inferred low-velocity fault zone layers are in good correspondence with the tabular damage zone around the narrow core slip regions discussed in section 2.1.

None of the geophysical imaging techniques can resolve dominant slip zones and fracture surfaces, and the tabular damaged zones imaged by the methods of sections 3.1–3.3 are probably blurred versions of the true structures. These tabular damaged zones may have important implications for fluid flow in the crust and a variety of related issues, however they are probably not the key mechanical structures accommodating the major long-term cumulative slip. To find evidence for properties of the key mechanical structure at seismogenic depth we discuss in the next section relations between various model predictions and earthquake observations.

4. Model Predictions

Since theoretical models of earthquakes and seismicity are predicated on an assumed GMM formulation, their success or failure at predicting observations bears directly on the problem of identifying the most suitable GMM framework. We now review several classes of theoretical models from this perspective. The spatial extent and resolution of the different classes depend on the observations used to validate model prediction and are not specified explicitly as done in sections 2, 3, and 5.

4.1. Seismic Radiation

A major triumph of modern seismology is the development in the last few decades of a quantitative ability to model successfully, with Euclidean surfaces and continuum-elastodynamics, seismic waves at all observed frequencies (e.g., AKI and RICHARDS, 1980). Some rupture and wave propagation models assume a self-similar collection of slip patches; however, those are modeled as occurring along planar surfaces in a continuum solid so mechanically the corresponding models still belong to the continuum-Euclidean framework. For example, ANDREWS (1980, 1981) pointed out that observed ω^{-2} spectral decay of ground acceleration and b-value of frequency-size (FS) earthquake statistics of about 1 can be generated by self-similar distributions of slip and stress on a planar fault. ZENG et al. (1994) and HERRERO and BERNARD (1994) simulated strong ground motion compatible with observed accelerograms with frequencies up to several tens of Hz by summing contributions from a self-similar distribution of slip patches, each radiating a dislocation pulse with a random phase, on a planar fault. It is of course possible that higher frequency waves that attenuate within the source region and do not reach the receivers are not compatible with slip on planar surfaces. To constrain the mechanical structure at the earthquake source region we must consider source phenomena that are observable at the earth surface. This leads us to predictions of mechanical models associated with spatio-temporal seismicity patterns.

4.2. Earthquake Triggering and Migration

Various studies have modeled observed changes of seismicity rates and space-time patterns of earthquake failure sequences in terms of stress transfer between frictional planar fault surfaces in a continuum solid (e.g., HARRIS, 1988; KING and COCCO, 2001, and references therein). As examples, SIMPSON and REASENBERG (1994) used such a framework to model changes of seismicity rates in the San Francisco Bay area following the 1989 Loma Prieta earthquake. STEIN *et al.* (1994) demonstrated that the main spatio-temporal characteristics of moderate to large earthquakes and their aftershock sequences in southern California over recent decades may be explained by interactions governed by elastic stress transfers between frictional planar faults. HARRIS and SIMPSON (1996) showed that space-time patterns of earthquakes in southern California in the 100 years following the great 1857 earthquake are compatible with elastic stress transfer generated by a continuum-Euclidean model of that event. Perhaps most impressively, STEIN *et al.* (1997) calculated with such a model elastic stress transfers associated with the remarkable 1942–1992 progression of ruptures along the North Anatolian fault. Significantly, the two largest subsequent earthquakes on the fault, the August and November 1999 M > 7 events, occurred along segments that fall within a region predicted to have high triggering stress by these calculations. NALBANT *et al.* (1998) modeled evolving seismicity in the more complex western end-region of the North Anatolian fault with elastic stress transfers between numerous frictional planar faults.

4.3. *Rupture along a Material Discontinuity Interface*

WEERTMAN (1980), ADAMS (1995), ANDREWS and BEN-ZION (1997) and others studied properties of dynamic rupture along a frictional planar interface between two different elastic solids. The studies indicate that a material discontinuity interface is a mechanically favored surface for rupture propagation. The sharp material contrast across the rupture plane leads to several additional model predictions. These include (1) a mode of rupture in the form of a narrow self-healing pulse associated with low generation of frictional heat, (2) asymmetric motion on the different sides of the fault, and (3) preferred direction of rupture propagation as that of the slip in the more compliant material (Fig. 8). Item (1) is compatible with inferred earthquake properties (e.g., HEATON, 1990; BRUNE *et al.*, 1993), however it may also be generated by a number of other mechanisms (see BEN-ZION, 2001, for a recent review).

►

Figure 8

(Top) Particle velocities at a given time for rupture along a material interface (thin horizontal line) separating elastic solids with 20% contrast of shear-wave velocities and mass densities. The slipping region is marked by the short thick segment on the fault and is propagating to the right. The existence of a material contrast across the fault produces an asymmetric motion in the different media that is especially prominent near the fault. (Bottom) A closer view around the fault (note changes in scales). Particle velocities in the more compliant material ($y > 0$) are larger than in the stiffer medium ($y < 0$). This leads to dynamic reduction of strength at the rupture front that enables slip to occur as a narrow wrinkle-like pulse. Spontaneous propagation occurs only in one direction, that of the slip in the more compliant medium. From BEN-ZION (2001).

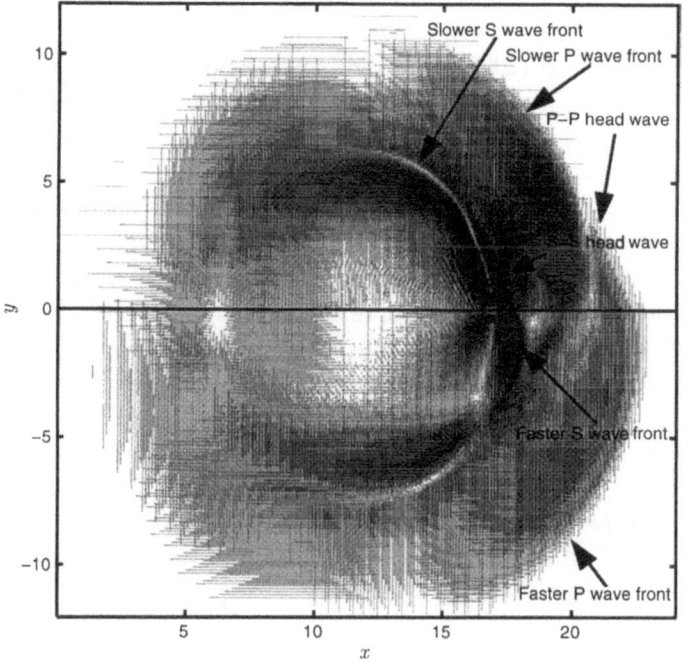

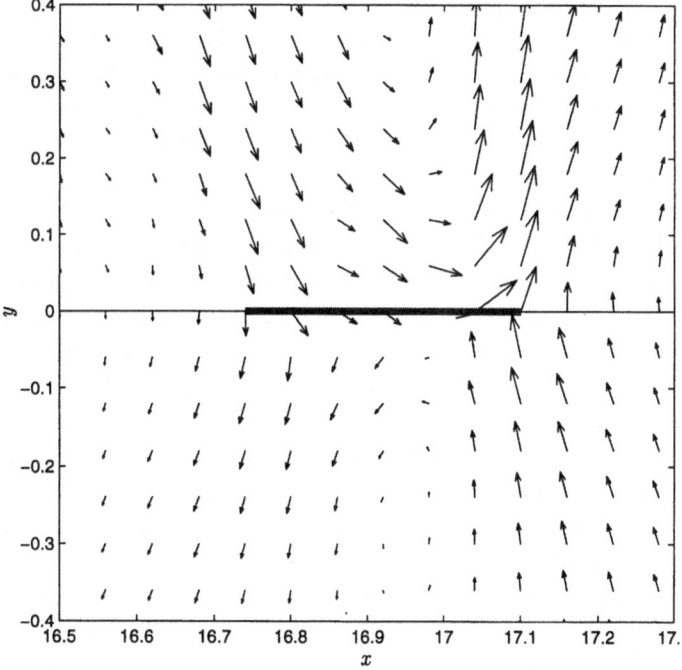

However, item (3) and to a lesser extent also item (2) are specific to rupture along a material interface.

RUBIN and GILLARD (2000) examined space-time properties of high-resolution locations of small earthquakes, obtained by a waveform cross-correlation technique, along the San Andreas fault north of San Juan Bautista, California. They found that the number of immediate aftershocks near the edges of prior ruptures to the northwest is more than double the number to the southeast. The strong asymmetry of aftershocks in the different directions was interpreted by RUBIN and GILLARD (2000) to be a manifestation of item (3), associated with rupture along a material contrast across the fault in that region. McGUIRE et al. (2000) analyzed source properties of about 30 global large earthquakes and found that rupture propagation of most events was predominantly unidirectional. This observation is compatible with a tendency of rupture to localize along a material interface between a relatively compliant fault zone layer and a stiffer surrounding rock, in agreement with the prediction of the above continuum-Euclidean models. Direct field studies also indicate that rupture commonly localizes near the boundary between a damaged fault zone layer and the host rock (MARTEL et al., 1988; BRUHN et al., 1994; SIBSON, 1999).

4.4. Statistics and Patterns of Earthquakes

KAGAN and KNOPOFF (1981) and KAGAN (1982) argued that earthquakes and faults are rough fractal processes and objects in space-time. Their argument is based on analysis of hypocenter locations discussed in the next section, and extrapolations of aftershock decay rates and strong ground motion to zero space and time distances from the generating sources, assuming strict self-similarity on trajectories having singularities at the sources. Since *in situ* measurements directly at the earthquake source are non-existent, such extrapolations are permissible in principle; however they are certainly not required by the available data. As noted in section 4.1, seismic radiation can be explained at all observed frequencies by heterogeneous distribution of dislocation sources on a planar fault. In continuum-Euclidean models, aftershock rates are truncated at short time due to initial build-up processes before instabilities associated with nonlinear rheology (e.g., DIETERICH, 1994; LOCKNER, 1998). This is reflected empirically by the positive constant c in the modified Omori law for aftershock rates, $\Delta n / \Delta t \sim (t + c)^{-p}$, where n is number of events, t is time, and the exponent p is close to 1. Observed values of c are derived based on incomplete recordings close in space-time to the mainshock and other uncertainties, so it is possible that c is essentially zero. However, UTSU et al. (1995) examined this possibility and concluded from high-resolution and corrected observations that actual c values are positive. DIETERICH (1994) showed that Omori's law can emerge directly from delayed nucleation associated with rate- and state-dependent friction observed in the laboratory, and does not require a fractal structure (see also LOCKNER, 1998).

KAGAN (1992, 1994) summarized various power-law distributions of earthquakes and concluded that the scale-invariant statistics and great irregularity of earthquake occurrence imply underlying fractal structures. Again, this is possible but not required since power-law statistics and complex space-time patterns can be generated by heterogeneous continuum-Euclidean models with many degrees of freedom, as well as by other models. For example, BEN-ZION (1996), FISHER et al. (1997), and DAHMEN et al. (1998) showed numerically and analytically that heterogeneous planar idealizations of segmented faults in an elastic solid can produce frequency-size, temporal, and spatial statistics of earthquakes that are compatible with observations covering broad ranges of space and time scales. In fact, for some parameter values these models generate fractal slip patterns (FISHER et al., 1997). Thus the complex earthquake phenomenology does not necessarily imply by itself any specific framework.

Another approach for imaging the geometry of faults is based on hypocenter distributions. Hypocenters provide an image of nucleation zones of brittle instabilities rather than entire fault zones, and they tend to cluster around geometrical and mechanical heterogeneities. Thus geometrical properties of hypocenters are probably more complex than those of earthquake ruptures and faults. Nevertheless, high resolution hypocenter locations appear to collapse on simple structures, as discussed in the next section.

5. Hypocenter Distributions

The spatial distribution of hypocenters helps to illuminate the geometry of the faults on which they occur. For individual faults, accurately located hypocenters should give a measure of the width of the slip localization zone and reveal evidence of spatial heterogeneity. For fault networks, hypocenters preferentially illuminate the most active elements and help to determine how the system accommodates regional strain.

5.1. Spatial Correlations among Routine Locations in Regional and Global Catalogs

These are regional studies and they have a resolution on the order of a few km. KAGAN and KNOPOFF (1980) and KAGAN (1981a, 1981b) estimated the spatial structure associated with seismicity patterns by calculating 2-, 3-, and 4-point correlation functions among hypocenters in the central California earthquake catalog. KAGAN (1981b) concluded in a summary of this set of works that the hypocenter locations reside on a self-similar volumetric branching structure. KAGAN (1991) measured the pair correlation of hypocenters in a global earthquake catalog to estimate the spatial fractal dimension of global seismicity. He obtained a correlation dimension of $D_2 = 2.1$–2.2 for shallow seismicity, 1.8–1.9 for intermediate depth events, and 1.5–1.6 for deep focus events. It is important to verify the results of these works by independent new studies using higher resolution data. We also note that a volumetric branching structure of hypocenters does not necessarily imply that the ruptures themselves do not reside, to a good approximation, on a collection of Euclidean surfaces.

5.2. Geometry and Analysis of Improved Locations

These are also regional studies but they have a resolution varying from a few hundred meters to about a km. ROBERTSON et al. (1995) used a 3-D box-counting algorithm to find the capacity dimension D_0 of seismicity in relocated aftershock sequences in southern California. They found the surprising result that the fractal dimension is only 1.8 in 3-D for those catalogs that had the smallest location errors, and a slightly larger value for catalogs with lower resolution. The low dimension was interpreted as evidence for localization of seismicity on the backbone of a critical percolation structure in 3-D, which is also 1.8. This interpretation is consistent with our hypothesis that most of the active deformation is localized on a small subset of the regional network, much of which is remnant from earlier phases of deformation. The results indicate that the entire fault network is not populated with earthquakes. Only the subset of faults that form a connected structure and allow finite shear (the percolation backbone) is active. More recently, RICHARDS-DINGER and SHEARER (2000) relocated the 1975–1998 seismicity in southern California using a distribution of station corrections that varies as a function of the earthquake positions. They have not attempted a fractal analysis, however visual inspection indicates that the relocated hypocenters tend to collapse, compared to the original locations, toward tabular Euclidean structures.

5.3. High-resolution Relocations of Local Network Catalogs

The highest-resolution locations are based on waveform cross correlation, sometimes done jointly with other methods (e.g., POUPINET et al., 1984; ITO, 1985; GOT et al., 1994; NADEAU et al., 1994; DODGE et al., 1995; SHEARER, 1997; RUBIN et al., 1999; WALDHAUSER et al., 1999; RUBIN and GILLARD, 2000). These results apply to specific fault segments or aftershock sequences and they have a resolution on the order of meters to tens of meters. In all these studies the hypocenters collapse toward regular Euclidean structures consisting of one or more planar subparallel segments. In no case do the high-resolution results show a fractal branching structure emerging from the original cloud of the standard locations.

The hypocenter locations on fault segments which have a large component of aseismic creep (NADEAU and MCEVILLY, 1997; NADEAU and JOHNSON, 1998; RUBIN et al., 1999; RUBIN and GILLARD, 2000) show interesting spatio-temporal patterns such as repeating events, linear streaks, clustering of immediate aftershocks along edges of

Mode II crack shapes, and preferred direction of rupture propagation. These patterns have been interpreted in terms of strength heterogeneity on a fault plane (SAMMIS *et al.*, 1999; SAMMIS and RICE, 2001; BEELER *et al.*, 2001), standard fracture mechanics (RUBIN and GILLARD, 2000), and rupture along a sharp material contrast across the fault (see section 4.3).

6. *Discussion*

We now return to our original questions: what is the best framework for modeling crustal deformation and how does the answer depend on scale? The multidisciplinary results reviewed in the paper suggest consistently that fault structures tend to evolve with cumulative slip (Fig. 9) toward geometrical simplicity and the continuum-Euclidean framework at all scales. The observed geometrical complexity appears to be largely a relict structure reflecting the superposition of many deformation phases in a long convoluted history. Structural complexity is generated primarily during initial deformation phases associated with strain hardening and low-slip organization phenomena that precede localization at various scales. Localization of deformation at a given scale is accompanied by a transition at that scale from strain hardening to weakening. Continuing slip leads to overall progressive structural regularization at the different scales.

Table 1 summarizes the evolution toward geometrical simplicity at four specific hierarchies. At scales of less than about 10 m representing internal fault zone structure, regularization of geometrical incompatibilities on the fault surface with progressive slip produces a tabular zone of granular rock that becomes fault gouge. Initially, shear deformation of the granular gouge zone is mostly accommodated by crushing particles, which is a hardening phenomenon and thus distributed across the layer. However, relatively small deformation leads to the development of a band-limited fractal distribution of grain sizes that suppresses further grain crushing and favors shear localization. The initial localization may be followed by several transient phases of delocalization, however the end result is localization onto a single surface within the gouge zone parallel to the direction of shear. Once this primary slip surface forms, the fractal granularity and oblique Riedel shear structures do not appear to play a significant role in subsequent slip. There is no evidence for persisting fluctuations between localized and delocalized deformation necessary to maintain active deformation on the entire fractal and/or granular structures.

A related evolution toward simplicity with increasing slip occurs on single fracture surfaces at scales on the order of a few cm or less. Fresh fractures with little or no slip have fractal surface roughness over broad ranges of scales. Often these are tensile fractures that are reactivated in shear, or shear fractures formed by the coalescence of many smaller shear fractures. With subsequent slip the surfaces

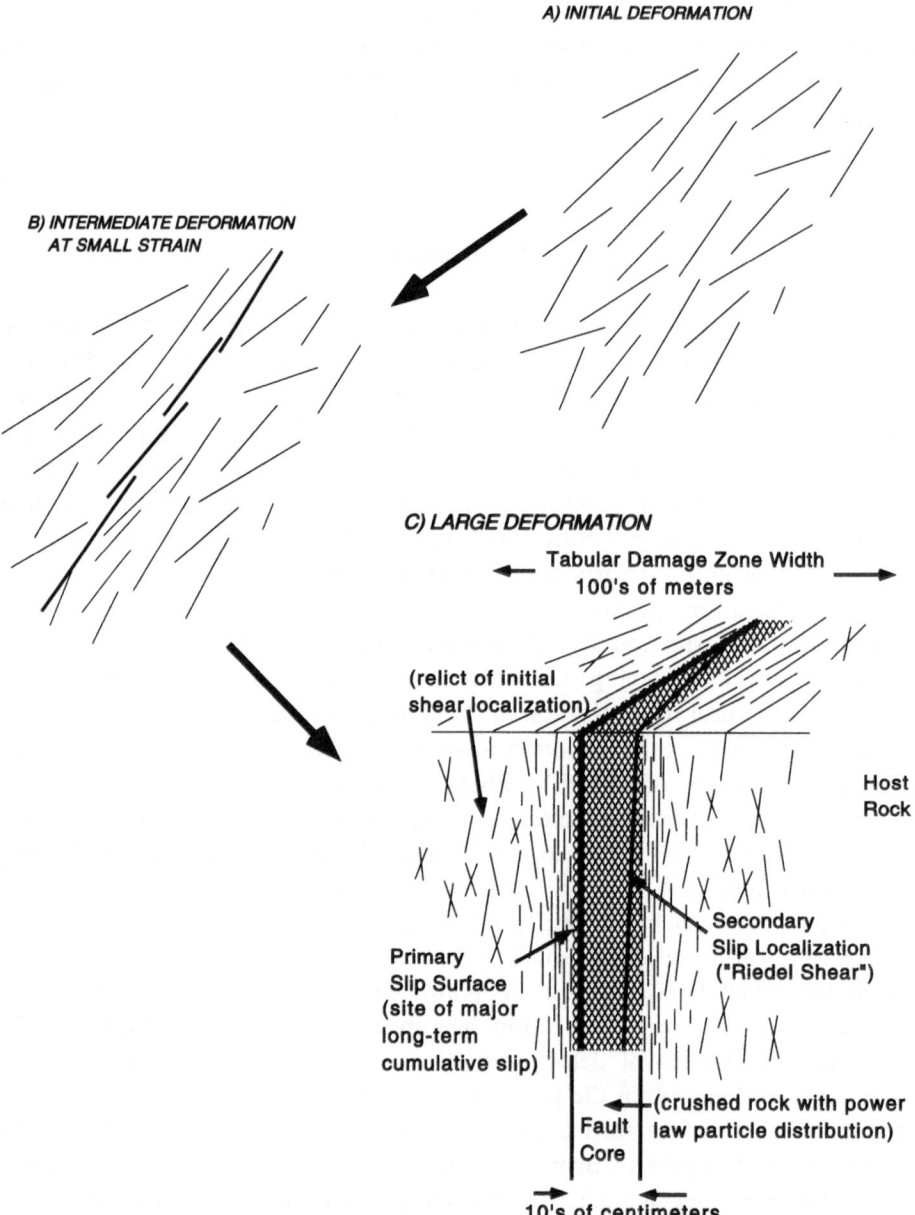

Figure 9
A conceptual representation of fault structures at three main evolutionary stages. (A) Initial deformation is associated with strain hardening. At this stage there is creation of granularity and band-limited fractal structures at several hierarchies. (B) After a relatively small initial strain there is localization to tabular primary slip zones accompanied by a transition to strain weakening. (C) Large deformation is dominated by strain weakening and overall evolution at different scales toward Euclidean geometry and progressive geometrical simplicity and regularization. The initial complex structure becomes largely passive at this stage.

Table 1

The Geometrical, Mechanical, and Mathematical (GMM) mode of deformation evolves at different scales from (1) to (2) to (3) with progressive slip

Dominant GMM Mode ⇒	Granular	Fractal	Continuum-Euclidean
SCALE ⇓			
Fracture Surface (< 1 cm)		(1)	(2)
Gouge Zone (< 10 m)	(1)	(2)	(3)
Fault System (> 1 km)		(1)	(2)
Fault Network "Plate Boundary" (> 100 km)	(1)	(2)	(3)

become more planar in the shear direction while becoming increasingly separated by a layer of crushed rock as discussed above. The same regularization of slip zone geometry can also be observed on a fault system at scales greater than one kilometer where jogs and bends in the fault zone are smoothed with increasing displacement. Finally, at plate boundary scales on the order of 100 km or more, a regional network of faults accommodates deformation. Whether such networks originate as shear localizations in a granular matrix of crustal blocks or by the nucleation, growth and interaction of individual fractures is not known. However, active seismicity does not appear to populate the entire network, but is rather limited to a lower dimension subset. Again, the implication is that the active elements of the network are evolving toward a simpler more Euclidean structure.

The results suggest that long-term accommodation of crustal deformation may be described to first order by the continuum-Euclidean framework. Episodes of hardening and distributed deformation may be triggered by geometrical incompatibility associated with bends, jogs, and growth of individual faults, or the finite strain rotation of Euclidean structures to unfavorable orientations. However, these are transient perturbations on an irreversible trend toward overall progressive weakening and geometrical simplicity. The ubiquitous observation of complex seismicity patterns is often interpreted as evidence for a correspondingly complex structure. However, modeling studies have found that strong geometrical and material heterogeneity on an ensemble of planar faults can produce complex spatio-temporal patterns of seismicity including fractal distributions, spatio-temporal clustering, repeating earthquakes, linear streaks of microearthquakes, and other observed features.

In many situations deformation is accommodated by (or partitioned among) several sets of active faults with different orientations. Prominent examples include transition zones between different tectonic domains (e.g., the western North Anatolian fault near the sea of Marmara and the southern San Andreas fault near the Salton sea), areas near large bends of faults (e.g., the Los Angeles basin), and places where different sets of faults are juxtaposed as a result of slip on other faults. These examples and others exhibit high apparent complexity manifested by a diversity of fault types and earthquake focal mechanisms. Such complexity, however, can still be modeled by a collection of Euclidean surfaces and zones in a continuum matrix, as is done for example in routine seismological derivation of fault plane solutions. The regularization processes summarized by Figure 9 and Table 1 should be understood in such situations as operating separately on each active set.

We note that while our focus here is on mechanics, structural properties of faults are important to a variety of other issues including fluid flow in the crust and propagation of seismic waves. In these applications the relative importance of various structures is essentially the reverse of that for the mechanics discussed above. The crushed gouge zone and shattered wall rock, which are relict structures from the mechanical point of view, are the controlling structures for permeability and seismic velocity. The primary shear surfaces, which appear to dominate the mechanics of all but the most immature faults, play little role in determining properties of seismic waves. However, the damage zones around the primary shear localizations can guide seismic waves, and the complex crack population in the bulk controls the velocity, attenuation and scattering of seismic waves in the crust. From the hydrological point of view, the gouge and shattered wall rock provide a conduit for fluid flow in the crust. However, the primary shear localization and core ultracataclasite layer, which are the main carriers of slip, create a barrier to flow normal to the fault zone.

The main result of our review of current information on the geometrical, mechanical, and mathematical properties of fault zones is that strain weakening, geometrical simplicity, and continuum-based description provide a long-term attractor for structural evolution at all scales. This gives an organizing principle that may be used to integrate the diverse and often conflicting reports on fault zone properties. Since fault zones at seismogenic depths are not accessible for direct observation, the available information on their character is still coarse and sparse. A better understanding of the GMM character of fault zones will require additional geological and geophysical studies employing high-resolution high-penetration techniques.

Acknowledgments

We thank Fred Chester and Yan Kagan for discussions. The manuscript benefited from comments by Geoff King, Chris Marone, Steve Martel, and Judith Sheridan. The studies were supported by the National Science Foundation (grants EAR-9725358 (YBZ), EAR-9903049 (YBZ) and EAR-9902901 (CGS)) and the Southern California Earthquake Center (based on NSF Cooperative Agreement EAR-8920136 and USGS Cooperative Agreement 14-08-0001-A0899).

REFERENCES

ADAMS, G. G. (1995), *Self-excited Oscillations of Two Elastic Half Spaces Sliding with Constant Coefficient of Friction,* J. Appl. Mech. *62,* 867–872.

AKI, K. and LEE, W. H. K. (1976), *Determination of Three-dimensional Velocity Anomalies under a Seismic Array Using First P Arrival Times from Local Earthquakes. 1. A Homogenous Initial Model,* J. Geophys. Res. *81,* 4381–4399.

AKI, K. and RICHARDS, P. G. *Quantitative Seismology: Theory and Methods.* (W. H. Freeman, San Francisco, California (1980)).

AN, L-J. and SAMMIS, C. G. (1994), *Particle Size Distribution of Cataclastic Fault Materials from Southern California,* Pure Appl. Geophys. *143,* 203–227.

AN, L-J. and SAMMIS, C. G. (1996), *A Cellular Automaton for the Development of Crustal Shear Zones,* Tectonophysics *253,* 247–270.

ANDREWS, D. J. (1980), *A Stochastic Fault Model 1. Static Case,* J. Geophys. Res. *85,* 3867–3877.

ANDREWS, D. J. (1981), *A Stochastic Fault Model 2. Time-dependent Case,* J. Geophys. Res. *86,* 10,821–10,834.

ANDREWS, D. J. (1989), *Mechanics of Fault Junctions,* J. Geophys. Res. *94,* 9389–9397.

ANDREWS, D. J. and BEN-ZION, Y. (1997), *Wrinkle-like Slip Pulse on a Fault Between Different Materials,* J. Geophys. Res. *102,* 553–571.

ARBOLEYA, M. L. and ENGELDER, T. (1995), *Concentrated Slip Zones with Subsidiary Shears: Their Development on Three Scales in the Cerro Brass Fault Zone, Appalachian Valley and Ridge,* J. Structural Geol. *17,* 519–532.

AVILES, C. A., SCHOLZ, C. H., and BOATWRIGHT, J. (1987), *Fractal Analysis Applied to Characteristic Segments of the San Andreas Fault,* J. Geophys. Res. 92, 331–344.

AYDIN, A. and JOHNSON, A. M. (1978), *Development of Faults as Zones of Deformation Bands and as Slip Surfaces in Sandstone,* Pure Appl. Geophys. *116,* 931–942.

AYDIN, A. and JOHNSON, A. M. (1983), *Analysis of Faulting in Porous Sandstones,* J. Structural Geol. 5, 19–31.

BEELER, N. M., LOCKNER, D. L., and HICKMAN, S. H. (2001), *A Simple Stick-slip and Creep-slip Model for Repeating Earthquakes and its Implication for Micro-earthquakes at Parkfield,* Bull. Seismol. Soc. Am., 91, 1797–1804.

BEN-ZION, Y. (1996), *Stress, Slip and Earthquakes in Models of Complex Single-fault Systems Incorporating Brittle and Creep Deformations,* J. Geophys. Res. *101,* 5677–5706.

BEN-ZION, Y. (1998), *Properties of Seismic Fault Zone Waves and their Utility for Imaging Low Velocity Structures,* J. Geophys. Res. *103,* 12,567—12,585.

BEN-ZION, Y. (2001), *Recent Results on Dynamic Rupture in Earthquake Fault Models,* J. Mech. Phys. Solids *49,* 2209–2244.

BEN-ZION, Y. and ANDREWS, D. J. (1998), *Properties and Implications of Dynamic Rupture Along a Material Interface,* Bull. Seismol., Soc. Am. *88,* 1085–1094.

BEN-ZION, Y., DAHMEN, K., LYAKHOVSKY, V., ERTAS, D., and AGNON, A. (1999), *Self-driven Mode Switching of Earthquake Activity on a Fault System,* Earth Planet. Sci. Lett. *172,* 11–21.

BEN-ZION, Y., KATZ, S., and LEARY, P. (1992), *Joint Inversion of Fault Zone Head Waves and Direct P Arrivals for Crustal Structure near Major Faults,* J. Geophys. Res. *97,* 1943–1951.

BEN-ZION, Y. and MALIN, P. (1991), *San Andreas Fault Zone Head Waves near Parkfield, California,* Science *251,* 1592–1594.

BEN-ZION, Y., OKAYA, D., PENG, Z., MICHAEL, A. J., SEEBER, L., ARMBRUSTER, J. G., OZER, N., BARIS, S., and AKTAR, M. (2000), *High Resolution Imaging of the Geometry and Seismic Properties of the Karadere-Duzce Branch of the North Anatolian Fault at Depth,* EOS Trans. Amer. Geophys. Union *81,* F1172.

BEN-ZION, Y. and RICE, J. R. (1993), *Earthquake Failure Sequences along a Cellular Fault Zone in a Three-dimensional Elastic Solid Containing Asperity and Nonasperity Regions,* J. Geophys. Res. *98,* 14,109–14,131.

BEN-ZION, Y. and RICE, J. R. (1995), *Slip Patterns and Earthquake Populations along Different Classes of Faults in Elastic Solids,* J. Geophys. Res. *100,* 12,959–12,983.

BIEGEL, R. L., SAMMIS, C. G., and DIETERICH, J. H. (1989), *The Frictional Properties of a Simulated Gouge Having a Fractal Particle Distribution*, J. Structural Geol. *11*, 827–846.

BOUR, O. and DAVY, P. (1999), *Clustering and Size Distributions of Fault Patterns: Theory and Measurements*, Geophys. Res. Lett. *26*, 2001–2004.

BROWN, S. R. (1995), *Simple Mathematical Models of Rough Fracture*, J. Geophys. Res. *100*, 5941–5952.

BRUHN, R. L., PARRY, W. T., YONKEE, W. A., and THOMPSON, T. (1994), *Fracturing and Hydrothermal Alteration in Normal Fault Zones*, Pure Appl. Geophys. *142*, 609–644.

BRUNE, J. N., BROWN, S., and JOHNSON, P. A. (1993), *Rupture Mechanism and Interface Separation in Foam Rubber Model of Earthquakes: A Possible Solution to the Heat Flow Paradox and the Paradox of Large Overthrusts*, Tectonophysics *218*, 59–67.

BURGMANN, R., ROSEN, P. A., and FIELDING, E. J. (2000), *Synthetic Aperature Radar Interferometry to Measure Earth's Surface Topography and its Deformation*, Ann. Rev. Earth and Plan. Sci. *28*,169–209.

CHESTER, F. M. and LOGAN, J. M. (1986), *Implications for Mechanical Properties of Brittle Faults from Observations of the Punchbowl Fault Zone, California*, Pure Appl. Geophys. *124*, 79–106.

CHESTER, F. M. and LOGAN, J. M. (1987), *Composite Planar Fabric of Gouge from the Punchbowl Fault, California*, J. Structural Geol. *9*, 621–634.

CHESTER, F. M., EVANS, J. P., and BIEGEL, R. L. (1993), *Internal Structure and Weakening Mechanisms of the San Andreas Fault*, J. Geophys. Res. *98*, 771–786.

CHESTER, F. M. and CHESTER, J. S. (1998), *Ultracataclasite Structure and Friction Processes of the Punchbowl Fault, San Andreas System, California*, Tectonophysics 295 199–221.

CHESTER, F. M. and KIRSCHNER, D. L. (2000), *Geochemical Investigation of Fluid Involvement in Exhumed Faults of the San Andreas System*, National Earthquake Hazards Reduction Program, Annual Project Summary, 41, U.S. Geological Survey.

DAHMEN, K., ERTAS, D., and BEN-ZION, Y. (1998), *Gutenberg Richter and Characteristic Earthquake Behavior in Simple Mean-Field Models of Heterogeneous Faults*, Phys. Rev. E *58*, 1494–1501.

DIETERICH, J. H., *A Model for the Nucleation of Earthquake Slip*. In *Earthquake Source Mechanics*. AGU Geophys. Mono. 37. (Washington, D.C., American Geophysical Union 1986) pp. 37–49.

DIETERICH, J. (1994), *A Constitutive Law for Rate of Earthquake Production and its Application to Earthquake Clustering*, J. Geophys. Res. *99*, 2601–2618.

DODGE, D., BEROZA, G. C., and ELLSWORTH, W. L. (1995), *Evolution of the 1992 Landers, California, Foreshock Sequence and its Implications for Earthquake Nucleation*, J. Geophys. Res. *100*, 9865–9880.

EBERHART-PHILLIPS, D. and MICHAEL, A. J. (1993), *Three-dimensional Velocity, Structure, Seismicity, and Fault Structure in the Parkfield Region, Central California*, J. Geophys. Res. *98*, 15,737–15,757.

EBERHART-PHILLIPS, D. and MICHAEL, A. J. (1998), *Seismotectonics of the Loma Prieta, California, Region Determined from Three-dimensional V_p, V_p/V_s, and Seismicity*, J. Geophys. Res. *103*, 21,099–21,120.

EBERHART-PHILLIPS, D., STANLEY, W. D., RODRIGUEZ, B. D., and LUTTER, W. J. (1995), *Surface Seismic and Electrical Methods to Detect Fluids Related to Faulting*, J. Geophys. Res. *97*, 12,919–12,936.

EVANS, J. P. and CHESTER, F. M. (1995), *Fluid-rock Interaction in Faults of the San Andreas System: Inferences from San Gabriel Fault Rock Geochemistry and Microstructures*, J. Geophys. Res. *100*, 13,007—13,020.

EVANS, J. P., SHIPTON, Z. K., PACHELL, M. A., LIM, S. J., and ROBESON, K. (2000), *The Structure and Composition of Exhumed Faults, and their Implication for Seismic Processes*, In Proc. of the 3rd Confer. on Tecto. problems of the San Andreas system, Stanford University.

FENG, R. and MCEVILLY, T. V. (1983), *Interpretation of Seismic Reflection Profiling Data for the Structure of the San Andreas Fault Zone*, Bull. Seismol. Soc. Amer. *73*, 1701–1720.

FISHER, D. S., DAHMEN, K., RAMANATHAN, S., and BEN-ZION, Y. (1997), *Statistics of Earthquakes in Simple Models of Heterogeneous Faults*, Phys. Rev. Lett. *78*, 4885–4888.

FREUND, R. (1974), *Kinematics of Transform and Transcurrent Faults*, Tectonophysics *21*, 93–134.

FUIS, G. S. and MOONEY, W. D. (1990), *Lithospheric Structure and Tectonics from Seismic-Refraction and other data*. In *The San Andreas Fault System, California* (ed. R. E. Wallace) U.S. Geol. Surv. Prof. Pap. *1515*, 207–238.

FUKAO, Y., HORI, S., and UKAWA, M. (1983), *A Seismological Constraint on the Depth of Basalt-Eclogite Transition in a Subducting Oceanic Crust*, Nature *303*, 413–415.

FUNG, Y. C., *A First Course in Continuum Mechanics* (2nd edition) (Prentice-Hall, Inc., New Jersey. 1977).

GALLAGHER, J. J., Jr. (1981), *Tectonics of China: Continental Style Cataclastic Flow, in Mechanical Behavior of Crustal Rocks*; The Handin Volume, Geophysical Monograph 24, pp. 259–273, American Geophysical Union, Washington, D.C.

GOT, J.-L., FRÉCHET, J., and KLEIN, F. W. (1994), *Deep Fault Plane Geometry Inferred from Multiplet Relative Relocation Beneath the South Flank of Kilauea*, J. Geophys. Res. *99*, 15,375–15,386.

HARRIS, R. A (1998), *Introduction to Special Section: Stress Triggers, Stress Shadows, and Implications for Seismic Hazard*, J. Geophys. Res. *103*, 24,347–24,358.

HARRIS, R. A. and DAY, S. M. (1993), *Dynamics of Fault Interactions: Parallel Strike-slip Faults*, J. Geophys. Res. *98*, 4461–4472.

HARRIS, R. A. and SIMPSON, R. W. (1996), *In the Shadow of 1857—the Effect of the Great Ft. Tejon Earthquake on Subsequent Earthquakes in Southern California*, Geophys. Res. Lett. *23*, 229–232.

HEATON, T. H. (1990), *Evidence for and Implications of Self-healing Pulses of Slip in Earthquake Rupture*, Phys. Earth and Plan. Interiors *64*, 1–20.

HERRERO, A. and BERNARD, P. (1994), *A Kinematic Self-similar Rupture Process for Earthquakes*, Bull. Seismol. Soc. Am. *84*, 1216–1228.

HIRATA, T. (1989), *Fractal Dimension of Fault Systems in Japan: Fractal Structure in Rock Fracture Geometry at Various Scales*, Pure Appl. Geophys. *131*, 157–170.

HOUGH, S. E., BEN-ZION, Y., and LEARY, P. (1994), *Fault-zone Waves Observed at the Southern Joshua Tree Earthquake Rupture Zone*, Bull. Seismol. Soc. Am., *84*, 761–767.

ITO, A. (1985), *High Resolution Relative Hypocenters of Similar Earthquakes by Cross-spectral Analysis Method*, J. Phys. Earth *33*, 279–294.

JAEGER, M., NAGEL, S. R., and BEHRINGER, R. P. (1996), *Granular Solids*, Liquids, and Gases, Rev. of Mod. Phys. *68*, 1259–1273.

KAGAN, Y. Y. (1981a), *Spatial Distribution of Earthquakes: The Tree-point Moment Function*, Geophys. J. R. Astron. Soc. *67*, 697–717.

KAGAN, Y. Y. (1981b), *Spatial Distribution of Earthquakes: The Four-point Moment Function*, Geophys. J. R. Astron. Soc. *67*, 719–733.

KAGAN, Y. Y. (1991), *Fractal Dimension of Brittle Fracture*, J. Nonlinear Sci. *1*, 1–16.

KAGAN, Y. Y. (1992), *Seismicity: Turbulence of Solids*, Nonlinear Sci. Today *2*, 1–13.

KAGAN, Y. Y. (1994), *Observational Evidence for Earthquakes as a Nonlinear Dynamic Process*, Physica D *77*, 160–192.

KAGAN, Y. Y. and KNOPOFF, L. (1980), *Spatial Distribution of Earthquakes: The Two-point Correlation Function*, Geophys. J. R. Astron. Soc. *62*, 303–320.

KEILIS-BOROK, V. I. and KOSSOBOKOV, V. G. (1990), *Premonitory Activation of Earthquake Flow: Algorithm M8*, Phys. Earth Planet. Inter. 61, 73–83.

KING, G. C. P. (1983), *The Accommodation of Large Strains in the Upper Lithosphere of the Earth and Other Solids by Self-similar Fault Systems: The Geometrical Origin of b-value*, Pure Appl. Geophys., *121*, 761–814.

KING, G. C. P. and COCCO, M. (2001), *Fault Interaction by Elastic Stress Changes: New Clues from Earthquake Sequences*, Adv. in Geophys. *44*, 1–38.

KUWAHARA, Y. and ITO, H. (2000), *Deep Structure of the Nojima Fault by Trapped Wave Analysis*, USGS, Open-file Report 00–129, 283–289.

LEARY, P. C. (1991), *Deep bore hole log evidence for fractal distribution of fractures in crystalline rock*, Geophys. J. Int., *107*, 615–627.

LEARY, P., LI, Y. G., and AKI, K. (1987), *Observations and Modeling of Fault Zone Fracture Anisotropy, I, P, SV, SH Travel Times*, Geophys. J. R. Astron. Soc. *91*, 461–484.

LEES, J. and MALIN, P. E. (1990), *Tomographic Images of P-wave Velocity Variation at Parkfield, California*, J. Geophys. Res. *95*, 21,793–21,804.

LI, Y. G., AKI, K., ADAMS, D., HASEMI, A., and LEE, W. H. K. (1994), *Seismic Guided Waves Trapped in the Fault Zone of the Landers, California, Earthquake of 1992*, J. Geophys. Res. *99*, 11705–11722.

LI, Y. G., AKI, K., VIDALE, J. E., and ALVAREZ, M. G. (1998), *A Delineation of the Nojima Fault Ruptured in the M7.2 Kobe, Japan, Earthquake of 1995 Using Fault-zone Trapped Waves*, J. Geophys. Res. *103*, 7247–7263.

LI, Y. G., ELLSWORTH, W. L., THURBER, C. H., MALIN, P. E., and AKI, K. (1997), *Fault-zone Guided Waves from Explosions in the San Andreas Fault at Parkfield and Cienega Valley*, Bull. Seismol. Soc. Am. *87*, 210–221.

LI, Y. G., LEARY, P., AKI, K., and MALIN, P. (1990), *Seismic Trapped Modes in the Oroville and San Andreas Fault Zones*, Science *249*, 763–766.

LOCKNER, D. (1998), *A Generalized Law for Brittle Deformation of Westerly Granite*, J. Geophys. Res. *103*, 5107–5123.

LOCKNER, D. A., BYERLEE, J. D., KUKSENKO, V., PONOMAREV, A., and SIDRIN, A., *Observations of Quasistatic Fault Growth from Acoustic Emissions*. In *Fault Mechanics and Transport Properties of Rocks* (eds. B. Evans and T.-f. Wong) pp. 3–31 (Academic, San Diego, Calif. 1992).

LUYENDYK, B. P. (1991), *A Model of Neogene Crustal Rotations, Transtension, and Transpression in Southern California*, Geol. Soc. Am. Bull. *103*, 1528–1536.

LYAKHOVSKY, V., BEN-ZION, Y., and AGNON, A. (1997), *Distributed Damage, Faulting, and Friction*, J. Geophys. Res. *102*, 27,635—27,649.

LYAKHOVSKY, V., BEN-ZION, Y., and AGNON, A. (2001), *Earthquake Cycle Faults, and Seismicity Patterns in Rheologically Layered Lithosphere*, J. Geophys. Res. *106*, 4103–4120.

MACKIE, R. L., LIVELYBROOKS, D. W., MADDEN, T. R., and LARSEN, J. C. (1997), *A Magnetotelluric Investigation of the San Andreas Fault at Carrizo Plain California*, Geophys. Res. Lett. *24*, 1847–1850.

MANDELBROT, B. B. (1983), *The Fractal Geometry of Nature* (3rd edition) (W. H. Freeman and Company, New York).

MARCO, S., STEIN, M., AGNON, A., and RON, H. (1996), *Long-term Earthquake Clustering: A 50,000 Year Paleoseismic Record in the Dead Sea Graben*, J. Geophys. Res. *101*, 6179–6192.

MARONE, C. and KILGORE, B. (1993), *Scaling of the Critical Slip Distance for Seismic Faulting with Shear Strain in Fault Zones*, Nature, *362*, 618–21.

MARONE, C. and SCHOLZ, C.H. (1989), *Particle-size Distribution and Microstructures within Simulated Fault Gouge*, J. Struct. Geology *11*, 799–814.

MARTEL, S. J. (1990), *Formation of Compound Strike-slip Fault Zones, Mount Abbot Quadrangle, California*, J. Struct. Geol. *12*, 869–881.

MARTEL, S. J., POLLARD, D. D., and SEGALL, P. (1988), *Development of Simple Fault Zones in Granitic Rock, Mount Abbot Quadrangle, Sierra Nevada, California*, Geol. Soc. of Am. Bull. *100*, 1451–1465.

McGUIRE, J. J., ZHAO, L., and JORDAN, T. H. (2000), *Predominance of Unilateral Rupture for a Global Distribution of Large Earthquakes*, EOS Trans. Am. Geophs. Union *81*, F1228.

MICHAEL, A. J. and BEN-ZION, Y. (2002), *Determination of Fault Zone Structure from Seismic Guided Waves by Genetic Inversion Algorithm and 2-D Analytical Solution: Application to the Parkfield Segment of the San Andreas Fault*, ms. in preparation.

MICHAEL, A. J. and EBERHART-PHILLIPS, D. (1991), *Relations Among Fault Behavior, Subsurface Geology, and Three-dimensional Velocity Models*, Science *253*, 651–654.

MICHELINI, A. and McEVILLY, T. V. (1991), *Seismological Studies at Parkfield, I, Simultaneous Inversion for Velocity Structure and Hypocenters Using Cubic B-spline Parameterization*, Bull. Seismol. Soc. Am. *81*, 524–552.

MOONEY, W. D. and BROCHER, T. M. (1987), *Coincident Seismic Reflection/Refraction Studies of Continental Lithosphere: A Global Review*, Rev. Geophys. *25*, 723–742.

MOONEY, W. D. (1989), *Seismic Methods for Determining Earthquake Source Parameters and Lithospheric Structure*, Mem. Geol. Soc. Am. *172*, 11–34.

NEAL, L. A., CHESTER, J. S., CHESTER, F. M., and Wintsch, R. P. (2000), *Internal Structure of the Kern Canyon Fault, California: A Deeply Exhumed Strike-slip Fault*, EOS Trans. Am. Geophys. Union *81*, F1145.

NADEAU, R., ANTOLIK, M., JOHNSON, P. A., FOXALL, W., and McEVILLY, T. V. (1994), *Seismological Studies at Parkfield III: Microearthquake Clusters in the Study of Fault-zone Dynamics*, Bull. Seismol. Soc. Am. *84*, 1247–263.

NADEAU, R. M. and JOHNSON, L. R. (1998), *Seismological Studies at Parkfield VI: Moment Release Rates and Estimates of Source Parameters for Small Repeating Earthquakes*, Bull. Seismol. Soc. Am. *88*, 790–814.

NADEAU, R. M. and McEVILLY, T. V. (1997), *Seismological Studies at Parkfield V: Characteristic Microearthquake Sequences as Fault-zone Drilling Targets*, Bull. Seismol. Soc. Am. *87*, 1463–1472.

NALBANT, S., HUBERT, A., and KING, G. C. P. (1998), *Stress Coupling in North West Turkey and the North Aegean*, J. Geophys. Res. *103*, B10, 24,469–24,486.

NISHIGAMI, K., ANDO, M., and TADOKORO, K. (2001), *Seismic Observations in the DPRI 1800 m Borehole Drilled into the Nojima Fault Zone, Southwest Japan, Island Arc, 10*, 288–295.

NUR, A., RON, H., and SCOTTI, O. (1989), *Kinematics and Mechanics of Tectonic Block Rotations*, Geophys. Mono. *49*, 31–46.

OKUBO, P. G. and AKI, K. (1987), *Fractal Geometry in the San Andreas Fault System*, J. Geophys. Res. *92*, 345–355.

OUILLON, G., CASTAING, C., and SORNETTE, D. (1996), *Hierarchical Geometry of Faulting*, J. Geophys. Res. *101*, 5477–5487.

PENG, Z., BEN-ZION, Y., and MICHAEL, A. J. (2000), *Inversion of Seismic Fault Zone Waves in the Rupture Zone of the 1992 Landers Earthquake for High Resolution Velocity Structure at Depth*, EOS Trans. Am. Geophys. Union *81*, F1146.

POWER, W. L. and TULLIS, T. E. (1991), *Euclidean and Fractal Models for the Description of Rock Surface Roughness*, J. Geophys. Res. *96*, 415–424.

POWER, W. L., TULLIS, T. E., and WEEKS, J. D. (1988), *Roughness and Wear During Brittle Faulting*, J. Geophys. Res. *93*, 15,268–15,278.

PRASHER, C., *Crushing and Grinding Process Handbook* (John Wiley and Sons Ltd., New York. 1987).

RICE, J. R. (1993), *Spatio-temporal Complicity of slip on a fault*, J. Geophys. Res., *98*, 9885–9907.

RICHARDS-DINGER, K. B. and SHEARER, P. M. (2000), Earthquake Locations in Southern California Obtained Using Source-specific Station Terms, J. Geophys. Res. *105*, 10,939–10,960.

ROBERTSON, M. C., SAMMIS, C. G., SAHIMI, M., and MARTIN, A. (1995), *The 3-D spatial Distribution of Earthquakes in Southern California with a Percolation Theory Interpretation*, J. Geophys. Res. *100*, 609–620.

ROCKWELL, T. K., LINDVALL, S., HERZBERG, M., MURBACH, D., DAWSON, T., and BERGER, G. (2000), *Paleoseismology of the Johnson Valley, Kickcapoo and Homestead Valley Faults of the Eastern California Shear Zone*, Bull. Seismol. Soc. Am. *90*, 1200–1236.

RUBIN, A. M. and GILLARD, D. (2000), *Aftershock Asymmetry/Rupture Directivity Among Central San Andreas Fault Microearthquakes*, J. Geophys. Res. *105*, 19,095–19,109.

RUBIN, A. M., GILLARD, D., and GOT, J.-L. (1999), *Streaks of Microearthquakes along Creeping Faults*, Nature *400*, 635–641.

SAMMIS, C. G., NADEAU, R. M., and JOHNSON, L. R. (1999), *How Strong is an asperity?*, J. Geophys. Res. *104*, 10,609–10,619.

SAMMIS, C. G., AN, L. and ERSHAGHI, I. (1992), *Determining the 3-D Fracture Structure of the Geysers Geothermal Reservoir*, Proc. 17th Workshop on Geothermal Reservoir Engineering, Stanford University, Stanford, CA, Jan. 29–31.

SAMMIS, C. G. and BIEGEL, R. (1989), *Fractals, Fault-gouge, and Friction*, Pure Appl. Geophys. *131*, 255–271.

SAMMIS, C. G. and STEACY, S. J. (1994), *The Micromechanics of Friction in a Granular Layer*, Pure Appl. Geophys. *142*, 777–794.

SAMMIS, C. G., KING, G. C. P. and BIEGEL, R. (1987), *The Kinematics of Gouge Deformation*, Pure Appl. Geophys. *125*, 777–812.

SAMMIS, C. G. and RICE, J. R. (2001), *Repeating Earthquakes as Low-stress-drop Events at a Border between Locked and Creeping Fault Patches*, Bull. Seismol. Soc. Am., *91*, 532–537.

SCHOLZ, C. H. (1991), *Earthquakes and faulting: Self-organized critical phenomena with a characteristic dimension*. In *Spontaneous Formation of Space-time Structures and Criticality* (eds. T. Riste and D. Sherrington) (Kluwer Acad., Norwell, Mass.) pp. 41–56.

SCHULZ, S. E. and EVANS, J. P. (2000), *Mesoscopic Structure of the Punchbowl Fault, Southern California and the Geologic and Geophysical Structure of Active Strike-slip Faults*, J. of Struct. Geol. *22*, 913–930.

SEGALL, P. and DAVIS, J. L. (1997), *GPS Applications for Geodynamics and Earthquake Studies*, Ann. Re. Earth and Planet. Sci. *25*, 301–336.

SEGALL, P. and POLLARD, D. D. (1983), *Nucleation and Growth of Strike-slip Faults in Granite*, J. Geophys. Res. *88*, 555–568.

SHAPIRO, N. M., CAMPILLO, M., SINGH, S. K., and PACHECO, J. (1998), *Seismic Channel Waves in the Accretionary Prism of the Middle America Trench*, Geophys. Res. Lett. *25*, 101–104.

SHEARER, P. (1997), *Improving Local Earthquake Locations Using the L1 Norm and Waveform Cross Correlation: Application to the Whittier Narrows, California, Aftershock Sequence*, J. Geophys. Res. *102*, 8269–8283.

SIBSON, R. H. (1986), *Brecciation Processes in Fault Zones: Inferences from Earthquake Rupturing*, Pure Appl. Geophys. *124*, 159–176.

SIBSON, R. H. (1999), *Thickness of the Seismogenic Slip Zone: Constraints from Field Geology*, EOS Trans. Amer. Geophys. Union *80*, F727.

SIMPSON, R. W. and REASENBERG, P. A. (1994), *Earthquake-induced Static Stress Changes on Central California Faults.* In *the Loma Prieta, California earthquake of October 17, 1989-tectonic processes and models*, (R. W. Simpson, ed.) U. S. Geological Survey Prof. Paper 1550-F.

SMITH, J. T. and BOOKER, J. R. (1991), *Rapid Inversion of Two- and Three-dimensional Magnetotelluric Data*, J. Geophys. Res. *96*, 3905–3922.

STEACY, S. J. and SAMMIS, C. G. (1991), *An Automaton for Fractal Patterns of Fragmentation*, Nature *353*, 250–252.

STEIN, R. S., BARKA, A. A., and DIETERICH, J. H. (1997), *Progressive Failure on the North Anatolian Fault since 1939 by Earthquake Stress Triggering*, Geophys. J. Int. *128*, 594–604.

STEIN, R. S., KING, G., and LIN, J. (1994), *Stress Triggering of the 1994 M-6.7 Northridge, California, Earthquake by its Predecessors*, Science *265*, 1432–1435.

STIERMAN, D. J. (1984), *Geophysical and Geological Evidence for Fracturing, Water Circulation, and Chemical Alteration in Granitic Rocks Adjacent to Major Strike-slip Faults*, J. Geophys. Res. *89*, 5849–4857.

STIRLING, M. W., WESNOUSKY, S. G., and SHIMAZAKI, K. (1996), *Fault Trace Complexity, Cumulative Slip, and the Shape of the Magnitude-frequency Distribution for Strike-slip Faults: a Global Survey*, Geophys. J. Int. *124*, 833–868.

TCHALENKO, J. S. (1970), *Similarities between Shear Zones of Different Magnitudes*, Bull. Geol. Soc. Am. *81*, 1625–1640.

UTSU T., OGATA, Y., and MATSU'URA, R. S. (1995), *The Centenary of the Omori Formula for a Decay Law of Aftershock Activity*, J. Phys. Earth *43*, 1–33.

UNSWORTH, M., EGBERT, G., and BOOKER, J. (1999), *High-resolution Electromagnetic Imaging of the San Andreas Fault in Central California*, J. Geophys. Res. *104*, 1131–1150.

WALD, D. J. (1992), *Rupture Characteristics of California Earthquakes*, Ph.D. Thesis, Caltech.

WALDHAUSER, F., ELLSWORTH, W. L., and COLE, A. (1999), *Slip-parallel Seismic Lineations Along the Northern Hayward Fault, California*, Geophys. Res. Lett. *26*, 3525–3528.

WANG, C. Y., RUI, F., ZHENGSHENG, Y., and XINGJUE, S. (1986), *Gravity Anomaly and Density Structure of the San Andreas Fault Zone*, Pure Appl. Geophys. *124*, 127–140.

WEERTMAN, J. (1980), *Unstable Slippage across a Fault that Separates Elastic Media of Different Elastic Constants*, J. Geophys. Res. *85*, 1455–1461.

WESNOUSKY, S. (1994), *The Gutenberg-Richter or Characteristic Earthquake Distribution, which is it?*, Bull. Seismol. Soc. Am. *84*, 1940–1959.

WESNOUSKY, S. (1988), *Seismological and Structural Evolution of Strike-slip Faults*, Nature *335*, 340–342.

WILLEMSE, E. J. M., PEACOCK, D. C. P., and AYDIN, A. (1997), *Nucleation and Growth of Strike-slip Faults in Limestones from Somerset, U. K.*, J. Struct. Geol. *19*, 1461–1477.

ZENG, Y., ANDERSON, J. G., and YU, G. (1994), *A Composite Source Model for Computing Realistic Synthetic Strong Ground Motion*, Geophys. Res. Lett. *21*, 725–728.

ZHU, L. and HELMBERGER, D. V. (1996), *Advancement in Source Estimation Techniques Using Broadband Regional Seismograms*, Bull. Seismol. Soc. Am. *86*, 1634–1641.

(Received, November 2, 2000, accepted April 6, 2001)

 To access this journal online:
http://www.birkhauser.ch

Pure appl. geophys. 160 (2003) 717–737
0033–4553/03/040717–21

| Pure and Applied Geophysics

Seismic Tomography of the Lithosphere with Body Waves

Clifford H. Thurber[1]

Abstract — A pair of papers in 1976 lead-authored by Kei Aki heralded the beginning of the field of seismic tomography of the lithosphere. The 1976 paper by Aki, Christoffersson, and Husebye introduced a simple and approximate yet elegant technique for using body-wave arrival times from teleseismic earthquakes to infer the three-dimensional (3-D) seismic velocity heterogeneities beneath a seismic array or network (teleseismic tomography). Similarly, a 1976 paper by Aki and Lee presented a method for inferring 3-D structure beneath a seismic network using body-wave arrival times from local earthquakes (local earthquake tomography). Following these landmark papers, many dozens of papers and numerous books have been published presenting exciting applications of and/or innovative improvements to the methods of teleseismic and local earthquake tomography, many by Aki's students.

This paper presents a brief review of these two types of tomography methods, discussing some of the underlying assumptions and limitations. Thereafter some of the significant methodological developments are traced over the past two and a half decades, and some of the applications of tomography that have reaped the benefits of these developments are highlighted. One focus is on the steady improvement in structural resolution and inference power brought about by the increased number and quality of seismic stations, and in particular the value of utilizing shear waves. The paper concludes by discussing exciting new scientific projects in which seismic tomography will play a major role — the San Andreas Fault Observatory at Depth (SAFOD) and USArray, the initial components of Earthscope.

Key words: Tomography, body waves, teleseismic, local earthquake.

Introduction

In 1974, Kei Aki visited NORSAR, and the availability of high-quality seismic data from a dense array combined with Aki's knowledge of inverse theory led to the "birth" of modern seismic tomography. Working together with A. Christoffersson and E. Husebye, the three developed what has become known as the ACH teleseismic tomography method. The method uses the arrival times of P waves at a seismic array or network to infer three-dimensional (3-D) velocity heterogeneities in the volume beneath the stations. The team published their analyses of data from 2 arrays and one regional network in 3 papers that appeared in 1976 and 1977 (AKI *et al.*, 1976, 1977; HUSEBYE *et al.*, 1976). As the story goes, the third paper in the series, which is the one most widely cited, was intended to be published first, but a sign error in the

[1]Department of Geology and Geophysics, University of Wisconsin-Madison, 1215 W. Dayton St., Madison WI 53706, U.S.A. E-mail: thurber@geology.wisc.edu

data delayed its publication (EVANS and ACHAUER, 1993). Then in 1975, Aki spent time at the U.S. Geological Survey in Menlo Park, California. Working together with Willie Lee, the pair extended the ACH tomography method to the case of local earthquakes. The local earthquake method utilizes P-wave arrival times from earthquakes beneath a seismic network to infer the 3-D velocity structure in the volume containing the earthquakes and stations. Their local earthquake tomography paper was published in 1976 (AKI and LEE, 1976). I will refer to their approach as the A&L method.

The author was fortunate to arrive at MIT in the fall of 1976. A number of graduate students, both of Aki's and of other faculty members, was either already or soon to be working on applications of seismic tomography (although we did not call it tomography at the time). Among this group were Bill Ellsworth, George Zandt, Steve Roecker, Steve Taylor and myself. Computers at MIT were slowly making the transition from punched cards to on-line disk storage and remote terminals, and color graphics were done with Exacto knives and colored acetate sheets. How times have changed!

The retirement of Kei Aki brings an opportunity to reflect on the brief history of seismic tomography using body waves, with a focus on work involving his students and their collaborators. The initial tomography methods made simplifying assump-tions that have become generally unnecessary due to advances in methodology and computer power. Local- and regional-scale applications were initially limited to places where seismic arrays or networks were in place. However, the creation of the Incorporated Research Institutions for Seismology (IRIS) portable seismic instru-ment program, the Program for Array Seismic Studies of the Continental Lithosphere (PASSCAL) (SMITH, 1986; FOWLER and PAVLIS, 1994), opened up virtually unlimited opportunities for carrying out tomography studies in regions of scientific interest. Now there may be the beginning of a new era of opportunity for seismic tomography studies in the U.S. with the impending establishment of the Earthscope program (HENYEY et al., 2000). With tomography components ranging from the fine-scale structure of the San Andreas fault in the SAFOD project to the broad-scale structure of the North American lithosphere in the USArray project, Earthscope promises to provide the field of seismic tomography with unprecedented data for imaging the internal structure of the earth.

Basic Tomography Methodology

Both teleseismic and local earthquake tomography make use of the arrival times of body waves (P and/or S) to infer the seismic velocity structure. In the case of teleseismic tomography, the earthquakes must be at relatively great distances from a localized cluster of seismic stations, so that the waves incident on the array can be treated as plane waves (Fig. 1a). With this assumption, it is only the structure immediately beneath the stations that contributes to arrival time perturbations (relative to a radially-symmetric model), and the source locations and origin times need not be precisely known. I note that this assumption has been questioned by MASSON and TRAMPERT (1997), who showed that travel times to the base of an ACH model through a representative global 3-D velocity model could not adequately be treated as a plane wave in some cases. However, this potential problem can be avoided using the approach of WIDIYANTORO and VAN DER HILST (1997) or BIJWAARD et al. (1998), by modeling coarse-scale global structure and fine-scale regional structure simultaneously. In contrast, for local earthquake tomography, the events and stations are in the same area (Fig. 1b). In this case, structure along the entire path from event to station contributes to arrival time perturbations, and the source location and origin time need to be treated formally as unknowns along with the structure model.

For teleseismic tomography, the equations are written to separate the contributions to arrival time from inside versus outside the model:

$$t = \bar{\tau} + \sum_{s=1}^{NL} \bar{t}_s \sum_{k=1}^{M} F_{sk}\left(1 + \frac{\Delta u_k}{u_s}\right) + \varepsilon \tag{1}$$

where t is the arrival time at the station, $\bar{\tau}$ is the arrival time at the base of the model (calculated from a standard earth model), $\bar{t}_s$ is the travel time through layer s in the unperturbed medium, F_{sk} is 1 if the ray in layer s has most of its length in block k and is otherwise zero, u_s is the slowness in layer s, Δu_k is the slowness perturbation in block k, and ε represents model errors and measurement errors. AKI et al. (1977) expressed this in matrix notation for a set of observations as

$$\mathbf{t} = \bar{\tau} - \mathbf{G}\bar{\mathbf{m}} + m_0\mathbf{i} + \varepsilon \tag{2}$$

where $\mathbf{t}$ is the vector of arrival times, $\bar{\tau}$ is the vector of arrival times at the base of the model, $\mathbf{G}$ is the relatively sparse matrix containing the appropriate $\bar{t}_s$ values, $\bar{\mathbf{m}}$ is the vector of fractional slowness deviations from the layer average in the blocks, m_0 is the travel time through the unperturbed model, $\mathbf{i}$ is a vector consisting of ones, and the vector ε contains the error terms. By rewriting equation (2) in terms of arrival time residuals and subtracting the average of the travel time residual over all stations for each event from the left-hand side, Aki and Lee arrived at the relation

$$\mathbf{t}^* = \mathbf{G}^*\bar{\mathbf{m}} + \varepsilon^* \tag{3}$$

where * indicates the vector or matrix value minus the corresponding average. Thus the ACH method inverts relative arrival time residuals for relative perturbations in fractional slowness. This approach effectively removes error contributions from inaccuracies in the standard earth model and in the source location and origin time.

For local earthquake tomography, the derivation of the equations is simpler because no averaging is required to remove the source and outside-the-model path

a)

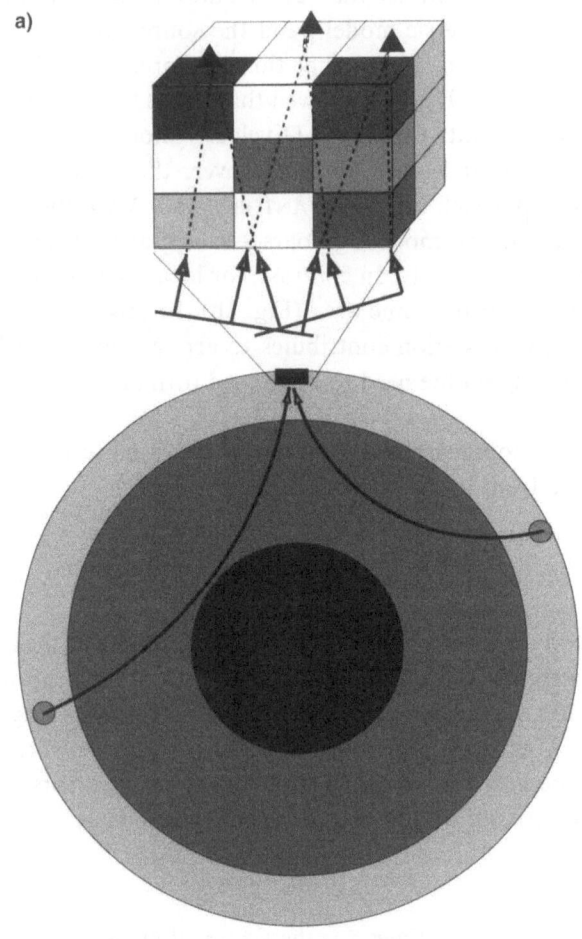

b)

LOCAL NETWORK

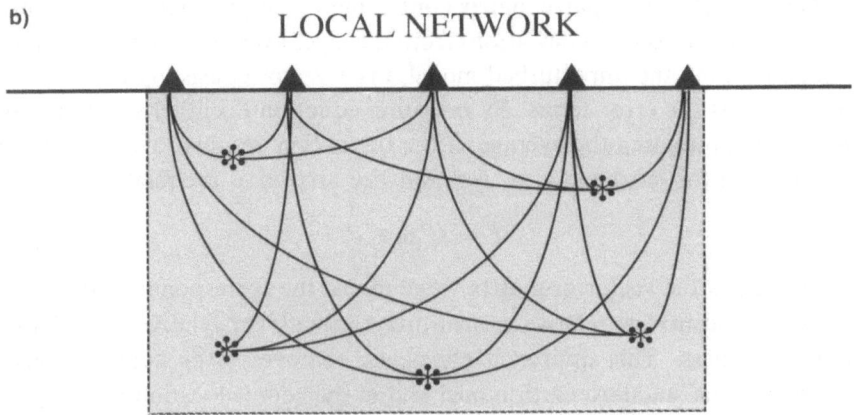

MODEL VOLUME

terms, although the equations themselves are more complicated. The equations relating residuals to model parameters can be written directly as

$$t_{ij}^{obs} - t_{ij}^{cal} = \sum_k T_{ij}^{(k)} \frac{\Delta u_k}{u_k} + \Delta T_j^o + \frac{\partial T_{ij}}{\partial x_j} \Delta x_j + \frac{\partial T_{ij}}{\partial y_j} \Delta y_j + \frac{\partial T_{ij}}{\partial z_j} \Delta z_j \qquad (4)$$

where the dependence of arrival time on source origin time and position is now an explicit part of the system to be solved, entering in the form of the perturbation to origin time (ΔT_j^o) and the partial derivatives of travel time with respect to source coordinates (e.g., $\partial T_{ij}/\partial x_j$) times the corresponding coordinate perturbations (e.g., Δx_j). For a starting model that is a homogeneous and isotropic medium, as assumed in the A&L method, the derivatives are proportional to the corresponding direction cosines of the vector connecting the source to the receiver. For the general case, the derivatives are proportional to the direction cosines of the ray direction at the source (THURBER, 1986).

The original ACH teleseismic and the A&L local earthquake methods both involved single-step linearized inversions for the model parameters. This approach allowed the use of linear inverse theory techniques for model estimation and model resolution and uncertainty analysis. AKI et al. (1977) compared models obtained using both a generalized inverse and a stochastic inverse. For the teleseismic case, the system of equations is always linearly dependent due to the trade off between average layer velocity and event origin times, consequently a simple least-squares solution was impossible. The generalized inverse approach uses the non-zero singular values of the $\mathbf{G}^*$ matrix (equation (3)) to compute a least-squares solution, while the stochastic inverse approach computes a damped least-squares solution using a damping value equal to the ratio of the data variance to the solution variance (AKI et al., 1977). AKI and LEE (1976) used just the stochastic inverse approach, because their matrix was nonsingular but contained very small eigenvalues that would amplify the effect of data errors on the model.

Advances in Seismic Tomography

There are a number of important aspects in which the methods and applications of body-wave seismic tomography have been improved over the last 25 years. Some of the advances I will discuss are improvements in the data, development of 3-D ray-tracing techniques, the use of iterative (nonlinear) inversion, the extension to active-

◀

Figure 1

(a) Geometry of the teleseismic tomography method. It is assumed that the P waves from teleseisms are incident on the base of the model approximately as plane waves and are recorded on the surface by a dense network. (b) Geometry of the local earthquake tomography method. The model volume contains both the earthquakes and the stations.

source applications, and multiple-method integration. Other improvements include the incorporation of interfaces (e.g., ZHAO *et al.*, 1992) and the use of anisotropic velocity models (e.g., HIRAHARA and ISHIKAWA, 1984).

Better data for tomography have arguably been the most important improvement, because without better data the other advances would have been of relatively limited use. Improvements in data have come in the form of additional instruments, the use of 3-component sensors, and more accurate arrival time picks. These factors allow for improved spatial resolution of structure, the ability to use *S* waves with reliability, and a sharpening of imaged structure with reduced uncertainty, respectively.

The development of the IRIS PASSCAL program in the U.S. and similar instrument pool developments in other countries has had a tremendously positive impact on the seismic tomography community. In the past, tomography studies were restricted to regions with existing seismic networks (with data access often limited to just those institutions operating the networks) or to a small number of institutions fortunate to have field instruments. Now, any U.S. scientist has the opportunity to propose and, if funded, carry out a seismic field project anywhere in the world involving the collection of seismic data using PASSCAL instruments.

One of the key features of the PASSCAL and other modern instrument pools is that 3-component instruments are the standard. As a result, *S* waves can be put to use to image structure with nearly the same capability as *P* waves, the main drawbacks being the presence of the *P*-coda noise interfering with precise *S*-wave arrival time picks and the possibility of shear wave splitting adding uncertainties to the data. The main value of *S* waves is the improved constraint on earthquake locations provided by the addition of *S* arrivals (GOMBERG *et al.*, 1990), which in turn improves the imaging of structure, and the increased constraints on model interpretation when 3-D models of both V_p and V_s (or V_p/V_s, equivalently Poisson's ratio) are available (EBERHART-PHILLIPS, 1990).

A study by ELLSWORTH (1977) demonstrated that a single-step inversion is adequate for teleseismic tomography when the size of velocity perturbations is relatively modest (less than 10%). The same is generally not true for local earthquake tomography. Beginning with THURBER (1981) and ROECKER (1982), a number of studies using synthetic models and data have shown that an iterative solution incorporating 3-D ray tracing is required for local earthquake tomography. The same is true for cross-borehole travel-time tomography, in which accounting for ray bending is widely recognized as being important. In the local earthquake case, it is even more important due to the coupling between structure and hypocenter locations (KISSLING, 1988; THURBER, 1992).

In the early stages of development, limited computer power made the use of exact ray-tracing techniques impractical. As a result, a variety of approximate ray-tracing techniques were developed (THURBER and ELLSWORTH, 1980; UM and THURBER, 1987; PROTHERO *et al.*, 1988). These methods achieved accuracies approaching that of reading error (a few tenths of a second or less) for moderate path lengths (up to 60 km), so they remain widely used even today. Over time, efficient exact ray-tracing (or more generally, travel time calculation) methods were developed. Examples include 3-D finite difference (VIDALE, 1988, 1990; PODVIN and LECOMTE, 1991), graph theory (MOSER, 1991), and perturbation methods (VIRIEUX et al., 1988; SNIEDER and SAMBRIDGE, 1992). See THURBER and KISSLING (2000) for a review.

It is important to recognize that even the "exact" methods have limited accuracy. HASLINGER and KISSLING (2001) and KISSLING et al. (2001) carried out extensive tests of both exact and approximate ray-tracing techniques, involving reciprocity tests on each method and comparing calculated travel times among methods. The basic conclusion of their analyses is that all methods have comparable accuracy variability for modest path lengths ($\sim$60 km), but the "exact" methods are superior for longer path lengths. They also noted that particular care must be taken regarding differences in the way the velocity model is parameterized in order to be able to make accurate comparisons among methods.

Seismic tomography was adapted to controlled-source problems in the 1980s. Non-linear solutions are generally required, as in the local earthquake case. Applications of controlled-source tomography include refraction, cross-borehole, and vertical seismic profiling (VSP). A review of these methods and their applications is beyond the scope of this paper.

Standard seismic tomography using first-arriving waves by itself provides only limited insight into the nature of the subsurface. By its nature, it provides a smoothed image of velocity structure, due to imperfect resolution, wavefront healing, etc. The use of information from secondary arrivals (reflections, conversions) provides one seismic approach for obtaining information about impedance structure, that is, identifying discontinuities. One example will be presented in the following section. Greater insight into the earth's structure can also be gained via multidisciplinary investigations. Two examples of methods that have proven useful in combination with seismic tomography are magnetotellurics (MT) and laboratory and down-hole measurements of seismic properties. EBERHART-PHILLIPS et al. (1990) compared velocity and MT models for a 2-D section crossing the San Andreas fault through the 1989 Loma Prieta main shock region (Fig. 2). High-V_p rocks between the San Andreas and Sargent faults were interpreted as mafic rocks due to their high resistivity. Low-V_p rocks in a wedge southwest of the San Andreas were interpreted as over-pressured marine sedimentary rocks due to their very high conductivity. LUTTER et al. (1999) compared a refraction-tomography image from the Los Angeles Region Seismic Experiment (LARSE) to well logs (from BROCHER et al., 1998) and laboratory measurements of seismic velocities of representative lithologies (from MCCAFFREE PELLERIN and CHRISTENSEN, 1998) (Fig. 3). The availability of well logs and laboratory data for representative lithologies permitted a detailed interpretation of the tomography image.

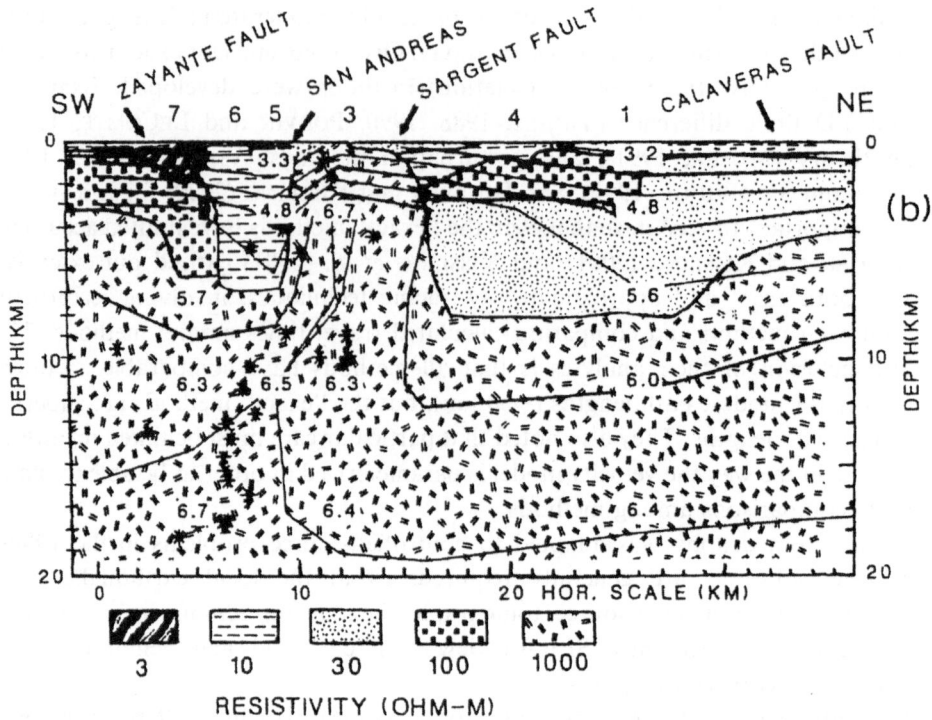

Figure 2

Comparison of a 2-D magnetotelluric model with the corresponding slice through the *P*-wave velocity
tomography model contours for the Loma Prieta area (modified from EBERHART-PHILLIPS *et al.*, 1990).
Note the excellent correspondence between the low-velocity basin and the low-resistivity units just SW of
the San Andreas fault (SAF) and the high-velocity "tongue" and the high resistivity unit between the SAF
and Sargent fault.

Applications of Seismic Tomography to Valles Caldera

Volcanic systems have been one of the key targets for seismic tomography studies
due to the desire to image the magma chambers at depth in the crust (IYER and
DAWSON, 1993). Aki's student Peter Roberts carried out a field project in Valles
Caldera in collaboration with Mike Fehler of LANL (also an Aki student). They
used six instruments staged in two deployments along a linear profile across the
caldera (ROBERTS *et al.*, 1991). Due to their limited teleseismic dataset, they used a
forward modeling approach, including *a priori* information on the shallow caldera
structure. Their final 2-D P-wave velocity model (Fig. 4) has a lens-shaped crustal
low-velocity zone (LVZ) with about a 35% velocity reduction compared to the
surrounding rock. The LVZ was interpreted to be the remnants of a cooling magma
chamber. The maximum width and minimum thickness of the LVZ are 17 km and 8
km, respectively.

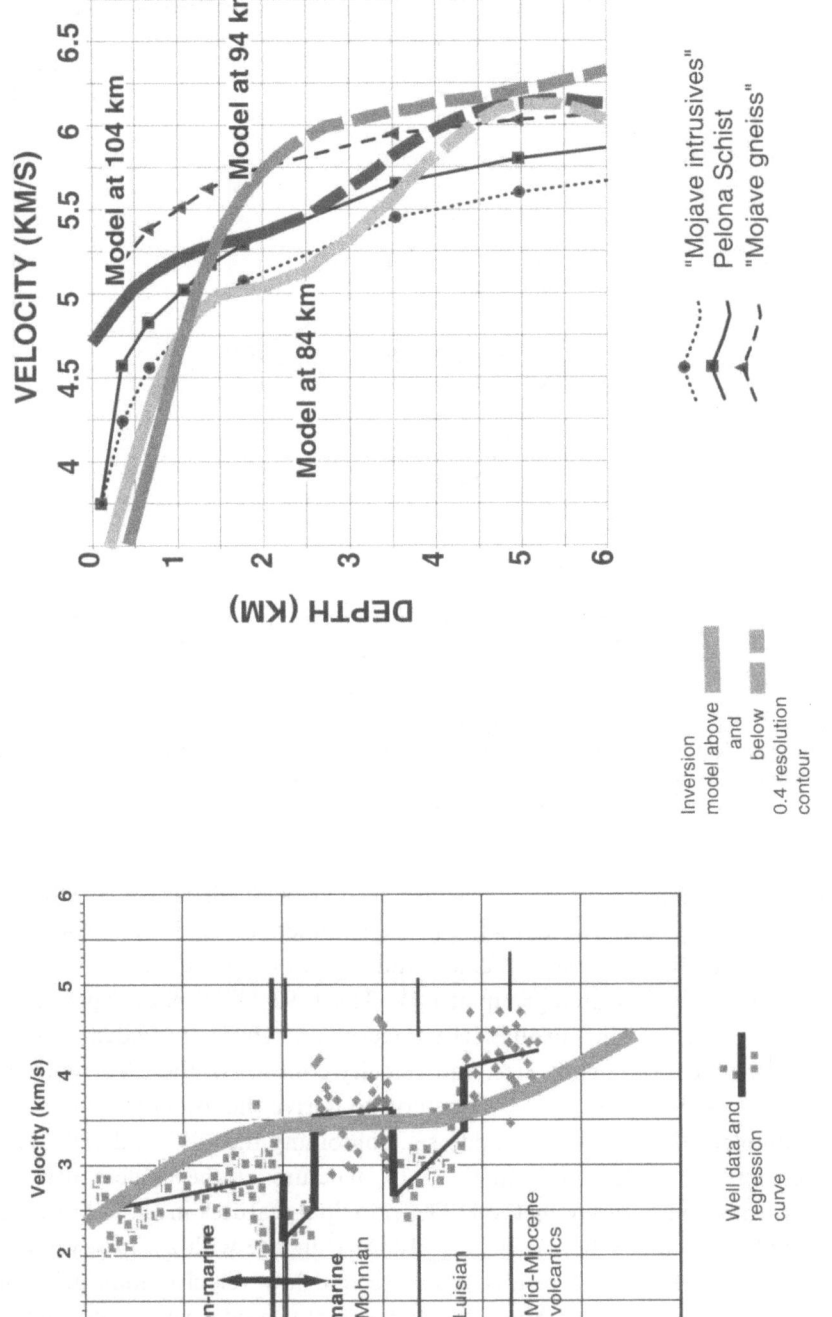

Figure 3

Comparison of velocity profiles through the 2-D LARSE Line 1 *P*-wave velocity model from refraction tomographic and the corresponding well log (left) and laboratory measurements (right) (from LUTTER *et al.*, 1999).

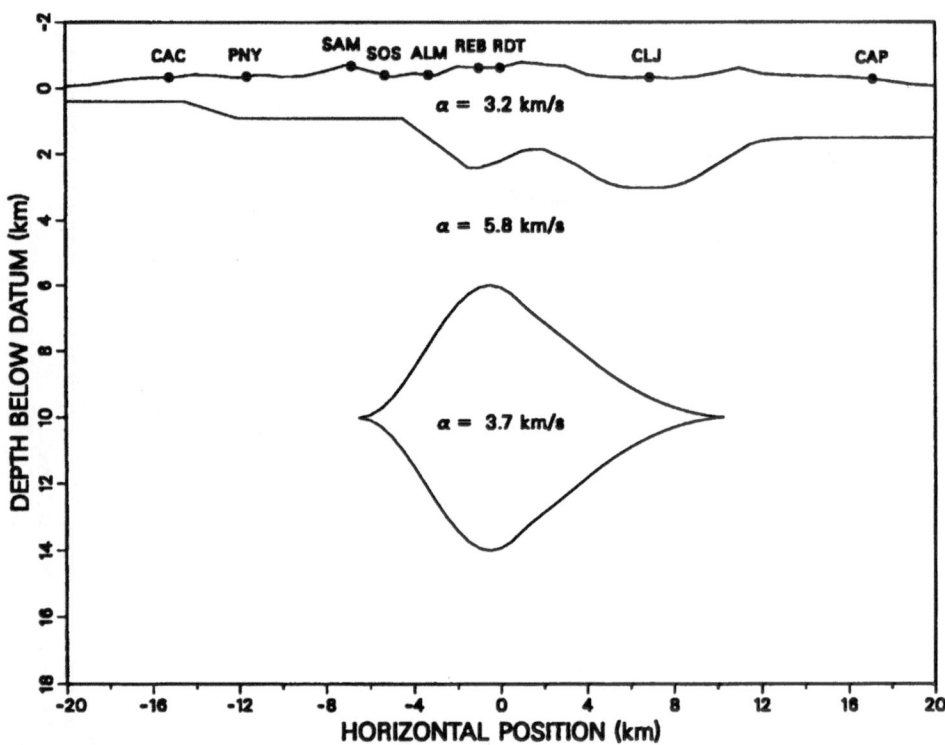

Figure 4
2-D P-wave velocity model for the structure beneath Valles caldera (from ROBERTS *et al.*, 1991). Note the lens-shaped low-velocity zone in the mid-crust ($Vp = 3.7$ km/s) and the low-velocity caldera fill ($Vp = 3.2$ km/s).

The results of ROBERTS *et al.* (1991) helped lead to a much larger PASSCAL teleseismic imaging project known as JTEX in the summers of 1993 and 1994. This project involved Peter Roberts, Mike Fehler, and several coworkers an LANL, as well as the author and his research group at UW. The 1993 JTEX pilot experiment, with 22 stations on 2 crossing profiles (LUTTER *et al.*, 1995), provided general confirmation of the location, depth, size, and velocity contrast of the LVZ reported by ROBERTS *et al.* (1991), using the ACH method to derive the velocity model.

With the complete JTEX teleseismic dataset, combining the 1993 and 1994 data (Fig. 5a), STECK *et al.* (1998) modeled the 3-D structure of the caldera using a nonlinear teleseismic tomography method. The principal features in the 3-D model are a mid-crustal low velocity body, roughly $(10 \text{ km})^3$ in size with a -35% velocity contract, consistent with the 2-D results, and a second low velocity body near the Moho (Fig. 5b). The authors combined the tomography model with geochemical information to conclude that Valles Caldera recently received a new pulse of magma from the upper mantle. Neither the 3-D geometry of the Valles magma chamber nor

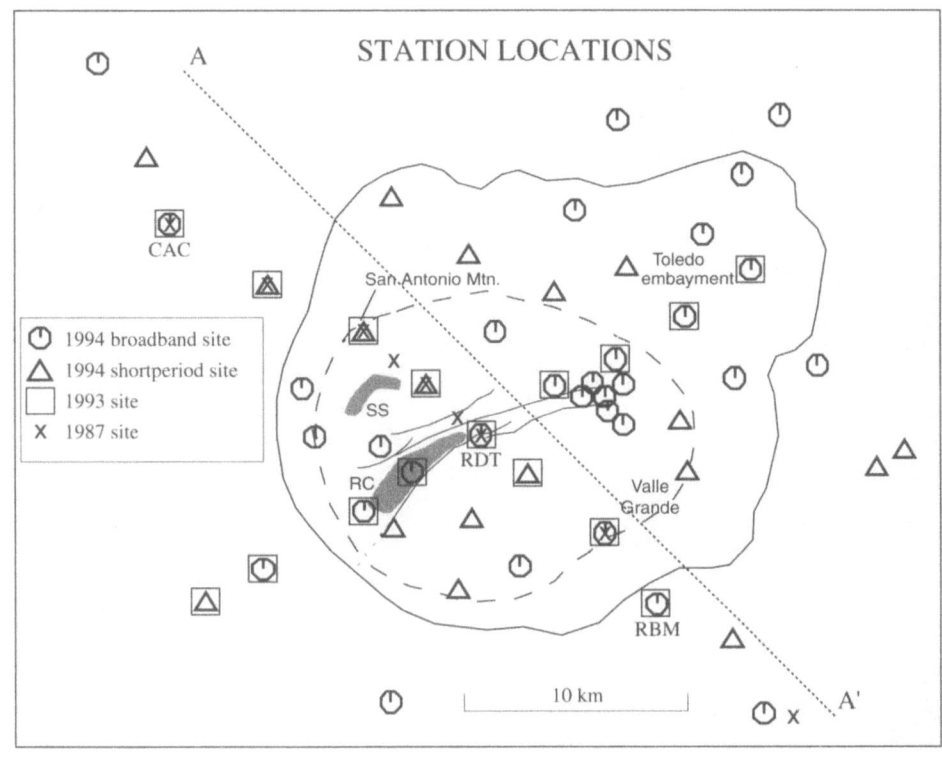

Figure 5a

Layout of the JTEX seismic arrays deployed in 1993 and 1994 (modified from STECK *et al.*, 1998). The solid line indicates the approximate rim of Valles caldera, the dashed line indicates its ring fracture system, and the gray areas indicate the two principle geothermal regions. Section line A-A' is shown in Figures 5b and c.

the presence of the second low velocity body near the Moho could have been resolved without a dense 2-D array of instruments.

Teleseismic first-arrival times are not useful for identifying structural discontinuities in the subsurface, so additional information from the seismograms is required to identify, for example, tops or bottoms of zones of magma. Possible approaches include the use of converted or scattered waves. APREA *et al.* (2002) employed a Kirchhoff migration approach to model the *P*-wave coda of teleseismic waveforms from the JTEX array. Their imaging approach used the surface reflection of the incident *P* wave that subsequently scattered off subsurface structures. This use of the surface-reflected phase for imaging allowed a separation of the target energy from scattered energy from the direct *P* coda. The imaging results (Fig. 5c) show a remarkable correspondence to the main features of the teleseismic tomography image. High amplitudes with opposite polarities are observed from the vicinity of the tops and bottoms of the low-velocity zones imaged by STECK *et al.* (1998).

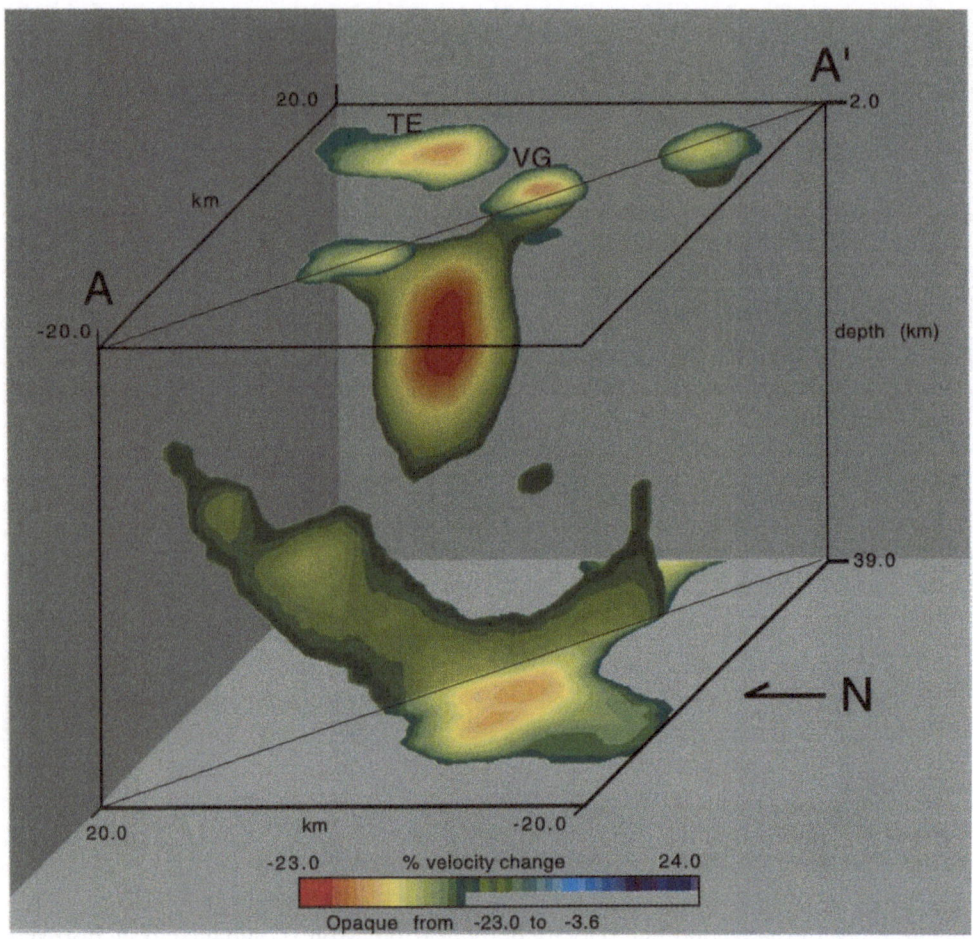

TE = Toledo Embayment, VG = Valle Grande

Figure 5b
Cut-away view of the 3-D model for Valles caldera, showing two low-velocity zones, one in the upper crust and another near the Moho (from STECK *et al.*, 1998).

Applications of Seismic Tomography to the San Andreas Fault

Fault zone studies are another important example for which improved data is vital to improving our understanding. One of the first 3-D local earthquake tomography studies was done by AKI and LEE (1976) along the Bear Valley segment of the San Andreas fault (SAF) in central California, part of the creeping section of the SAF. The data were from the USGS regional network plus a very short-term deployment of portable instruments (long enough to capture 32 local earthquakes),

3-D MIGRATED IMAGE

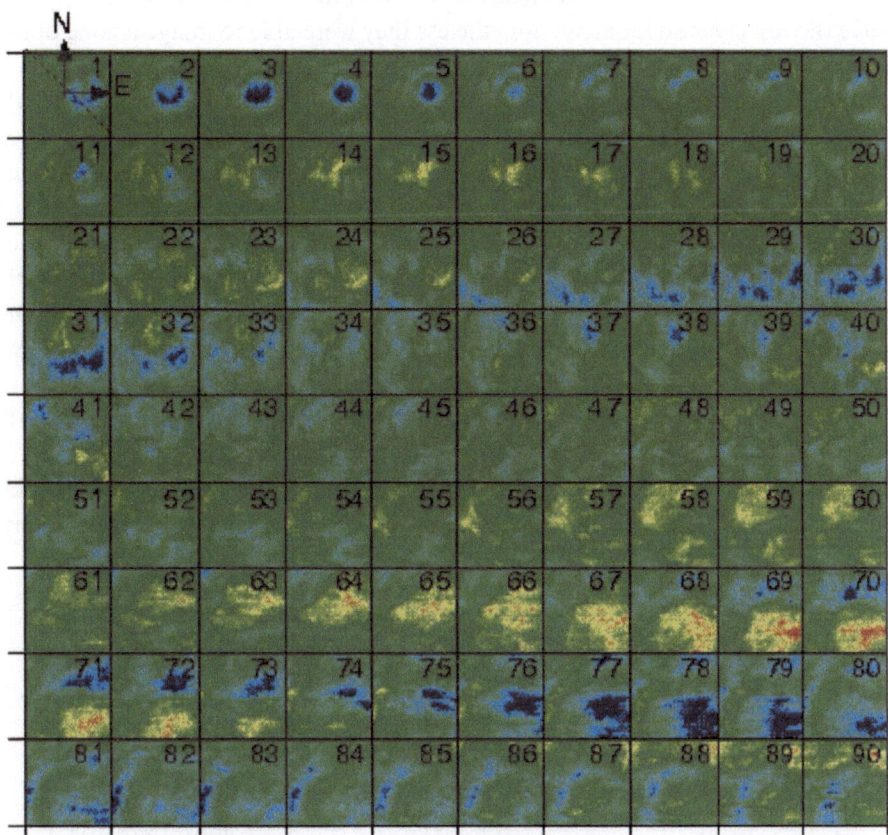

The 3-D MIGRATED IMAGE is presented as 90 horizontal-slices. Each square is a 30x30 km2 horizontal slice of the 3-D image oriented as indicated in the first square in the top-left. Each of the slices are dz=0.5 km apart. The first to the top-left is at 0.5 km below sea-level. The 3-D image is normalized to [-1,1], extremes (blues or reds in this case) may be representative of reflectors.

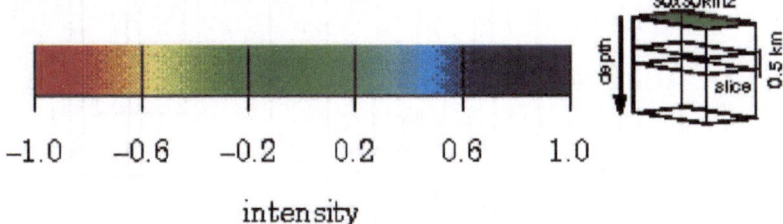

intensity

Figure 5c

Horizontal slices through the migrated image from the JTEX waveforms; blues and reds indicate opposite polarity (modified from APREA *et al.*, 2001). There are anomalies at depths roughly corresponding to the tops and bottoms of the low-velocity zones shown in Figure 5b. Section line A-A' is indicated with a dashed line in the upper left panel (although it extends about 7 km beyond each corner of the panel).

all of which were single-component instruments. The *P*-wave model that could be obtained was very coarse, with block size $3 \times 4 \times 5$ km. The authors were only able to resolve the top layer adequately, nonetheless they were able to image a zone of low velocity paralleling the SAF, with higher velocities on either side (Fig. 6a).

The most puzzling finding of the AKI and LEE (1976) study concerned the earthquake locations. Conventional wisdom held that the SAF is a simple, vertical strike-slip fault in this area, therefore the epicenters should lie along the fault trace. However, whether AKI and LEE (1976) started their inversion with hypocenters aligned on the fault or located several km southwest of the fault, the inversion result had the earthquakes several km southwest of the fault trace (Fig. 6b). Thus either the inversion method is flawed or the SAF is unexpectedly complex in Bear Valley. One possibility is that first-arriving fault zone head waves (BEN-ZION and MALIN, 1991) may be being modeled as direct waves, resulting in location bias. BEN-ZION *et al.* (1992) demonstrate how to use fault zone head waves in an inversion for laterally heterogeneous fault zone structure.

A recent study of a nearby segment of the SAF (also in the central creeping section) indicates how much more information can be obtained with a longer-term

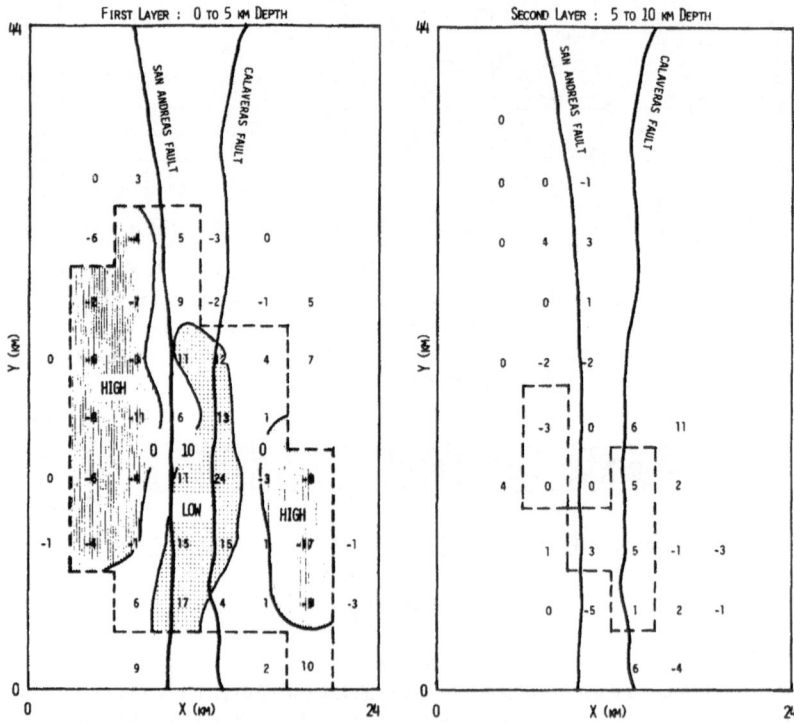

Figure 6a

3-D crustal model for the Bear Valley region of central California, showing a low-velocity fault zone sandwiched between faster basement on either side (modified from AKI and LEE 1976).

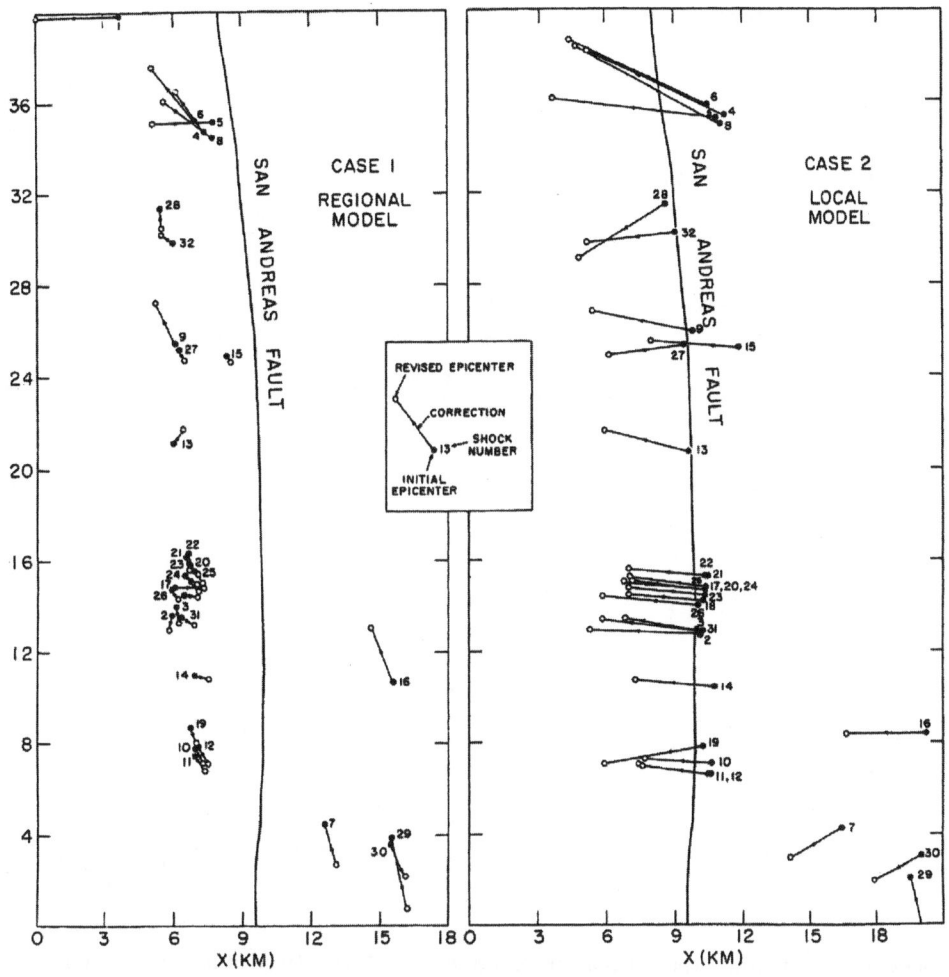

Figure 6b

Earthquake relocation results starting from the "regional" (left) and "local" (right) locations (modified
from AKI and LEE 1976). Starting and ending locations are the solid and open circles, respectively. Note
the general agreement in the ending locations for the two different sets of starting locations.

deployment of three-component instruments combined with a small active-source
experiment (Fig. 7a). The author's research group at UW along with Steve
Roecker's group at RPI and Bill Ellsworth (another Aki student) and co-workers
at the USGS-Menlo Park deployed 50 PASSCAL instruments for 7 months in a
20×15 km area near Hollister, California, recording hundreds of local
earthquakes (THURBER et al., 1997). Their best resolution was along a 2-D
section normal to the fault, using a grid size as fine as 1 km. Figure 7b shows
their models for V_p and V_p / V_s structure, with seismicity superimposed. They
were able to resolve a deep low V_p and high V_p / V_s anomaly in the fault zone,

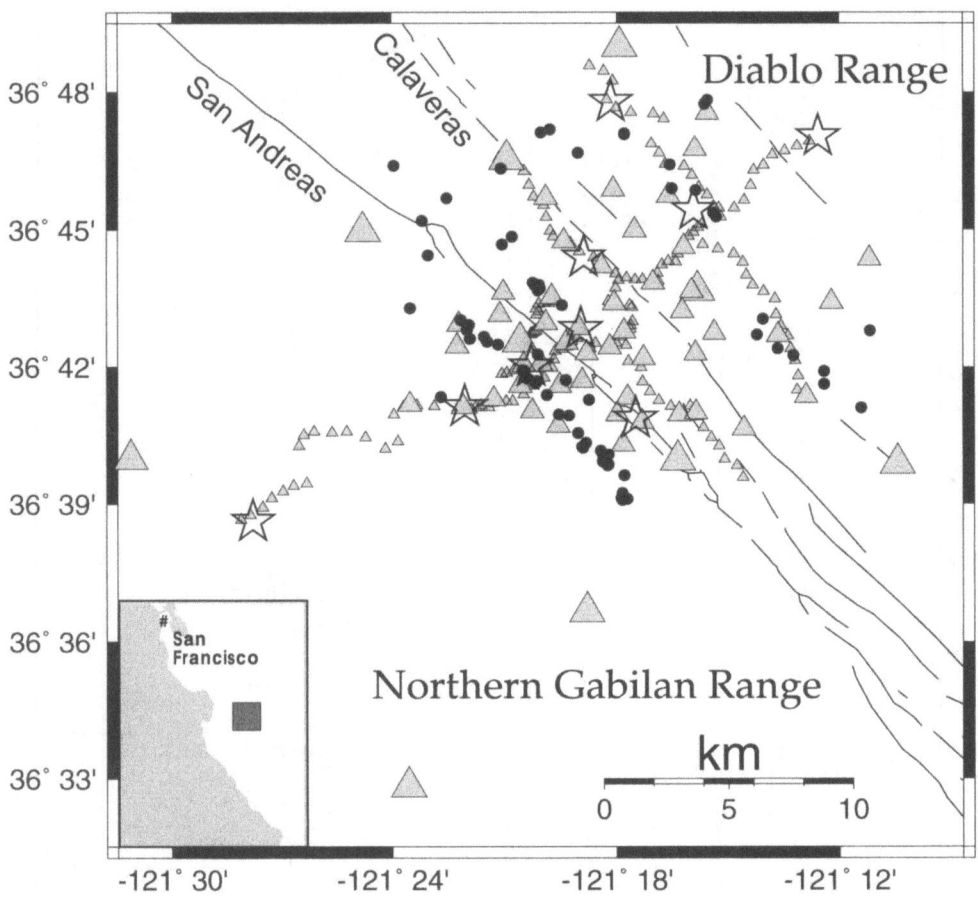

Figure 7a
Layout of the seismic array deployed along the San Andreas fault near Hollister in 1994–1995 (modified from THURBER *et al.*, 1997).

indicative of a highly fractured and probably fluid-rich region. Interestingly, the *P*-wave velocity contrasts across the fault are similar between this study and the original AKI and LEE (1976) study, although they were done along different sections of the SAF. Note how the seismicity is concentrated on the boundary between the high and low V_p regions in Fig. 7a, and how regions of high V_p / V_s are virtually aseismic in Fig. 7b. For this section of the SAF, it appears that the fault dips to the southwest at about 70°.

Other Applications

Although a thorough review of the applications of body-wave seismic tomography would require a book (for example, IYER and HIRAHARA (1993)), it is of value

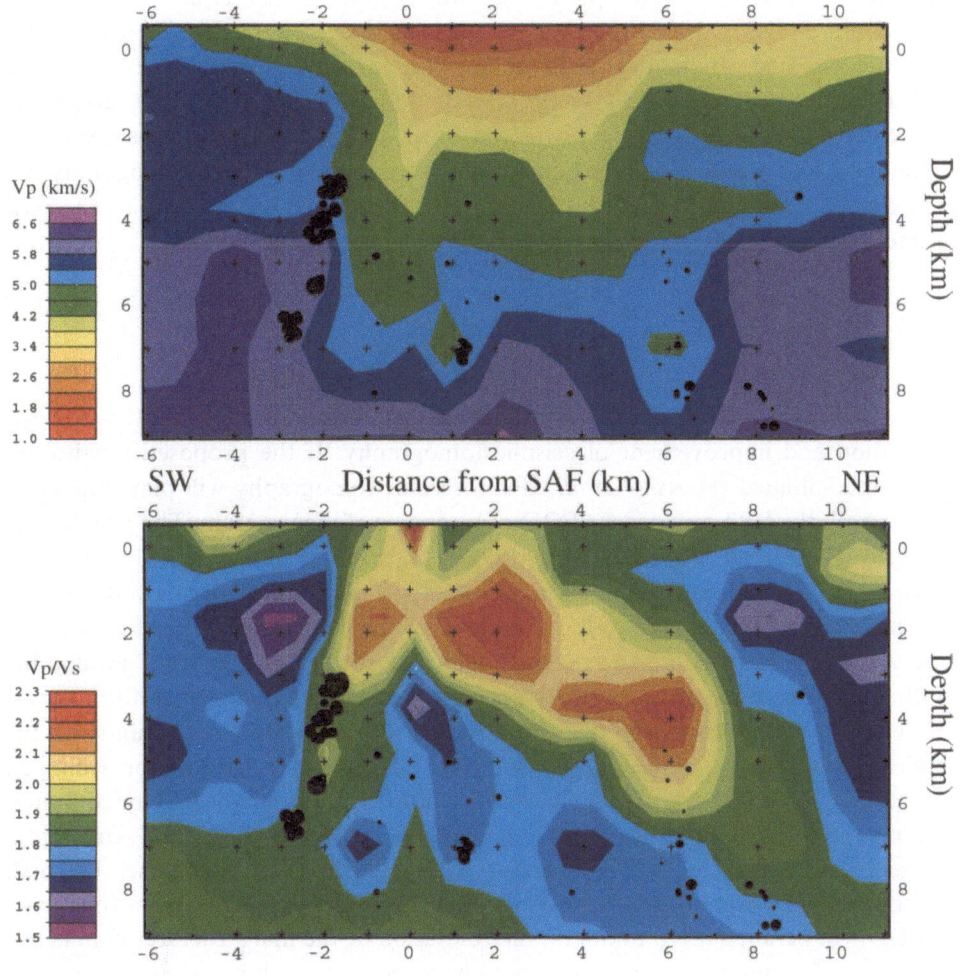

Figure 7b

2-D models of *Vp* (top) and *Vp/Vs* (bottom) for the San Andreas fault zone in the Cienega Valley region of central California. The approximate surface trace of the Calaveras fault is about 4 km NE of the SAF. Note the dipping low-velocity/high-*Vp/Vs* fault zone, and the concentrations of seismicity adjacent to high-*Vp* zones (modified from THURBER *et al.*, 1997).

to mention some of the highlights achieved over recent decades. Foremost among them is probably the successful imaging of the high-velocity anomalies of down-going (subducted) slabs, for example by HIRAHARA (1981). Substantial improvement in slab imaging has been obtained by the use of ocean-bottom seismometers (OBS) to augment arrays on land (e.g., HUSEN *et al.*, 2000). OBS controlled-source and local earthquake data has been used to provide remarkable images of mid-ocean ridge structure (e.g., TOOMEY *et al.*, 1990), helping to refine our concept of mid-ocean ridge magma chamber structure and magma propagation. Seismic tomography has also provided a view of the mantle plume beneath Iceland (WOLFE *et al.*, 1997; ALLEN, 2001). On a global scale, seismic tomography has imaged apparent fossil slabs and other deep-seated anomalies that likely play a fundamental role in global geodynamics (VAN DER HILST *et al.*, 1997). Seismic tomography has also been adapted successfully to surface waves (on local, regional, and global scales), free oscillations, and attenuation (for body and surface waves). Seismic tomography has been one of the most widely and successfully used geophysical tools of the past decade.

Future of Tomography

As we begin the new millennium, there are exciting opportunities for the application and improvement of seismic tomography. If the proposed Earthscope program is initiated (HENYEY *et al.*, 2000), seismic tomography will play important roles in both the USArray and SAFOD components of the program. The plan for the USArray (LEVANDER *et al.*, 1999; MELTZER *et al.*, 1999) is to deploy 400 broadband instruments at about 60–70 km spacing, initially in the southwestern United States. This array would cover about 1/8 of the area of the continental United States at a time. About a year after the array is fully deployed, it will begin to "roll" around the continental United States, eventually providing complete, spatially uniform coverage. The dataset resulting from this effort will provide an exciting opportunity for the application of tomography and other imaging and analysis techniques, providing an unprecedented view of the architecture of the lithosphere beneath the United States. The plan for SAFOD is to drill a deviated borehole to about 3.5 km depth, intersecting the SAF in the immediate vicinity of the rupture patches of small earthquakes. The combination of surface and downhole seismic observations from SAFOD will permit superb tomographic resolution of the fine structure of an active fault zone. The author and Steve Roecker deployed a 15-station PASSCAL array in July 2000, around the proposed drill site for improving earthquake locations, to start the preparations for drilling at Parkfield.

Acknowledgements

The author is grateful to a number of people and government agencies for advice and support, respectively, over the last nearly 25 years of my work on seismic tomography. The list of people starts with Kei Aki, and includes Bill Ellsworth, Steve Roecker, Donna Eberhart-Phillips, Edi Kissling, Gary Pavlis, Steve Taylor, Bob Nowack, Bill Prothero, John Evans, Rob Comer, George Zandt, Wim Spakman, Florian Haslinger, and Stefan Husen. I also thank Florian Haslinger for creating Figure 1a and for a careful reading of the manuscript. I appreciate the constructive reviews of Tom Parsons, Donna Eberhart-Phillips, and Associate Editor Yehuda Ben-Zion. The list of government agencies supporting my tomography research includes: the National Science Foundation (currently via awards EAR-9814192 and EAR-9814359), with special thanks to Leonard Johnson; the U.S. Geological Survey (currently via award 00HQGR0053); and the Defense Threat Reduction Agency (currently via contract DTRA01-01-C-0085 the content does not necessarily reflect the position or the policy of the U.S. Government, and no official endorsement should be inferred). Finally, I thank Jim Fowler and the staff of the IRIS PASSCAL instrument centers for their dedicated efforts in support of my PASSCAL field projects over the past decade. The facilities of the IRIS Consortium are supported by the National Science Foundation under Cooperative Agreement EAR-0004370.

REFERENCES

AKI, K. (1982), *Three dimensional Inhomogeneities in the Lithosphere and Asthenosphere: Evidence for Decoupling in the Lithosphere and Flow in the Asthenosphere*, Rev. Geophys. Space Phys. *20*, 161–170.

AKI, K., CHRISTOFFERSSON, A., and HUSEBYE, E.S. (1976), *Three-dimensional Seismic Structure of the Lithosphere under Montana LASA*, Bull. Seism. Soc. Am. *66*, 501–524.

AKI, K., CHRISTOFFERSSON, A., and HUSEBYE, E.S. (1977), *Determination of the 3-dimensional Seismic Structure of the Lithosphere*, J. Geophys. Res. *82*, 277–296.

AKI, K., and LEE, W.H.K. (1976), *Determination of Three-dimensional Anomalies under a Seismic Array Using First P Arrival Times from Local Earthquakes, 1. A Homogeneous Initial Model*, J. Geophys. Res. *81*, 4381–4399.

ALLEN, R.M. (2001), *The Mantle Plume beneath Iceland and its Interaction with the North-Atlantic Ridge: A Seismological Investigation*, Ph.D. thesis, Princeton University, 184 pp.

APREA, C.M., HILDEBRAND, S., FEHLER, M., STECK, L. BALDRIDGE, W.S., ROBERTS, P., THURBER, C.H., and LUTTER, W.J. (2002), *3-D Kirchhoff Migration: Imaging of the Jemez Volcanic Field Using Teleseismic Data*, J. Geophys. Res., in press.

BEN-ZION, Y., KATZ, S., and LEARY, P. (1992), *Joint Inversion of Fault Zone Head Waves and Direct P Arrivals for Crustal Structure near Major Faults*, J. Geophys. Res. *97*, 1943–1951.

BEN-ZION, Y., and MALIN, P. (1991), *San Andreas Fault Zone Head Waves near Parkfield, California*, Science *251*, 1592–1594.

BIJWAARD, H., SPAKMAN, W., and ENGDAHL, E.R. (1998), Closing the gap between regional and global travel time tomography, J. Geophys. Res. 103, 30,055–30,078.

BROCHER, T.M., RUBEL, A.L., WRIGHT, T.L., and OKAYA, D.A. (1998), *Compilation of 20 Sonic and Density Logs from 12 Oil Test Wells along LARSE Lines 1 and 2 in the Los Angeles Basin, California*, U.S. Geol. Surv. Open File Rep. *98-366*, 53 pp.

EBERHART-PHILLIPS, D. (1990), *Three-dimensional P and S Velocity Structure in the Coalinga Region, California*, J. Geophys. Res. *95*, 15,343–15,363.

EBERHART-PHILLIPS, D., LABSON, V.F., STANLEY, W.D., MICHAEL, A.J., and RODRIGUEZ, B.D. (1990), *Preliminary Velocity and Resistivity Models of the Loma Prieta Earthquake Region*, Geophys. Res. Lett. *17*, 1235–1238.

ELLSWORTH, W.L. (1977), *Three-dimensional Structure of the Crust and Mantle beneath the Island of Hawaii*, Ph.D. Thesis, M.I.T. 327 pp.

ENGDAHL, E.R. and LEE, W.H.K. (1976), *Relocation of Local Earthquakes by Seismic Ray Tracing*, J. Geophys. Res. *81*, 4400–4406.

EVANS, J.R. and ACHAUER, U. (1993), *Teleseismic velocity tomography using the ACH method: theory and application to continental–scale studies*. In Seismic Tomography (Iyer, H.M., and Hirahara, K., eds.) (Chapman and Hall, London, 1993), pp. 319–360.

FOWLER, J. and PAVLIS, G. (1994), *PASSCAL; A Facility for Portable Seismological Instrumentation*, EOS, Trans. Am. Geophys. Un. Suppl. *75*, 66.

GOMBERG, J.S., SHEDLOCK, K.M., and ROECKER, S.W. (1990), *The Effect of S-wave Arrival Times on the Accuracy of Hypocenter Estimation*, Bull. Seismol. Soc. Am. *80*, 1605–1628.

HASLINGER, F. and KISSLING, E. (2001), *Investigating Effects of 3-D Ray-tracing Methods in Local Earthquake Tomography*, Phys. Earth Plan. Int., in press.

HENYEY, T. and the EARTHSCOPE WORKING GROUP (2000), *Earthscope: A Look into our Continent*, Geotimes *45*, 5 and 40.

HIRAHARA, K. (1981), *Three-dimensional Seismic Structure beneath the Japan Islands and its Tectonic Implications*, J. Phys. Earth *28*, 221–241.

HIRAHARA, K. and ISHIKAWA, Y. (1984), *Travel-time Inversion for Three-dimensional P-wave Velocity Anisotropy*, J. Phys. Earth *32*, 197–218.

HUSEBYE, E.S., CHRISTOFFERSSON, A., AKI K. and POWELL, C. (1976), *Preliminary Results of the 3-dimensional Seismic Structure of the Lithosphere under the USGS Central California Seismic Array*, Geophys. J. Roy. Astron. Soc. *46*, 319–340.

HUSEN, S., KISSLING, E., and FLUEH, E.R. (2000), *Local Earthquake Tomography of Shallow Subduction in North Chile: A Combined Onshore and Offshore Study*, J. Geophys. Res. *105*, 28,183–28,198.

IYER, H.M. and DAWSON, P.B. (1993), *Imaging volcanoes using teleseismic tomography*, In *Seismic Tomography*, (Iyer, H.M., and Hirahara, K. eds.), (Chapman and Hall, London, 1993), pp. 466–492.

IYER, H.M. and HIRAHARA, K. (eds.), *Seismic Tomography* (Chapman and Hall, London, 1993), 842 pp.

KISSLING, E. (1988), *Geotomography with Local Earthquake data*, Rev. Geophys. *26*, 659–698.

KISSLING, E., HASLINGER, F., and HUSEN, S. (2001), *Model Parameterization in Seismic Tomography: A Choice of Consequence for the Solution Quality*, Phys. Earth Plan. Int., in press.

LEVANDER, A., HUMPHREYS, E., EKSTRÖM, G., MELTZER, A., and SHEARER, P. (1999), *Proposed Project Would Give Unprecedented Look under North America*, EOS, Trans. Am. Geophys. Un. *80*, 245, 250–251.

LUTTER, W.J., FUIS, G.S., THURBER, C.H., and MURPHY, J.R. (1999), *Tomographic Images of the Upper Crust from the Los Angeles Basin to the Mojave Desert: Results from the Los Angeles Region Seismic Experiment*, J. Geophys. Res. *104*, 25,543–25,565.

LUTTER, W.J., ROBERTS, P.M., THURBER, C.H., STECK, L.K., FEHLER, M.C., STAFFORD, D.G., BALDRIDGE, W.S., and ZEICHERT, T.A. (1995), *Teleseismic P–wave Image of Crust and Upper Mantle Structure beneath the Valles Caldera, New Mexico: Initial Results from the 1993 JTEX Passive Array*, Geophys. Res. Lett. *22*, 505–508.

MASSON, F. and TRAMPERT, J. (1997), *On ACH, or How Reliable is Regional Teleseismic Delay Time Tomography?*, Phys. Earth Planet. Int. *102*, 21–32.

MCCAFFREE, C.L. and CHRISTENSEN, N.I. (1998), *Interpretation of Crustal Seismic Velocities in the San Gabriel-Mojave Region, Southern California*, Tectonophysics *286*, 252–273.

MELTZER, A., RUDNICK, R., ZEITLER, P., LEVANDER, A., HUMPHREYS, G., KARLSTROM, K., EKSTRÖM, G., CARLSON, C., DIXON, T., GURNIS, M., SHEARER, P., and VAN DER HILST, R. (1999), *USArray initiative*, GSA Today *9*, 8–10.

MOSER, T.J. (1991), *Shortest Path Calculation of Seismic Rays*, Geophysics *56*, 59–67.

PAVLIS, G.L. and BOOKER, J.R. (1980), *The Mixed Discrete Continuous Inverse Problem: Application to the Simultaneous Determination of Earthquake Hypocenters and Velocity Structure*, J. Geophys. Res. *85*, 4801–10.

PODVIN, P. and LECOMTE, I. (1991), *Finite Difference Computation of Travel Times in Very Contrasted Velocity Models: A Massively Parallel Approach and its Associated Tools*, Geophys. J. Int. *105*, 271–284.

PROTHERO, W.A., TAYLOR, W.J., and EICKEMEYER, J.A. (1988), *A Fast, Two-point, Three-dimensional Ray-tracing Algorithm Using a Simple Step Search Method*, Bull. Seismol. Soc. Am. *78*, 1190–1198.

ROBERTS, P.M., AKI, K., and FEHLER, M.C. (1991), *A Low-velocity Zone in the Basement Beneath the Valles Caldera, New Mexico*, J. Geophys. Res. *96*, 21,583–21,596.

ROECKER, S.W. (1982), *Velocity Structure of the Pamir–Hindu Kush Region: Possible Evidence of Subducted Crust*, J. Geophys. Res. *87*, 945–959.

SMITH, S.W. (1986), *IRIS: A Program for the Next Decade*, EOS, Trans. Am. Geophys. Un. *67*, 213–219.

SNIEDER, R. and SAMBRIDGE, M. (1992), *Ray Perturbation Theory for Travel Times and Ray Paths in 3-D Heterogeneous Media*, Geophys. J. Int. *109*, 294–322.

SPENCER, C. and GUBBINS, D. (1980), *Travel-time Inversion for Simultaneous Earthquake Location and Velocity Structure Determination in Laterally Varying Media*, Geophys. J. R. Astron. Soc. *63*, 95–116.

STECK, L., THURBER, C., FEHLER, M., LUTTER, W., ROBERTS, P., BALDRIDGE, S., STAFFORD, D., and SESSIONS, R. (1998), *Crust and Upper Mantle P-wave Velocity Structure Beneath the Valles Caldera, New Mexico: Results from the JTEX Teleseismic Experiment*, J. Geophys. Res. *103*, 24,301–24,320.

THURBER, C.H. (1981), *Earth Structure and Earthquake Locations in the Coyote Lake Area, Central California*, Ph.D. Thesis, M.I.T. 331 pp.

THURBER, C.H. (1986), *Analysis Methods for Kinematic Data from Local Earthquakes*, Rev. Geophys. *24*, 793–805.

THURBER, C.H. (1992), *Hypocenter-velocity Structure Coupling in Local Earthquake Tomography*, Phys. Earth Planet. Int. *75*, 55–62.

THURBER, C.H. and ELLSWORTH, W.L. (1980), *Rapid Solution of Ray-tracing Problems in Heterogeneous Media*, Bull. Seismol. Soc. Am. *70*, 1137–48.

THURBER, C.H. and KISSLING, E., *Advances in travel-time calculations for three-dimensional structures*. In *Advances in Seismic Event Location*, (Thurber, C., and Rabinowitz, N., eds.), (Kluwer Academic Publishers, Dordrecht, Netherlands 2000) pp. 71–99.

THURBER, C., ROECKER, S., ELLSWORTH, W., CHEN, Y., LUTTER, W., and SESSIONS, R. (1997), *Two-dimensional Seismic Image of the San Andreas Fault in the Northern Gabilan Range, Central California: Evidence for Fluids in the Fault Zone*, Geophys. Res. Lett. *24*, 1591–1594.

TOOMEY, D.R., PURDY, G.M., SOLOMON, S.C., and WILCOCK, W.S.D. (1990), *The Three-dimensional Seismic Velocity Structure of the East Pacific Rise near Latitude 9° 30' N*, Nature *347*, 639–645.

UM, J. and THURBER, C.H. (1987), *A Fast Algorithm for Two-point Seismic Ray Tracing*, Bull. Seismol. Soc. Am. *77*, 972–986.

VAN DER HILST, R.D., WIDIYANTORO, S., and ENGDAHL, E.R. (1997), *Evidence for Deep Mantle Circulation from Global Tomography*, Nature *386*, 578–584.

VIDALE, J.E. (1988), *Finite-difference Travel-Time Calculation*, Bull. Seis. Soc. Am. *78*, 2062–2076.

VIDALE, J.E. (1990), *Finite-difference Calculation of Travel Times in Three Dimensions*, Geophysics *55*, 521–526.

VIRIEUX, J., FARRA, V., and MADARIAGA, R. (1988), *Ray Tracing for Earthquake Location in Laterally Heterogeneous Media*, J. Geophys. Res. *93*, 6585–6599.

WIDIYANTORO, S., and VAN DER HILST, R. (1997), *Mantle Structure beneath Indonesia Inferred from High-resolution Tomographic Imaging*, Geophys. J. Int. *130*, 167–182.

WOLFE, C.J., BJARNASON, I.T., VAN DECAR, J.C., and SOLOMON, S.C. (1997), *Seismic Structure of the Iceland Mantle Plume*, Nature, *385*, 245–247.

ZHAO, D., HASEGAWA, A., and HORIUCHI, S. (1992), *Tomographic Imaging of P- and S- wave Velocity Structure beneath Northeastern Japan*, J. Geophys. Res. *97*, 19,909–19,928.

(Received July 1, 2000, accepted January 31, 2001)

Pure appl. geophys. 160 (2003) 739–788
0033–4553/03/040739–50

❙ Pure and Applied Geophysics

Volcano Seismology

BERNARD CHOUET[1]

Abstract — A fundamental goal of volcano seismology is to understand active magmatic systems, to characterize the configuration of such systems, and to determine the extent and evolution of source regions of magmatic energy. Such understanding is critical to our assessment of eruptive behavior and its hazardous impacts. With the emergence of portable broadband seismic instrumentation, availability of digital networks with wide dynamic range, and development of new powerful analysis techniques, rapid progress is being made toward a synthesis of high-quality seismic data to develop a coherent model of eruption mechanics. Examples of recent advances are: (1) high-resolution tomography to image subsurface volcanic structures at scales of a few hundred meters; (2) use of small-aperture seismic antennas to map the spatio-temporal properties of long-period (LP) seismicity; (3) moment tensor inversions of very-long-period (VLP) data to derive the source geometry and mass-transport budget of magmatic fluids; (4) spectral analyses of LP events to determine the acoustic properties of magmatic and associated hydrothermal fluids; and (5) experimental modeling of the source dynamics of volcanic tremor. These promising advances provide new insights into the mechanical properties of volcanic fluids and subvolcanic mass-transport dynamics. As new seismic methods refine our understanding of seismic sources, and geochemical methods better constrain mass balance and magma behavior, we face new challenges in elucidating the physico-chemical processes that cause volcanic unrest and its seismic and gas-discharge manifestations. Much work remains to be done toward a synthesis of seismological, geochemical, and petrological observations into an integrated model of volcanic behavior. Future important goals must include: (1) interpreting the key types of magma movement, degassing and boiling events that produce characteristic seismic phenomena; (2) characterizing multiphase fluids in subvolcanic regimes and determining their physical and chemical properties; and (3) quantitatively understanding multiphase fluid flow behavior under dynamic volcanic conditions. To realize these goals, not only must we learn how to translate seismic observations into quantitative information about fluid dynamics, but we also must determine the underlying physics that governs vesiculation, fragmentation, and the collapse of bubble-rich suspensions to form separate melt and vapor. Refined understanding of such processes—essential for quantitative short-term eruption forecasts—will require multidisciplinary research involving detailed field measurements, laboratory experiments, and numerical modeling.

Key words: Long-period events, fluid-filled crack model, hydrothermal systems, magmatic transport, high-resolution tomography, very-long-period seismicity, laboratory experiments with analog fluids, Q, Sompi method.

[1] U. S. Geological Survey, 345 Middlefield Road, MS 910, Menlo Park, California 94025, U.S.A. E-mail: chouet@chouet.wr.usgs.gov

Introduction

A central goal of volcano seismology is to understand the nature and dynamics of seismic sources associated with the injection and transport of magma and related hydrothermal fluids. Its ultimate objective is the elaboration of a comprehensive theory of seismic sources associated with volcanic activity that can be applied to a quantitative interpretation of elastic wavefields associated with magmatic and/or hydrothermal transport in volcanoes. To reach these objectives, volcano seismologists must tackle wave emission and radiation problems in both elastodynamics and mutiphase-fluid dynamics.

The course of volcano seismology was fundamentally altered in the late 1970s with the publication of a landmark paper by AKI et al. (1977a) and companion papers by AKI and LEE (1976), AKI et al. (1976), HUSEBYE et al. (1976), and AKI (1977), in which Aki and his colleagues developed a method for the determination of the three-dimensional seismic structure of the lithosphere based on an inversion of earthquake travel-time data. Tomography techniques have since become the most diagnostic approaches to image complex heterogeneous volcanic structures and detect and map magma reservoirs.

In another breakthrough, AKI et al. (1977b) used the standard elastodynamics theory to develop a seminal model of the dynamics of a fluid-driven crack, which for the first time provided a quantitative basis for the study of the source of volcanic tremor and long-period (LP) events. A few years later, AKI and KOYANAGI (1981) published a study of the kinematics of magma injection and ascent beneath Kilauea Volcano, Hawaii, which provided a quantitative estimate of the magma transport budget beneath this volcano based on seismic observations. Another study by AKI (1984) discussed magmatic activity beneath Long Valley Caldera, California, and proposed a quantitative basis for an estimation of the energetics of volcanic tremor.

Concurrently with these theoretical advances, AKI et al. (1978) conducted a series of key field experiments at Kilauea Iki, Hawaii, in which they demonstrated that the use of multiple seismic methods is essential for a determination of a complex seismic structure such as found in a cooling, partially solidified basaltic lava lake (RICHTER and MOORE, 1966; RICHTER et al., 1970; HELZ, 1993; BARTH et al., 1994). In active and passive experiments conducted at Kilauea Iki, Aki and associates used a combination of P waves and surface waves from explosions detonated in the lake crust to determine the velocity structure of the lava lake, relied on teleseismic S waves transmitted through the residual lens of melt to further constrain the shear velocity structure, and obtained an estimate of the lateral dimensions of the still-molten magma lens from the spatial distribution of seismic events originating within the cooling crust.

These early efforts, along with parallel studies inspired by Aki's work (FEHLER and AKI, 1978; CHOUET, 1979, 1981, 1982, 1985; FEHLER and CHOUET, 1982; FEHLER, 1983), established a sound physical basis for the interpretation of volcanic phenomena and provided the impetus that transformed volcano seismology from a qualitative into an increasingly quantitative science. As quantitative volcano seismology ushered in a new era in our understanding of volcanic processes and structures, concurrent technological developments occurred that further advanced the field. With the emergence of portable broadband seismic instrumentation, availability of digital networks with wide dynamic range, and development of new powerful analysis techniques made possible by greatly increased computer capacity, volcano seismology has now reached a mature stage where insights are rapidly being gained on the role played by magmatic and hydrothermal fluids in the generation of seismic waves.

A critically important area of research in volcano seismology today is aimed at the quantification of the source properties of LP events and tremor. The origins of LP events and tremor are critical to understand because this type of seismicity commonly precedes and inevitably accompanies volcanic eruptions and, therefore, can be used to assess the eruptive state or near-term eruptive potential of a volcano (CHOUET et al., 1994; CHOUET, 1996a). Monitoring of precursory seismicity at restless volcanoes is the most reliable, diagnostic, and widely used technique in volcano monitoring (SCARPA and TILLING, 1996). Other key areas of research include the elaboration of the three-dimensional velocity structure of volcanic edifices using high-resolution tomography, and imaging of magma conduits and quantification of magma transport from very-long-period (VLP) seismic data (CHOUET, 1996b). High-resolution tomography is critical to studies of source processes in volcanoes, because the precision and accuracy achieved in the identification of forces operating in a volcano depend on the degree of resolution achieved for the volcanic structure. High-resolution tomography based on iterative inversions of seismic travel-time data can image three-dimensional structures at scales of a few hundred meters provided adequate local short-period earthquake data are available. Thus, temporal variations in forces in a volcano are potentially resolvable for periods as short as a few seconds. Knowledge of the distribution of these forces, their amplitudes, and time histories represents the first step toward assessing the fluid-pathway geometry and mass-transport budget in a volcano.

Once the source space-time functions of the forces associated with fluid transport are known, the next step is to interpret these forces in terms of a physically realistic model of the fluid dynamics. A better understanding of the mechanisms driving fluid-induced seismicity can only be obtained from detailed analyses of the dynamic interactions between gas, liquid, and solid. A logical step toward achieving this goal is to rely on observations derived from well-controlled laboratory experiments with fluids with physical properties analogous to those in active volcanic conduits. The different flow regimes and associated characteristic pressure-oscillation behavior observed in expanding gas-liquid flows produced under experimental conditions may then be modeled numerically. If similarities between seismic and laboratory observations can be demonstrated, the resulting models then may be applied to quantitatively interpret the observed seismic data. Laboratory experiments conducted to date strongly suggest that pressure oscillations are fundamental phenomena resulting from dynamic flow processes involving the gas and liquid phases. The detailed documentation of the behavior of such flows and of their similarities to observed ground responses to flow within volcanic conduits is a prerequisite to translating seismic information into flow processes and modeling these processes. Such laboratory investigations represent another fundamental area of research in volcano seismology that is critical to the interpretation of magmatic and hydrothermal processes.

The present article offers a brief review of the state-of-the-art in volcano seismology and addresses basic issues in the quantitative interpretation of processes operative in active volcanic systems. The article covers volcano structure (via tomography) and the physics and quantitative analysis of source processes for volcanic earthquakes with examples based on work by the author and his colleagues. This discussion is not intented to be a comprehensive review of the entire field, however. Additional perspectives on other aspects of the field may be found in papers by McNUTT (2000a, 2000b), which focus more on empirical properties of volcanic earthquake swarms and on case histories for some well-known eruptions.

Volcanic Structures Imaged from High-resolution Three-dimensional Travel-time Tomography

The three-dimensional velocity structure of a volcanic edifice can be imaged through the inversion of first-arrival times from local earthquakes. The technique uses crisscrossing ray paths to separate the integrated effects of slowness (the reciprocal of velocity) on travel times and derive an image of the velocity structure. The arrival time of a seismic wave from an earthquake at a receiver is expressed by the nonlinear relationship

$$t = \tau + \int_{\ell[u(\vec{r})]} u(\vec{r}) \, d\ell, \tag{1}$$

where t is the arrival time, τ is the earthquake origin time, $u(\vec{r})$ is the slowness, $d\ell$ is the differential length along the ray, and $\ell[u(\vec{r})]$ is the ray path, a function of the earthquake location and velocity structure. Linearization of eq. (1) about a starting reference model and earthquake location then yields the relation (HOLE, 1992; BENZ et al., 1996)

$$r_{ij} = \sum_{k=1}^{3} \frac{\partial T_{ij}}{\partial x_k} \Delta x_k + \Delta \tau_i + \int_{\text{raypath}} \delta u(\vec{r}) \, d\ell, \tag{2}$$

where r_{ij} is the arrival time residual for the i-th earthquake and j-th receiver, T_{ij} is the travel time from the i-th earthquake to the j-th receiver, $\partial T_{ij}/\partial x_k$ are the partial derivatives of travel time with respect to the spatial earthquake coordinates, x_k, Δx_k and $\Delta \tau_i$ are perturbations to the starting earthquake location and origin time, respectively, and $\delta u(\vec{r})$ is the slowness perturbation to the reference model. The arrival time residual is the difference between the observed travel time and calculated travel time based on the starting earthquake location and reference slowness model.

In this approach, the initial reference slowness model is discretized by a uniformly spaced set of grid points in three dimensions; the first arrival times from a source location to each node in the model are calculated using finite-difference operators based on the Eikonal equation assuming a constant slowness within each cell (VIDALE, 1988, 1991; PODVIN and LECOMTE, 1991). Once the travel times are known everywhere in the model for a given source, rays are found by tracing from the receivers back to the source through the travel-time field. The ray within each cell is approximated by a straight segment and the ray direction is taken along the average travel-time gradient across the cell. Once a ray is traced, the length increments $d\ell$ within all cells crossed by the ray are known, and slowness and hypocenter perturbations are calculated by minimizing the differences between observed and calculated travel times. The model is then updated by the addition of the model perturbations, and the resulting new reference model is used for another linearized inversion. Iterations are stopped when the root-mean-square error in travel times between data and model reaches a stable minimum value.

High-resolution velocity models (0.5 km resolution) of the Kilauea caldera region were obtained by DAWSON et al. (1999) by tomographic inversion of P- and S-wave arrival times from local earthquakes. The data used for their inversion were recorded on a network of 67 stations, including 16 permanent vertical-component seismometers from the Hawaiian Volcano Observatory (HVO) short-period network, and 51 temporary three-component seismometers deployed as part of a joint Japan-U.S. seismic experiment (McNUTT et al., 1997) (Fig. 1). The average receiver spacing within 5 km of the center of the caldera is about 650 m. The data include 4695 P-wave and 3195 S-wave arrivals from 206 earthquakes during a 20-day period from January 15 through February 3, 1996.

Map views of the P-wave velocity and V_P/V_S models are shown for four depths in Figure 2. High P-wave velocities are observed along the southwest and east rift zones and within the northeast and southwest sectors of the caldera in the shallowest 1 km of structure. Low-velocity zones are observed in the first kilometer of structure in the center and along the south edge of the caldera. One kilometer below the surface, a marked low P-wave anomaly with lateral dimensions on the order of the size of the caldera is resolved. This anomaly is centered on the southeastern edge of the caldera and extends 3 km south-southeast of the caldera. The anomaly extends in depth to 4 km below the surface and has a maximum velocity contrast of about 10% with the surrounding region. Using a 5% reduction in velocity contrast from the initial model, the volume of the anomaly is estimated to be approximately 27 km^3. Corresponding V_p/V_s ratios within the first kilometer show high values along the south and

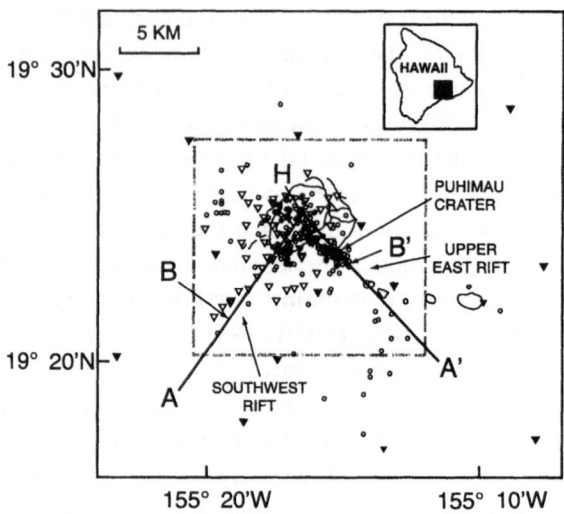

Figure 1

Map of the Kilauea caldera region with inset showing the island of Hawaii and model boundary (black box). The map boundary outlines the horizontal extent of the initial models. The square bounded by dashed, shaded perimeter delimitates the horizontal extent of the P-wave velocity model shown in Figures 2a–d and V_P/V_S model shown in Figures 2e–h. Fractures and pit craters are indicated by solid lines. Solid inverted triangles mark vertical-component stations of the HVO permanent network, open triangles indicate temporary three-component stations, and open circles mark earthquake locations. Lines A–A' and B–B' mark the locations of the vertical sections shown in Figures 3a–b. H, which marks the apex of these two profiles, is located 1 km northeast of the Halemaumau pit crater. (Modified from DAWSON et al., 1999.)

southeastern outline of the caldera. Below 1 km depth, a high V_p/V_s ratio occupies the same region as the low P-wave anomaly, but separates into two distinct zones (Fig. 2g), one centered on the southern portion of the caldera, the other beneath the upper east rift.

Of interest in the shallow (<1 km depth) structure beneath Kilauea are the marked low-velocity zones to the south of the caldera and near the Halemaumau pit crater. The first zone correlates well with the caldera fracture system and a region of thick tephra and ash deposits to the south of the caldera. The second zone within the caldera may be attributed to reduced P-wave velocities due to hydrothermal activity at Halemaumau. The high P-wave velocities in the southwestern and northeastern portions of the caldera are attributed to accumulated prehistoric and historic pahoehoe flows.

The V_P/V_S model clearly shows that the shear-wave velocity in the upper 5 km beneath the southern portion of the caldera and the upper east rift is anomalously low. The corresponding Poisson ratio in the anomalous region approaches values near 0.37, suggestive of the presence of either highly fractured material and/or a significant fraction of partial melt. Theoretical calculations of seismic wave velocities for different melt geometries (MAVKO, 1980) and constitutive relations of solid-liquid

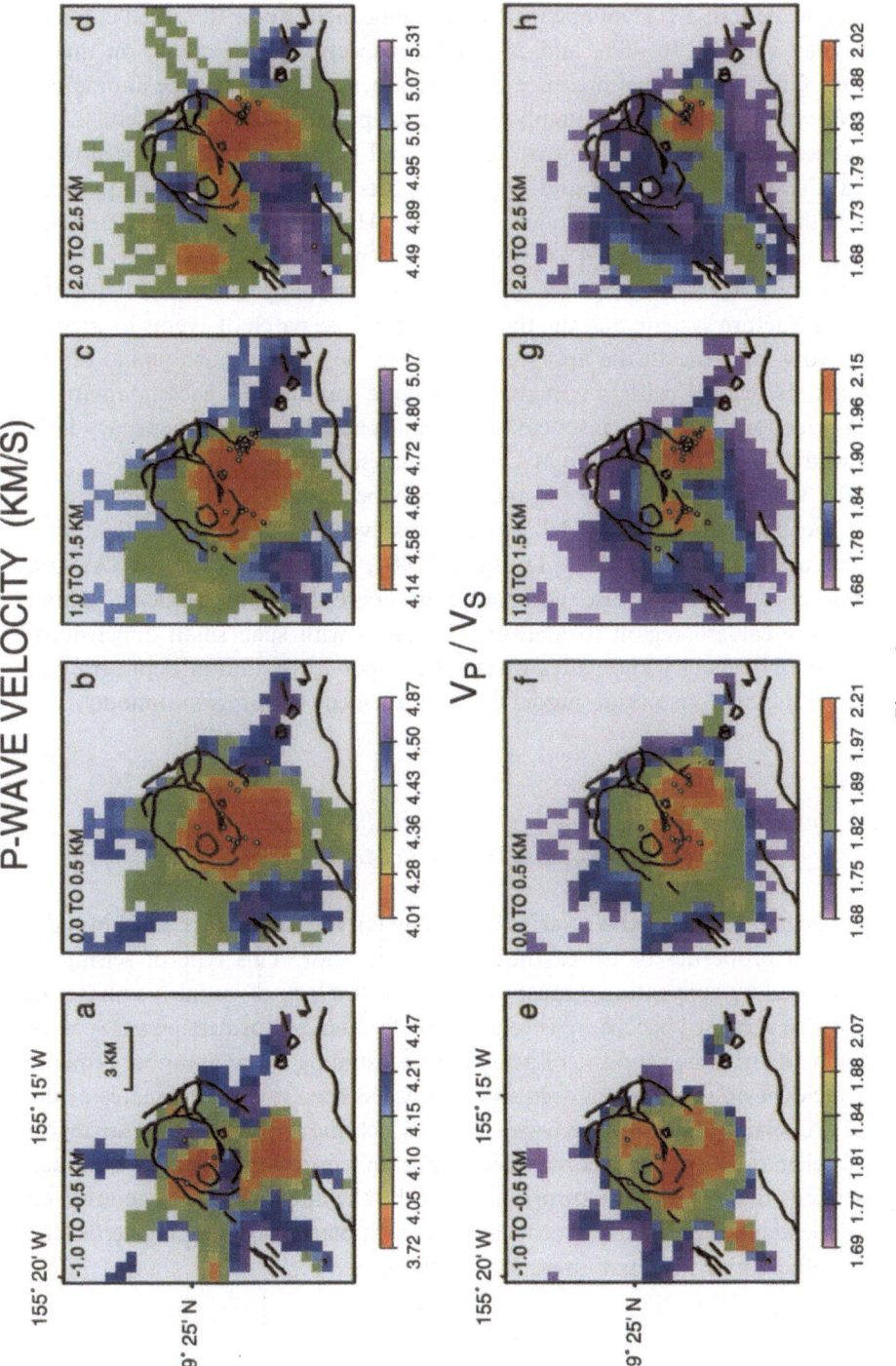

Figure 2

Map views of the *P*-wave velocity (a–d) and V_P/V_S velocity ratio (e–h) at four depths. Zero depth indicates sea level and negative depths indicate elevations above sea level. The spatial resolution is 500 m. Pit craters and fractures are indicated by solid black lines. White circles indicate earthquake hypocenters within each layer. (From DAWSON *et al.*, 1999, after correction for erroneous color scales in the original V_P/V_S panels.)

composites (TAKEI, 1998) suggest an approximately 1% perturbation in P-wave velocity is equivalent to a 1% volume fraction of mafic melt. For a 10% melt content, V_P and V_S may vary by 10–40% and 20–100%, respectively, depending on model assumptions. Given the known presence of magma beneath Kilauea, the anomalous P and V_P/V_S volumes can be reasonably ascribed to presence of partial melt.

Vertical cross sections of the P-wave velocity and V_p/V_s structures taken along a transect parallel to the southwest and upper east rifts further illustrate the primary features of the two structural models (Fig. 3). Also shown are the relocated hypocenters of events within a 4 km wide zone centered on the sections. Three clusters of events are observed, two of which are spatially correlated with the southern ring fracture system, and the third a more diffuse patch of events located at depths of 1 to 4 km beneath the upper east rift (Fig. 3b). These earthquakes reflect stress release associated with an increased magma production accompanying a strong, 4.5-hour-long inflation episode of the Kilauea summit in early February 1996 that produced a 22 μrad tilt change at Uwekahuna (see Fig. 13a).

It should be noted that the low P-wave anomaly beneath Puhimau crater (Fig. 3) imaged in the high-resolution model was not resolved in earlier regional models (ELLSWORTH and KOYANAGI, 1977; THURBER, 1984, 1987; ROWAN and CLAYTON, 1993; OKUBO et al., 1997), because these former models did not have sufficient spatial sampling in the caldera region to identify anomalies with such small dimensions. However, the high-velocity anomalies within the upper 1 km, and at depth around the caldera in the high-resolution model are in agreement with regional models.

Spatio-temporal Properties of Long-period Seismicity Mapped with Small-Aperture Seismic Antennas

Small-aperture seismic arrays (seismic antennas) are useful for tracking the spatio-temporal properties of LP earthquakes and tremor. This type of seismicity cannot be located by conventional means because of the lack of impulsive phases in the records. The standard procedure is to compute frequency-slowness power spectra over successive short-time windows of array data. Assuming a finite number of plane waves are incident on the antenna, frequency-slowness spectra yield estimates of the directional properties of these plane waves, from which the locations of the strongest sources of seismic energy may be inferred using an appropriate structure model. Time-dependent changes in signal properties seen by the antenna are thus converted into time-dependent changes in source locations. Each plane wave moving across the antenna is defined by its ray parameter, P, where

$$P = \sqrt{S_x^2 + S_y^2}$$

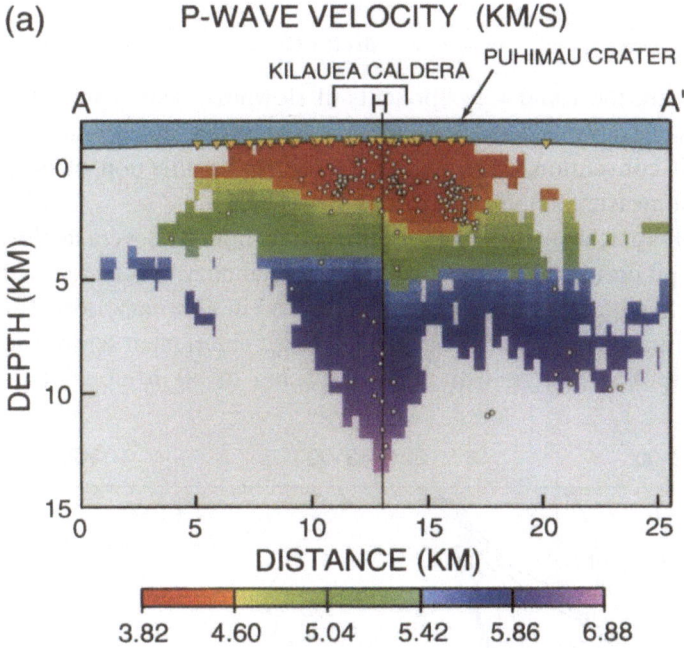

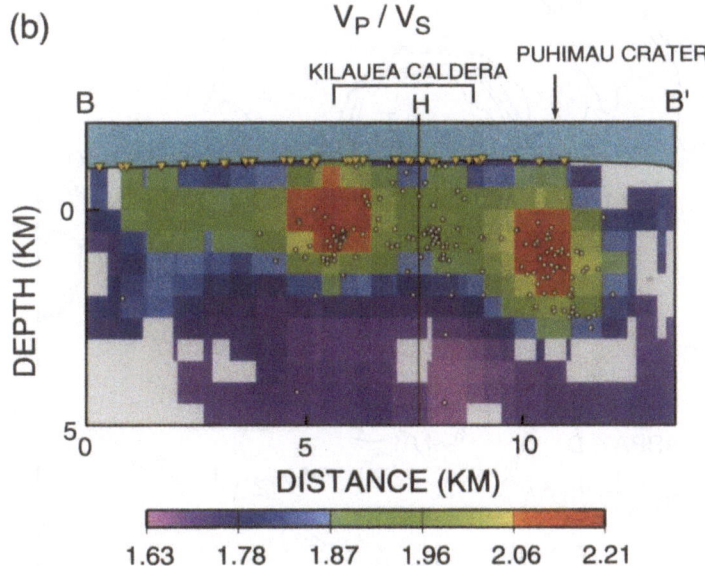

Figure 3

Northeast-southeast cross sections for (a) *P*-wave velocity and (b) V_P/V_S models. The profile locations are shown in Figure 1; *H* and vertical line beneath indicate sharp bend in section (see Fig. 1). White circles indicate earthquake hypocenters located within ± 2 km of the profiles. Yellow inverted triangles mark the projections of stations onto the profiles. (From DAWSON *et al.*, 1999, after correction for an erroneous color scale in the original V_P/V_S panel.)

and azimuth, Φ, where

$$\Phi = \pi/2 - \arctan(S_y/S_x),$$

and S_x and S_y are the x and y components of slowness, respectively. The Cartesian coordinates are usually selected with the y axis pointing north and x axis pointing east. With this convention, Φ represents the apparent direction of propagation or signal azimuth measured clockwise from north.

Three small-aperture arrays of short-period seismometers were deployed near the Halemaumau pit crater and along the southern boundary of Kilauea caldera as part of a Japan-U. S. cooperative experiment conducted in February 1997 (Fig. 4). Array D has an aperture of 400 m and consists of 41 three-component sensors deployed in a semi-circular spoked pattern with station spacing of 50 m along the spokes and

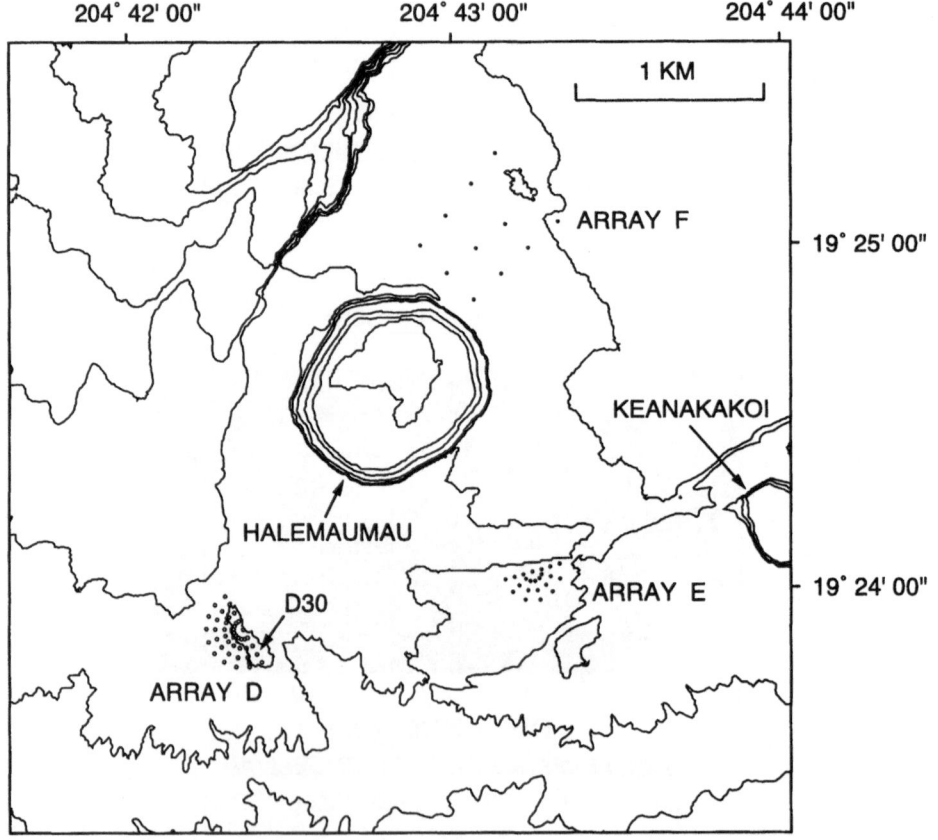

Figure 4

Map of the southern extent of the Kilauea caldera region showing locations of pit craters, main topographic features, and seismic antennas deployed during the 1997 Japan–U. S. experiment. Solid dots mark locations of vertical-component sensors and open circles indicate three-component sensors. Station D30 was used to obtain the seismograms displayed in Figure 5.

angular spacing of 20° between spokes. Array E, with an aperture of 300 m, includes 22 vertical-component sensors deployed in a similar pattern as array D with station spacing of 50 m along spokes but with a wider angular spacing of 30° between spokes. Array F is a polygonal array with approximate dimensions of 400 by 600 m and consists of three rows of 4 vertical-component receivers spaced roughly 200 m apart. The semi-circular geometry of arrays D and E was selected to enable both detailed frequency-slowness analyses of the wavefield and a quantification of the wavefield properties using the correlation method of AKI (1957, 1959, 1965), which is not discussed in this article. Details of the method can be found in AKI (1957) and in studies by FERRAZZINI et al. (1991) and CHOUET et al. (1998).

All the temporary stations deployed during the experiment used 16-bit, four-channel, Hakusan dataloggers. Three-component stations were equipped with Mark Products L22-3D sensors, and vertical-component stations were equipped with Mark Products L11-4A sensors. The seismometers had a natural frequency of 2 Hz and sensitivity of 0.5 V/cm/s, and were sampled at 100 samples/s/channel. All the array instruments used a common Global Positioning System (GPS) time base with an accuracy of 5 μs among all the channels.

The three arrays were deployed for the specific purpose of tracking LP events and tremor originating beneath Kilauea caldera. Figure 5 shows an 11-hour-long sample of vertical-component data recorded at one of the receivers in array D. This sample is representative of the seismicity of Kilauea at the time. Frequency-slowness analyses of the vertical-component data from the three arrays were performed using the Multiple Signal Classification (MUSIC) method (SCHMIDT, 1981, 1986; GOLD-STEIN and ARCHULETA, 1987, 1991a, b; GOLDSTEIN, 1988; GOLDSTEIN and CHOUET, 1994; CHOUET et al., 1997). The advantages of this method over more conventional array-processing techniques have been extensively discussed by GOLDSTEIN (1988) and GOLDSTEIN and ARCHULETA (1991a, b). Power spectra were computed for successive 2.56-s windows with 2.36 s of overlap spanning the array records. For data from arrays D and E, the power spectra were calculated over a bandwidth from 1 to 10 Hz using slowness stacking (SPUDICH and OPPENHEIMER, 1986). Both components of slowness were allowed to range over a 4 s km^{-1} window to insure the inclusion of both body and surface waves. Data from array F were processed in a similar manner over a reduced bandwidth of 1–6 Hz and narrower slowness range of 2 s km^{-1} because of this array's lack of resolution for frequencies above 6 Hz. Use of this array was restricted to the detection of body waves with ray parameters smaller than 1 s km^{-1}. The precision of ray parameter measurements, estimated according to eq. (12) in GOLDSTEIN and ARCHULETA (1991a), is approximately 0.05 s km^{-1} on array D, 0.09 s km^{-1} on array E, and 0.04 s km^{-1} on array F.

Figure 6 shows the temporal distributions of power, azimuth, and ray parameter for the vertical component of motion derived from the stacked slowness power spectra for a LP event recorded on array D. Errors were estimated by measuring the spread in azimuths and slownesses at a level of 90% of the peak spectral amplitude.

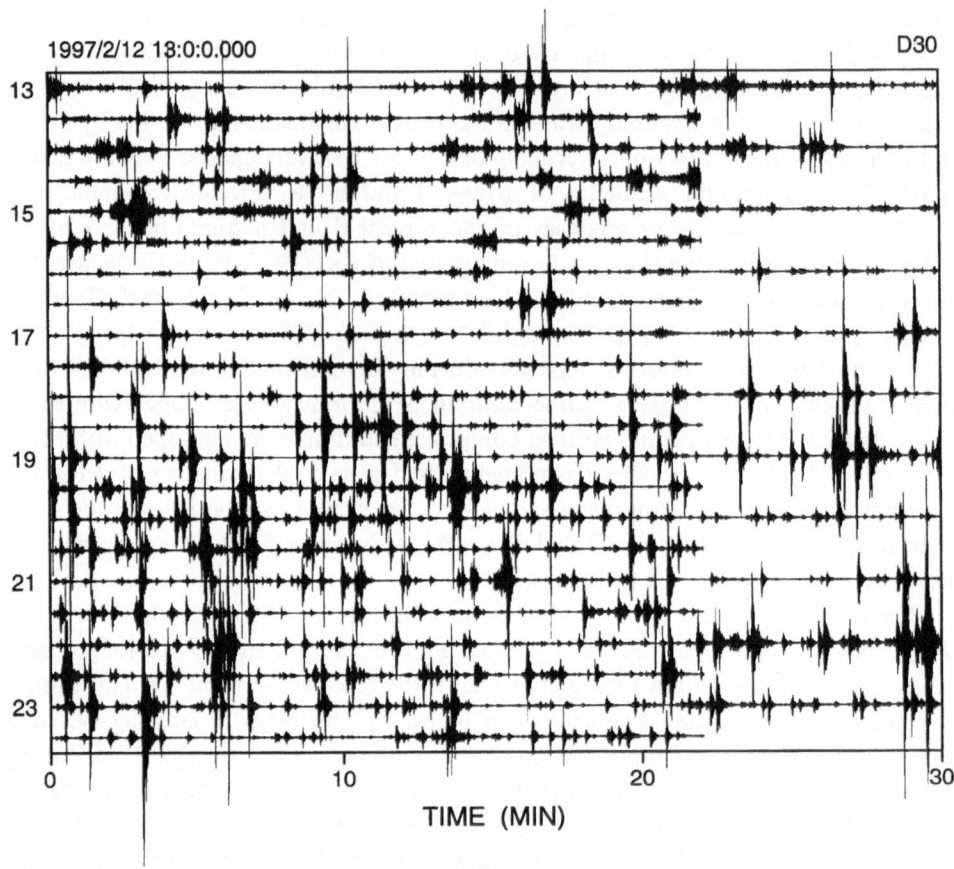

Figure 5

Typical vertical-component record (velocity) of LP seismicity selected for frequency-slowness analyses. The record is from station D30 of array D (see Fig. 4). Date and time (UT) at the start of the record are listed at the upper left of the plot. Numbers in the left margin indicate the start of individual hours in this recording period (beginning 1300, 12 February, 1997). Gaps in the record represent periods used by the data logger to perform GPS clock synchronization.

This method of error estimation is appropriate when there is a well-defined dominant spectral peak as seen near the onset of the LP event depicted. However, the error estimates obtained in such manner are less useful when dealing with the coda of the LP event, where multiple scattered arrivals with distinct azimuths and ray parameters are observed with roughly equal power above the 90% power level. This latter situation can lead to significant overestimation of the error as observed in the tail section of the record (Fig. 6). The 4.6-s time interval characterized by enhanced spectral power and stable azimuths was used to determine the directions of primary wave arrivals for this event.

Figure 7 shows the ranges of backazimuths for waves detected during 2–5 s bracketing the onset of the LP event. The backazimuths are obtained from

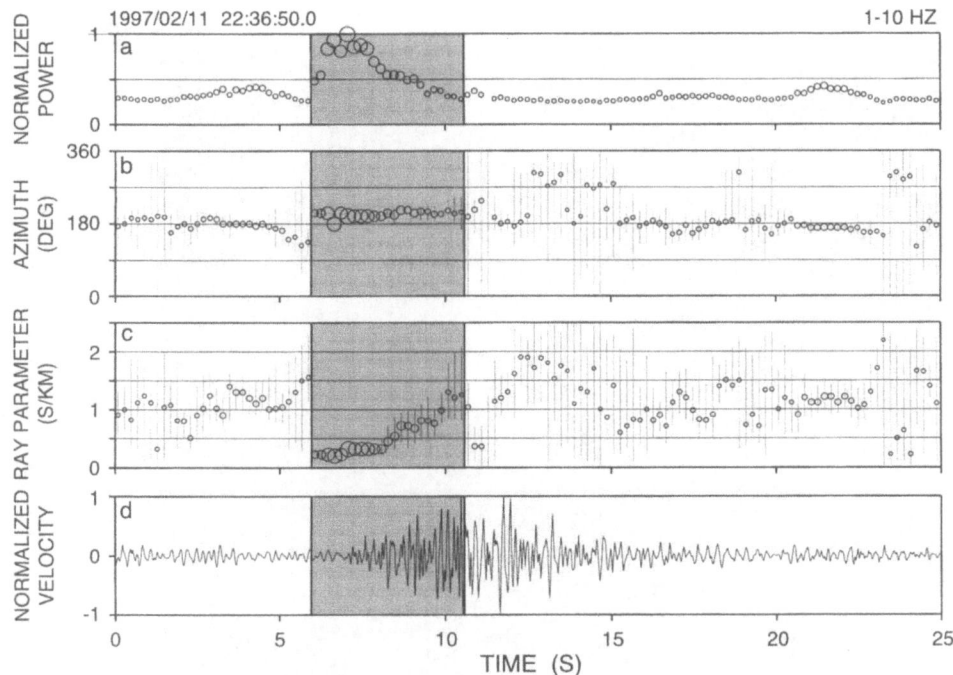

Figure 6

Peak spectral power, azimuth and ray parameter versus time obtained from frequency-slowness power spectra calculated for successive 2.56-s windows with 2.36 s of overlap ranging over 25 s of data for a LP event recorded on February 11, 1997. The plots show the results obtained for the vertical components of ground velocity recorded on array D. The bandwidth is indicated at the upper right, and the date and time at the start of the record are listed at the upper left. The vertical band of shading marks the time interval selected for measurements of azimuths and ray parameters associated with direct arrivals from the event shown in the bottom panel. (a) Power level of the dominant peak in individual stacked slowness spectra normalized by its maximum value reached during the entire interval depicted. (b) Azimuth, and (c) ray parameter versus time derived from the dominant peak in the stacked slowness power spectra calculated for individual windows. The sizes of the open circles in (a), (b), and (c) are proportional to the peak spectral power. (d) Vertical-component velocity seismogram from one of the receivers in array D.

frequency-slowness analyses performed simultaneously on data recorded with the three antennas. The intersection of the three colored wedges marks the most probable source region. First arrivals for 1129 LP events and 147 samples of tremor were detected by this method over a 23-hour period of swarm activity. With the exception of a few more distant events, this activity appears to originate from a region beneath and northeast of Halemaumau.

The geometrical method illustrated in Figure 7 provides only a rough idea of the epicentral source region because it does not take into account the distortion of the wavefield due to the volcanic structure. To improve the resolution of the source hypocenter, ray parameter and azimuth information from the three antennas must be interpreted in the context of a realistic model of the medium.

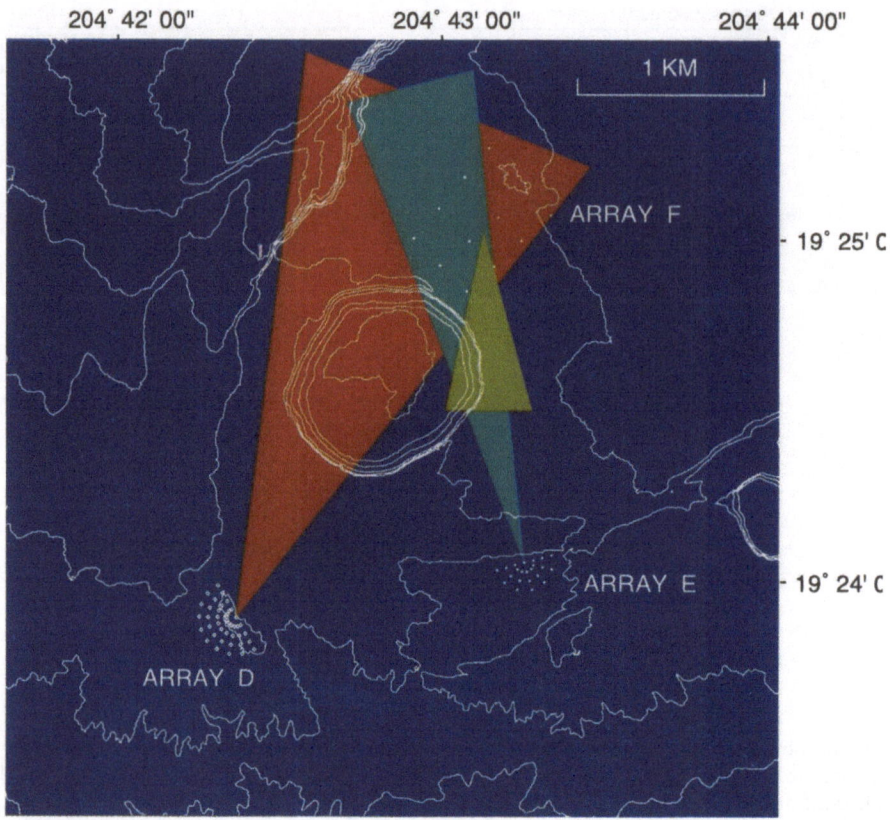

Figure 7

Map of the southern part of Kilauea caldera showing main topographic features, small-aperture arrays (D, E and F), and spread of backazimuths for direct waves propagating from a LP source (see event in Fig. 6d) detected by the three arrays. The domain defined by the intersections of the three wedges near the northeast boundary of Halemaumau encloses the most probable source epicenter.

A better estimate of the spatial extent of the source region of LP seismicity may be obtained by modeling the wavefield in the 3-D velocity structure and 3-D topography of Kilauea. Figure 8 illustrates the effect of topography on waves propagating from an isotropic point source located at a depth of 520 m below the northeast edge of Halemaumau. The medium is homogeneous with compressional and shear-wave velocities of 4 km/s and 2 km/s, respectively. The amplitude of the source dipole moment is 10^{12} N m and the source time function is a cosine-shaped pulse with period of 0.5 s. The snapshots (Fig. 8) represent the vertical component of ground displacement calculated by the finite-difference method of OHMINATO and CHOUET (1997). Diffraction of the incident P wave by the Halemaumau pit crater is observed as a distinct red patch at the base of Halemaumau and orange arcs extending from Halemaumau at $t = 0.5$ s. Wave diffraction by Halemaumau is also apparent in the snapshot at $t = 0.9$ s where it manifests as a notch of lighter blue in

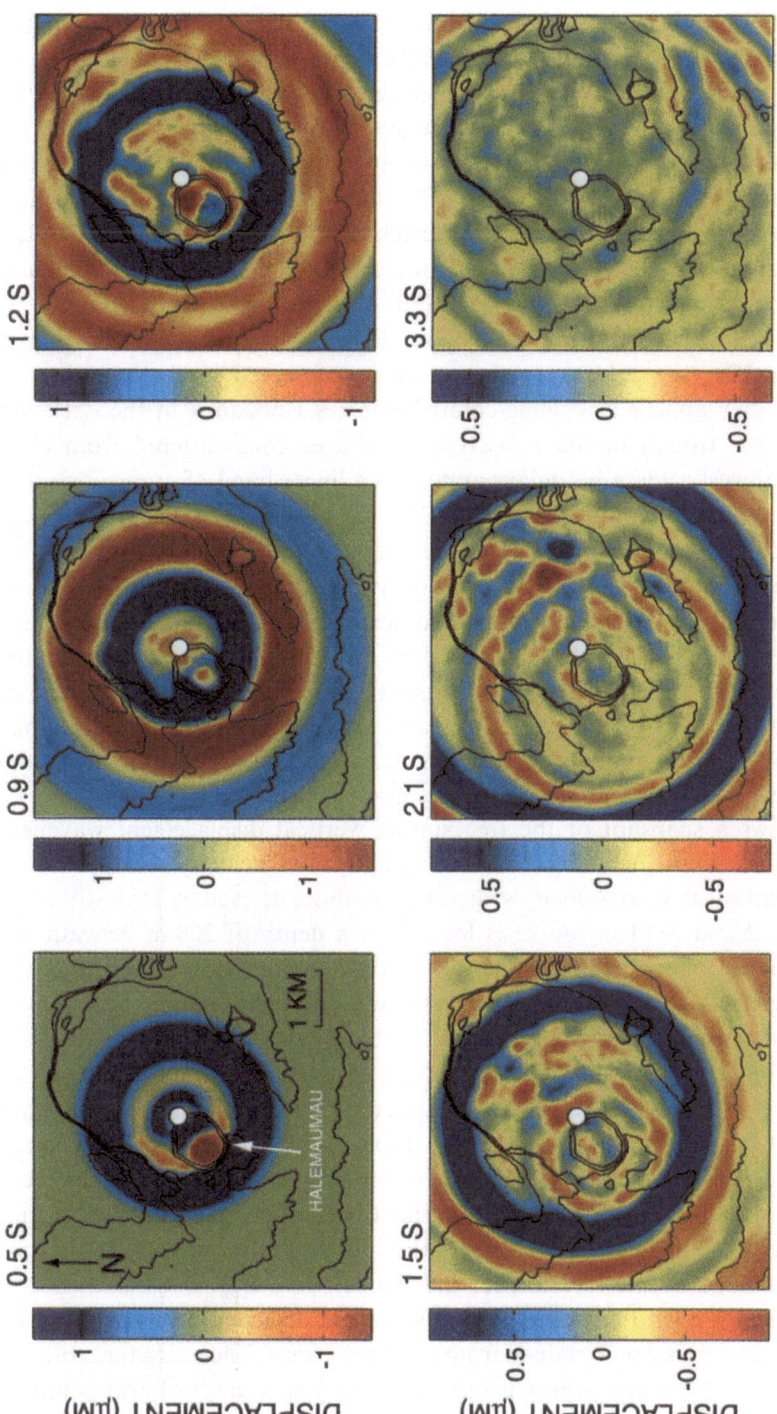

Figure 8

Snapshots showing wave propagation in the topography model of the Kilauea caldera region. Displayed is the vertical component of displacement at the surface of the medium. An isotropic point source is embedded at a depth of 520 m below the northeast edge of Halemaumau (see text for details). The white circle marks the source epicenter. The snapshots are superimposed on a map of Kilauea caldera (black contour lines with 50-m interval) to emphasize the effect of topography on waves propagating from the source. Colors indicate displacement amplitudes (in μ meters) according to the scale at the left of each snapshot. The medium is homogeneous and the observed wavefield complexity only reflects the effect of topography.

the dark blue ring marking the second peak of the wavefront near the southwest edge of the pit crater. The bright triangular orange patch observed near the source epicenter in this snapshot identifies converted SV waves reflected from the wall of Halemaumau facing the source. Waves are generated at the circular bottom corner of Halemaumau by the incident P wave. These waves travel along the floor of the pit crater where they interfere constructively to create a pattern of standing circular waves. These trapped waves are observed as a bright patch of orange near the center of Halemaumau at $t = 0.9$ s and appear as a bright orange ring coincident with the walls of Halemaumau at $t = 1.2$ s. Rayleigh waves scattered by the pit crater are observed as orange-colored rings propagating outward from Halemaumau at $t = 1.5$ s and $t = 2.1$ s. Wave diffractions by a smaller pit crater east of Halemaumau and by cliffs flanking the caldera to the north are observed in the snapshot at $t = 0.9$ s, where they appear as orange-colored notches embedded in the dark red ring identifying the trough in the P wavefront. Waves backscattered from cliffs marking the northwest caldera boundary appear as a linear band of orange oriented parallel to the topographic contours north of Halemaumau at $t = 1.2$ s. These waves propagate in a clockwise direction through the northern sector of the caldera where they interfere with the Rayleigh waves scattered by Halemaumau to produce a wave pattern identified by the orange patches located near the northern caldera boundary at $t = 1.5$ s and $t = 2.1$ s. Waves backscattered by cliffs flanking the northern sector of the caldera are also observed propagating southward past the southern edge of Halemaumau at $t = 3.3$ s. The mottled pattern of yellow on the caldera floor observed in this snapshot represents decaying standing waves resulting from the interference of waves backscattered by the topography of the caldera.

Figure 9 shows a snapshot of the free surface vertical displacement wavefield calculated for a medium including both topography and velocity structure. The compressional and shear-wave velocity structures are those derived by DAWSON et al. (1999) (see Figs. 2 and 3). The source is located at a depth of 200 m beneath the northeast edge of Halemaumau. The effect of velocity structure is a distortion of the features observed in the presence of topography alone. The arrows in Figure 9 represent the slowness vectors determined at three synthetic antennas whose locations and configurations simulate the setups of the three arrays deployed at Kilauea in February 1997 (see Fig. 4). The apparent slownesses and propagation azimuths of the waves are different in the model including topography only, compared to the model including both topography and structure. The differences between the slowness vectors derived from the two models are quite large and indicate that the combined effects of topography and structure must be taken into account to obtain an accurate location for the source.

The results from forward modelling of the wavefield depicted in Figure 9 can be used to invert the slowness data obtained from the three arrays. The procedure consists in a discretization of the source region using a uniformly spaced set of grid points in three dimensions, and computation of the free-surface responses produced by

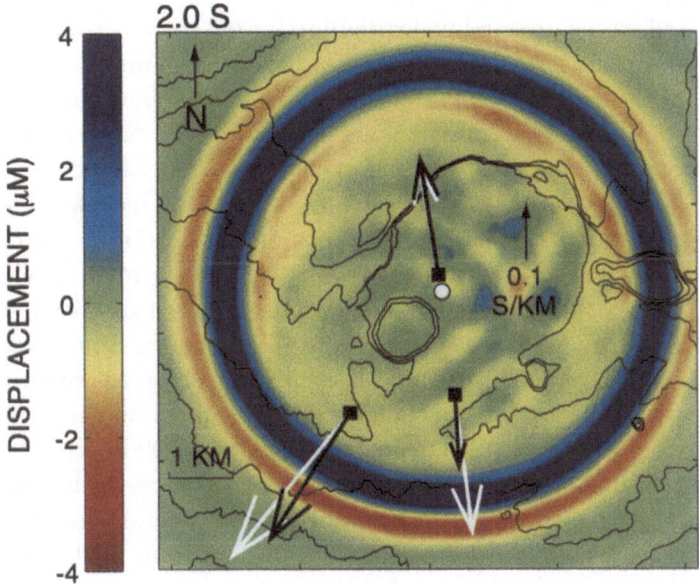

Figure 9

Snapshot of the vertical component of the free surface displacement wavefield 2.0 s after the origin time of a synthetic event. The medium includes both the topography and 3D velocity structure of Kilauea. An isotropic point source is embedded at a depth of 200 m below the northeast edge of Halemaumau. The source parameters are the same as those used in Figure 8 (see text for details). A white circle marks the source epicenter and black squares mark the locations of arrays D, E, and F (see Fig. 4). The arrows represent the slowness vectors estimated at the three arrays for a homogeneous medium including topography only (white arrows) and a medium including both topography and structure (black arrows). Differences are relatively large and affect both the azimuth and ray parameter. These results emphasize the importance of considering both topography and structure in the definition of a synthetic slowness vector model. Colors indicate displacement amplitudes (in μ meters) according to the scale at the left of the snapshot.

isotropic sources located at each grid node. Synthetic seismograms are calculated at three synthetic antennas that simulate the three arrays shown in Figure 4, and frequency-slowness analyses of the synthetic data yield estimates of ray parameters and azimuths at the three arrays for each point source. In this way, a three-dimensional slowness vector model is generated at each array from the grid of source positions.

The source locations of LP events and tremor are obtained in a probabilistic sense. First, a probability is assigned to every point in the source domain by comparing the results from the frequency-slowness analyses to the slowness vector model obtained for the domain. The statistical distributions of slownesses observed at each array for each individual LP event or tremor sample determine the shape of the probability functions. To each value in the slowness vector model corresponds, for each of the arrays, a given probability that can be mapped. The source location is obtained as the point corresponding to the maximum probability of the combined distributions for azimuths and ray parameters from the three arrays. As a final step, the source location probability distributions for individual events are stacked to obtain an overall spatial probability distribution for the ensemble of events analyzed (see ALMENDROS et al., 2001a, b, for details). The distributions of probabilities for the locations of the sources of LP events and tremor are shown in Figure 10. The source regions of the two types of signals are defined as the volumes contained within the surfaces corresponding to 5% of the maximum of the stacked source location probability. The choice of 5% is conservative enough to include all the located hypocenters and provides maximum constraints on the sizes of the source regions.

The data illustrated in Figure 10 represent 83% of the total activity recorded during the 23 hours considered, hence these results provide a good representation of the overall behavior of the LP swarm. Figure 10a shows the source region of LP events. At least three different clusters of events are observed within this source volume. The most active cluster includes 857 hypocenters and is centered about 200 m northeast of Halemaumau at depths shallower than 200 m beneath the caldera floor (region A in Figs. 10b, c). A second cluster of 132 events is located at depths near 400 m beneath the northeast quadrant of Halemaumau (region B in Fig. 10c), and a third less dense cluster of 131 events is located at depths shallower than 200 m in an elongated zone extending southeastward from the northeast quadrant of Halemaumau. This latter region is seen in Figure 10a as the arm extending southeastward from Halemaumau; the northwest end of this arm is visible in the cutaway view in Figure 10c (region C). The individual clusters are not completely separated, and a few events do occur between these clusters, suggesting a connection between the different zones of high activity and providing a rough glimpse of the three-dimensional structure of the overall source region. Of the entire LP data set considered, only 9 events were found to be located outside of the source domain considered for analysis and these are not considered to be part of the LP source region. For the tremor the method reveals a single source region (Figs. 10d, e), which coincides with the most active source region of LP events northeast of Halemaumau

▶

Figure 10

Location and extent of the source region of long-period seismicity recorded at Kilauea Volcano in February 1997. (a) 3-D view of the source region of LP events. The source region is estimated as the region inside the surface corresponding to 5% of the maximum of the stacked source location probability obtained for 1120 LP events considered in the analysis. The horizontal projection of the source region and rim of the Halemaumau pit crater shown on the bottom of the cube provide additional information about the horizontal extent of this seismicity. (b) Cutaway view of the source region showing the most active source zone (A) located northeast of Halemaumau. (c) Secondary clusters of hypocenters (B, C) are also present at two depths beneath the crater. The maximum of the color scale is 0.3 in this panel. Accordingly, the probability is saturated in the main source region shown in (b). (d) 3-D view of the source region of tremor obtained from the surface of 5% of the maximum stacked probability corresponding to the 147 tremor samples considered. (e) Cutaway view of the stacked source location probability of tremor showing that there are no subregions and that the maximum probability occurs in a region (A') which coincides with the most active zone of LP event generation. (f) General view of the summit region. The blue cube shows the boundaries of the region extracted in (a)–(e). (Reproduced from ALMENDROS et al., 2001b.)

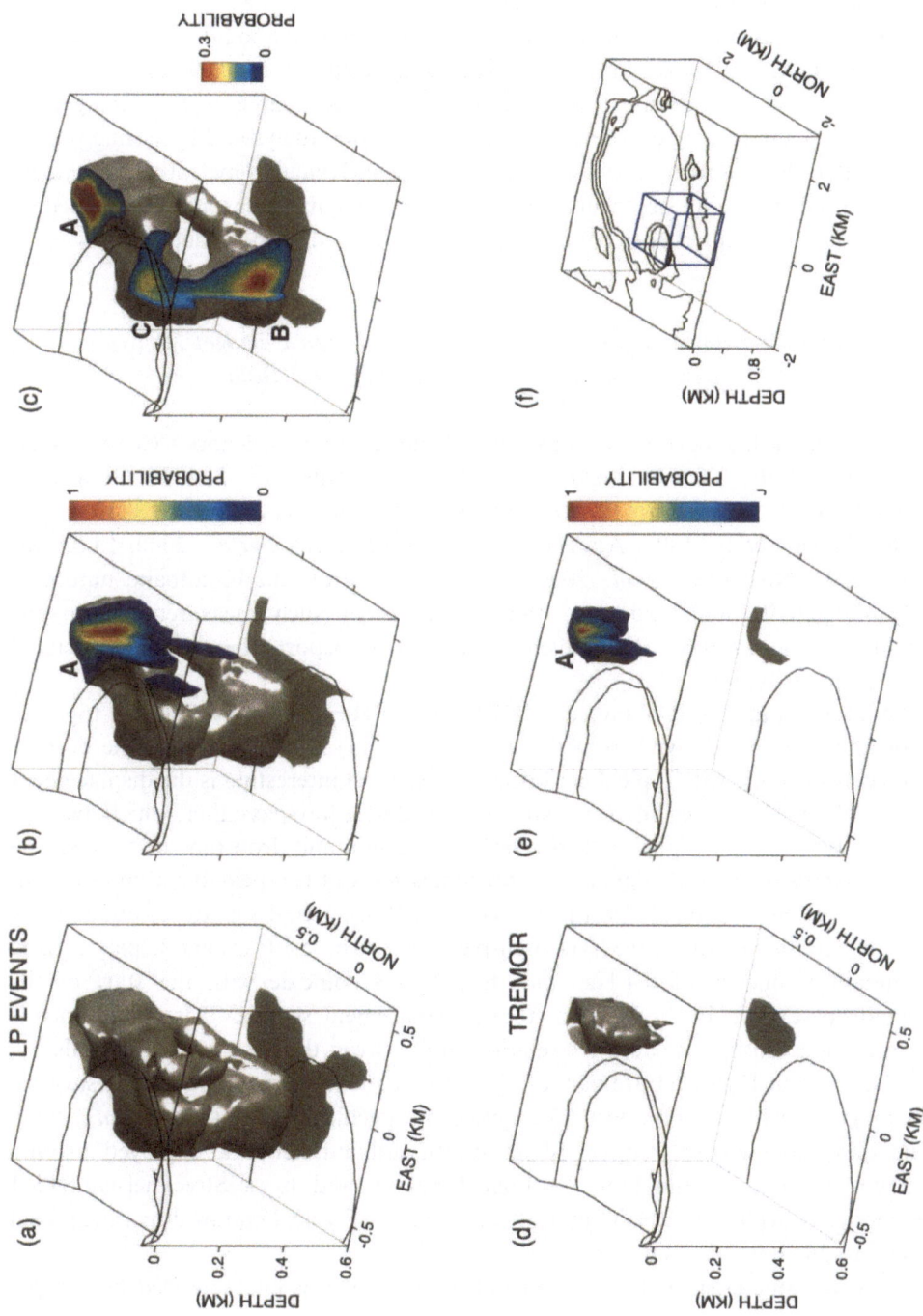

(compare A and A' in Fig. 10). This result demonstrates the strong relation that exists between the two types of activities. The dimensions of the source regions in the east-west, north-south, and depth directions are roughly $0.6 \times 1.0 \times 0.5$ km for LP events and $0.2 \times 0.5 \times 0.2$ km for tremor, with volumes of 0.09 and 0.01 km^3, respectively. The locations, depths, and sizes of the source regions imaged in Figure 10 point to a hydrothermal origin for all the analyzed LP seismicity and suggest that this seismicity may reflect the response of Kilauea's hydrothermal system to enhanced degassing associated with increased magma transport in the deeper magma conduit. This deeper conduit system is discussed in more detail below.

Magma Pathway Geometry and Magma Volumetric Budget Derived from Moment Tensor Inversion of VLP Data

Broadband seismic observations carried out at many volcanoes (KAWAKATSU et al., 1992, 1994, 2000; NEUBERG et al., 1994; KANESHIMA et al., 1996; OHMINATO and EREDITATO, 1997; OHMINATO et al., 1998; DAWSON et al., 1998; ROWE et al., 1998; CHOUET et al., 1999; ARCINIEGA et al., 1999; FUJITA et al., 2000; LEGRAND et al., 2000; NISHIMURA et al., 2000) clearly demonstrate the broadband nature of volcanic activity and undescore the usefulness of such measurements in the quantification of source mechanisms and mass transport phenomena associated with eruptive activity.

A broadband record obtained at Kilauea on February 1, 1996, shown together with filtered traces derived from this record (Fig. 11) is dominated by the oceanic microseismic noise with typical periods of 3–7 s. Most interesting is the displacement record obtained by filtering the signal with a 0.125-Hz low-pass filter, which displays a repetitive sawtooth signal with rise time of 2–3 min and drop time of 5–10 s. The repetitiveness of the VLP signal is a clear indication of the repeated action of a non-destructive source. Notice also the characteristic long-period signatures obtained by filtering the signal with a 0.33-Hz high-pass filter. These LP events display a fixed dominant frequency of 0.4 Hz, and their onsets coincide with the start of the downdrop segment in the VLP sawtooth displacement signals. This coincidence is strongly suggestive of a causative relationship between the VLP and LP signals.

The data in Figure 11 are part of a seismic sequence observed during a surge in the magma flow feeding the east rift eruption of the volcano (THORNBER et al., 1996). The data were recorded by a 10-station broadband network deployed around Kilauea Caldera by the U.S. Geological Survey and by a Streckheisen STS-1 seismometer deployed at HVO by Caltech (Fig. 12). Typical bandwidths range from 0.04 s to 30, 100, 120, and 300 s.

The volcanic crisis at Kilauea on February 1, 1996 is characterized by a rapid summit inflation and increasing seismic activity starting at about 0800 Hawaiian Standard Time (HST). Figure 13a shows the tilt record obtained at the Uwekahuna

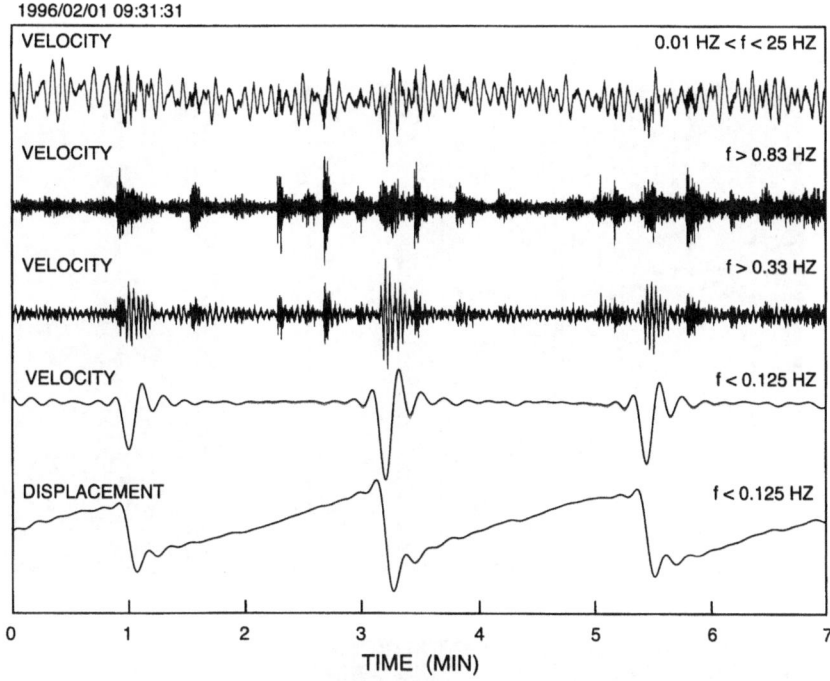

1996/02/01 09:31:31

Figure 11

Broadband record and associated filtered signals obtained at station GU9 of the Kilauea broadband network (Fig. 12) during a volcanic crisis on February 1, 1996. The broadband signal from the vertical component of velocity is filtered in various frequency bands to produce five records for the same 7-min time interval. The top trace shows the broadband signal ($0.01 < f < 25$ Hz), which is dominated by the oceanic wave-action microseism with periods in the range of 3 to 7 s. The second trace shows the signal after a high-pass filter has been applied ($f > 0.83$ Hz). The result is equivalent to a typical short-period record and shows a series of events superimposed on a background of tremor. The third trace also has a high-pass filter applied ($f > 0.33$ Hz); long-period (LP) signals with a dominant frequency of about 0.4 Hz are enhanced in this record. The fourth trace shows the signal when a low-pass filter is applied ($f < 0.125$ Hz); a repetitive very-long-period (VLP) signal consisting of pulses with period of about 20 s is observed. The fifth trace is the corresponding displacement record with a low-pass filter applied ($f < 0.125$ Hz), showing a repetitive sawtooth pattern with rise time of 2–3 min and drop time of 5–10 s. Notice that the onset of the LP signal observed above coincides with the onset of the downdrop in the VLP displacement. (Reproduced from DAWSON *et al.*, 1998.)

vault near HVO. Summit inflation lasts approximately 4.5 hours, following which the summit deflates back to its original state over a 3-day period. Figure 13*b* shows details of the tilt signal and ground displacement at HVO observed during a 6-hour period starting at 0700 HST. The tilt reverses from inflation to deflation around 1230 HST. Figure 13*c* shows the details of the sawtooth displacement waveforms recorded at HVO and GU9 during a 15-minute period during inflation. Particle motions associated with the displacements observed on the broadband network are essentially linear and point to a source region beneath the northeast edge of Halemaumau (OHMINATO *et al.*, 1998).

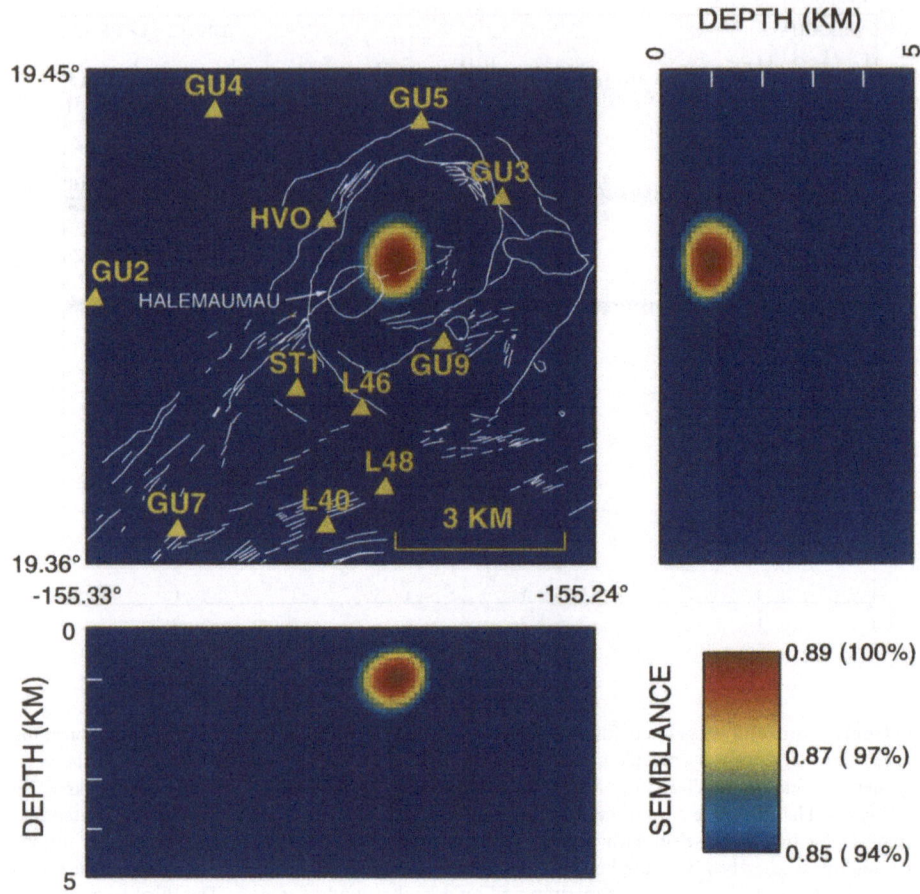

Figure 12

Map view and vertical cross sections of semblance distribution for the very-long-period signals observed at Kilauea on February 1, 1996 (see Fig. 11). Colors indicate semblance levels according to the scale at the bottom right. The map view shows the broadband seismic network at Kilauea. Solid triangles show station locations. Summit caldera, pit craters, faults, and fractures are shown by thin white lines. (Reproduced from OHMINATO et al., 1998.)

A detailed analysis of these data was performed by OHMINATO et al. (1998). To locate the source OHMINATO et al. (1998) used the extended definition of semblance of MATSUBAYASHI (1995), which includes a penalty function that accounts for departures of the signal from perfect rectilinearity (see Appendix of OHMINATO et al., 1998, for details). Figure 12 shows vertical and horizontal cross sections of semblance distribution for the VLP velocity pulses in Figure 11. The location of the peak of semblance points to a source hypocenter at a depth of roughly 1 km beneath the northeast edge of Halemaumau. OHMINATO et al. (1998) used this result

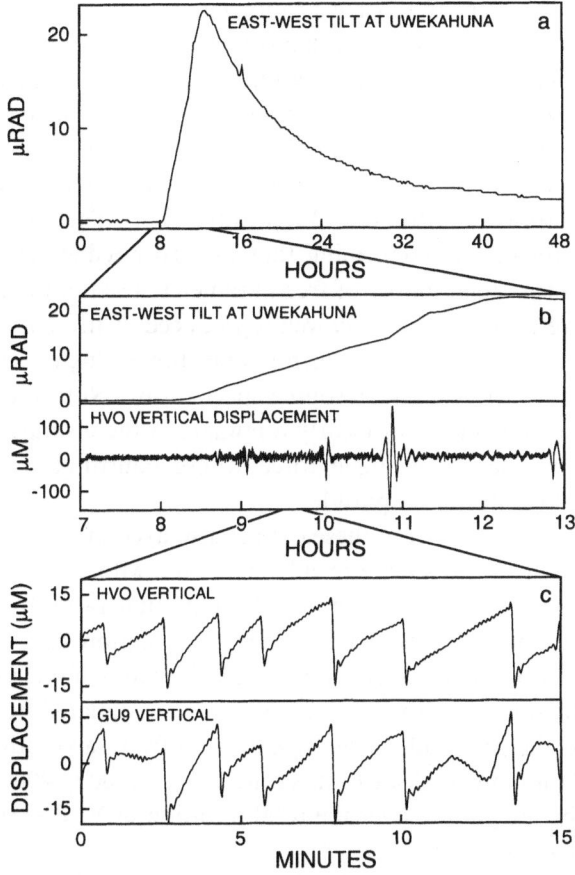

Figure 13

Tilt and displacements observed at Kilauea on February 1, 1996. (a) 48-hour-long record of tilt observed at the Uwekahuna vault near HVO. (b) Tilt at Uwekahuna and displacement at HVO observed over a 6-hour period starting at 0700 HST. (c) Sawtooth displacement waveforms recorded at HVO and GU9 (Fig. 12). Only the vertical component is shown as the radial displacement displays essentially the same waveform owing to the linearity of particle motion. (Reproduced from OHMINATO et al., 1998.)

as an initial source location in their inversion of the observed waveforms for source mechanism.

OHMINATO et al. (1998) carried out their inversion assuming a point source embedded in a homogeneous half space with compressional wave velocity $V_P = 3$ km/s, shear-wave velocity $V_S = V_P/\sqrt{3}$ km/s, and density $\rho = 2.7$ g/cm^3. They considered three cases including (1) a purely volumetric source, (2) a source including six moment tensor components but no single force, and (3) a source including six moment tensor components and three single force components. Inversions performed under the assumption of purely volumetric tensor components M_{xx}, M_{yy} and M_{zz} resulted in poor fits of the observed waveforms. The waveform match was significantly improved when six moment tensor components, but no single force, were included in the source mechanism. However, residual errors were still found to be high, suggesting that some stations were not well explained. Only in the last case, in which six moment tensor components and three single force components were used to express the source mechanism, were the residual errors observed to be significantly reduced. OHMINATO et al. (1998) evaluated each model by calculating the Akaike Information Criterion (AIC) (AKAIKE, 1974). They found the minimum value of the AIC for case 3, confirming that the error reduction was not solely a result of the increase in the number of free parameters used in the model. Figure 14 shows that the waveforms are generally well reproduced by inversion of the VLP data assuming six moment tensor components and three single force components. Figure 15 shows the corresponding source mechanism. Note that the volumetric components of the moment tensor clearly dominate in these solutions. There is also evidence of the presence of a single force whose contribution to the observed waveforms is roughly 10% in amplitude.

The source model derived from these data is displayed in Figure 16. The solution shows a compression of the source in all directions, with a dominant eigenvector slightly inclined from the vertical direction. The amplitude ratios for the three axes of the moment tensor are roughly 1:1:2, hence the solution can be interpreted as representative of a sub-horizontal crack or sill-like structure if one assumes a Poisson ratio $v = 1/3$ in the source region. The volume change estimated from the moment is about 3000 m^3. As the amplitude of the signal varies from sawtooth to sawtooth, the volume estimated for individual sawtooths ranges from 1000 to 4000 m^3. The volume budget obtained by integration over the total duration of VLP seismic activity is on the order of 500,000 m^3.

As indicated above, the inversion results also provide evidence for the presence of a single force in the source mechanism of the VLP events at Kilauea. A single force can be generated by a linear exchange of momentum between the source volume and the rest of the earth (TAKEI and KUMAZAWA, 1994). For example, a vertical single force may be generated as a consequence of the release of gravitational energy in the source volume. Such situation occurs during the ascent of a slug of gas in a column of liquid. As the slug rises toward the surface, liquid moves downward to fill the void left behind by the ascending gas. The sinking of dense liquid associated with the ascent of less dense gas changes the density structure of the fluid column and releases gravitational energy. Another example of vertical single force is the recoil force generated by a volcanic jet during an eruption (KANAMORI et al., 1984; CHOUET et al., 1997). The shear traction due to the flow of viscous liquid in a conduit provides yet another means to generate a single force (CHOUET, 1996b). In the inversion results shown in Figure 15, the force is synchronous with the compression of the crack, consistent with a discharge of fluid from the crack. OHMINATO et al. (1998) attributed the origin of this force to the drag force generated by the flow of viscous magma through a narrow constriction in the flow path.

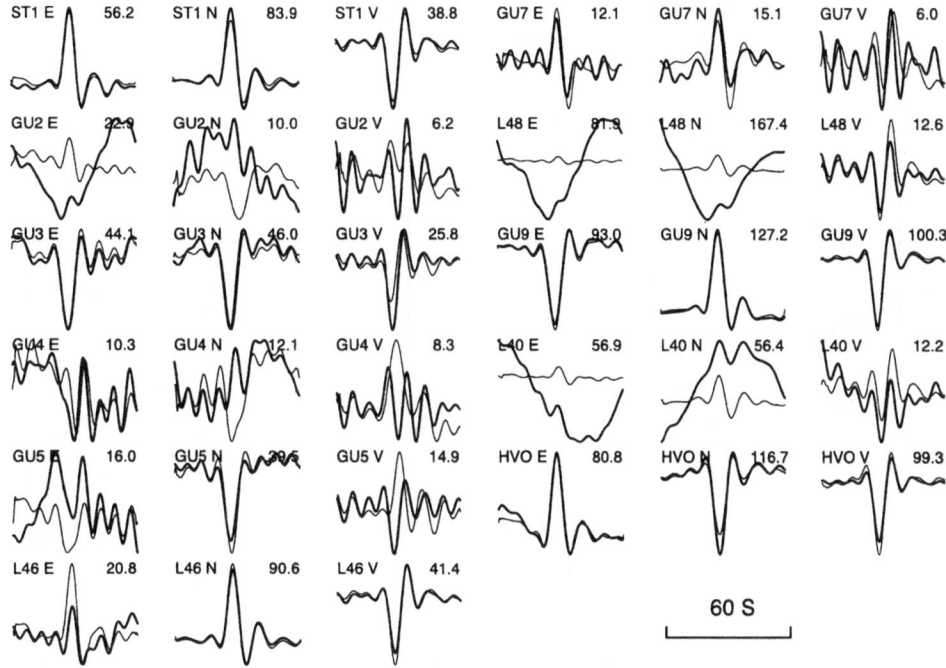

Figure 14

Waveform match obtained for a point source with mechanism consisting of six moment tensor components and three single forces. Horizontal components at GU2, L48, and L40 were not included in the inversion because these traces are contaminated by noise. The source is embedded at a depth of 1 km in a homogeneous half space. Thin and bold lines represent synthetics and data, respectively. The station code and component of motion are indicated at the upper left of each seismogram. Numbers at the upper right of the seismograms indicate the peak to peak amplitudes of the seismograms in units of 10^{-7} m/s. (Reproduced from OHMINATO et al., 1998.)

Momentum conservation in the overall source-earth system requires that the drag force generated by viscous fluid flow in the conduit must be counterbalanced by a force in the opposite direction. Such restoring force is not apparent in the F_z component in Figure 15. One possible explanation for this may be that the compensating force is generated by a slow process operating over a time scale that is beyond the responses of the broadband seismometers monitoring Kilauea.

A conceptual model which may explain both seismic data and total lava flow is shown in Figure 17. The mechanism involves the injection of a large slug of gas from the summit magma reservoir into the crack. Both liquid magma and gas flow through a narrow constriction at the crack outlet. Initially, liquid magma occupies the entire outlet cross section and gas accumulates in the crack, deforming the crack walls as the back pressure of gas increases in the crack. Eventually, the gas slug deforms and rapidly squeezes past the constriction along with the liquid, triggering the rapid deflation of the crack in the process. This sequence is repeated with successive slugs. In this model, the restraining force of the liquid on the gas slug acts like a self-

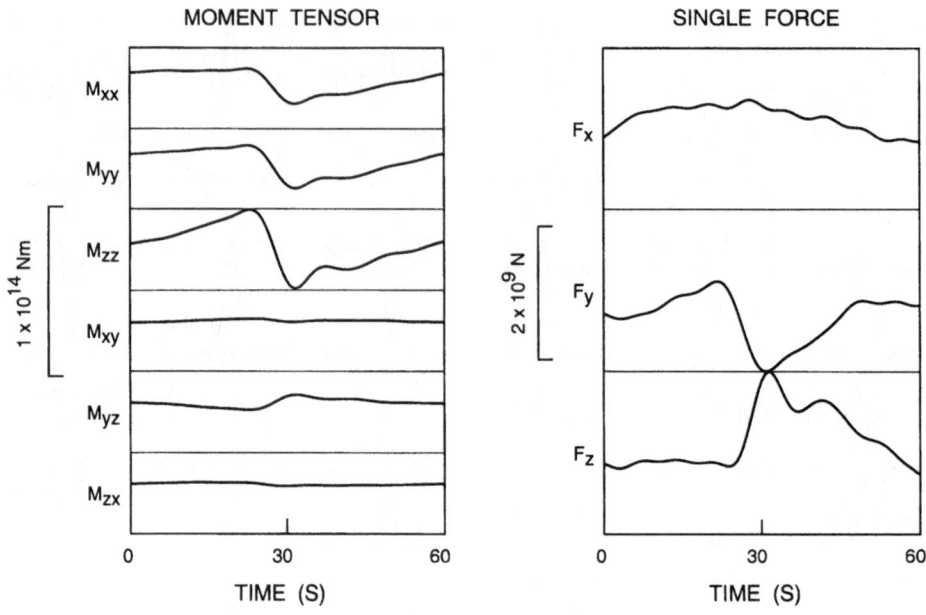

Figure 15
Source time functions obtained for a point source with mechanism consisting of six moment tensor components and three single forces. (Reproduced from OHMINATO et al., 1998.)

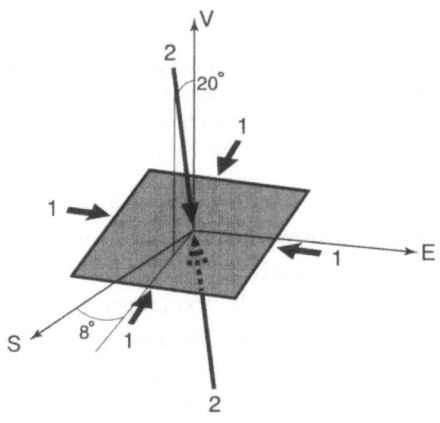

Figure 16
Source model obtained from inversion of VLP data at Kilauea. The amplitude ratios for the three axes of the moment tensor are 1:1:2. This solution is representative of a subhorizontal crack if one assumes a Poisson ratio $v = 1/3$ in the source region. (Reproduced from OHMINATO et al., 1998.)

activated viscosity-controlled valve. In the stratified flow through the nozzle, the liquid moves at a steady slow pace on the order of m/s but the gas flow itself is choked in the narrow opening between the liquid/gas interface and the upper wall

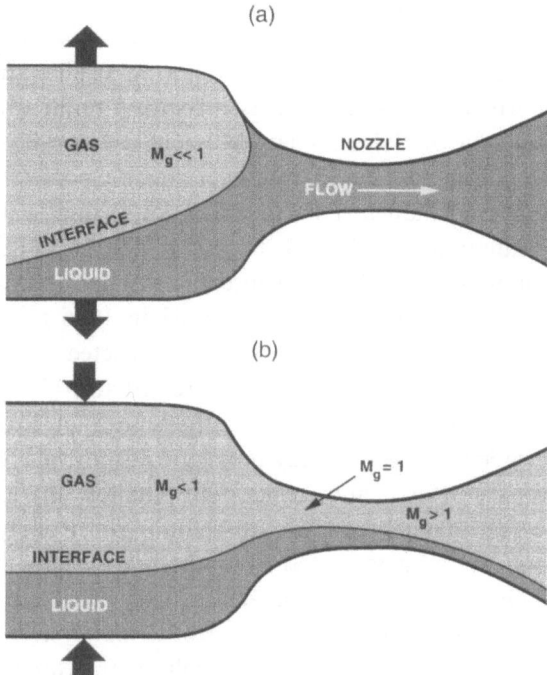

Figure 17

Conceptual model of separated gas-liquid flow through a converging-diverging nozzle under choked conditions (after WALLIS, 1969, pp. 71–74). (a) Inflation phase showing gas accumulating upstream of the nozzle and building excess pressure and deforming the crack as a result. The gas slug is essentially stationary and the Mach number of the gas is $M_g \ll 1$. This phase coincides with the upgoing ramp in the observed sawtooth displacement signals. (b) Separated gas-liquid flow through the nozzle under compound choked conditions. The gas flow is choked ($M_g = 1$) in the nozzle defined by the liquid-gas interface and upper solid wall, and is supersonic ($M_g > 1$) immediately downstream of the nozzle. There is a maximum possible gas flow rate which is fixed by the rate of liquid discharge for given upstream conditions and fixed throat geometry. In the limit, liquid fills the duct entirely, and there is no gas flow. (Reproduced from OHMINATO et al., 1998.)

(WALLIS, 1969). The formation of a shock associated with compound choking of the flow may act as a trigger of acoustic oscillations of the liquid/gas mixture which may be at the origin of the LP event with characteristic resonant frequency of 0.4 Hz observed in conjunction with the rapidly downgoing part of the sawtooth displacement signals shown in Figure 11.

Distinct VLP waveforms were again observed during a reinflation of the Kilauea summit following a 30 μrad deflation episode on January 29–30, 1997 (unpublished data, USGS Hawaiian Volcano Observatory). During February 1997, periods of inflation and minor deflation were observed, and each time a sharp transition from deflation to inflation occurred, VLP waveforms with periods of 30–40 s were detected by the Kilauea broadband network (Fig. 29a). As seen in the volcanic crisis of February 1, 1996, the VLP velocity waveforms observed in February 1997 share common features among each other, pointing again to a process involving the repeated activation of a fixed source. Particle motions associated with each pulse are approximately linear, exhibiting compressional motion at all stations, and analyses of semblance and particle motion are consistent with a point source located 1 km beneath the northeast edge of Halemaumau. Moment tensor inversions of these data are well explained by a transport mechanism operating on a crack linking the summit reservoir to the east rift of Kilauea. The crack location is similar to that of the crack activated during the inflation episode on February 1, 1996 (see Fig. 12), although the transport mechanism involved in the two sources is different (CHOUET and DAWSON, 1997). The proximity of the two sources observed in 1996 and 1997 points to a reactivation of a permanent plexus of cracks in a restricted source zone below the northeast edge of Halemaumau. As seen in Figures 10 and 12, the magma conduit imaged from VLP data lies directly beneath the source of hydrothermal energy imaged from LP seismicity. The unsteady magma transport accompanying summit reinflation in February 1997, and concurrent shallow hydrothermal activity immediately above the magma conduit, therefore are strongly suggestive of a dynamic link between the magmatic and hydrothermal systems beneath Halemaumau. Accordingly, enhanced hydrothermal activity may be viewed as the result of an increased flow of heat into the base of the hydrothermal system resulting from an increase in the discharge of magmatic gases through the fractured rock mass capping the magma conduit.

Acoustic Properties of Magmatic and Hydrothermal Fluids Derived from Spectral Analyses of LP Events

Long-period seismicity, including LP events and tremor, has been widely observed in relation to magmatic and hydrothermal activities in volcanic areas and has been recognized as a precursory phenomenon for eruptive activity (CHOUET et al., 1994; CHOUET, 1996a). The waveform of the LP event is characterized by simple decaying harmonic oscillations except for a brief time interval at the event onset (Fig. 18). This characteristic signature is commonly interpreted as oscillations of a fluid-filled resonator in response to a time-localized excitation. By the same token, tremor may be viewed as oscillations of the same resonator in response to a sustained excitation. LP events are particularly important in the quantification of volcanic and hydrothermal processes, because the properties of the resonator system at the source of this event may be inferred from the complex frequencies of the decaying harmonic oscillations in the tail of the seismogram. The damped oscillations in the LP coda may be characterized by two parameters, f and Q. The constant f is the frequency of the dominant mode of oscillation, and the parameter Q represents the quality factor of the oscillatory system. The observed Q of LP events may be expressed as

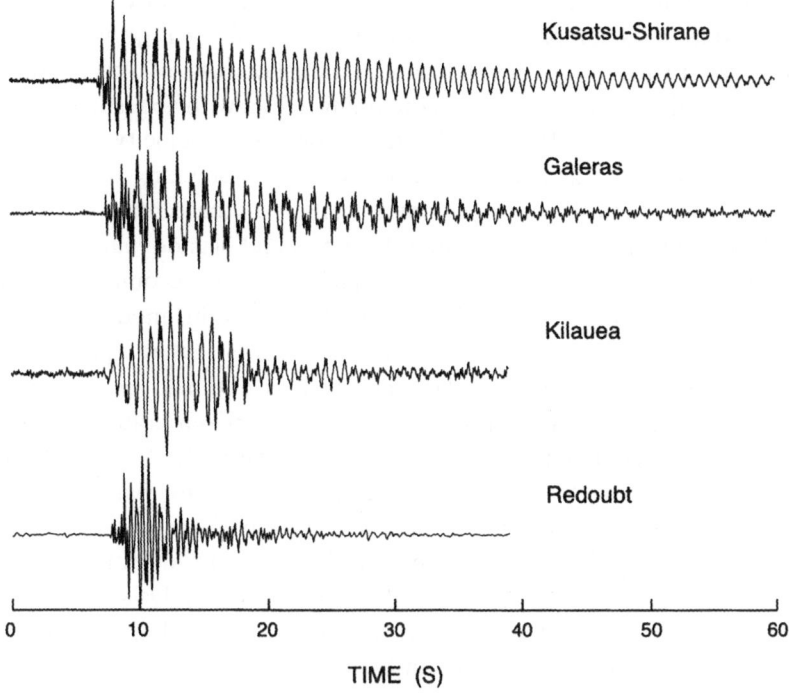

Figure 18

Typical signatures of long-period events observed at Kusatsu-Shirane, Galeras, Kilauea and Redoubt volcanoes. The signatures are all characterized by a harmonic coda following a signal onset enriched in higher frequencies. (Reproduced from KUMAGAI and CHOUET, 1999.)

$$Q^{-1} = Q_r^{-1} + Q_i^{-1}, \tag{3}$$

where Q_r^{-1} and Q_i^{-1} represent the radiation and intrinsic losses, respectively (AKI, 1984). Typical frequencies observed for LP events are in the range 0.5–5 Hz (CHOUET, 1996a), and typical observed Q range from values near 1 to values larger than 100. For example, the events at Kilauea and Redoubt in Figure 18 are representative of values of Q between 20 and 50, and the events at Kusatsu-Shirane and Galeras are characterized by values of Q higher than 100.

To compare the complex frequencies of LP events with those predicted by a resonator model we use a crack geometry, which is appropriate for mass transport conditions beneath a volcano. The fluid-filled crack model was originally proposed by AKI et al. (1977b) and has been extensively studied by CHOUET (1986, 1988, 1992) using a more detailed description of the coupling between fluid and solid. Chouet's studies showed that the fluid-filled crack generates a very slow wave propagating along the crack wall, which he called the "crack wave." The asymptotic behavior of this wave for a crack of infinite length was investigated by FERRAZZINI and AKI (1987) in an analytical study of normal modes trapped in a liquid layer sandwiched between two solid half spaces. The crack wave leads to more realistic estimates of the size of the resonator as compared to a resonator with spherical geometry (CROSSON and BAME, 1985; FUJITA *et al.*, 1995). Chouet's model consists of a rectangular crack with length L, width W, and aperture d, containing a fluid with acoustic velocity a and density ρ_f, embedded in an elastic solid with compressional velocity α and density ρ_s (Fig. 19). The crack is excited into resonance by a pressure transient applied symmetrically over small areas of the crack walls near the center of the crack. A solution of the coupled equations of fluid dynamics and elastodynamics is obtained by finite differences. Note that the crack model does not account for dissipation mechanisms within the fluid so that simulations done with this model predict quality factors due to the radiation loss only. Intrinsic losses are treated separately.

Once the space-time motion of the crack wall is obtained in response to the pressure transient, the ground response due to a fluid-driven crack embedded in a homogeneous or layered half space can be easily synthesized using the discrete wavenumber method (BOUCHON, 1979, 1981) and propagator-based formalism of CHOUET (1987). Figure 20*a* shows a vertical velocity waveform calculated at the free surface in the near field of a resonating fluid-filled crack embedded in a homogeneous half space. Figure 20*b* shows the results from a Sompi analysis of the synthetic waveform in the form of a frequency–growth rate $(f - g)$ diagram (KUMAZAWA *et al.*, 1990). Densely populated regions in the $f - g$ diagram represent the signal for which the complex frequencies are stably determined for different autoregressive (AR) orders (KUMAZAWA *et al.*, 1990), while scattered points represent incoherent noise. Figure 20*c* shows the corresponding normalized FFT spectrum. As can be

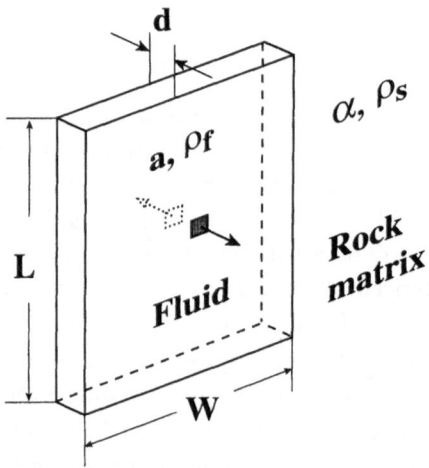

Figure 19
Geometry of the fluid-filled crack model. The crack has length L, width W, and aperture d, and contains a fluid with sound speed a and density ρ_f. The crack is embedded in a solid with compressional wave velocity α and density ρ_s. Excitation of the crack is provided by a pressure transient applied symmetrically on both walls over the small areas indicated by the grey patches near the center of the crack.

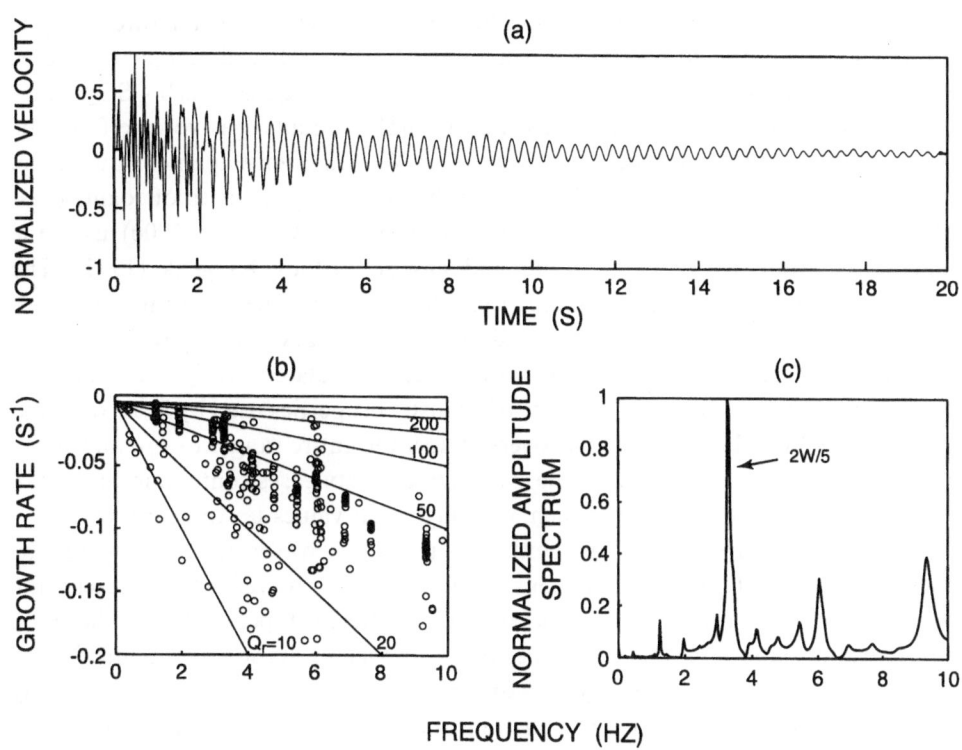

Figure 20
(a) Synthetic velocity waveform observed at an epicentral distance of 500 m and azimuth $\phi = 45°$ measured from the crack trace for a vertical crack (vertical extent $L = 150$ m and horizontal extent $W = 75$ m) buried at a depth of 500 m. The aspect ratio is $L/d = 10^4$, and the parameters of the fluid and solid are $a = 300$ m/s, $\rho_f = 2120$ kg/m^3, and $\alpha = 4500$ m/s, $\rho_s = 2650$ kg/m^3. (b) Plots of the complex frequencies of wave elements for all trial AR orders estimated for the waveform in (a). The straight lines represent lines along which the Q_r factor is constant. (c) Normalized amplitude spectrum corresponding to the waveform in (a). (Reproduced from KUMAGAI and CHOUET, 2000.)

readily seen in Figures 20*b* and 20*c*, there are four dominant peaks and other minor peaks in the 0–10 Hz frequency range, which show an almost constant Q_r factor due to radiation near 50. The peak at frequency near 3.5 Hz corresponds to the transverse mode with wavelength $2W/5$. This mode was used by KUMAGAI and CHOUET (2000) to examine the dependence of the complex frequency on the parameters α/a and ρ_f/ρ_s.

Using the Sompi method KUMAGAI and CHOUET (2000) estimated the factor Q_r and dimensionless frequency $v = fL/\alpha$ of the mode $2W/5$ for α/a ranging from 5 to 45 and ρ_f/ρ_s ranging from 0.05 to 1. The results of their study are shown in Figure 21. The factor Q_r monotonically increases with increasing α/a, and slightly increases with decreasing ρ_f/ρ_s. Thus, Q_r increases with increasing impedance contrast $Z = \alpha\rho_s/(a\rho_f)$, although Q_r more strongly depends on α/a than on Z. The dimensionless frequency v decreases with increasing α/a and ρ_f/ρ_s, and depends equally on both of these parameters. The fluids considered are mixtures of gas, liquid, and solid, including gas-gas mixtures ($H_2O - CO_2$), liquid-gas mixtures (water $- H_2O$, basalt $- H_2O$), and dusty and misty gases (ash $- SO_2$, water droplet $- H_2O$). Sound speeds and densities of bubbly liquids (gas-volume fractions less than 10%) were estimated using the Van Wijngaarden-Papanicolaou model (COMMANDER and PROSPERETTI, 1989). For the acoustic properties of foams (gas-volume fractions between 10 and 90%), KUMAGAI and CHOUET (2000) used the adiabatic equation of state for liquid and gas derived by KIEFFER (1977). The acoustic properties of gas-gas mixtures were derived using the ideal mixing theory (e.g., MORRISSEY and CHOUET, 2001), and those of dusty and misty gases (ash $- SO_2$ gas and water droplet $- H_2O$ gas mixtures) were calculated using the model of TEMKIN and DOBBINS (1966). The range of acoustic properties used by KUMAGAI and CHOUET (2000) for the crack model covers almost the entire ranges of α/a and ρ_f/ρ_s for the various types of fluids, with the exception of dusty gases containing very small weight fractions of pure gas for which α/a may reach values up to nearly 60.

Figure 22 shows examples of synthetic far-field P wavetrains that would be generated in an infinite homogeneous medium. These waveforms represent the impulse response of a fluid-filled crack containing different types of fluids. The crack geometry and spatio-temporal properties of the applied excitation are identical in all the examples shown. Each signal shows distinct resonance characteristics, suggesting that LP signatures may be used to constrain the fluid composition at the source.

A comparison of Q_r^{-1} with Q_i^{-1} in bubbly fluids is shown in Figure 23. The estimates of Q_i^{-1} represent the contributions from three damping mechanisms, namely, viscosity, heat transfer, and acoustic radiation. In bubbly water as in bubbly basalt, Q_i^{-1} becomes comparable to or larger than Q_r^{-1} in fluids containing bubbles whose radii are larger than 1 mm. Intrinsic absorption becomes trivial in fluids containing bubbles with radii smaller than 0.3 mm so that Q becomes essentially equivalent to Q_r in such fluids.

Figure 24 shows a comparison of Q_r^{-1} with Q_i^{-1} in dusty and misty gases. The estimates of Q_i^{-1} represent the combined loss mechanisms of viscosity and thermal diffusion. The factor Q_i^{-1} in dusty gases containing particles with radii larger than 10 μm is comparable or larger than Q_r^{-1}. For solid particles with sizes 1 μm or less intrinsic absorption is trivial. Similar results are observed in a misty gas where intrinsic

▶

Figure 21

Plots of α/a versus ρ_f/ρ_s curves for various types of fluids, and Q_r and ν contours for the crack model. Solid dots mark points where the synthetic waveforms are calculated. The contour plots of Q_r and ν (yellow curves) are obtained by fitting quadratic polynomials to the calculated Q_r or ν values along profiles with constant ρ_f/ρ_s. Dashed, solid, and dash-dotted lines indicate α/a and ρ_f/ρ_s calculated for fluids at pressures of 5, 10, and 20 or 25 MPa, respectively, using $\alpha = 2700$ m/s, $\rho_s = 2500$ kg/m^3 at 5 MPa pressure, $\alpha = 4000$ m/s, $\rho_s = 2650$ kg/m^3 at 10 MPa pressure, and $\alpha = 5000$ m/s, $\rho_s = 2700$ kg/m^3 at 20 and 25 MPa pressure. Thermodynamic properties of the fluids can be found in KUMAGAI and CHOUET (2000). (Reproduced from KUMAGAI and CHOUET, 2000.)

absorption becomes trivial in the presence of water droplets with sizes 10 μm or less, but may dominate when water droplets of 100 μm or larger are present. Thus, dusty gases containing small-size particles or misty gases containing small-size droplets can both generate long-lasting oscillations characterized by large values of Q_r ranging up

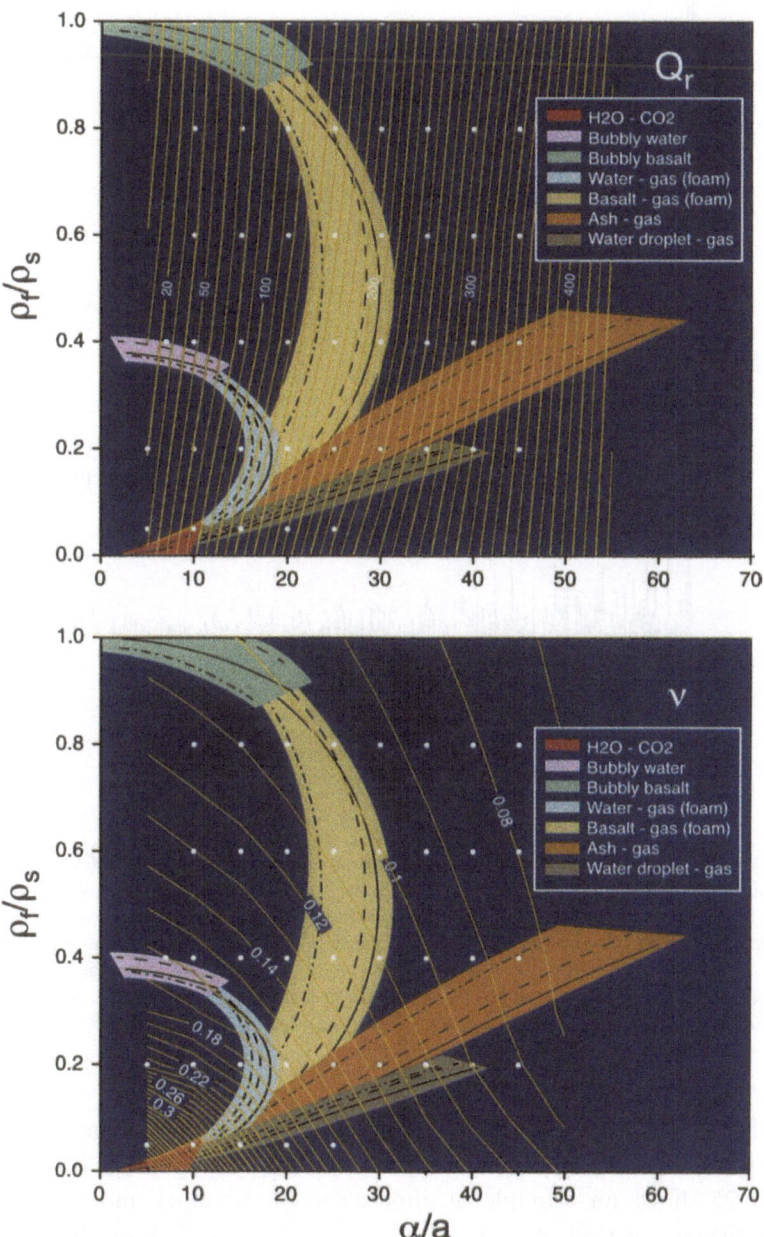

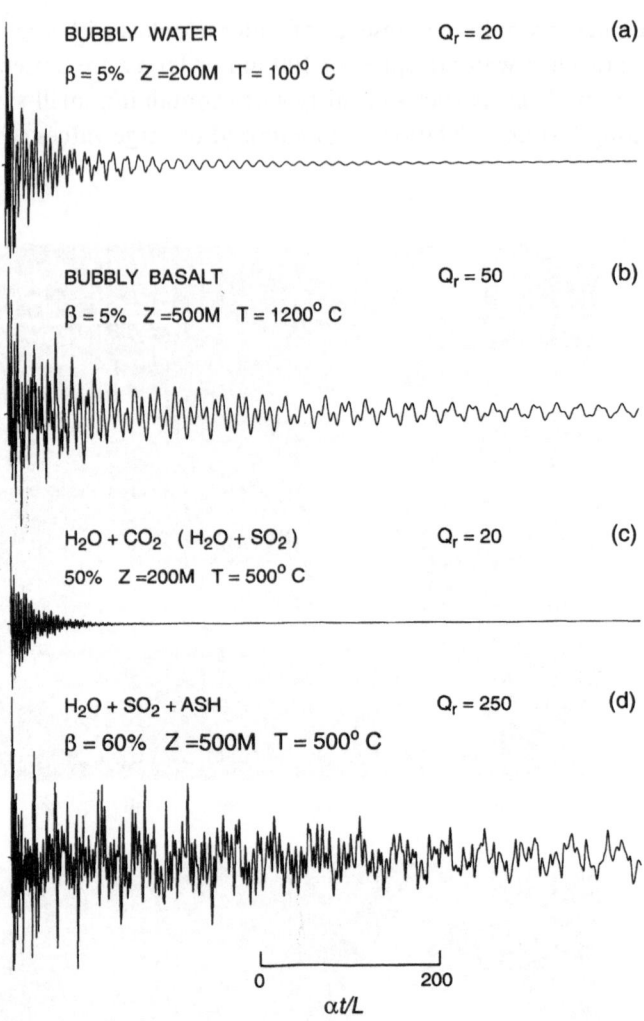

Figure 22
Synthetic far-field compressional waveforms produced by the excitation of a crack containing different types of fluids. The crack is embedded in a rock with fixed properties $\alpha = 2700$ m/s, $\rho_s = 2500$ kg/m^3. (a) Waveform for a crack at depth of 200 m containing bubbly water with 5% gas-volume fraction at a temperature of 100°C. (b) Waveform for a crack at depth of 500 m containing a bubbly basalt with 5% gas-volume fraction at a temperature of 1200°C. (c) Waveform for a crack at depth of 200 m containing a 50% mixture of either $H_2O - CO_2$ or $H_2O - SO_2$ gases at a temperature of 500°C. (d) Waveform for a crack a depth of 500 m containing an ash-laden gas with gas-volume fraction of 60% at a temperature of 500° C.

to several hundreds. KUMAGAI and CHOUET (2000) found that intrinsic absorption is trivial in gas-gas mixtures. No reliable estimates of Q_i are available for foams.

Figure 25 shows an example of application of the crack model to LP events recorded during a period of renewed activity at Kusatsu-Shirane Volcano, Japan.

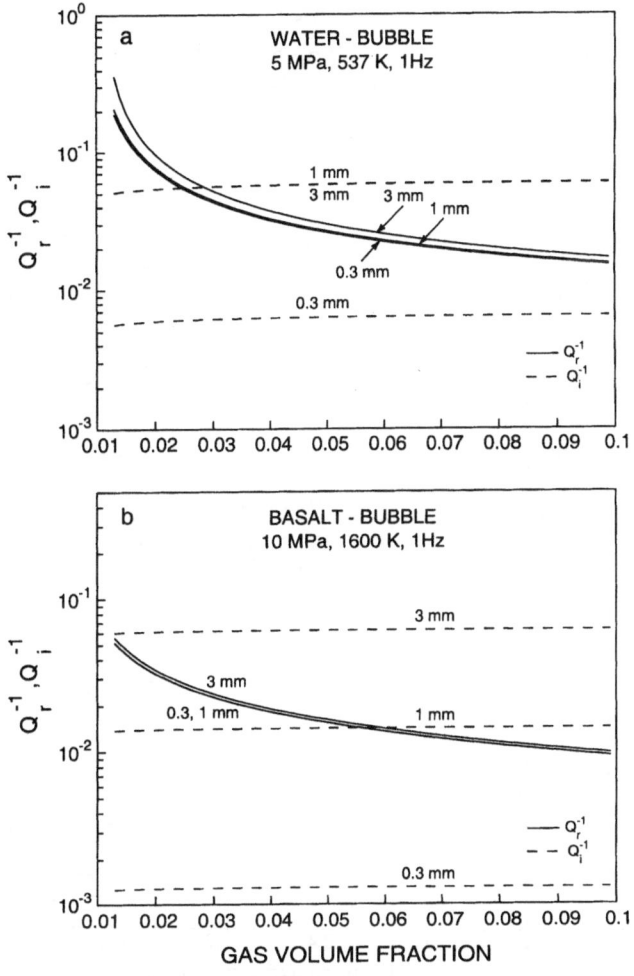

Figure 23

Comparison of Q_r^{-1} (solid lines) with Q_i^{-1} (dashed lines) for (a) bubbly water at 5 MPa and 537° K, and (b) bubbly basalt at 10 MPa and 1600° K as a function of gas-volume fraction for the three bubble radii 0.3, 1, and 3 mm, and frequency of the traveling wave of 1 Hz. (Reproduced from KUMAGAI and CHOUET, 2000.)

During this activity, LP events were frequently observed with hypocenters located several hundred meters below the summit crater (NAKANO et al., 1998). High-quality records of this LP seismicity were obtained in September–November 1992 from a network of seismometers surrounding the source region (NAKANO et al., 1998; KUMAGAI et al., 2002; NAKANO et al., 2002). The recorded events display a dominantly monochromatic signature (Fig. 18) with dominant frequency and factor Q both varying gradually with time (Fig. 25). Measurements of the dominant frequency and factor Q obtained by KUMAGAI et al. (2002) for individual events recorded over a 70-day period are shown together with fits obtained from the crack

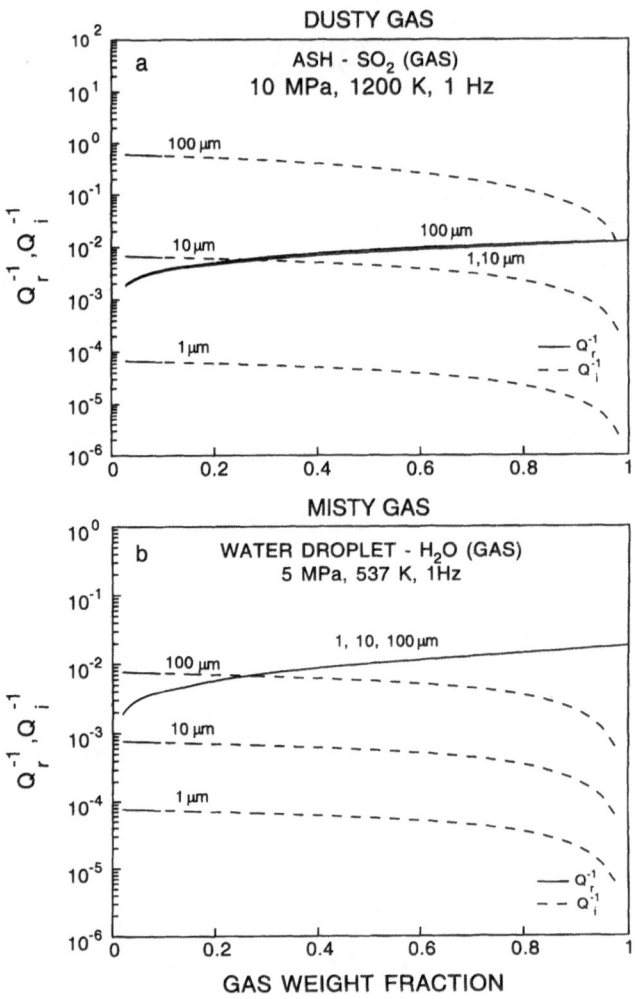

Figure 24
Comparison of Q_r^{-1} (solid lines) with Q_i^{-1} (dashed lines) for (a) ash-SO_2 gas mixture (dusty gas) at 10 MPa and 1200° K, and (b) water droplet-H_2O gas mixture (misty gas) at 5 MPa and 537° K for the three particle radii 1, 10, and 100 μm, and frequency of the traveling wave of 1 Hz. (Reproduced from KUMAGAI and CHOUET, 2000.)

model assuming that the fluid in the crack is a misty gas. The gas starts off fairly wet with 10% gas-weight fraction and ends up dry with 100% gas-weight fraction over the course of 70 days, suggesting a gradual heating of the LP source region. As the dominant mode of oscillation cannot be identified in these data, fits are obtained for different assumed modes. The fit in Figure 25a assumes that the mode $2W/5$ is the dominant mode and yields a crack length of 560 m. If one assumes instead that the dominant frequency represents the mode $2W/3$, one obtains essentially the same fit with a shorter crack length of 220 m (Fig. 25b).

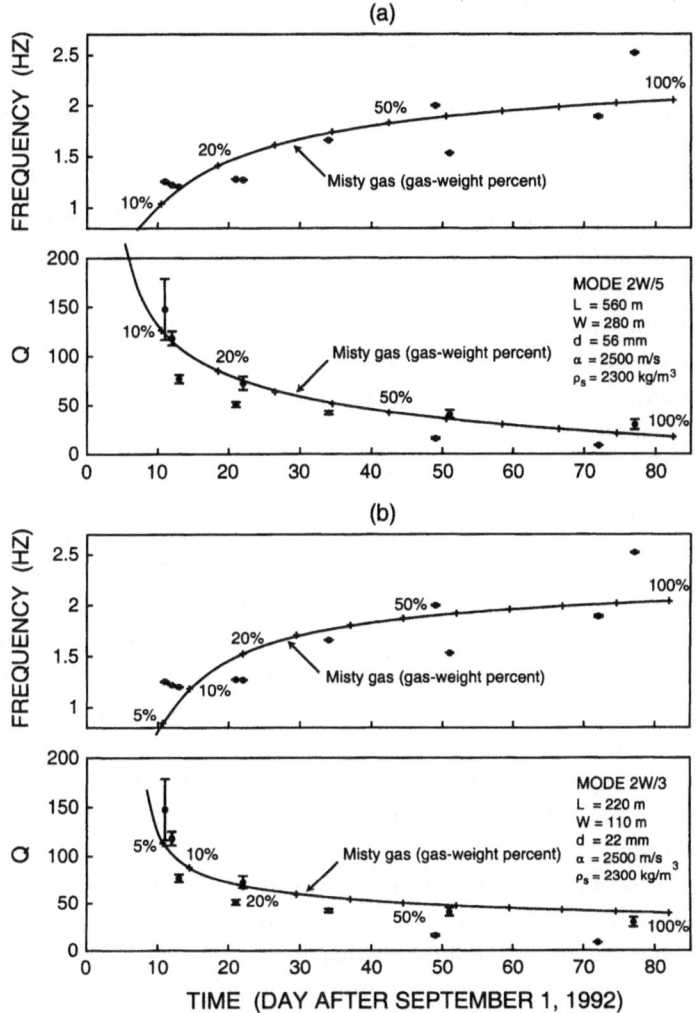

Figure 25
Temporal variation of Q and dominant frequency (solid dots with error bars) observed for LP events at Kusatsu-Shirane Volcano, Japan, in 1992. The solid line is the fit to these data obtained from the crack model assuming a crack filled with a misty gas. (a) Fit obtained assuming the observed dominant frequency represents the mode $2W/5$. (b) Fit obtained assuming the mode $2W/3$ dominates. (From KUMAGAI *et al.*, 2002.)

Experimental Modelling of the Source Dynamics of Tremor

Although a good understanding of the resonance properties of resonators in the earth has been achieved, we still have a relatively poor understanding of the actual fluid dynamics and attendant pressure fluctuations underlying the excitation of LP events and tremor (JULIAN, 1994; CHOUET, 1996b; MORRISSEY and CHOUET, 1997). As flows through volcanic conduits are not directly observable *in situ*, a useful approach to achieve a better understanding of such flows is to use analogue laboratory experiments. Expanding gas-liquid flows, designed to be analogous to those in volcanic conduits, were generated in the laboratory by LANE *et al.* (2001) using organic gas-gum rosin (natural pine resin) mixtures expanding in a narrow vertical tube of borosilicate glass connected to a vacuum chamber (Fig. 26). The tube is isolated from the vacuum chamber by a thin plastic diaphragm and an eruption is triggered by the rupture of the diaphragm. Depending on the level of volatile supersaturation, different flow regimes are observed, ranging from gentle unfragmented to violent fragmenting flows. The photograph in Figure 26 illustrates the growth of foam from the liquid interface and the unfragmented and fragmented regions of flow.

Figure 27 shows the pressure traces obtained at various heights in the tube during a fragmenting diethyl ether driven flow. Colors represent different flow regimes identified in video and still-camera images, and in optical and pressure sensor data. Figure 28 gives a schematic representation of the flow regimes and pressure oscillations in the fragmenting foam flow based on the positions of the flow regions 0.3 s into the flow (see Fig. 27). Region R4, shaded yellow, represents the source liquid. Region R3 in green shading, represents the foam growing from the surface of the source liquid. Region R2 in shades of pink and red, identifies a region where the foam becomes detached from the wall. This region is subdivided into a lower region R2b (pink in Fig. 28), in which gas pockets are developing at the wall, and an upper region R2a (red in Fig. 28) of gas-at-wall annular flow. Region 1, shaded blue, consists of foam fragments dispersed in gas. Region R1 is further subdivided into R1b (light blue in Fig. 28) and R1a (dark blue in Fig. 28) on the basis of the pressure oscillation behavior.

With the exception of R4, each region displays a characteristic pressure oscillation behavior (Fig. 27). Pressure oscillations are not detectable in R4 despite oscillations being present in the flow regions above this region. This implies that no oscillations are generated within the source liquid and also that no pressure disturbances can penetrate the liquid source region from the foam regions above. The reason for this probably lies in the strong impedance contrast between liquid and foam. This foam layer apparently acts like an acoustic barrier that prevents the pressure disturbances from propagating down into the source liquid.

Pressure oscillations in R3 display dominant frequencies in the range 10–30 Hz with pressure fluctuations $\Delta P/P$ of 2–10%, interpreted by LANE *et al.* (2001) as longitudinal oscillation modes of the foam column. Low-amplitude oscillations in the 80–100 Hz range are also present and may represent weak radial oscillation modes. A 50% increase in pressure oscillation amplitude is observed across the boundary from R3 to R2b. Pressure oscillations in region R2 have dominant frequencies of 10–30 Hz similar to those observed in region R3 and also show more prominent spectral peaks in the 80–100 Hz band. These oscillations were interpreted by LANE *et al.* (2001) to reflect both longitudinal and radial resonances in the detached section of the flow.

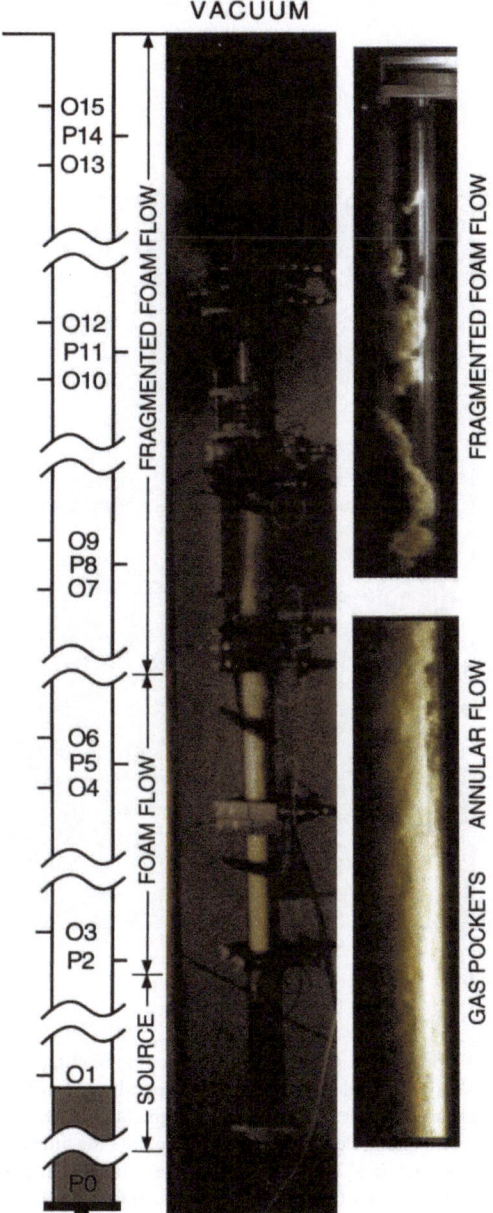

Figure 26

Experimental set up used by LANE *et al.* (2001) in their investigation of the dynamics of expanding gas-liquid flows. A 1.5-m-long tube with internal diameter of 3.8 cm is connected to a vacuum chamber with volume of $0.4\,m^3$. Optical (**O**) and pressure (**P**) transducers monitor the flow conditions along the tube. A still film camera image shows a fragmenting diethyl ether driven foam flow in progress, illustrating the growth of foam from the liquid interface and the unfragmented and fragmented regions of the flow. Digital video frames (1/400 s) show typical fragmented, gas-at-wall annular, and gas-pocket-at-wall regions of the flow. (Reproduced from LANE *et al.*, 2001.)

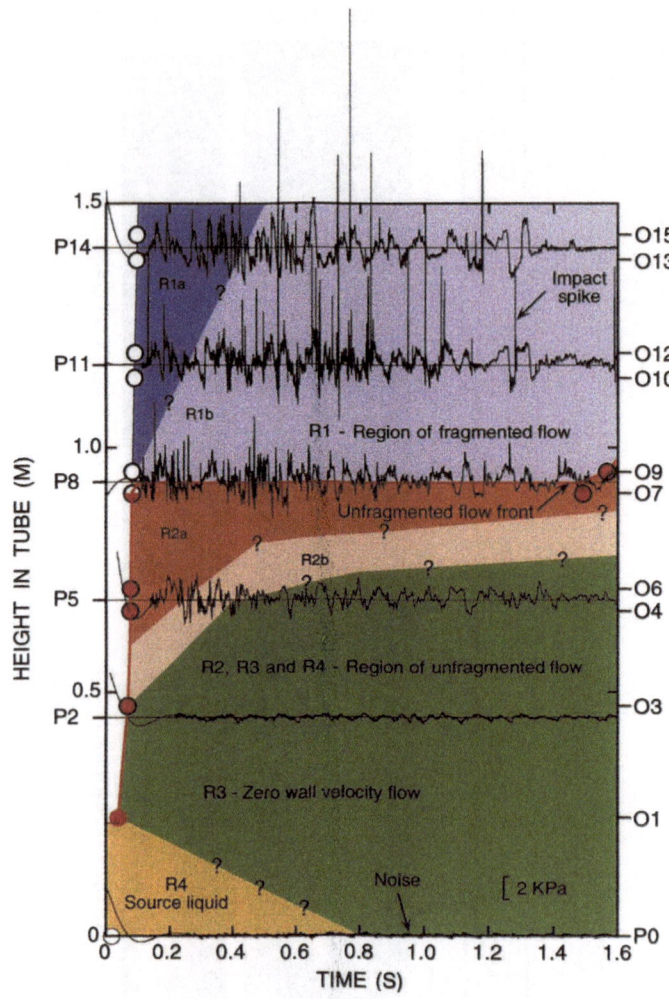

Figure 27

Height versus time plot of optical and pressure data for a fragmenting diethyl ether driven flow. The positions of the pressure (**P**) and optical (**O**) transducers are indicated at the left and right, respectively. Open circles mark the times at which the first 270 foam fragments are detected, and filled circles indicate the onset of unfragmented flow. The circle at zero height represents the time at which **P14** detected pressure falling below 99 kPa on decompression. The pressure data have been high-pass filtered at 5 Hz to remove transducer drift. The noise-free section of pressure data between 0.1 and 0.3 s is synthetic data included to reduce high-pass filter end effects. The pressure range (2 kPa) is indicated by the scale bar. Shading identifies regions in which different flow regimes are observed (see text for details). (Reproduced from LANE *et al.*, 2001.)

Wave packets with frequencies in the range 0.5–5 kHz are commonly observed in Regions R1b and R2a, rarely present in regions R1a and R2b, and never present in regions R3 and R4. Thus, wave packets are dominantly present when the flow is fully

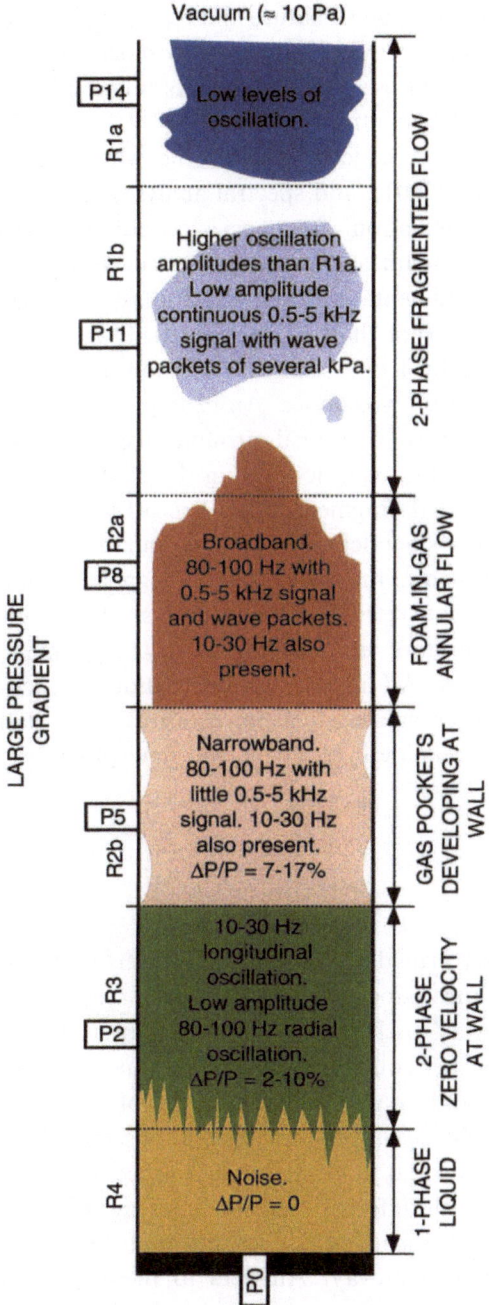

Figure 28
Schematic representation of flow regimes and pressure oscillations observed in a diethyl ether driven
fragmenting foam flow. The positions of the pressure transducers are indicated at the left and the flow regions
are representative of the conditions 0.3 s into the flow (see Fig. 27). (Reproduced from LANE et al., 2001.)

detached from the tube wall but still interacting with the wall on occasion. Some wave packets have a broadband signature and represent impacts of gum rosin on the tube walls, while others display a spectrum encompassing a series of narrow-band peaks attributed to resonant oscillation of the gas between foam slugs. Region R1a is marked by low levels of oscillations.

Pressure oscillations were detected over the whole range of flow conditions explored by LANE *et al.* (2001), and spectral analyses of these oscillations indicated that the dominant oscillation frequencies could be attributed to resonant oscillations in both the foam and open gas spaces within the experimental tube. Flows in which foam fragmentation occurred displayed a broader, higher frequency and more energetic spectrum than unfragmented flows. Interestingly, some of the pressure data obtained by LANE *et al.* (2001) show remarkable similarities to LP and VLP data recorded for volcanoes. An example of such similarity is shown in Figure 29 which compares a vertical ground displacement record obtained at Kilauea with pressure pulses recorded during the diethyl ether driven flow near the point of optically detected fragmentation. The VLP events at Kilauea occurred during a transition from slow deflation to rapid inflation of the volcano. There are marked similarities between the two types of signals. Although the reasons for these similarities are not yet clear, they suggest that similar fluid mechanical processes may be operating in both laboratory experiments and in the fluids moving beneath Kilauea. Additional documentation of such similarities between laboratory and field observations is a prerequisite to translating seismic information into a refined understanding of volcanic flow processes and modeling those processes.

Conclusions

Volcano seismology provides a unique and powerful tool to study the dynamics of active magmatic and associated hydrothermal systems, determine the *in situ* mechanical properties of such systems, and interpret the nature, evolution and extent of magmatic energy source regions, all of which advance our understanding of eruptive behavior and enable the assessment of volcanic hazards. Three key questions summarize the situation with seismic observations regarding the nature and distribution of fluids and fluid mechanical processes beneath a volcano. These are: (1) can we distinguish different types of fluids by their LP signatures; (2) what are the excitation mechanisms of LP events and tremor; and (3) what is the geometry of the magma pathway? Answers to these questions might potentially contribute to improved prediction capability and improved scientific responses to volcanic crises to avert volcanic disasters, provided that the information can be obtained and applied in near real time (CHOUET, 1996a). The promise of a forecasting strategy based on LP seismicity is quite evident in its successful

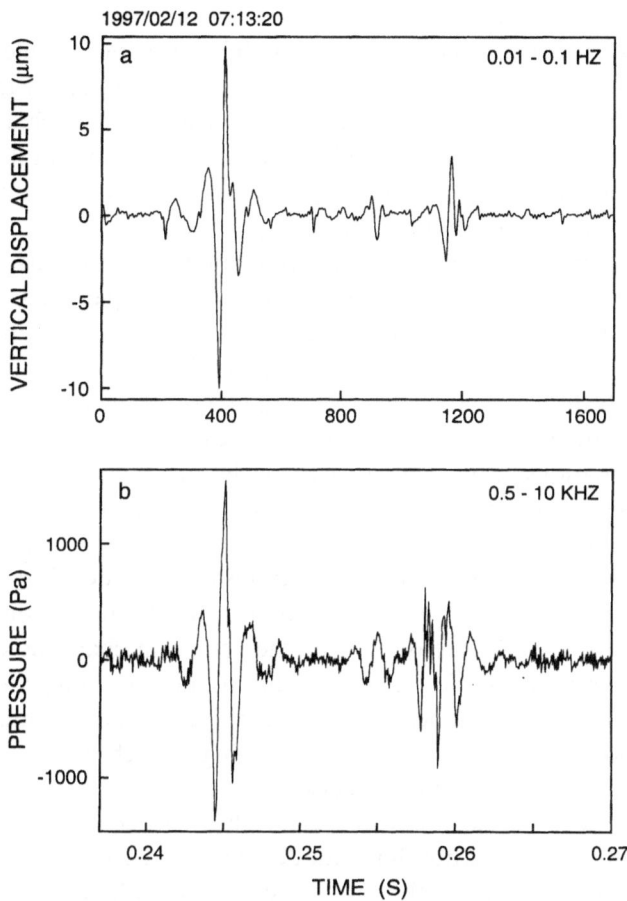

Figure 29

Comparison between (a) VLP displacement waveforms recorded at Kilauea Volcano on 12 February, 1997, during a transition from slow deflation to rapid inflation, and (b) pressure waveforms obtained at the boundary between regions R1 and R2 (see Figs. 27 and 28) in the diethyl ether driven fragmenting foam flow. The record in (a) has been band-pass filtered in the 0.01–0.1 Hz band, and the record in (b) has been similarly filtered in the 0.5–10 kHz band. Date and time at the start of the Kilauea record are indicated at the upper left. Notice the different time scales.

application to precursory swarms of LP events associated with the pressurization sequences leading to dome-destroying eruptions at Redoubt Volcano, Alaska (CHOUET et al., 1994; MORRISSEY and CHOUET, 1997), and its potential application to precursory LP events at volcanoes such as Galeras, Colombia (FISCHER et al., 1994; GIL CRUZ and CHOUET, 1997; STIX et al., 1997; NARVÁEZ et al., 1997), Soufriere Hills, Monserrat (WHITE et al., 1998), Popocatepetl, Mexico (ARCINIEGA-CEBALLOS et al., 2000), and Pinatubo, Philippines (WHITE et al., 1992; WHITE, 1996; RAMOS et al., 1999).

The acoustic properties of fluids may be inferred from the spectrum of LP seismicity. A caveat is that a crack containing different types of fluids can produce similar values of Q, so that interpretations of fluid types based on the observed Q may be manifold and equivocal. Uncertainties in intrinsic loss mechanisms and rock matrix properties and in the pressure and temperature conditions at the source of LP events further complicate the task of interpreting their source properties. Geological and geochemical constraints on possible fluid types beneath volcanoes are necessary to further constrain the seismological interpretation of LP events. In particular, further advances in real-time geochemical monitoring capabilities are needed to provide robust continuous data on volcanic fluids. Improvements in the resolution of 3-D velocity structures of volcanoes and improved locations for LP seismicity are also required to better fix the rock properties and thermodynamical conditions at the source. With additional constraints from geochemical and geological data, the spatial and temporal variations in the complex frequencies of LP events recorded on dense networks of seismometers offer great potential to image magmatic and hydrothermal systems beneath volcanoes.

As the resolution of the 3-D velocity structures of volcanoes sharpens, leading to more precise locations for LP events and improving the resolution of the forces operating at the source, the resolution of the source mechanisms of shallow LP events improves as well. Seismic source mechanisms, however, only yield clues about the fluid dynamics; thus, well-designed laboratory experiments simulating natural eruptive conditions, along with numerical modeling of such flows, are required for the identification of plausible fluid processes.

Although a direct link has been well established between shallow LP seismicity and eruptive activity, the relationship with deeper LP events is less clear. LP events are regularly observed at depths of 30–60 km beneath Kilauea, where their occurrence appears to be more directly related to deep magma supply dynamics than surface activity (AKI and KOYANAGI, 1981; SHAW and CHOUET, 1991). A low background of deep LP events is not uncommon, and may indeed be the rule rather than the exception as accumulating evidence now increasingly suggests (UKAWA and OHTAKE, 1987; HASEGAWA et al., 1991; PITT and HILL, 1994). The difficulty here has to do with our lack of knowledge concerning the character and dynamics of deep-seated fluid transport under volcanoes. The origin of deep LP seismicity still remains a major challenge to our understanding of eruptive phenomena.

Moment tensor inversions of VLP signals associated with variations in magma transport rate may be used to map the magma conduit structure (location, geometry, orientation, size) and to resolve the mass transport budget of magmatic fluids beneath a volcano. However, laboratory and numerical experiments of fluid mechanical processes analogous to those in volcanoes are again required to interpret flow processes compatible with the source mechanisms derived from seismic data.

Recent laboratory studies (LANE *et al.*, 2001) have elucidated self-excitation mechanisms inherent to fluid nonlinearity which are leading to a new understanding of the mechanisms underlying volcanic tremor. In parallel to these advances, numerical studies of multiphase flows are starting to shed light on both the micro- and macrophysics of these flows (e.g., VALENTINE and WOHLETZ, 1989; DOBRAN *et al.*, 1993), while new advances with computer models coupling the fluid dynamics with the elastodynamics in the solid (NISHIMURA and CHOUET, 1998) constitute the first steps toward a quantification of fluid processes in volcanoes and elastic radiation resulting from such processes. A comprehensive synthesis of seismic data derived by different methods such as discussed in this article, together with petrological, geochemical and geological data, and laboratory experiments, will be required to achieve a full understanding of the dynamics of active magmatic systems. Such endeavor is a prerequisite to accurate assessments of volcanic hazards and to mitigate the human and socio-economic loss from destructive eruptions.

Acknowledgements

I am grateful to Phil Dawson and Javier Almendros for their assistance in drafting figures. I am also indebted to David Hill, Robert Tilling, Raoul Madariaga, and Hitoshi Kawakatsu for critical reviews.

REFERENCES

AKAIKE, H. (1974), *A New Look at the Statistical Model Identification*, IEEE Trans. Autom. Contr. *AC-9*, 716–723.

AKI, K. (1957), *Space and Time Spectra of Stationary Stochastic Waves, with Special Reference to Microtremors*, Bull. Earthquake Res. Inst. Tokyo Univ. *25*, 415–457.

AKI, K. (1959), *Correlational Study of Near Earthquake Waves*, Bull. Earthquake Res. Inst. Tokyo Univ. *37*, 207–232.

AKI, K. (1965), *A Note on the Use of Microseisms in Determining the Shallow Structures of the Earth's Crust*, Geophysics, *30*, 665–666.

AKI, K. (1977), *Three-dimensional Seismic Velocity Anomalies in the Lithosphere*, J. Geophys. *43*, 235–242.

AKI, K. (1984), *Evidence for Magma Intrusion during the Mammoth Lakes Earthquakes of May 1980 and Implications of the Absence of Volcanic (Harmonic) Tremor*, J. Geophys. Res. *89*, 7,689–7,696.

AKI, K. and LEE, W. H. K. (1976), *Determination of Three-dimensional Velocity Anomalies under a Seismic Array Using first P Arrival Times from Local Earthquakes, 1. A Homogeneous Initial Model*, J. Geophys. Res. *81*, 4,381–4,399.

AKI, K., CHRISTOFFERSSON, A., and HUSEBYE, E. S. (1976), *Three-dimensional Seismic Structure of Lithosphere under Montana LASA*, Bull. Seismol. Soc. Am. *66*, 501–524.

AKI, K., CHRISTOFFERSSON, A., and HUSEBYE, E. S. (1977a), *Determination of the Three-dimensional Seismic Structure of the Lithosphere*, J. Geophys. Res. *82*, 277–296.

AKI, K., FEHLER, M., and DAS, S. (1977b), *Source Mechanism of Volcanic Tremor: Fluid-driven Crack Models and their Application to the 1963 Kilauea Eruption*, J. Volcanol. Geotherm. Res. *2*, 259–287.

AKI, K., CHOUET, B., FEHLER, M., ZANDT, G., KOYANAGI, R., COLP, J., and HAY, R. G. (1978), *Seismic Properties of a Shallow Magma Reservoir in the Kilauea Iki by Active and Passive Experiments*, J. Geophys. Res. *83*, 2,273–2,282.

AKI, K. and KOYANAGI, R. (1981), *Deep Volcanic Tremor and Magma Ascent Mechanism under Kilauea, Hawaii*, J. Geophys. Res. *86*, 7,095–7,109.

ALMENDROS, J., CHOUET, B., and DAWSON, P. (2001a), *Spatial Extent of a Hydrothermal System at Kilauea Volcano, Hawaii, Determined from Array Analyses of Shallow Long-period Seismicity, 1: Method*, J. Geophys. Res. *106*, 13,565–13,580.

ALMENDROS, J., CHOUET, B., and DAWSON, P. (2001b), *Spatial Extent of a Hydrothermal System at Kilauea Volcano, Hawaii, Determined from Array Analyses of Shallow Long-period Seismicity, 2: Results*, J. Geophys. Res. *106*, 13,581–13,597.

ARCINIEGA-CEBALLOS, A., CHOUET, B. A., and DAWSON, P. (1999), *Very Long-period Signals Associated with Vulcanian Explosions at Popocatepetl Volcano, Mexico*, Geophys. Res. Lett. *26*, 3,013–3,016.

ARCINIEGA-CEBALLOS, A., VALDES-GONZALEZ, C., and DAWSON, P. (2000), *Temporal and Spectral Characteristics of Seismicity Observed at Popocatepetl Volcano, Central Mexico*, J. Volcanol. Geotherm. Res. *102*, 207–216.

BARTH, G. A., KLEINROCK, M. C., and HELZ, R. T. (1994), *The Magma Body at Kilauea Iki Lava Lake: Potential Insights into Mid-ocean Ridge Magma Chambers*, J. Geophys. Res. *99*, 7,199–7,217.

BENZ, H. M., CHOUET, B. A., DAWSON, P. B., LAHR, J. C., PAGE, R. A., and HOLE, J. A. (1996), *Three-dimensional P- and S-Wave Velocity Structure of Redoubt Volcano, Alaska*, J. Geophys. Res. *101*, 8,111–8,128.

BOUCHON, M. (1979), *Discrete Wave Number Representation of Elastic Wave Fields in Three-space Dimensions*, J. Geophys. Res. *84*, 3,609–3,614.

BOUCHON, M. (1981), *A Simple Method to Calculate Green's Functions for Elastic Layered Media*, Bull. Seismol. Soc. Am. *71*, 959–971.

CHOUET, B. (1979), *Sources of Seismic Events in the Cooling Lava Lake of Kilauea Iki, Hawaii*, J. Geophys. Res. *84*, 2,315–2,330.

CHOUET, B. (1981), *Ground Motion in the Near Field of a Fluid-driven Crack and its Interpretation in the Study of Shallow Volcanic Tremor*, J. Geophys. Res. *86*, 5,985–6,016.

CHOUET, B. (1982), *Free Surface Displacements in the Near Field of a Tensile Crack Expanding in Three Dimensions*, J. Geophys. Res. *87*, 3,868–3,872.

CHOUET, B. (1985), *Excitation of a Buried Magmatic Pipe: A Seismic Source Model for Volcanic Tremor*, J. Geophys. Res. *90*, 1,881–1,893.

CHOUET, B. (1986), *Dynamics of a Fluid-driven Crack in Three Dimensions by the Finite Difference Method*, J. Geophys. Res. *91*, 13,967–13,992.

CHOUET, B. (1987), *Representation of an Extended Seismic Source in a Propagator-based Formalism*, Bull. Seismol. Soc. Am. *77*, 14–27.

CHOUET, B. (1988), *Resonance of a Fluid-driven Crack: Radiation Properties and Implications for the Source of Long-period Events and Harmonic Tremor*, J. Geophys. Res. *93*, 4,375–4,400.

CHOUET, B., *A Seismic Model for the Source of Long-period Events and Harmonic Tremor*. In *Volcanic Seismology* (eds. P. Gasparini, R. Scarpa, and K. Aki) (Springer-Verlag, New York. 1992) pp. 133–156.

CHOUET, B. (1996a), *Long-Period Volcano Seismicity: Its Source and Use in Eruption Forecasting*, Nature *380*, 309–316.

CHOUET, B. (1996b), *New Methods and Future Trends in Seismological Volcano Monitoring*, In *Monitoring and Mitigation of Volcano Hazards*, (eds. R. Scarpa and R. Tilling) (Springer-Verlag, New York. 1996b) pp. 23–97.

CHOUET, B. A., PAGE, R. A., STEPHENS, C. D., LAHR, J. C., and POWER, J. A. (1994), *Precursory Swarms of Long-period Events at Redoubt Volcano (1989–1990), Alaska: Their Origin and Use as a Forecasting Tool*, J. Volcanol. Geotherm. Res. *62*, 95–135.

CHOUET, B., SACCOROTTI, G., MARTINI, M., DAWSON, P., DELUCA, G., MILANA, G., and SCARPA, R. (1997), *Source and Path Effects in the Wavefields of Tremor and Explosions at Stromboli Volcano, Italy*, J. Geophys. Res. *102*, 15,129–15,150.

CHOUET, B. A. and DAWSON, P. B. (1997), *Observations of Very-long-period Impulsive Signals Accompanying Summit Inflation of Kilauea Volcano, Hawaii, in February 1997*, EOS (Trans. Am. Geophys. Union), *78*, no. 46 (Supplement), F429–F430.

CHOUET, B., DELUCA, G., MILANA, G., DAWSON, P., MARTINI, M., and SCARPA, R. (1998), *Shallow Velocity Structure of Stromboli Volcano, Italy, Derived from Small-aperture Array Measurements of Strombolian Tremor*, Bull. Seismol. Soc. Am. *88*, 653–666.

CHOUET, B., SACCOROTTI, G., DAWSON, P., MARTINI, M., SCARPA, R., DELUCA, G., MILANA, G., and CATTANEO, M. (1999), *Broadband Measurements of the Sources of Explosions at Stromboli Volcano, Italy*, Geophys. Res. Lett. *26*, 1,937–1,940.

COMMANDER, K. W. and PROSPERETTI, A. (1989), *Linear Pressure Waves in Bubbly Liquids: Comparison between Theory and Experiments*, J. Acoust. Soc. Am. *85*, 732–746.

CROSSON, R. S. and BAME, D. A. (1985), *A Spherical Source Model for Low Frequency Volcanic Earthquakes*, J. Geophys. Res. *90*, 10,237–10,247.

DAWSON, P. B., DIETEL, C., CHOUET, B. A., HONMA, K., OHMINATO, T., and OKUBO, P. (1998), *A Digitally Telemetered Broadband Seismic Network at Kilauea Volcano, Hawaii*, U.S. Geol. Surv. Open File Report *98–108*, 1–121.

DAWSON. P. B., CHOUET, B. A., OKUBO, P. G., VILLASEÑOR, A., and BENZ, H. M. (1999), *Three-dimensional Velocity Structure of Kilauea Caldera, Hawaii*, Geophys. Res. Lett. *26*, 2,805–2,808.

DOBRAN, F., NERI, A., and MACEDONIO, G. (1993), *Numerical Simulation of Collapsing Volcanic Columns*, J. Geophys. Res. *98*, 4,231–4,259.

ELLSWORTH, W. L. and KOYANAGI, R. Y. (1977), *Three-dimensional Crust and Mantle Structure of Kilauea Volcano, Hawaii*, J. Geophys. Res. *82*, 5,379–5,394.

FEHLER, M. (1983), *Observations of Volcanic Tremor at Mount St. Helens Volcano*, J. Geophys. Res. *88*, 3,476–3,484.

FEHLER, M. and AKI, K. (1978), *Numerical Study of Diffraction of Plane Elastic Waves by a Finite Crack with Application to Location of a Magma Lens*, Bull. Seismol. Soc. Am. *68*, 573–598.

FEHLER, M. and CHOUET, B. (1982), *Operation of a Digital Seismic Network on Mt. St. Helens Volcano and Observations of Long-period Seismic Events that Originate under the Volcano*, Geophys. Res. Lett. *9*, 1,017–1,020.

FERRAZZINI, V. and AKI, K. (1987), *Slow Waves Trapped in a Fluid-filled Infinite Crack: Implication for Volcanic Tremor*, J. Geophys. Res. *92*, 9,215–9,233.

FERRAZZINI, V., AKI, K., and CHOUET, B. (1991), *Characteristics of Seismic Waves Composing Hawaiian Volcanic Tremor and Gas-piston Events Observed by a Near-source Array*, J. Geophys. Res. *96*, 6,199–6,209.

FISCHER, T. P., MORRISSEY, M. M., CALVACHE, M. L., GÓMEZ, V. D., TORRES, M. R., STIX, C. J., and WILLIAMS, S. N. (1994), *Correlations between SO₂ Flux and Long-period Seismicity at Galeras Volcano*, Nature, *368*, 135–137.

FUJITA, E., IDA, Y., and OIKAWA, J. (1995), *Eigen Oscillation of a Fluid Sphere and Source Mechanism of Harmonic Volcanic Tremor*, J. Volcanol. Geotherm. Res. *69*, 365–378.

FUJITA, E., FUKAO, Y., and KANJO, K. (2000), *Strain Offsets with Monotonous Damped Oscillations during the 1986 Izu-Oshima Volcano Eruption*, J. Geophys. Res. *105*, 443–462.

GIL CRUZ, F. and CHOUET, B. A. (1997), *Long-period Events, the Most Characteristic Seismicity Accompanying the Emplacement and Extrusion of a Lava Dome in Galeras Volcano, Colombia, in 1991*, J. Volcanol. Geotherm. Res. *77*, 121–158.

GOLDSTEIN, P. (1988), *Array Measurements of Earthquake Rupture*, Ph. D. Thesis, University of California Santa Barbara, Santa Barbara.

GOLDSTEIN, P. and ARCHULETA, R. J. (1987), *Array Analysis of Seismic Signals*, Geophys. Res. Lett. *14*, 13–16.

GOLDSTEIN, P. and ARCHULETA, R. J. (1991a), *Deterministic Frequency-wavenumber Methods and Direct Measurements of Rupture Propagation during Earthquakes Using a Dense Array: Theory and Methods*, J. Geophys. Res. *96*, 6,173–6,185.

GOLDSTEIN, P. and ARCHULETA, R. J. (1991b), *Deterministic Frequency-wavenumber Methods and Direct Measurements of Rupture Propagation during Earthquakes Using a Dense Array: Data Analysis*, J. Geophys. Res. *96*, 6,187–6,198.

GOLDSTEIN, P. and CHOUET, B. (1994), *Array Measurements and Modeling of Sources of Shallow Volcanic Tremor at Kilauea Volcano, Hawaii*, J. Geophys. Res. *99*, 2,637–2,652.

HASEGAWA, A., ZHAO, D., HORI, S., YAMAMOTO, A., and HORIUCHI, S. (1991), *Deep Structure of the Northeastern Japan Arc and its Relationship to Seismic and Volcanic Activity*, Nature, *352*, 683–689.

HELZ, R. T. (1993), *Drilling Report and Core Logs for the 1988 Drilling of Kilauea Iki Lava Lake, Kilauea Volcano, Hawaii, with Summary Descriptions of the Occurrence of Foundered Crust and Fractures in the Drill Core*, U. S. Geological Survey Open-file Report 93-15, pp. 1–57.

HOLE, J. A. (1992), *Nonlinear High-resolution Three-dimensional Seismic Travel Time Tomography*, J. Geophys. Res. *97*, 6,553–6,562.

HUSEBYE, E. S., CHRISTOFFERSSON, A., AKI, K., and POWELL, C. (1976), *Preliminary Results on the Three-dimensional Seismic Structure of the Lithosphere under the USGS Central California Seismic Array*, Geophys. J. R. Astr. Soc. *46*, 319–340.

JULIAN, B. R. (1994), *Volcanic Tremor: Nonlinear Excitation by Fluid Flow*, J. Geophys. Res. *99*, 11,859–11,877.

KANAMORI, H., GIVEN, J. W., and LAY, T. (1984), *Analysis of Seismic Body Waves Excited by the Mount St. Helens Eruption of May 18, 1980*, J. Geophys. Res. *89*, 1,856–1,866.

KANESHIMA, S., KAWAKATSU, H., MATSUBAYASHI, H., SUDO, Y., TSUTSUI, T., OHMINATO, T., ITO, H., UHIRA, K., YAMASATO, H., OIKAWA, J., TAKEO, M., and IIDAKA, T. (1996), *Mechanism of Phreatic Eruptions at Aso Volcano Inferred from Near-field Broadband Seismic Observations*, Science, *273*, 642–645.

KAWAKATSU, H., OHMINATO, T., ITO, H., KUWAHARA, Y., KATO, T., TSURUGA, K., HONDA, S., and YOMOGIDA, K., (1992), *Broadband Seismic Observation at the Sakurajima Volcano, Japan*, Geophys. Res. Lett. *19*, 1,959–1,962.

KAWAKATSU, H., OHMINATO, T., and ITO, H. (1994), *10-s-period Volcanic Tremors Observed over a Wide Area in Southwestern Japan*, Geophys. Res. Lett. *21*, 1,963–1,966.

KAWAKATSU, H., KANESHIMA, S., MATSUBAYASHI, H., OHMINATO, T., SUDO, Y., TSUTSUI, T., UHIRA, K., YAMASATO, H., ITO, H., and LEGRAND, D. (2000), *Aso94: Aso Seismic Observation with Broadband Instruments*, J. Volcanol. Geotherm. Res. *101*, 129–154.

KIEFFER, S. W. (1977), *Sound Speed in Liquid-gas Mixtures: Water-air and Water-steam*, J. Geophys. Res. *82*, 2,895–2,904.

KUMAGAI, H. and CHOUET, B. A. (1999), *The Complex Frequencies of Long-period Seismic Events as Probes of Fluid Composition Beneath Volcanoes*, Geophys. J. Int. *138*, F7–F12.

KUMAGAI, H. and CHOUET, B. A. (2000), *Acoustic Properties of a Crack Containing Magmatic or Hydrothermal Fluids*, J. Geophys. Res. *105*, 25,493–25,512.

KUMAGAI, H., CHOUET, B. A., and NAKANO, M. (2002), *Temporal Evolution of a Hydrothermal System in Kusatsu-Shirane Volcano, Japan, Inferred from the Complex Frequencies of Long-period Events*, J. Geophys. Res., (in press).

KUMAZAWA, M., IMANISHI, Y., FUKAO, Y., FUROMOTO, M., and YAMAMOTO, A. (1990), *A Theory of Spectral Analysis Based on the Characteristic Property of a Linear Dynamic system*, Geophys. J. Int. *101*, 613–630.

LANE, S. J., CHOUET, B. A., PHILLIPS, J. C., DAWSON, P., RYAN, G. A., and HURST, E. (2001), *Experimental Observations of Pressure Oscillations and Flow Regimes in an Analogue Volcanic System*, J. Geophys. Res. *106*, 6,461–6,476.

LEGRAND, D., KANESHIMA, S., and H. KAWAKATSU (2000), *Moment Tensor Analysis of Broadband Waveforms Observed at Aso Volcano, Japan*, J. Volcanol. Geotherm. Res. *101*, 155–169.

MATSUBAYASHI, H. (1995), *The Source of the Long-period Tremors and the Very-long Period Events Preceding the Mud Eruption at Aso Volcano, Japan (in Japanese with English abstract)*, M. A. Thesis, Tokyo University, Tokyo.

MAVKO, G. M. (1980), *Velocity and Attenuation in Partially Molten Rocks*, J. Geophys. Res. *85*, 5,173–5,189.

McNUTT, S. R. (2000a), *Volcanic Seismicity*. In *Encyclopedia of Volcanoes* (H. Sigurdsson, B. Houghton, S. R. McNutt, H. Rymer, and J. Stix eds.), pp. 1,015–1.033.

McNUTT, S. R. (2000b), *Seismic Monitoring*. In *Encyclopedia of Volcanoes* (H. Sigurdsson, B. Houghton, S. R. McNutt, H. Rymer, and J. Stix eds.), pp. 1,095–1,119.

McNutt, S. R., Ida, Y., Chouet, B. A., Okubo, P., Oikawa, J., and Saccorotti, G. (1997), *Kilauea Volcano Provides Hot Seismic Data for Joint Japanese-U. S. Experiment*, EOS (Trans. Am. Geophys. Union), *78*, 105, 111.

Morrissey, M. M. and Chouet, B. A. (1997), *A Numerical Investigation of Choked Flow Dynamics and its Application to the Triggering Mechanism of Long-period Events at Redoubt Volcano, Alaska*, J. Geophys. Res. *102*, 7,965–7,983.

Morrissey, M. M. and Chouet, B. A. (2001), *Trends in Long-period Seismicity Related to Magmatic Fluid Compositions*, J. Volcanol. Geotherm. Res., *108*, 265–281.

Nakano, M., Kumagai, H., Kumazawa, M., Yamaoka, K., and Chouet, B. A. (1998), *The Excitation and Characteristic Frequency of the Long-period Volcanic event: An approach Based on an Inhomogeneous Autoregressive Model of a Linear Dynamic System*, J. Geophys. Res. *103*, 10,031–10,046.

Nakano, M., Kumagai, H., and Chouet, B. A. (2002), *Source Mechanism of Long-period Events at Kusatsu-Shirane Volcano, Japan, Inferred from Waveform Inversion of the Effective Excitation Functions*, J. Volcanol. Geotherm. Res., submitted.

Narváez, M. L., Torres, R. A., Gómez., C. D. M., Cortés, G. P. Cepeda, J. H., and Stix, J. (1997), *'Tornillo'-type Seismic Signals at Galeras Volcano, Colombia, 1992–1993*, J. Volcanol. Geotherm. Res. *77*, 159–171.

Neuberg, J., Luckett, R., Ripepe, M., and Braun, T., (1994), *Highlights from a Seismic Broadband Array on Stromboli Volcano*, Geophys. Res. Lett. *21*, 749–752.

Nishimura, T., Nakamichi, H., Tanaka, S., Sato, M., Kobayashi, T., Ueki, S., Hamaguchi, H., Ohtake, M., and Sato, H. (2000), *Source Process of Very Long Period Seismic Events Associated with the 1998 Activity of Iwate Volcano, Northeastern Japan*, J. Geophys. Res. *105*, 19,135–19,147.

Nishimura, T. and Chouet, B. (1998), *Numerical Simulations of Seismic Waves and Magma Dynamics Excited by a Volcanic Eruption*, EOS (Trans. Am. Geophys. Union) *79*, no. 45 (Supplement), F595.

Ohminato, T. and Chouet, B. A. (1997), *A Free-surface Boundary Condition for Including 3D Topography in the Finite Difference Method*, Bull. Seismol. Soc. Am. *87*, 494–515.

Ohminato, T. and Ereditato, D. (1997), *Broadband Seismic Observations at Satsuma-Iwojima Volcano, Japan*, Geophys. Res. Lett. *24*, 2,845–2,848.

Ohminato, T., Chouet, B. A., Dawson, P. B., and Kedar, S. (1998), *Waveform Inversion of Very-long-period Impulsive Signals Associated with Magmatic Injection beneath Kilauea Volcano, Hawaii*, J. Geophys. Res. *103*, 23,839–23,862.

Okubo, P. G., Benz, H. M., and Chouet, B. A. (1997), *Imaging the Crustal Magma Source beneath Mauna Loa and Kilauea Volcanoes, Hawaii*, Geology *25*, 867–870.

Pitt, A. M. and Hill, D. P. (1994), *Long-period Earthquakes in the Long Valley Caldera Region, Eastern California*, Geophys. Res. Lett. *21*, 1,679–1,682.

Podvin, P. and Lecomte, I. (1991), *Finite Difference Computation of Traveltimes in Very Contrasted Velocity Models: A Massively Parallel Approach and its Associated Tools*, Geophys. J. Int. *105*, 271–284.

Ramos, E. G., Hamburger, M. W., Pavlis, G. L., and Laguerta E. P. (1999), *The Low-frequency Earthquake Swarms at Mount Pinatubo, Philippines: Implications for Magma Dynamics*, J. Volcanol. Geotherm. Res. *92*, 295–320.

Richter, D. H. and Moore, J. G. (1966), *Petrology of the Kilauea Iki Lava Lake, Hawaii*, U. S. Geological Survey Professional Paper 537-B, pp. B1–B26.

Richter, D. H., Eaton, J. P., Murata, K. J., Ault, W. U., and Krivoy, H. L., (1970), *Chronological Narative of the 1959–60 eruption of Kilauea Volcano, Hawaii*, U. S. Geological Survey Professional Paper 537-E, pp. E1–E73.

Rowan, L. R. and Clayton, R. W. (1993), *The Three-dimensional Structure of Kilauea Volcano, Hawaii, from Travel-time Tomography*, J. Geophys. Res. *98*, 4,355–4,375.

Rowe, C. A., Aster, R. C., Kyle, P. R., Schlue, J. W., and Dibble, R. R. (1998), *Broadband Recording of Strombolian Explosions and Associated Very-long-period Seismic Signals on Mount Erebus Volcano, Ross Island, Antarctica*, Geophys. Res. Lett. *25*, 2,297–2,300.

Scarpa, R. and Tilling, R. I., *Monitoring and Mitigation of Volcano Hazards*, (Springer-Verlag, New York (1996)), 841 pp.

Schmidt, R. O. (1981), *A Signal Subspace Approach to Multiple Emitter Location and Spectral Estimation*, Ph. D. Thesis, Stanford University, Palo Alto.

SCHMIDT, R. O. (1986), *Multiple Emitter Location and Signal Parameter Estimation*, IEEE Trans. Antennas Propagation, *34*, 276–280.

SHAW, H. R. and CHOUET, B. (1991), *Fractal Hierarchies of Magma Transport in Hawaii and Critical Self-organization of Tremor*, J. Geophys. Res. *96*, 10,191–10,207.

SPUDICH, P. and OPPENHEIMER, D. (1986), *Dense Seismograph Array Observations of Earthquake Rupture Dynamics*. In *Earthquake Source Mechanics*, Geophys. Monogr. Ser. *37*, (eds. S. Das, J. Boatwright and C. D. Scholz), 285–296, AGU, Washington.

STIX, J., TORRES, R., NARVÁEZ, C. L., CORTÉS, M. G. P., RAIGOSA, J. J., GÓMEZ, A. D. and CASTONGUAY, R. (1997), *A Model of Vulcanian Eruptions at Galeras Volcano, Colombia*, J. Volcanol. Geotherm. Res. *77*, 285–303.

TAKEI, Y. and KUMAZAWA, M. (1994), *Why Have the Single Force and Torque been Excluded from Seismic Source Models?* Geophys. J. Int. *118*, 20–30.

TAKEI, Y. (1998), *Constitutive Mechanical Relations of Solid-liquid Composites in Terms of Grain-boundary Contiguity*, J. Geophys. Res. *103*, 18,183–18,203.

TEMKIN, S. and DOBBINS, R. A. (1966), *Attenuation and Dispersion of Sound by Particulate-relaxation Processes*, J. Acoust. Soc. Am. *40*, 317–324.

THORNBER, C. R., HELIKER, C. C., REYNOLDS, J. R., KAUAHIKAUA, J., OKUBO, P., LISOWSKI, M., SUTTON, A. J., and CLAGUE, D. (1996), *The Eruptive Surge of February 1, 1996: A Highlight of Kilauea's Ongoing East Rift Zone eruption*, EOS (Trans. Am. Geophys. Union), *77*, no. 46 (Supplement), F798.

THURBER, C. H. (1984), *Seismic Detection of the Summit Magma Complex of Kilauea Volcano, Hawaii*, Science, *223*, 165–167.

THURBER, C. H. (1987), *Seismic Structure and Tectonics of Kilauea Volcano*, U. S. Geological Survey Professional Paper *1350*, 919–934.

UKAWA, M. and OHTAKE, M. (1987), *A Monochromatic Earthquake Suggesting Deep-seated Magmatic Activity Beneath the Izu-Ooshima Volcano, Japan*, J. Geophys. Res. *92*, 12,649–12,663.

VALENTINE, G. A. and WOHLETZ, K. H. (1989), *Numerical Models of Plinian Eruption Columns and Pyroclastic Flows*, J. Geophys. Res. *94*, 1,867–1,887.

VIDALE, J. (1988), *Finite-difference Calculations of Traveltimes*, Bull. Seismol. Soc. Am. *78*, 2,062–2,076.

VIDALE, J. (1991), *Finite-difference Calculation of Traveltimes in Three Dimensions*, Geophysics, *55*, 521–526.

WALLIS, G. B., *One-Dimensional Two-Phase Flow*, McGraw-Hill, New York. (1969), 408 pp,.

WHITE, R. A. (1996), *Precursory Deep Long-period Earthquakes at Mount Pinatubo, Philippines: Spatio-temporal Link to a Basalt Trigger*. In *Fire and Mud: Eruptions and Lahars of Mount Pinatubo, Philippines*, (Newhall, C., and Punongbayan, R., eds.), PHIVOLCS and Univ. Washington Press, pp. 307–328.

WHITE, R., HARLOW, D., and CHOUET, B. (1992), *Long-period Earthquakes Preceding and Accompanying the June 1991 Mount Pinatubo Eruptions*, EOS (Trans. Am. Geophys. Union) *73*, no. 43 (Supplement), 347.

WHITE, R. A., MILLER, A. D., LYNCH, L., and POWER, J., (1998), *Observations of Hybrid Seismic Events at Soufriere Hills Volcano, Montserrat: July 1995 to September 1996*, Geophys. Res. Lett. *25*, 3,657–3,660.

(Received July 23, 2000, accepted May 4, 2001)

 To access this journal online:
http://www.birkhauser.ch

Pure appl. geophys. 160 (2003) 789–807
0033–4553/03/040789–19

❙ Pure and Applied Geophysics

Seismic Detection and Characterization of the Altiplano-Puna Magma Body, Central Andes

GEORGE ZANDT[1], MARK LEIDIG[1], JOSEF CHMIELOWSKI[1],
DAVID BAUMONT[1] and XIAOHUI YUAN[2]

Abstract — The Altiplano-Puna Volcanic Complex (APVC) in the central Andes is the product of an ignimbrite "flare-up" of world class proportions (DE SILVA, 1989). The region has been the site of large-scale silicic magmatism since 10 Ma, producing 10 major eruptive calderas and edifices, some of which are multiple-eruption resurgent complexes as large as the Yellowstone or Long Valley caldera. Seven PASSCAL broadband seismic stations were operated in the Bolivian portion of the APVC from October 1996 to September 1997 and recorded teleseismic earthquakes and local intermediate-depth events in the subducting Nazca plate. Both teleseismic and local receiver functions were used to delineate the lateral extent of a regionally pervasive ~20-km-deep, very low-velocity layer (VLVL) associated with the APVC. Data from several stations that sample different parts of the northern APVC show large amplitude Ps phases from a low-velocity layer with $Vs \leq 1.0$ km/s and a thickness of ~1 km. We believe the crustal VLVL is a regional sill-like magma body, named the Altiplano–Puna magma body (APMB), and is associated with the source region of the Altiplano–Puna Volcanic Complex ignimbrites (CHMIELOWSKI *et al.*, 1999).

Large-amplitude P–SH conversions in both the teleseismic and local data appear to originate from the top of the APMB. Using the programs of LEVIN and PARK (1998), we computed synthetic receiver functions for several models of simple layered anisotropic media. Upper-crustal, tilted-axis anisotropy involving both Vp and Vs can generate a "split Ps" phase that, in addition to the Ps phase from the bottom of a thin isotropic VLVL, produces an interference waveform that varies with backazimuth. We have forward modeled such an interference pattern at one station with an anisotropy of 15%–20% that dips 45° within a 20-km-thick upper crust. We develop a hypothesis that the crust above the "magma body" is characterized by a strong, tilted-axis, hexagonally symmetric anisotropy. We speculate that the anisotropy is due to aligned, fluid-filled cracks induced by a "normal-faulting" extensional strain field associated with the high elevations of the Andean Puna.

Key words: PASSCAL, receiver functions, anisotropy, magma body, Altiplano-Puna, APVC.

Introduction

The detection and characterization of magma bodies in the crust and upper mantle has been a long-standing goal in seismology, and a topic of interest for Kei Aki throughout his career (e.g., AKI, 1968; AKI *et al.*, 1978; ROBERTS, AKI,

[1] Department of Geosciences, Gould-Simpson Building, University of Arizona, Tucson, Arizona 85721-0077, U.S.A. E-mail: zandt@geo.arizona.edu
[2] GeoForschungsZentrum, Potsdam, Germany.

and FEHLER, 1991). Early studies of magma bodies were often based on the fortuitous observation of "exotic" phases within waveforms recorded in volcanic regions (PEPPIN, 1987). Aki and his colleagues pioneered the formulation of the seismic travel-time inversion technique now commonly called "tomography" (AKI et al., 1977). Among the first targets of the new technique taken up by his students was detection of crustal magma bodies and mantle asthenosphere in Hawaii, Yellowstone, and California (ELLSWORTH and KOYANAGI, 1977; IYER et al., 1981; ZANDT and FURLONG, 1982). IYER (1984) summarized the evidence for the locations, shapes, and sizes of magma bodies beneath regions of Quaternary volcanism based on seismic tomography as well as other geophysical techniques.

The difficulties of geophysical imaging of a crustal magma body are exemplified by the many studies conducted at Long Valley caldera, California (RUNDLE and HILL, 1988). Although a low-velocity zone beneath the caldera had been found as early as 1976 by a teleseismic P-wave delay study by STEEPLES and IYER (1976), the precise depth, shape and velocity reduction has been difficult to accurately assess (STECK and PROTHERO, 1994). The difficulties of imaging a magma body are caused by a combination of factors: complex structure in volcanic regions, especially with near-surface heterogeneity, the tendency of seismic waves to refract around low-velocity zones, and strong attenuation and anisotropy effects often associated with magmatic structures in the crust. Among the best-resolved magma chambers are mid-crustal, tabular-shaped bodies exemplified by the Socorro magma body delineated by microearthquake studies (SANDFORD et al., 1973), seismic reflection surveys (BROWN et al., 1980), and converted phase observations (SHEETZ and SCHLUE, 1992; SCHLUE et al., 1996).

In this contribution, we describe another well-constrained observation of a regionally extensive, sill-like magma body at mid-crustal depth. We will present data for the existence of highly developed anisotropy in the upper 20 km of the crust in the Altiplano-Puna Volcanic Complex (APVC), a major silicic volcanic field straddling the political triple junction of Bolivia, Chile, and Argentina (Fig. 1). The anisotropic upper crust appears to be associated with the areal extent of the magma body. The existence of the magma body is based on the observations of large-amplitude, negative-polarity Ps phases observed on receiver functions from PASSCAL stations that we had deployed in the APVC during 1996–1997 (CHMIELOWSKI et al., 1999). Matching the large amplitudes of these Ps phases requires a $Vs \leq 1$ km/s within a $\sim$1-km-thick layer. A purely isotropic model cannot explain the observation of significant azimuthal variations of the Ps amplitude on the radial receiver functions, nor the existence of corresponding large-amplitude phases on the transverse component. Based on preliminary forward modeling of the waveform data, we suggest that the upper crust is characterized by 15%–20% anisotropy with a tilted symmetry axis.

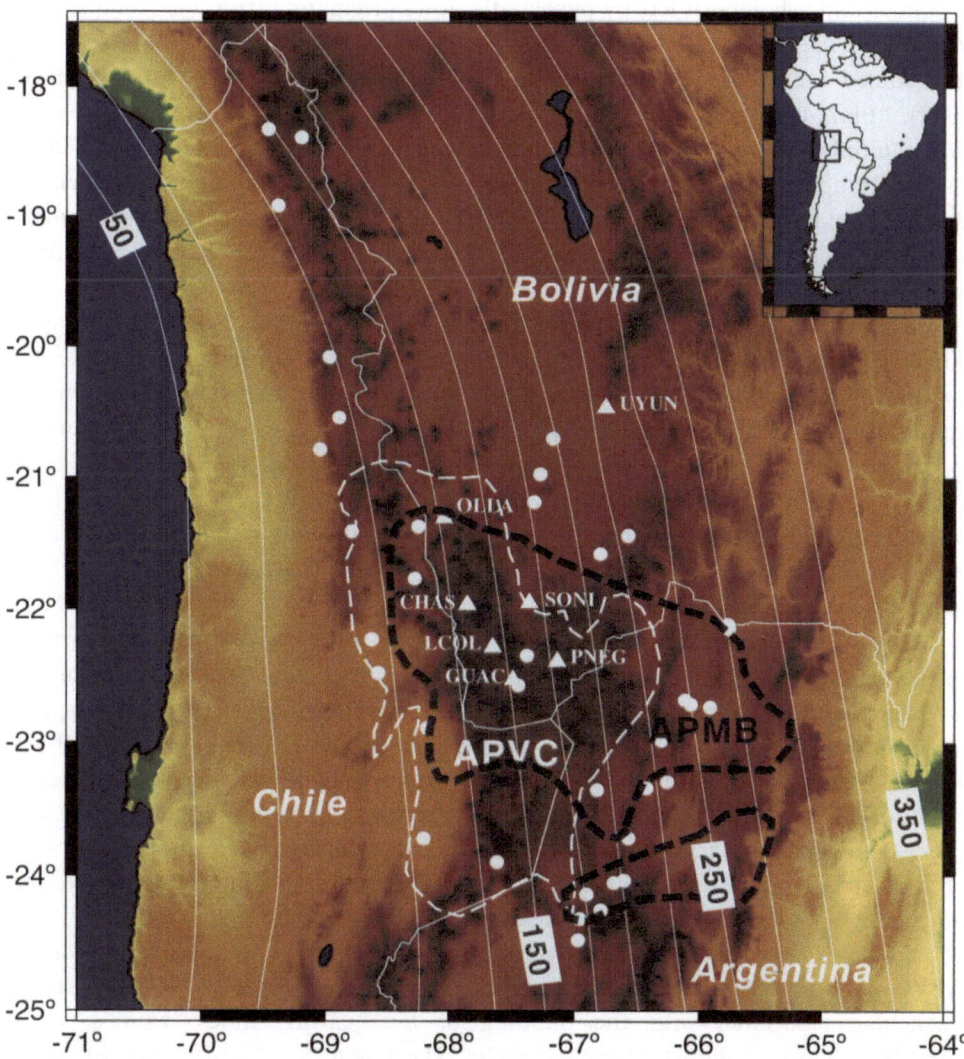

Figure 1

Location map of the Altiplano-Puna Volcanic Complex (APVC, enclosed by the dashed white line) and the one-year deployment of PASSCAL broadband stations (triangles). Dashed black line encloses the approximate extent of the mid-crustal Altiplano-Puna magma body (APMB) determined in this study. Solid circles are epicenters of intermediate-depth earthquakes utilized in this study. Thin lines are the contours of depth (km) to the Wadati-Benioff zone from CAHILL and ISACKS (1992).

APVC Background

Since the late Miocene, an ignimbrite "flare-up" has produced a major silicic volcanic province, the Altiplano–Puna Volcanic Complex (APVC), in the central Andes of South America (DE SILVA, 1989) (Fig. 1). An ignimbrite (or ash-flow tuff) is a pumice and ash deposit left by pyroclastic flows originating from the collapse of a vesiculated magmatic eruption column (FRANCIS, 1993). The term "flare-up" was first coined for the mid-Tertiary eruptions of the western United States (CONEY, 1972). "Flare-ups" result when large calderas occur in clusters or complexes such that multiple eruptions produce voluminous ignimbrite cover. Examples of large caldera complexes within the United States include the San Juan field in the central Rocky Mountains and the Indian Peak volcanic field in southern Nevada (MAUGHAN *et al.*, 2000). The APVC is a youthful, active analogue of these western U.S. caldera complexes.

Several young ignimbrite fields occur in the Central Volcanic Zone of the central Andes (BAKER, 1981; DE SILVA and FRANCIS, 1991). The Altiplano-Puna volcanic complex is the youngest of these and covers approximately 50,000 km^2 between 21° and 24°S. The APVC constitutes the largest ignimbrite concentration in the Central Volcanic Zone of the Andes (Fig. 1) and one of the largest in the world (DE SILVA, 1989). The APVC is located at the southward transition between the ~4-km-high plateau of the Altiplano and the ~5-km-high Puna, a change that may be associated with a rapid southward thinning of the South American lithosphere (WHITMAN *et al.*, 1996).

The APVC is located 100–250 km above the Nazca plate where it subducts with a relatively uniform ~30° dip (CAHILL and ISACKS, 1992) (Fig. 1). Crustal seismicity levels are very low in the APVC crust, as they are throughout the central Andes, due to the unusually thick and weak crust (BECK *et al.*, 1996; ZANDT *et al.*, 1996; BILLS *et al.*, 1994). However, a concentration of larger-magnitude intermediate-depth earthquakes with focal depths between 150 to 300 km (the Jujuy cluster) is located beneath the eastern margin of the APVC. This cluster of strong earthquakes located almost directly beneath the APVC presents a rare opportunity to probe the lithosphere of a major caldera complex.

Studies of the voluminous ignimbrites by DE SILVA (1989) and others suggest the APVC flare-up is due primarily to crustal melting in response to tectonic crustal thickening associated with the building of the central Andes, with a lesser contribution from subduction-related melts. The region has been the site of large-scale silicic magmatism since 10 Ma, with several major caldera-forming eruptions producing >15,000 km^3 of ignimbrites (DE SILVA, 1989; DE SILVA and FRANCIS, 1991). The APVC ignimbrites are part of a distinct end-member class called the monotonous intermediate type characterized by large volumes of a limited compositional range that probably reflects a pre-eruption evolution in a slab-like magma chamber (MAUGHAN *et al.*, 2000). Late Pleistocene to Recent volcanic activity in the form of large silicic lava flows and domes and two major geothermal fields indicate that the APVC province remains magmatically active (DE SILVA 1989; DE SILVA, *et al.*, 1994).

Discovery of the APMB

Recently we published a study of teleseismic receiver functions for 14 events recorded by seven PASSCAL broadband stations deployed for one year in the northeast portion of the APVC (CHMIELOWSKI et al., 1999). We used the method of receiver functions to probe the crustal stratigraphy. The basis of this technique is that a P wave propagating up to a seismic station produces converted S waves (Ps) at any seismic impedance contrast (e.g., the Moho or a magma body). A teleseismic P wave, or a P wave generated by a subcrustal local event, is steeply incident and dominates the vertical component of ground motion, whereas the Ps conversions are contained mostly on the horizontal components. Receiver functions are seismic waveforms computed by deconvolving the vertical component from the radial component to isolate and highlight the Ps converted phases within the coda of the P waves (ZANDT et al., 1995). We computed all of our receiver functions, for teleseismic and local events, using a new time-domain, iterative deconvolution algorithm (LIGORRIA and AMMON, 1999). The method entails doing a cross correlation between the vertical and radial components of the seismograms to find the lag and amplitude of the first and largest spike (Gaussian) in the receiver function. Then the current estimate of the receiver function is convolved with the vertical component and subtracted from the radial component, and the procedure is repeated to estimate other spike lags and amplitudes. With each additional spike in the receiver function the misfit between the convolution of the vertical component and the receiver function and the radial component is reduced, and the iteration halts when the misfit falls below a prescribed threshold. Long-period stability is insured *a priori* by construction of the deconvolution as a sum of Gaussian pulses. This method has the advantage of finding the "simplest" receiver function with high-frequency content but without side-lobe artifacts or long-period instability often associated with frequency-domain and some other time-domain techniques.

The teleseismic radial receiver functions for all our stations within the APVC have a large-amplitude, negative-polarity Ps conversion at approximately 2.5 s after the direct P wave and a smaller positive Ps conversion between 3 and 5 s after the direct P wave (Fig. 2). We suggested from forward modeling that the first conversion is produced by the top of a very low-velocity layer (VLVL) at a depth of 19 km, and we interpreted the second Ps as the bottom of the VLVL. We found the VLVL must have a $Vs \leq 1.0$ km/s and a thickness of ~1 km to match the amplitudes and timing from several stations that sample different parts of the northern APVC. We interpreted this crustal VLVL as a regional sill-like magma body associated with the Altiplano–Puna Volcanic Complex, and we named it the Altiplano–Puna magma body (APMB) (CHMIELOWSKI et al., 1999).

To test the resolving capability of relatively long-period receiver functions on thin, crustal low-velocity zones, we computed reflectivity synthetics for a suite of models varying both the thickness and velocity of a low-velocity layer (Fig. 3). The results

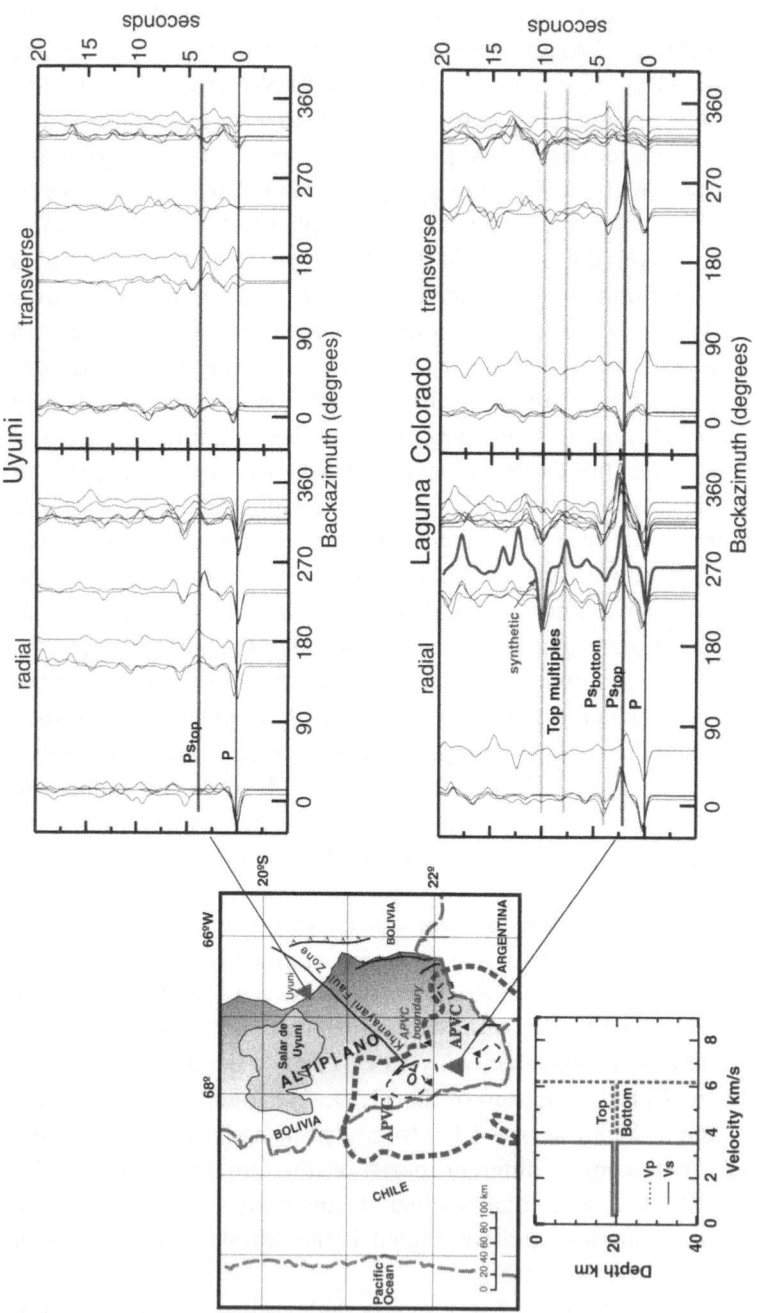

Figure 2

Comparison of teleseismic receiver functions recorded outside the APVC (Uyuni) and within the APVC (Laguna Colorado). The receiver functions at Uyuni are unremarkable except for the evidence of a relatively subdued lower crustal low-velocity zone at a depth of approximately 30–35 km. In contrast, radial receiver functions at Laguna Colorado show a very large amplitude, negative polarity arrival at ~2.5 seconds, consistent with a *Ps* conversion from the top of a very low-velocity layer at a depth of ~20 km. A synthetic receiver function from a simple model (lower left) with a 1-km-thick layer with V_s ~0.5 km/s can fit the main features of the radial data. However, the variations in the amplitude of the radial *Ps* phase and the corresponding arrivals on the transverse data showing reversals in polarity cannot be explained by isotropic models but could be explained by the presence of highly developed anisotropy between the converter interface and the surface.

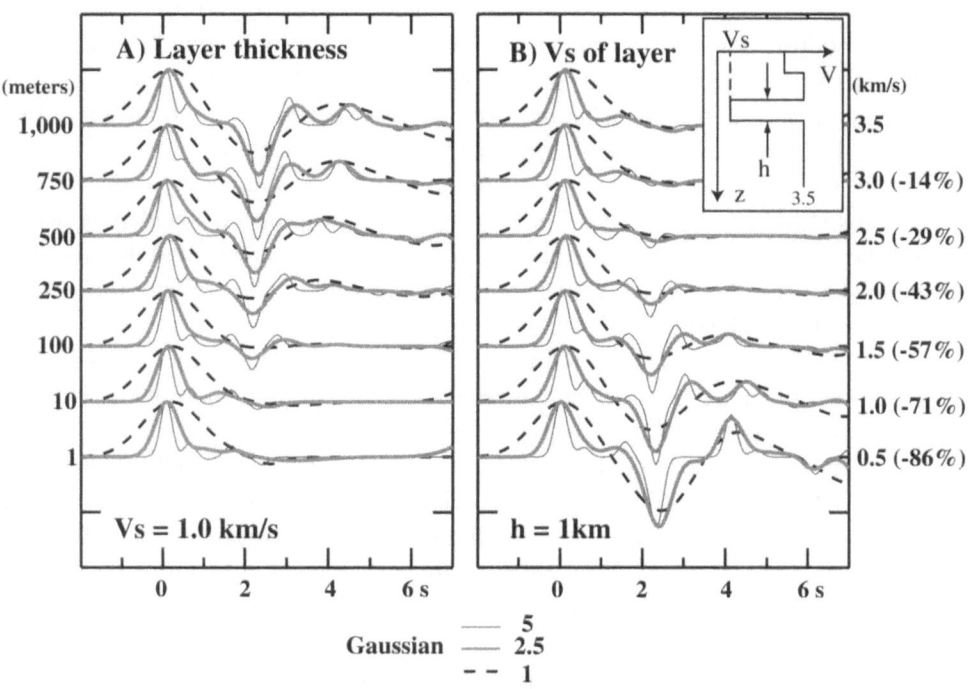

Figure 3

Resolution test for low-velocity-layer (lvl) thickness, *h* and velocity, *Vs* as a function of frequency. The three Gaussian widths of 5, 2.5, and 1 correspond to Gaussian low-pass filters with 90% cutoff at frequencies of 2.4, 1.2, and 0.5 Hz, respectively. (A) Synthetic receiver functions for varying lvl thickness with constant velocity of 1 km/s. Note that for Gaussians ≥ 2.5 the lvl is detectable for layers as thin as 100 m, although the full amplitude is not developed until the thickness approaches 1 km. (B) Synthetic receiver functions for varying lvl velocity with constant thickness of 1 km. Note that for Gaussians ≥ 2.5 the lvl is detectable for layers with velocity < ~2 km/s, although the full amplitude is not developed until the velocity approaches 1 km/s.

demonstrate that with a nominal frequency content of an earthquake-source receiver function, a crustal low-velocity layer with $Vs = 1$ km/s can be detected if it is greater than ~100 m thick. The full amplitude is developed as the thickness approaches 1 km. Conversely, a 1-km-thick low-velocity layer is detectable for velocities less than ~2 km/s, and the full amplitude is developed as the velocity approaches 1 km/s. In contrast to the case of travel-time tomography where the ray bending effects of low-velocity zones impede resolution, the low velocity aids resolution of a thin layer by creating greater separation of the converted phases from the top and bottom of the layer.

Lateral Extent of the APMB

Our PASSCAL deployment was largely confined to the northeast portion of the Altiplano–Puna volcanic complex. We were able to greatly expand our analysis to include most of the APVC using data from the SFB-267 project (Deformazion in den Andes) in cooperation with researchers from the GFZ-Potsdam and Freie University-Berlin. This large multi-year project occupied ~100 sites with short-period instruments deployed in three separate, overlapping experiments of about 3 months each (Fig. 4). Combining data from the three GFZ-FU deployments, we computed a total of 88 receiver functions using the same processing techniques as in the APVC study (CHMIELOWSKI *et al.*, 1999). The relatively short duration of the deployments limited the number of events available for analysis, however the spatial density and broad coverage of the German instruments was essential in constraining the lateral boundaries of the APMB.

We considered a receiver function to contain evidence for the APMB if a negative-polarity conversion arrived between 1.5 and 3.0 s after the *P* wave with an amplitude >50% of that of the direct *P* wave. If the amplitude of the *Ps* conversion was between 30% and 50% of that of the direct *P* wave, we considered the presence of the APMB ambiguous. Finally, if the amplitude of the *Ps* conversion was <30% of the direct *P* wave amplitude, or if its timing was not between 1.5 and 3.0 s, we considered the APMB absent beneath the station. The resulting map of the lateral extent of the APMB covers an area about 3° in longitude and about 2° in latitude, or approximately 60,000 km².

An important observation about the APMB is that it most closely correlates with the distribution of the <7 Ma ignimbrite complexes and not with the line of Quaternary arc volcanoes (Fig. 4). This spatial correlation strengthens the argument that this seismic structure is associated with the magmatic plumbing of the APVC and not necessarily with the active arc volcanoes.

Evidence for Strong Anisotropy

A unique advantage in studying the seismic structure of the APVC compared to many other caldera complexes is the presence of a very active Wadati–Benioff zone in the subducting slab (Fig. 1). The subcrustal earthquakes are abundant and many have unusually large magnitudes. We combined global data for the largest events ($M > 5.8$) and many a selection of additional smaller magnitude earthquakes distributed around our APVC network and performed a joint hypocenter relocation on the resulting set of 52 events. From the relocated events, we selected 38 from which to compute receiver functions using the iterative deconvolution technique of LIGORRIA and AMMON (1999). From among the best receiver functions, we stacked those with similar backazimuths for each station.

After low-pass filtering (corner of 0.9 Hz), the stacked local receiver functions were plotted by back-azimuth with the teleseismic receiver functions. Where we have both teleseismic and local receiver functions for comparable backazimuths, the waveforms are similar, giving us confidence in the fidelity of the local receiver

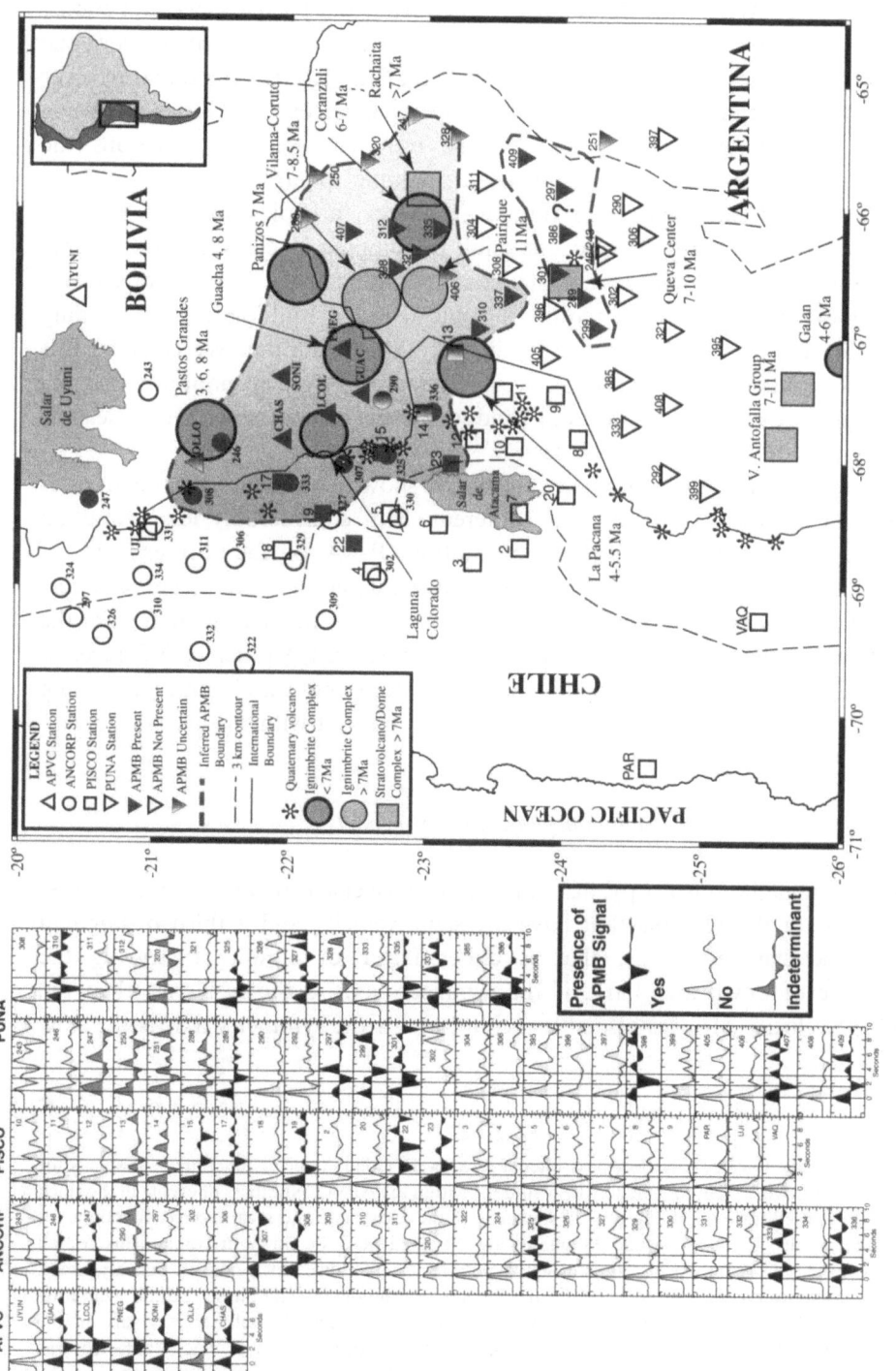

Figure 4

Areal extent of the Altiplano-Puna Magma Body (APMB) as mapped out by stations of the APVC, ANCORP, PISCO, and PUNA deployments. Note that it underlies the high elevations (>3 km) of the Altiplano–Puna transition and correlates most closely with the distribution of the caldera complexes and not with the occurrence of Quaternary volcanoes.

functions (Fig. 5). With the additional coverage from the local receiver functions, significant azimuthal variations in the radial waveform, only hinted at from the teleseismic data, become obvious in the APVC data (Fig. 5a). On the radial receiver functions, the *Ps* from the top of the low-velocity layer exhibits large-amplitude variations and timing perturbations. On the transverse receiver functions the corresponding phases are large and have multiple polarity reversals. These characteristics, especially the large transverse energy, are diagnostic of either a dipping layer or anisotropy. We do not think the dipping layer explanation works here because the large amplitudes and similarity of timing of the *Ps* phase at stations tens of kilometers apart would require a coincidence of multiple steeply dipping interfaces located at about the same depth beneath each station. We consider this scenario geologically unreasonable. A decidedly more likely explanation is that the variations are due to *P–SH* conversions produced at the top of the APMB by seismic anisotropy (LEVIN and PARK, 1997, 1998).

Just as importantly, the combined data clearly show that the receiver functions at stations UYUN and OLLA are quite different, with considerably less azimuthal variations and transverse energy (Fig. 5b). Station OLLA, on the northern edge of the APVC, is within a few kilometers of the active Quaternary arc volcano Ollague. While the OLLA data do display negative polarity arrivals, neither the timing nor the amplitude are similar to the large-amplitude *Ps* phase on the other APVC receiver functions. This contrast supports the idea that both the VLVL and the strong anisotropy are features associated with the APVC and not with the arc volcanoes.

Preliminary Anisotropy Model

Hexagonal symmetry anisotropy, also known as transverse isotropy, is characterized by two orthogonal directions that have the same velocity and a third orthogonal direction, the symmetry axis, that is either faster or slower. Following the nomenclature of Levin and Park, if the symmetry axis is the fast axis, then the equal phase velocity surfaces resemble melons; and if the symmetry axis is the slow

▶

Figure 5(a)

Combined teleseismic (dashed) and low-pass filtered local event (solid) receiver functions and stacks for both radial and transverse components for the 5 stations within the APVC. Numbers of events in the stack are shown above each trace. Note the general similarity of the teleseismic and local receiver functions for similar backazimuths. Large-amplitude *Ps* phases are observed at 2–4 s on both components for all 5 stations. On the radial component a large negative polarity *Ps* is followed by a positive polarity phase. Corresponding phases are usually observed on the transverse component. The phases exhibit azimuthal variations in amplitude and timing on the radial component and multiple polarity reversals on the transverse component. These characteristics are diagnostic of tilted-axis hexagonal symmetry anisotropy as described by LEVIN and PARK (1998). The timing lines are based on station LCOL and repeated on the other stations for comparison.

axis, then the equal phase velocity surfaces resemble pumpkins. This leads to "melon" and "pumpkin" models of anisotropy (LEVIN and PARK, 1998). The former is most often associated with lattice-preferred orientation (LPO) of mineral crystals

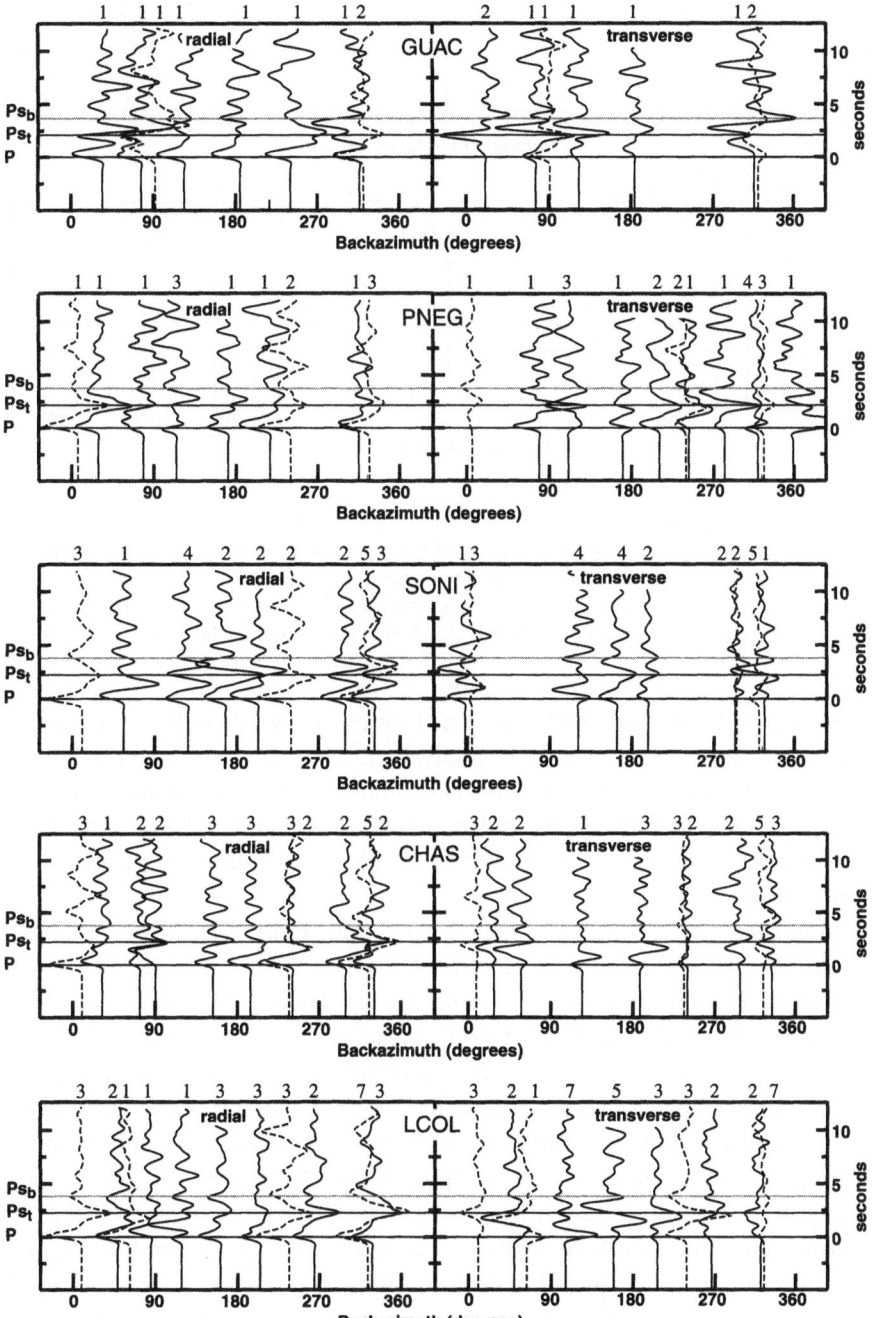

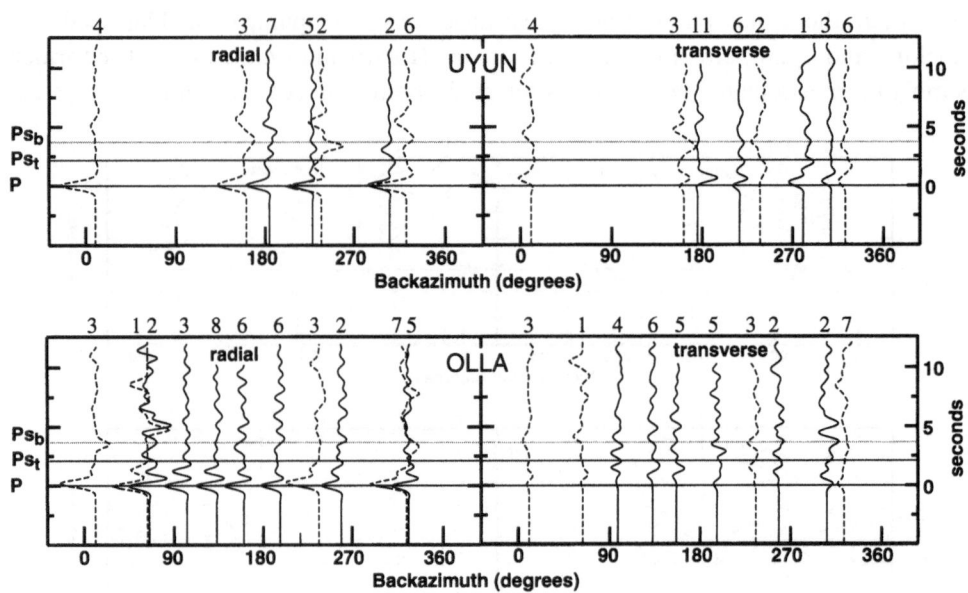

Figure 5(b)

Combined teleseismic (dashed) and low-pass-filtered local event (solid) receiver functions and stacks for both radial and transverse components for stations OLLA and UYUN. Numbers of events in the stack are shown above each trace. Note the lack of similarity of the UYUN and OLLA data to the other APVC stations (Fig. 5a) both in the absence of the large-amplitude *Ps* phases at 2–4 s and less azimuthal variability. The timing lines are based on station LCOL and repeated on the other stations for comparison.

and the latter with aligned cracks or thin layering perpendicular to the symmetry axis. Most previous converted-phase analysis of anisotropy assumed a horizontal axis of symmetry. LEVIN and PARK (1997) formulated the problem with an arbitrary orientation of the symmetry axis and demonstrated that tilted-axis anisotropy in simple flat-layered models can lead to complex waveforms, especially on the transverse component.

We followed the "cookbook" of LEVIN and PARK (1998) to analyze the anisotropy at station Laguna Colorado (LCOL). *P–SH* conversions occur at interfaces where one or both layers are anisotropic. To generate anisotropic features in the *Ps* phase from the top of the APMB, the anisotropy can be in the upper crust, in the low-velocity layer, or in both. The presence of initial *P*-wave energy on the transverse receiver functions is diagnostic of upper-crustal anisotropy. Consequently we first concentrated on this case, assuming the rest of the crust is isotropic. The effect of upper-crustal anisotropy is to "split" the *Ps* phase from the top of the low-velocity zone, generating two phases with variable amplitudes on the radial component and with multiple polarity reversals on the transverse component (Fig. 6, top). The *Ps* phase from the bottom of the low-velocity zone is not split but exhibits significant azimuthal timing perturbations and develops an interference

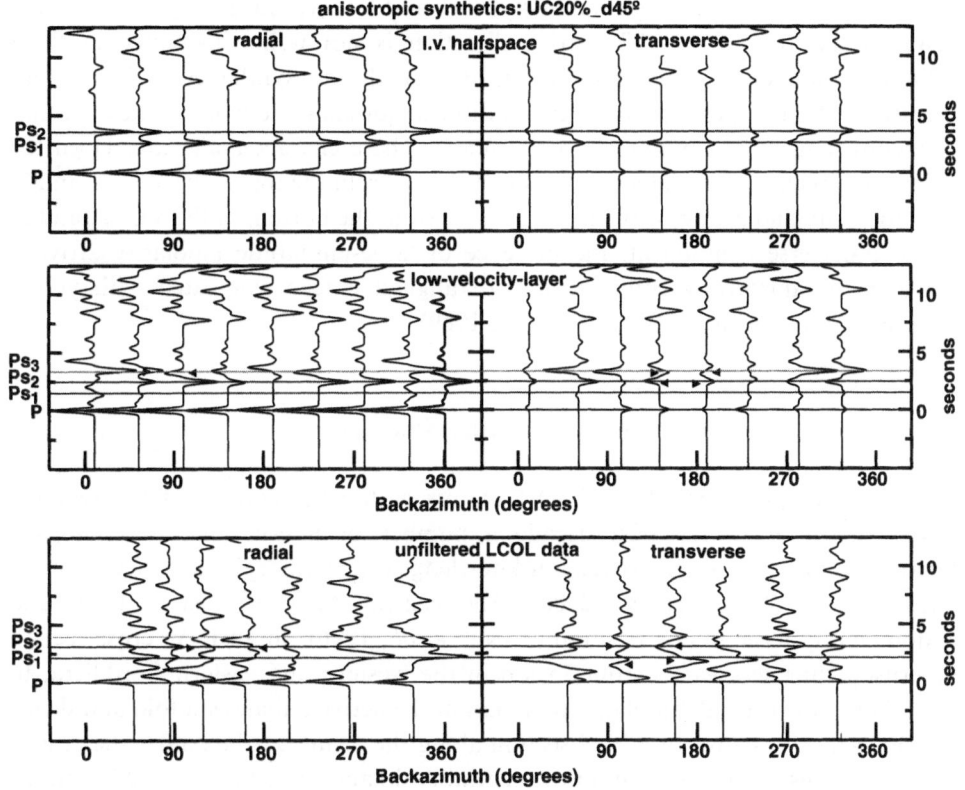

Figure 6

Top panel is Park-Levin synthetic receiver functions for a layer with 20% peak-to-peak "pumpkin" style anisotropy with 45° dip over an isotropic low-velocity halfspace. Note how the negative polarity *Ps* phase from the top of the low-velocity halfspace is "split", with variable amplitudes on the radial component and polarity reversals on the transverse component. Second panel shows synthetics for a model with the same 20% upper crust anisotropy but overlying an isotropic low-velocity layer (lvl) only 1 km thick. The effect of the bottom of the low-velocity layer is a third *Ps* phase with positive polarity and variable arrival time that interferes in a complicated pattern with the slow "split-*Ps*." An azimuth independent synthetic for a completely isotropic low-velocity layer model is plotted at 360° as a heavy line. The bottom panel shows the unfiltered local earthquake receiver functions at LCOL. Comparison of the data with synthetics indicates that anisotropy can explain several major characteristics of the data that are inconsistent with a purely isotropic model.

pattern with the second "split *Ps*" (Fig. 6, second panel). The net effect is a complicated waveform that varies in both timing and amplitude as a function of backazimuth, which in a general way mimics the behavior of our APVC data.

Based on these initial observations, we computed synthetics for a range of upper-crust anisotropy parameters, varying the amount of *P* and *S* anisotropy and the inclination of the symmetry axis for both "melon" and "pumpkin" anisotropy. On the basis of visual inspection we found a certain class of models that explains some of the specific waveform complexities observed in the LCOL data (Fig. 6, bottom two panels). These models have "pumpkin-style" *P*- and *S*-wave anisotropy of 20% and a symmetry axis tilt of 45°. The strike direction is not well constrained in these preliminary models. Comparison of the data with the synthetics indicates that ansiotropy of the upper crust can explain three major characteristics of the data that are inconsistent with a purely isotropic model. These characteristics are: 1) on the radial component, the variable amplitude of the *Ps* from the top of the VLVL, and 2) the complexity and changing polarity of the *Ps* from the bottom of the VLVL, and 3) on the transverse component, the presence of large-amplitude bipolarity arrivals. This type of anisotropy could be due to aligned cracks associated with young volcanic features in the APVC (DE SILVA, 1989).

Regional Crustal Structure

An important question is how the VLVL fits within the context of the regional crustal structure. In the past decade a number of international seismological experiments have greatly improved our knowledge of the lithospheric structure of the central Andes (YUAN *et al.*, 2000). The central Andean crust is thick and predominantly felsic in composition (ZANDT *et al.*, 1996). In an E–W cross section the crustal thickness varies from 75 km in the western and eastern Cordillera, to 60–65 km in the central Altiplano, to 40 km in the active Subandean fold and thrust belt (BECK *et al.*, 1996). In a N–S section along the center of the Andes the crustal thickness varies from 60–65 km in the central Altiplano to as thin as ~50 km in the Puna (YUAN *et al.*, 2000). In conjunction with this crustal thinning, the already low bulk crustal seismic velocity also decreases southward. A regionally prevalent crustal low-velocity layer characterizes the central and southern Altiplano, the western flank of the Eastern Cordillera, and the Puna. The top of the VLVL appears to link with the basal detachment of the fold-thrust belt under the Eastern Cordillera (YUAN *et al.*, 2000). This depth may represent the brittle-ductile transition and be the decoupling depth between upper-crustal imbrication and lower-crustal ductile flow. The APMB is apparently localized at this boundary.

On the APVC receiver functions, the large converted waves and multiples from the VLVL effectively hide the Moho *Ps*. To investigate the larger-scale structure we measured interstation surface-wave dispersion and used a stochastic inversion technique to find models that satisfy the data within the error bounds (BAUMONT *et al.*, 2001). Within the APVC the short interstation distances and large lateral velocity variations contributed to substantial errors in the phase velocity measurements. We improved the results greatly by solving simultaneously for the phase velocity and backazimuth (Fig. 7). Significant deviations from the theoretical backazimuth are observed, and accounting for these deviations greatly improved the precision of the phase velocity measurements. The dispersion data cannot resolve a thin VLVL; nonetheless they are sensitive to the total crustal travel time, which is

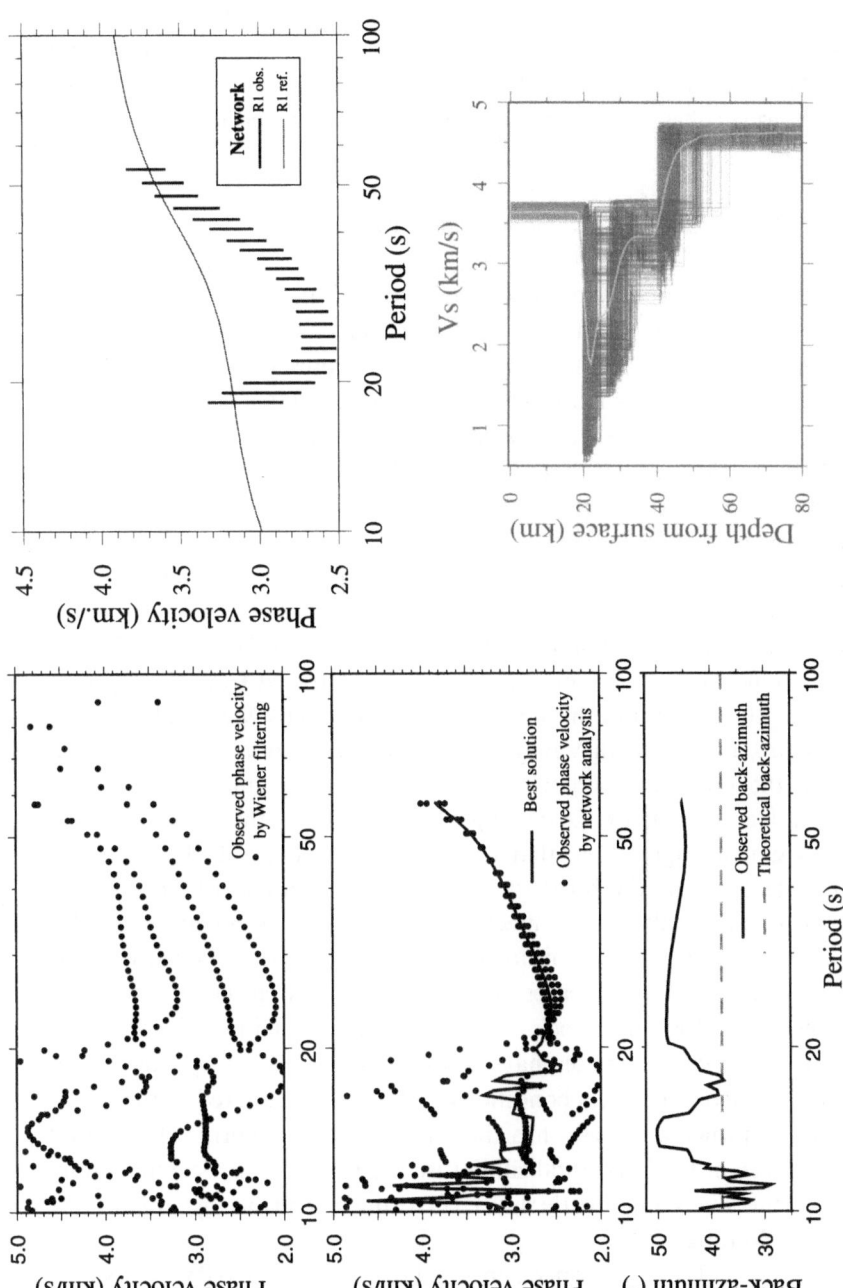

Figure 7

Interstation Rayleigh wave dispersion measurements and inversion results for the region within the APVC network (excluding stations UYUN and OLLA). The left panels show the improvements in the phase velocity measurements by using the APVC network to simultaneously estimate the phase velocity and backazimuth of the surface waves. The final phase velocity measurements and estimated error bounds are shown in the upper right compared with a reference curve for the ZANDT et al. (1996) model for the central Altiplano. Lower right is a preliminary model for the APVC from stochastic inversion of the dispersion data with constraints on the 20-km-deep low-velocity layer.

significantly affected by the presence of the VLVL. We ran a series of inversions, including ones with *a priori* constraints on the presence of a thin low-velocity layer based on the receiver functions. The resulting models have extremely low velocities (approaching 0.5 km/s) in the low-velocity layer and Moho depths between 40 and 55 km (Fig. 7). The relatively thin crust is consistent with the regional trends described above, however these preliminary results still require verification.

Magmatism and Tectonics in the Central Andes

The coexistence of a pervasive horizontal sill and inclined aligned cracks might at first seem paradoxical in light of the local stress field. Yet a number of well-documented examples of exactly this situation occur in the present and past geologic record. Two present-day examples from the western United States occur in the Rio Grande Rift near Socorro (New Mexico) and in the Death Valley area of the southern Great Basin (California). In both areas a mid-crustal magma body has been inferred from seismic reflection data, and the dominant tectonic strain field is extensional. The tabular magma bodies are interpreted to lie at a major rheological boundary between a brittle upper crust extending by faulting and a ductile lower crust stretching by penetrative flow and intrusion (DE VOOGD *et al.*, 1988). In Death Valley the upper-crustal faults are planar, with moderate dips of 20° to 35°. In one spectacular seismic reflection record, normal faults can be traced from the inferred magma body at a depth of 15 km to the surface location of a 0.69 Ma basaltic cinder cone (DE VOOGD *et al.*, 1988). Death Valley is also the location of a Miocene pluton that intruded into the mid-crust and which has since been exhumed by extreme extension as part of a core complex. Detailed geologic studies of this pluton and the surrounding rocks convincingly show that entrapment was controlled by a high-strength barrier represented by the brittle-ductile transition (HOLM, 1995).

PARSONS *et al.* (1992) analyzed the mechanics of the emplacement of tabular intrusions into a rheologically stratified lithosphere under extension. The basic idea is that a vertically intruding dike increases the horizontal least principal stress by pushing out on the walls of the host rock. This effect is magnified in a ductile zone, such as the lower crust, where stresses have been partially relaxed by flow. Then subsequent intrusions encounter stress conditions that have been altered to the extent that the local least principal stress has become vertical, favoring horizontal intrusions. The inflation of the horizontal sill in turn increases the vertical maximum principal stress in the overlying brittle layer, favoring extension by either dike intrusion or normal faulting.

All the examples cited above are from extensional regions, yet the APVC is located in one of the largest compressional mountain belts in the world. Perhaps the APVC, with average elevations above 4.4 km, represents a portion of the Andes that has recently undergone the transition from vertical thickening to horizontal, orogen-parallel extension. It has been postulated that the onset of ignimbrite eruptions may mark this temporal transition in the central Andes (RILLER et al., 1999), while collapse apparently followed ignimbrite volcanism in the western United States (BEST and CHRISTIANSEN, 1991). One interesting possibility was recently published by BOTT (1999), who modeled localized crustal isostasy caused by ductile flow in the lower crust in response to emplacement of a tabular low-density "granitic" body in the upper crust. Both the preliminary indication of crustal thinning within the APVC and geodetic evidence for northward tectonic tilting of the southern Altiplano (BILLS et al., 1994) are possible indications of this type of on-going process. We believe that further study of these crustal sill-like magma bodies will contribute to our understanding of the interplay of tectonics and magmatism in compressional orogens.

Acknowledgments

U.S. funding was provided by National Science Foundation EAR-9505816 and funding to GFZ-Potsdam and FU-Berlin was provided by the SFB 267 project. The PASSCAL program of IRIS provided instrumentation and logistical support. Observatorio San Calixto and the U.S. Embassy in La Paz provided in-country logistical support. We thank Bernard Chouet and an anonymous reviewer for their constructive reviews, and to Norm Meader for manuscript preparation. One of us (GZ) thanks K. Aki for showing him the fun in deciphering what all of those wiggles mean and dedicates this work to the memory of his parents (Donald Zandt and Sumie Sawamura Zandt).

REFERENCES

AKI, K. (1968), Seismological Evidence for the Existence of Soft Thin Layers in the Upper Mantle under Japan, J. Geophys. Res. 73, 585–594.

AKI, K., CHOUET, B., FEHLER, M., ZANDT, G., KOYANAGI, R., COLP, J., and HAY, R.C. (1978), Seismic Properties of a Shallow Magma Reservoir in Kilauea Iki by Active and Passive Experiments, J. Geophys. Res. 83, 2273–2282.

AKI, K., CHRISTOFFERSSON, A., and HUSEBYE, E. S. (1977), Determination of the Three-dimensional Seismic Structure of the Lithosphere, J. Geophys. Res. 82, 277–296.

BAKER, M.C.W. (1981), The Nature and Distribution of Upper Cenozoic Ignimbrite Centres in the Central Andes, J. Volcanology Geothermal Res. 11, 293–315.

BAUMONT, D., PAUL, A., BECK, S., ZANDT, G., and PEDERSEN, H. (2001), Lithospheric Structure of the Central Andes Based on Surface Wave Dispersion, revised for J. Geophys. Res.

BECK, S. L., ZANDT, G., MYERS, S. C., WALLACE, T. C., SILVER, P. G., and DRAKE, L. (1996), Crustal Thickness Variations in the Central Andes, Geology 24, 407–410.

BEST, M. G. and CHRISTIANSEN, E. H. (1991), Limited Extension During Peak Tertiary Volcanism, Great Basin of Nevada and Utah, J. Geophys. Res. 96, 13509–13528.

BILLS, B. G., DE SILVA, S. L., CURREY, D. R., EMENGER, R. S., LILLQUIST, K. D., DONNELLAN, A., and WORDEN, B. (1994), Hydro-isostatic Deflection and Tectonic Tilting in the Central Andes: Initial Results of a GPS Survey of Lake Minchin Shorelines, Geophys. Res. Lett. 21, 293–296.

BOTT, M. H. P. (1999), *Modeling Local Crustal Isostasy Caused by Ductile Flow in the Lower Crust*, J. Geophys. Res. *104*, 20349–20359.

BROWN, L., CHAPIN, C., SANFORD, A., KAUFMAN, S., and OLIVER, J. (1980), *Deep Structure of the Rio Grande Rift from Seismic Reflection Profiling*, J. Geophys. Res. *85*, 4773–4800.

CAHILL, T. and ISACK, B.L. (1992), *Seismicity and Shape of the Subducted Nazca Plate*, J. Geophys. Res. *97*, 17 503–17 529.

CHMIELOWSKI, J., ZANDT, G., and HABERLAND, C. (1999), *The Central Andean Altiplano–Puna Magma Body*, Geophys. Res. Lett. *26*, 783–786.

CONEY, P. J. (1972), *Cordilleran Tectonics and North American Plate Motion*, Am. J. Sci. *272*, 602–628.

DE SILVA, S. L. (1989), *Altiplano–Puna Volcanic Complex of the Central Andes*, Geology *17*, 1102–1106.

DE SILVA, S. L. and FRANCIS, P. W., *Volcanoes of the Central Andes* (Springer-Verlag, Berlin 1991).

DE SILVA, S. L., SELF, S., FRANCIS, P. W., DRAKE, R. E., and CARLOS, R.R. (1994), *Effusive Silicic Volcanism in the Central Andes: The Chao Dacite and Other Young Lavas of the Altiplano–Puna Volcanic Complex*, J. Geophys. Res. *99*, 17805–17825.

DE VOOGD, B., SERPA, L., and BROWN, L. (1988), *Crustal Extension and Magmatic Processes: COCORP Profiles from the Death Valley and Rio Grande Rift*, Geol. Soc. Am. Bull. *100*, 1550–1567.

ELLSWORTH, W. L. and KOYANAGI, R. Y. (1977*)*, *Three-Dimensional Crust and Mantle Structure of the Kilauea Volcanoe, Hawaii*, J. Geophys. Res. *82*, 5379–5394.

FRANCIS, P., *Volcanoes, a Planetary Perspective* (Clarendon Press, Oxford 1993).

HOLM, D. K. (1995), *Relation of Deformation and Multiple Intrusion in the Death Valley Extended Region, California, with Implications for Magma Entrapment Mechanism*, J. Geophys. Res. *100*, 10495–10505.

IYER, H. M. (1984), *Geophysical Evidence for the Location, Shapes and Sizes, and Internal Structures of Magma Chambers Beneath Regions of Quaternary Volcanism*, Phil. Trans. R. Soc. Lond., *A310*, 473–510.

IYER, H. M., EVANS, J. R., ZANDT, G., STEWART, R. M., COAKLEY, J. M., and ROLOFF, J. N. (1981), *A Deep Low Velocity Body under the Yellowstone Caldera, Wyoming; Delineation Using Teleseismic P-wave Residuals and Tectonic Interpretation: Summary*, Bull. Geol. Soc. Am. *92*, part 1, 792–798.

LEVIN, V. and PARK, J. (1997), *P–SH Conversions in a Flat-layered Medium with Anisotropy of Arbitrary Orientation*, Geophys. J. Int. *131*, 253–266.

LEVIN, V. and PARK, J. (1998), *P-SH Conversions in Layered Media with Hexagonally Symmetric Anisotropy: A Cook Book*, Pure Appl. Geophys. *151*, 669–697.

LIGORRIA, J. P. and AMMON, C. J. (1999), *Iterative Deconvolution and Receiver-function Estimation*, Bull. Seismol. Soc. Am. *89*, 1395–1400.

MAUGHAN, L. L., CHRISTIANSEN, E. H., BEST, M. G., GROMME, C. S., DEINO, A., and TINGEY, D. G. (2002), *The Oligocene Lund Tuff, Great Basin, USA: A Voluminous Monotonous Intermediate*, J. Volcan. Geothermal Res. *113*, 129–157.

PARSONS, T., SLEEP, N. H., and THOMPSON, G. A. (1992), *Host Rock Rheology Controls on the Emplacement of Tabular Intrusions: Implications for Underplating of Extending Crust*, Tectonics *11*, 1348–1356.

RILLER, U., ONCKEN, O., PETRINOVIC, I., and STRECKER, M. (1999), *Late Cenozoic Tectonism and Caldera Formation in the Central Andes*, Eos Trans. AGU *80*, F1061 (1999 Fall AGU Meeting, San Francisco).

ROBERTS, P. M., AKI, K., and Fehler, M. C. (1991), *A Low-Velocity Zone in the Basement Beneath the Valles Caldera, New Mexico*, J. Geophys. Res. *96*, 21,583–21,596.

RUNDLE, J. B. and HILL, D. P. (1988), *The Geophysics of a Restless Caldera – Long Valley, California*, Ann. Rev. Earth Planet. Sci. *16*, 251–271.

SANDFORD, A. R., ALPTEKIN, O., and TOPPOZADA, T. R. (1973*)*, *Use of Reflection Phases on Microearthquake Seismograms to Map an Unusual Discontinuity Beneath the Rio Grande Rift*, Bull. Seismol. Soc. Am. *63*, 2021–2034.

SCHLUE, J. W., ASTER, R. C., and MEYER, R. P. (1996), *A Lower Crustal Extension to a Midcrustal Magma Body in the Rio Grande Rift, New Mexico*, J. Geophys. Res. *101*, 25,283–25,291.

SHEETZ, K. E. and SCHLUE, J. W. (1992*)*, *Inferences for the Socorro Magma Body from Teleseismic Receiver Functions*, Geophys. Res. Lett. *19*, 1867–1870.

STECK, L. K. and PROTHERO, W. A., Jr. (1994), *Crustal Structure Beneath Long Valley Caldera from Modeling Teleseismic P-wave Polarizations and Ps Converted Waves*, J. Geophys. Res. *99*, 6881–6898.

STEEPLES, D. W. and IYER, H. M. (1976), *Low-velocity Zone under Long Valley as Determined from Teleseismic Events*, J. Geophys. Res. *81*, 849–860.

WHITMAN, D., ISACKS, B. L., and KAY, S. M. (1996), *Lithospheric Structure and Along-strike Segmentation of the Central Andean Plateau: Topography, Tectonics, and Timing*, Tectonophysics *259*, 29–40.

YUAN, X., SOBOLEV, S. V., KIND, R., ONCKEN, O., and the ANDES SEISMOLOGY GROUP (2000), *Subduction and Collision Processes in the Central Andes Constrained by Converted Seismic Phases*, Nature *408*, 958–961.

ZANDT, G. and FURLONG, K. P. (1982*)*, *Evolution and Thickness of the Lithosphere Beneath Coastal California*, Geology *10*, 376–381.

ZANDT, G., MYERS, S. C., and WALLACE, T. C. (1995), *Crust and Mantle Structure Across the Basin and Range-Colorado Plateau Boundary at 37°N Latitude and Implications for Cenozoic Extensional Mechanism*, J. Geophys. Res. *100*, 10 529–10 548.

ZANDT, G., BECK, S. L., RUPPERT, S. R., AMMON, C. J., ROCK, D., MINAYA, E., WALLACE, T. C., and SILVER, P. G. (1996), *Anomalous Crust of the Bolivian Altiplano, Central Andes: Constraints from Broadband Regional Seismic Waveforms*, Geophys. Res. Lett. *23*, 1159–1162.

(Received August 15, 2000, accepted May 15, 2001)

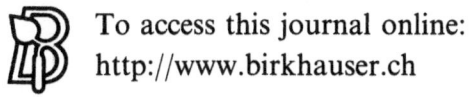

To access this journal online:
http://www.birkhauser.ch